Causality

WILEY SERIES IN PROBABILITY AND STATISTICS

Established by WALTER A. SHEWHART and SAMUEL S. WILKS

A complete list of the titles in this series appears at the end of this volume.

Causality

Statistical Perspectives and Applications

Edited by

Carlo Berzuini • Philip Dawid • Luisa Bernardinelli
Statistical Laboratory, Centre for Mathematical Sciences
University of Cambridge, Cambridge, UK

WILEY
A John Wiley & Sons, Ltd., Publication

This edition first published 2012

Registered office
John Wiley & Sons Ltd, The Atrium, Southern Gate, Chichester, West Sussex, PO19 8SQ, United Kingdom

For details of our global editorial offices, for customer services and for information about how to apply for permission to reuse the copyright material in this book please see our website at www.wiley.com.

Library of Congress Cataloging-in-Publication Data

Berzuini, Carlo.
Causality : statistical perspectives and applications / Carlo Berzuini, Philip Dawid, Luisa Bernardinelli.
p. cm. – (Wiley series in probability and statistics)
Includes bibliographical references and index.
ISBN 978-0-470-66556-5 (hardback)
1. Estimation theory. 2. Causation. 3. Causality (Physics) I. Dawid, Philip. II. Bernardinelli, Luisa. III. Title.
QA276.8.B475 2012
519.5′44–dc23

2011049795

A catalogue record for this book is available from the British Library.

ISBN: 978-0-470-66556-5

Set in 10/12pt Times by Aptara Inc., New Delhi, India.

Printed and bound in Malaysia by Vivar Printing Sdn Bhd

1 2012

Contents

List of contributors

Elja Arjas
Department of Mathematics and Statistics
University of Helsinki
Helsinki, Finland

Luisa Bernardinelli
Statistical Laboratory
Centre for Mathematical Sciences
University of Cambridge
Cambridge, UK

Carlo Berzuini
Statistical Laboratory
Centre for Mathematical Sciences
University of Cambridge
Cambridge, UK

Clive G. Bowsher
School of Mathematics
University of Bristol
Bristol, UK

Simon Cousens
Centre for Statistical Methodology
London School of Hygiene and Tropical Medicine, London, UK

D.R. Cox
Nuffield College
University of Oxford
Oxford, UK

Rhian Daniel
Centre for Statistical Methodology
London School of Hygiene and Tropical Medicine, London, UK

Philip Dawid
Statistical Laboratory
Centre for Mathematical Sciences
University of Cambridge
Cambridge, UK

Bianca De Stavola
Centre for Statistical Methodology
London School of Hygiene and Tropical Medicine, London, UK

Vanessa Didelez
Department of Mathematics
University of Bristol
Bristol, UK

Graham Dunn
Health Sciences Research Group
School of Community Based Medicine
The University of Manchester
Manchester, UK

Michael Eichler
Department of Quantitative Economics
Maastricht University
Maastricht, The Netherlands

Richard Emsley
Health Sciences Research Group
School of Community Based Medicine
The University of Manchester
Manchester, UK

Krista Fischer
MRC Biostatistics Unit, Cambridge, UK
and
Estonian Genome Center
University of Tartu, Estonia

Luisa Foco
Department of Applied Health Sciences
University of Pavia
Pavia, Italy

Sander Greenland
Department of Epidemiology and
Department of Statistics
University of California
Los Angeles, California, USA

Miguel A. Hernán
Department of Epidemiology
Harvard School of Public Health
Boston, Massachusetts, USA

Miles Parkes
Department of Gastroenterology
Addenbrookes Hospital
Cambridge, UK

Roberta Pastorino
Department of Applied Health Sciences
University of Pavia
Pavia, Italy

Judea Pearl
Computer Science Department
University of California
Los Angeles, California, USA

Roland R. Ramsahai
Statistical Laboratory
Centre for Mathematical Sciences
University of Cambridge
Cambridge, UK

Paul R. Rosenbaum
Wharton School
University of Pennsylvania
Philadelphia, Pennsylvania, USA

Michael Rutter
MRC SGDP Centre
Institute of Psychiatry
London, UK

Ilya Shpitser
Department of Epidemiology
Harvard School of Public Health
Boston, Massachusetts, USA

Arvid Sjölander
Department of Medical Epidemiology
and Biostatistics
Karolinska Institutet
Stockholm, Sweden

Tyler J. VanderWeele
Departments of Epidemiology and
Biostatistics
Harvard School of Public Health
Boston, Massachusetts, USA

Stijn Vansteelandt
Ghent University, Ghent, Belgium
and
London School of Hygiene and Tropical
Medicine
London, UK

Ian R. White
MRC Biostatistics Unit
Cambridge, UK

Hu Zhang
Department of Gastroenterology
Addenbrookes Hospital
Cambridge, UK

An overview of statistical causality

Many statistical analyses aim at a causal explanation of the data. The early observational studies on the risks of smoking (Cornfield *et al.*, 1959), for example, aimed at something deeper than to show the poorer prognosis of a smoker. The hoped-for interpretation was *causal*: those who smoked would, on average, have had a better health had they not done so and, consequently, any future intervention against smoking will, at least in a similar population, have a positive impact on health. Causal interpretations and questions are the focus of this present book. They underpin many statistical studies in a variety of empirical disciplines, including natural and social sciences, psychology, and economics. The case of epidemiology and biostatistics is noted for a traditionally cautious attitude towards causality. Early researchers in these areas did not feel the need to use the word 'causal'. Emphasis was on the requirement that the study be 'secure': that its conclusions should not rely on special assumptions about the nature of uncontrolled variation, something that is ideally only achieved in experimental studies. In the work of Fisher (1935), security was achieved largely by using randomization within an experimental context. This ensures that, when we form contrasts between the treatment groups, we are comparing 'like with like', and thus there are no systematic pre-existing differences between the treatment groups that might be alternative explanations of the observed difference in response.

Another idea originated by Fisher (1932) and later developed by Cochran (1957), Cox (1960), and Cox and McCullagh (1982) is the use of supplementary variables to improve the efficiency of estimators and of instrumental variables to make a causal effect of interest identifiable. The use of supplementary and instrumental variables in causal inference is discussed in Chapter 16 of this book, 'Supplementary variables for causal estimation' by Roland Ramsahai.

Early advances in the theory of experimental design, largely contributed by Rothamsted researchers, are discussed in Chapter 1, 'Statistical causality: some historical remarks', by David Cox. Also discussed in this chapter are some implications of the 'Rothamsted view' (and of the controversies that arose around it) for the current discussion on causal inference. A technical discussion of the problems of causal inference in randomized experiments in medicine is given in Chapter 21, 'Causal inference in clinical trials', by Krista Fischer and Ian White.

The 1960s witnessed the early development of a theory of causal inference in observational studies, a notable example being the work of Bradford Hill (1965). Hill proposed a set of guidelines to strengthen the case for a causal interpretation of the results of a given observational study. One of these guidelines, the presence of a dose–response relationship, is discussed in depth in Chapter 19, 'Nonreactive and purely reactive doses in observational studies', by Paul Rosenbaum. Hill's guidelines are informal, and they do not provide a definition of 'causal'. During the 1990s, a wider community of researchers, gathered from such disciplines as statistics, philosophy, economics, social science, machine learning, and artificial intelligence, proposed a more aggressive approach to causality, reminiscent of the long philosophers' struggle to reduce causality to probabilities. These researchers transformed cause–effect relationships into objects that can be manipulated mathematically (Pearl, 2000). They attempted to formalize concepts such as confounding and to set up various formal frameworks for causal inference from observational and experimental studies. In a given application, such frameworks allow us (i) to define the target causal effects, (ii) to express the causal assumptions in a clear way, and determine whether they are sufficient to allow estimation of the target effects from the available data, (iii) to identify analysis modalities and algorithms that render the estimate feasible, and (iv) to identify observations and experiments that would render the estimate feasible, or assumptions under which the conclusions of the analysis have a causal interpretation.

In retrospect, such effort came late. Many of the tools for conceptualizing causality had been available for some time, as in the case of the potential outcomes representation (Rubin, 1974), which Rubin adapted from experimental (Fisher, 1935) to observational studies in the early 1970s. Potential outcomes are discussed in Chapter 2, 'The language of potential outcomes', by Arvid Sjölander, and are used in several other chapters of this book. In this representation, any individual is characterized by a notional response Y_k to each treatment T_k, regarded as fixed even before the treatment is applied. In Chapter 10, 'Cross-classifications by joint potential outcomes', Arvid Sjölander discusses the idea of a 'principal stratification' of the individuals, on the basis of the joint values of several potential outcomes of the same variable (so that each stratum specifies exactly how that variable would respond to a variety of different settings for some other variable). A sometimes serious limitation of such an approach is that typically there is no way of telling which individual falls into which stratum. In Chapter 10, principal stratification is used to bound nonidentifiable causal effects and to deal with problems due to imperfect observations. Principal stratification is also used in Chapter 21 to deal with problems of protocol nonadherence and of contamination between treatment arms, in the context of randomized clinical trials.

Another legacy from the past is the use of graphical representations of causality, predated by Wright's work on path diagrams (Wright, 1921, 1934) and later advocated by Cochran (1965). This area is currently dominated by the Non Parametric Structural Equations Models (NPSEMs) discussed in Chapter 3, 'Structural equations, graphs and interventions', by Ilya Shpitser. Shpitser emphasizes a conceptual symbiosis between NPSEMs and potential outcomes, where NPSEMs contribute a transparent language for expressing assumptions in terms of conditional independencies implied by the structure of a causal graph (Dawid, 1979; Geiger *et al.*, 1990). Constructive criticism of NPSEMs is given in Chapter 5, 'Causal inference as a prediction problem: assumptions, identification and evidence synthesis', by Sander Greenland. A strong interpretation of an NPSEM regards each node in the model as associated with a fixed collection of potential responses to the various possible configurations of interventions on the set of parents of that node. This interpretation sheds light on some of the problems of nonstochasticity and nonidentifiability that Greenland mentions in relation to NPSEMs.

In the light of these problems, some researchers have set aside potential outcomes and NPSEMs in favour of approaches that fully acknowledge the stochastic nature of the world. One of these is the *decision theoretic* approach of Chapter 4. Focus here is on the assumptions under which an inference of interest, which we would ideally obtain from an experiment, can be drawn from a given set of observational data. In general, inferences are not transportable between an observational and an experimental regime of data collection, the reason being that the distributions of the domain variables in the two regimes may be completely different. The decision-theoretic approach considers special 'regime indicator' variables and uses them to formalize (in terms of conditional independence relationships) those conditions of invariance between regime-specific distributions that make cross-regime inference possible. In some problems of causal inference, the decision-theoretic approach leads to the same conclusions one reaches by using potential outcomes, but relaxing the strong assumptions of the latter. Chapters 8, 14, 15, and 16 further illustrate the use of a decision-theoretic formalism in combination with explicit graph representations of the assumed data-generating mechanism. A theme for future research is a comparison of different formulations of statistical causality, in relation to real data analysis situations and in terms of their ability to clarify the assumptions behind the validity of an inference.

As noted in Chapter 5 by Greenland, potential outcomes can be ambiguous in the absence of a well-defined physical mechanism that determines the value of the causal factors. Difficulties then arise with causal factors that one cannot conceivably (let alone technologically) manipulate. In fact, the definition of a causal effect of variables like sex, say, remains a veritable conundrum. The problem is discussed in Chapter 9, 'Causal effects and natural laws: towards a conceptualization of causal counterfactuals for nonmanipulable exposures, with application to the effects of race and sex', by Tyler VanderWeele and Miguel Hernán. The authors of this chapter argue that the notion of causal effect for nonmanipulable properties is defensible from a potential outcomes perspective if these can be assumed to affect the outcome through an underlying deterministic 'natural law', of the kind we encounter in physics. In the 'sex and employment' example of Chapter 12, Judea Pearl deals with the sex effect by invoking a hypothetical intervention that prevents the employer from being informed about the applicant's sex.

Much causal inference literature, in both observational and experimental contexts, conceptualizes causality in terms of assessing the consequences of a (future or hypothetical) intervention. This will typically involve estimating, from data, effects that are formally defined in terms of a relevant intervention distribution, induced by the intervention of interest. A given set of data will be sufficient to estimate the causal effects of interest if the relevant intervention distribution can be expressed as a function of the distribution from which the data have been obtained. A formal check of this property can be performed with the aid of the 'do-calculus' described in Chapter 6, 'Graph-based criteria of identifiability of causal questions', by Ilya Shpitser. The need for such formal machinery becomes acute in highly structured causal problems, notably in the presence of multiple causal paths (such as in Chapters 11 and 12), and of complexly structured confounding, and in the identification of dynamic plans.

The usual conception of 'intervention' involves fixing variables to a particular configuration of values. This definition may prove inadequate in the presence of temporal precedence relationships between exposures or actions. In such cases, interventions may be more appropriately described as *plans*, the archetypal example being a plan of treatment of a medical patient, where periodic decisions have to be made in the light of the outcomes of previous treatment actions. The evaluation will often be based on information gleaned from records of the performance of past decision makers. Problems of this kind are discussed in Chapters 7, 8, and 17 of this book. In Chapter 7, 'Causal inference from observational data: a Bayesian predictive approach', Elja Arjas approaches the problem from an entirely probabilistic point of view, and clarifies important links with mainstream analysis of time-dependent events.

In Chapter 8, 'Assessing dynamic treatment strategies', Carlo Berzuini, Philip Dawid, and Vanessa Didelez tackle the problem from a decision-theoretic point of view, and in the context of examples in the treatment of an HIV virus infection and of an aortic aneurism. In Chapter 17, 'Time-varying confounding: some practical considerations in a likelihood framework', Rhian Daniel, Bianca De Stavola, and Simon Cousens tackle the problem from a potential outcomes point of view, and look more closely into aspects of statistical estimation and computation, with the aid of a simulation experiment.

Problems of the above kind typically involve biases introduced by post-treatment variables, which cannot be adjusted for within a standard regression analysis framework. Recent research in causal inference has developed algorithms to deal with a wide class of problems of this kind (albeit often only by introducing suitable parametric assumptions) and, by so doing, overcome the limits of standard regression. A notable example is Robins's G-computation algorithm (Robins, 1986). Use of this algorithm for the evaluation of dynamic plans, and the conditions under which this leads to valid causal inferences, are discussed in Chapters 8 and 17. The same algorithm is discussed in Chapter 11 in relation to the analysis of mediation.

How does causal inference connect with science? One might argue that the randomization philosophy at the basis of many experimental designs (e.g. clinical trials) is a very poor example of 'science'. Randomization may make the study conclusions more secure, but it does little to unravel the biological, or physical, or psychological processes operating behind an observed causal effect. Such a limitation may be negligible in those areas of application whose exclusive aim is to predict the consequences of a future intervention, or to identify the most promising treatment, such as often occurs in clinical trials. But there are study areas which are largely driven by the wish to understand an underlying mechanism. One example are studies of genetic association to elucidate a molecular disease mechanism, or, say, a studies to investigate whether anxiety plays a role in the response of an individual to a specific stimulus. In these and other application areas, there is a need for statistical concepts and tools that help us to unravel the mechanism behind a causal effect.

Different interpretations of the concept of mechanism will inspire different statistical approaches to mechanistic inference. One class of methods of inference about mechanism is inspired by the *sufficient cause* conceptualization discussed in Chapter 13, 'The sufficient cause framework in statistics, philosophy and the biomedical and social sciences', by Tyler VanderWeele. This models the outcome event in terms of a collection of triggers, each involving a set of component events, such that occurrence of all the component events of at least one trigger always produces the outcome event (Mackie, 1965). The above conceptualization provides a basis for considering two causal factors as *interacting mechanistically* – with respect to the outcome of interest – if they are jointly involved in at least one trigger. The importance of this concept in the study of mechanisms is illustrated by the deep interest of genetic epidemiologists in methods for assessing epistasis (mechanistic interaction between the effects of genes). Methodological development in this area owes much to the work of several researchers, including Rothman (1976) and Skrondal (2003). VanderWeele and Robins (2008, 2009) have provided a potential outcomes formalization of the framework and have established empirical tests for mechanistic interaction in an observational context. In Chapter 14, 'Analysis of interaction for identifying causal mechanisms', Carlo Berzuini, Philip Dawid, Hu Zhang, and Miles Parkes justify such tests from a decision-theoretic point of view, in a formulation that embraces continuous causal factors. They use the method to identify components of the autophagy pathway which may cause susceptibility to Crohn's Disease. In Chapter 15, 'Ion channels as a possible mechanism of neurodegeneration in multiple sclerosis', Luisa

Bernardinelli, Carlo Berzuini, Luisa Foco, and Roberta Pastorino discuss the use of family-structured genetic association data to detect patterns of statistical interaction where the value of one causal factor affects the sign of the effect of another factor. These patterns can be reasonably interpreted in terms of mechanistic interaction, because they cannot be eliminated by variable or response transformation.

Other conceptualizations of mechanism exist. In a variety of empirical disciplines, the term 'causal mechanism' designates the process through which the treatment causally affects the outcome (Salmon, 1984). It is often helpful to decompose this into an *indirect* effect, which operates through an observed mediator of interest, and a *direct* effect, which includes all other possible mechanisms (Robins and Greenland,1992; Pearl, 2005). By allowing separate estimation of these two components, analysis of mediation enables the researcher to explore the role of a specific causal pathway in the transmission of an effect of interest. Many questions of scientific interest can be formulated in terms of direct and indirect effects. One example is discussed in Chapter 20, 'Evaluation of potential mediators in randomised trials of complex interventions (psychotherapies)' by Richard Emsley and Graham Dunn. The study described in this chapter is motivated by the question of whether the quality-of-life improvement induced by cognitive behaviour therapy in depressed patients is mediated by a therapy-induced beneficial change of beliefs. Any evidence of a direct effect of the therapy, unmediated by that change, would prompt scientific hypotheses concerning the possible involvement of further mediating mechanisms, and thus point to specific directions for future experimental investigation. Problems of analysis of mediation arise in different disciplines. In the natural sciences, direct effects reveal how Nature works. In the social sciences, they allow prediction of the effect of a normative intervention that blocks specific causal paths (for example the effect of sex discrimination on employment).

Thanks to the ability to formalize such notions as 'holding the mediating variables fixed' (as distinct from probabilistic conditionalization), researchers in causal inference have been able to provide analysis of mediation with a sound framework. Two chapters in this book contribute significant advances in this area. The first, Chapter 11, entitled 'Estimation of direct and indirect effects', is contributed by Stijn Vansteelandt. The second, Chapter 12, written by Judea Pearl, is entitled 'The mediation formula: a guide to the assessment of causal pathways in nonlinear models'.

Causal inference research has produced important advances in the analysis of mediation. First, the traditional framework involving only linear relationships has been generalized to allow arbitrary, nonparametric, relationships. Second, it has been recognized that, in a nonlinear system, the concept of direct effect can be defined in different ways, not all of which support a meaningful definition of a corresponding indirect effect. Third, the conditions under which well-established mediation analysis methods produce causally meaningful estimates (Baron and Kenny, 1986) have been formalized.

Our earlier distinction between inference about the consequences of an intervention (ICI) and inference about mechanism (IAM) deserves further elaboration. In statistical ICI, pre-existing scientific/mechanistic knowledge about the studied system may be helpful, but will rarely be essential to the design, analysis and interpretation of a study. The validity of a clinical trial, for example, does not typically depend (at least not in an essential way) on the availability of a deep scientific understanding of the studied process. By contrast, in statistical IAM, the causal validity of the analysis conclusions will often require assumptions justified by a deep scientific understanding of the involved mechanisms, as illustrated by the genetic study example of Chapter 14.

An interesting illustration of the potential interplay between statistical inference about mechanism and scientific knowledge is given in Chapter 23, 'Dynamic molecular networks and mechanisms in the biosciences: a statistical framework' by Clive Bowsher. This chapter illustrates the general principle that causality operates in time, rather than instantaneously, in this specific case being governed by a continuous-time jump process called a *stochastic kinetic model*. In Bowsher's chapter, a priori biological knowledge and experimental evidence from previous studies are combined into a causal graph representation of the process. Considerations based on this graph will suggest experimental interventions that may reliably improve our knowledge of the system, and lead to further, incremental, refinements of the model.

The conceptualization of causality in terms of the consequences of an intervention (e.g. of the effect of aspirin in terms of the consequences I observe when I administer it to a patient) has not been universally accepted. In a time-series context, for example, the traditional view of causality is in terms of dependence not explained away by other appropriate explanatory variables, with no reference to intervention. This idea, found, for example, in the work of Granger (1969) and Schweder (1970), although understandable in disciplines (like economics) where experimental interventions are difficult to implement, misses the requirement that causal relationships should persist if certain aspects of the system are changed. Recent work by Eichler and Didelez (2010) remedies this. In Chapter 22 of this book, 'Causal inference in time series analysis', Michael Eichler reconciles the two views, by incorporating the idea of intervention into causal graph representations of time-series models.

The benefits of randomization are not exclusive to controlled experiments. They are also available in 'natural experiments', where certain variables that causally affect the outcome can (in certain conditions) be regarded as generated by Nature in a random way, unaffected by potential sources of confounding. In Chapter 18, 'Natural experiments as a means of testing causal inferences', Michael Rutter offers a discussion of this opportunity, and illustrates it with the aid of a wealth of scientific examples. Just one example is where an individual's genotype G, at a specific DNA locus, may be regarded as generated, by Nature, by a random draw. Assuming that G does not affect the outcome except via changes to an intermediate phenotype M, the method called *Mendelian randomization* provides a vehicle for an unconfounded assessment of the causal effect of M on the outcome. This will often be possible only after appropriate conditioning, e.g. on a (natural or adoption) family or environment.

In many situations no randomization, be it achieved via experimental control or through the benevolence of Nature, can be invoked. One of these is discussed in Chapter 19, 'Nonreactive and purely reactive doses in observational studies', by Paul Rosenbaum. In this chapter, observational data is used to investigate a possible beneficial effect of higher doses of chemotherapy in ovarian cancer. The question is a difficult one, since most of the variation in treatment dose is a response to variation in the health of patients. The method suggested deals with potential confounding by capitalizing on that portion of the observed variation in treatment dose introduced by doctors with different views on dose prescription. This chapter illustrates the effectiveness of matching techniques and the usefulness of rigorous causal inference reasoning in the context of a scientific study.

In a recent paper on statistical causality, David Cox (2004) reminds us that deep conclusions often require synthesis of evidence of different kinds. Perhaps it is in this sense that we should interpret the Fisher's famous aphorism that in order to make observational studies more like randomized experiments we should 'make our theories elaborate'. This idea is illustrated in Chapter 15 by Luisa Bernardinelli, Carlo Berzuini, Luisa Foco, and Roberta Pastorino. The question whether ion channels are a possible cause of multiple sclerosis is here tackled by integrating different sources of statistical and experimental evidence into a coherent hypothesis about the role of specific molecular dysfunctions in susceptibility to multiple sclerosis.

Besides being an important topic in natural science disciplines, causality is important also in such normative areas as Forensic Law and Legal Philosophy. Although this book does not extensively cover this area, we should ask: How do the assumptions, methods, and requirements involved in the empirical evaluation of causal hypotheses in these normative areas differ from those relevant to the natural sciences?

In scientific applications we are typically interested in generic causal relationships. However, the focus in legal contexts will more often be on assigning causal responsibility in an individual case. Put otherwise, scientific causality is forward-looking and concerned with understanding the unfolding of generic causal processes, while legal causality is retrospective, concerned with unpicking the causes of specific observed effects. In other words, scientific causality is concerned with questions about the *effects of causes*, whereas legal causality is concerned about the *causes of effects*. This volume largely confines attention to the former enterprise; its contents amply attest that there is currently much debate and disagreement about how to formalize and analyse questions about the effects of causes. However, these already tricky philosophical and conceptual difficulties and disputes become greatly magnified when we turn to trying to understand the causes of observed effects, especially in a stochastic world (Dawid, 2011). Much work in this area tiptoes around the quicksands by ignoring stochasticity and assuming deterministic dependence of outcomes on sufficiently many inputs (unobserved as well as observed). Formalisms such as Mackie's INUS criterion are then applicable, and have proved popular with legal philosophers. Very close to INUS is a formalism that lawyers often refer to using the acronym NESS (Miller, 2011), which stands for 'Necessary Element of a Sufficient Set'. However, when, more realistically, we wish to allow genuine indeterminism, it turns out that it may simply be impossible, even with the best data in the world, to estimate causes of effects at the individual level without making arbitrary and empirically untestable additional assumptions. However, we can sometimes extract empirically meaningful interval bounds for relevant causal quantity, as illustrated, using potential outcome methods, in Chapter 10.

This book arose out of the International Meeting on 'Causal Inference : The State of the Art', that we organized in 2009 at the University of Cambridge, UK, with support from the European MolPage Project and the Cambridge Statistics Initiative, and under the auspices of the Royal Statistical Society. The book project has graced us with an enriching and stimulating experience. Profound thanks are due to the chapter authors for their commitment and enthusiasm. We also express our gratitude to the Wiley series team, first and foremost to Ilaria Meliconi for her early support for this project, to Richard Davies for his indefatigable and competent assistance, and to Prachi Sinha Sahay and Baljinder Kaur for their competent management of the copyediting.

Carlo Berzuini, Philip Dawid and Luisa Bernardinelli

References

Baron, R.M. and Kenny, D.A. (1986). The moderator–mediator variable distinction in social psychological research: Conceptual, strategic, and statistical considerations. *Journal of Personality and Social Psychology*, **51**, 1173–1182.

Cochran, W.G. (1957) Analysis of covariance: its nature and uses. *Biometrics*, **13**, 261–281.

Cochran, W.G. (1965) The planning of human studies of human populations (with discussion). *Journal of the Royal Statistical Society, Series A*, **128**, 234–266.

Cornfield, J., Haenszel, W., Hammond, E.C., Lilienfeld, A.M., Shimkin, B. and Wynder, E.L. (1959) Smoking and lung cancer: recent evidence and a discussion of some principles. *Journal of the National Cancer Institute*, **22**, 173–204.

Cox, D.R. (1960) Regression analysis when there is prior information about supplementary variables. *Journal of the Royal Statistical Society, Series B*, **22**, 172–176.

Cox, D.R. (2004) Causality: a statistical view. *International Statistical Review*, **72** (3), 285–305.

Cox, D.R. and P. McCullagh, P. (1982) Some aspects of analysis of covariance. *Biometrics*, **38**, 541–561.

Dawid, A.P. (1979) Conditional independence in statistical theory. *Journal of the Royal Statistical Society, Series B*, **41**, 1–31.

Dawid, A.P. (2011) The role of scientific and statistical evidence in assessing causality. In Richard Goldberg, editor, *Perspectives on Causation* (ed. R. Goldberg). Oxford: Hart Publishing, pp. 133–147.

Eichler, M. and Didelez, V. (2010) On Granger-causality and the effect of interventions in time series. *Life Time Data Analysis*, **16**, 332.

Fisher, R.A. (1932) *Statistical Methods for Research Workers*. Edinburgh: Oliver and Boyd.

Fisher, R.A. (1935 and also subsequent editions) *Design of Experiments*. Edinburgh: Oliver and Boyd.

Geiger, D., Verma, T. and Pearl, J. (1990) Identifying independence in Bayesian networks. *Networks*, **20** (5), 507–534.

Granger, C.W.J. (1969) Investigating causal relations by econometric models and cross-spectral method. *Econometrica*, **3** (3).

Hill, A.B. (1965) The environment and disease: association or causation? *Proceedings of the Royal Society of Medicine*, **58**, 295–300.

Mackie, J. (1965) Causes and conditions. *American Philosophical Quarterly*, **2**, 245–264.

Miller, C. (2011) NESS for beginners. In Richard Goldberg, editor, Perspectives on Causation (ed. R. Goldberg). Oxford: Hart Publishing, pp. 323–337.

Pearl, J. (2000) *Causality: models, reasoning, and inference*. Cambridge: Cambridge University Press, March 2000.

Pearl, J. (2005) Direct and indirect effects. In *Proceedings of the American Statistical Association Joint Statistical Meetings*, Technical Report R-273. MIRA Digital Publishing, pp.1572–1581.

Robins, J.M. (1986) A new approach to causal inference in mortality studies with a sustained exposure period: applications to control of the healthy worker survivor effect. *Mathematical Modeling*, **7**, 1393–1512.

Robins, J.M. and S. Greenland, S. (1992) Identifiability and exchangeability for direct and indirect effects. *Epidemiology*, **3**, 143–155.

Rothman, K.J. (1976) Causes. *American Journal of Epidemiology*, **104**, 587–592.

Rubin, D.B. (1974) Estimating causal effects of treatments in randomized and nonrandomized studies. *Journal of Educational Psychology*, **66**, 688–701.

Salmon, W. (1984) *Scientific Explanation and the Causal Structure of the World*. Princeton: Princeton University Press.

Schweder, T. (1970) Composable Markov processes. *Journal of Applied Probability*, **7**, 400–410.

Skrondal, A. (2003) Interaction as departure from additivity in case-control studies: a cautionary note. *American Journal of Epidemiology*, **158** (3).

VanderWeele, T.J. and Robins, J.M. (2008) Empirical and counterfactual conditions for sufficient cause interactions. *Biometrika*, **95** (1), 49–61.

VanderWeele, T.J. and Robins, J.M. (2009) Minimal sufficient causation and directed acyclic graphs. *Annals of Statistics*, **37** (3), 1437–1465.

Wright, S. (1921) Correlation and causation. *Journal of Agricultural Research*, **20**, 557–585.

Wright, S. (1934) The method of path coefficients. *Annals of Mathematical Statistics*, **5** (3), 161–215.

1

Statistical causality: Some historical remarks

D.R. Cox
Nuffield College, University of Oxford, UK

1.1 Introduction

Some investigations are essentially descriptive. Others are concerned at least in part with probing the nature of dependences.

Examples of the former are studies to estimate the number of whales in a portion of ocean, to determine the distribution of particle size in a river bed and to find the mortality rates of smokers and of nonsmokers. Examples of the second type of investigation are experiments to compare the effect of different levels of fertilizer on agricultural yield and investigations aimed to understand any apparent differences between the health of smokers and nonsmokers, that is to study whether smoking is the explanation of differences found. Also much, but not all, laboratory work in the natural sciences comes in this category.

Briefly the objectives of the two types are respectively to describe the world and in some sense to understand it. Put slightly more explicitly, in the agricultural field trial the object is essentially to understand how the yield of a plot would differ if this level of fertilizer were used rather than that level or, in smoking studies, how the outcomes of subjects who smoke compare with what the outcomes would have been had they not smoked. These are in some sense studies of causality, even though that word seems to be sparingly used by natural scientists and until recently by statisticians.

Neyman (1923) defined a basis for causal interpretation, using a working and not directly testable assumption of unit-treatment additivity specifying the way a particular experimental unit would respond to various treatments, only one of which could actually be used for a specific unit. In the absence of randomization specific consequences were unclear. His

Causality: Statistical Perspectives and Applications, First Edition. Edited by Carlo Berzuini, Philip Dawid and Luisa Bernardinelli.

later more ambitious work (Neyman *et al.*, 1935) is discussed below. The landmark paper on observational studies by Cochran (1965) did deal with causal interpretation and moreover in the discussion Bradford Hill outlined his considerations pointing towards stronger interpretation. Both authors emphasized the difficulties of interpretation in experiments and in observational contexts.

1.2 Key issues

Key issues are first what depth of understanding is involved in claims of causality, that is just what is meant by such a claim. Then there is the crucial matter of the security with which such causality can be established in any specific context. It is clear that defining models aimed at causal interpretation is desirable, but that calling such models causal and finding a good empirical fit to data is in some specific instances far short of establishing causality in a meaningful subject-matter sense. Cox and Wermuth (1996, pp. 219–) suggested *potentially causal* as a general term for such situations. This was criticized as overcautious by Pearl (2000). Indeed, overcaution in research is in general not a good idea. In the present situation, however, especially in a health-related context, caution surely is desirable in the light of the stream of information currently appearing, suggesting supposedly causal and often contradictory interpretations about diet and so forth.

1.3 Rothamsted view

The key principles of experimental design, developed largely at Rothamsted (Fisher, 1926, 1935; Yates, 1938, 1951), contain the essential elements of what is often needed to achieve reasonably secure causal conclusions in an experimental context, even though, as far as I can recall, the word cause itself is rarely if ever used in that period. An attempt to review that work largely in nontechnical terms (Cox, 1958a) centred the discussion on:

- a distinction between on the one hand experimental units and their properties and on the other hand treatments;
- a working assumption of unit-treatment additivity;
- a general restriction of supposedly precision-enhancing adjustments to features measured before randomization;
- a sevenfold classification of types of measurement with the above objectives in mind.

In more modern terminology the third of these points requires conditioning on features prior to the decision on treatment allocation and marginalization over features between treatment and final outcome. It presupposes, for example, the absence of what in a clinical trial context is called *noncompliance* or *nonadherence*. More generally, any intervention in the system between treatment allocation and response should either be independent of the treatment or be reasonably defined as an intrinsic part of the treatment. A possibly apocryphal agricultural example is that of a particular fertilizer combination so successful that the luxuriant growth of crop on the relevant plots attracted birds from afar who obtained their food solely from those plots. Consequently, these plots ended up showing a greatly depressed yield. Thus in a so-called intention-to-treat analysis the treatment was a failure, a conclusion correct in one sense but misleading both scientifically and in terms of practical implication.

If some measurements had been available plot by plot on the activity of the birds, the situation might have been partly rescued, although by making strong assumptions. In general, causal interpretation of randomized experimental designs may be inhibited by unobserved, uncontrolled and unwanted interventions in the period between implementing the randomization and the measurement of outcome. This consideration is of especial concern if there is an appreciable time gap between randomization and measurement of outcome.

Causal interpretation of observational studies, on the other hand, is handicapped primarily by the possibility of unobserved confounders, that is unobserved explanatory variables. When the effect measured as a relative risk is large some protection against the distortions induced by such unobserved variables is provided by Cornfield's inequality (Cornfield *et al.*, 1959).

1.4 An earlier controversy and its implications

Neyman and colleagues read to the Royal Statistical Society (Neyman *et al.*, 1935) an account of work in which the notion of unit-treatment additivity, stemming from Neyman (1923) and in a sense implicitly underpinning the Rothamsted work on design, was replaced by a much more general assumption in which different experimental units had different treatment effects. A provisional conclusion of the analysis was that the standard estimate of error for a Latin square design is inappropriate. This suggestion led to a vehement denunciation by R.A. Fisher, recorded in appreciable detail in the published version of the discussion. The general issues of the role of randomization were further discussed in the next few years, mostly in *Biometrika*, with contributions from Student, Yates, Neyman and Pearson, and Jeffreys. With the exception of Student's contribution, which emphasized the role of randomization in escaping biases arising from personal judgement, the discussion focused largely on error estimation. Interestingly, the aspect most stressed in current writing, namely the decoupling from the estimation of treatment effects of unobserved systematic explanatory features, is not emphasized.

The specific issue of the Latin square was developed in more detail nearly 20 years later by Kempthorne and his colleagues, confirming (Kempthorne and Wilk, 1957) the bias of the error estimate in the Latin square analysis. Cox (1958b) pointed out the unreality of the null hypothesis being tested, namely that in spite of unexplained variation in treatment effect the null hypothesis considered was that the treatment effects balanced out *exactly* over the finite set of units in a trial. When a more realistic null hypothesis was formulated the biases disappeared, restoring the respectability of the Latin square analysis, as argued by Fisher.

While the status of error estimation in the Latin square may seem a rather specialized or even esoteric matter in the present context, the relevance for current discussions is this. It is common to define an average causal effect in essentially the way that Neyman *et al.* did, that is without any assumption that the effect is identical for all units of study. Of course, if the variation in treatment effect between units of study can be explained in substantive terms that is preferable, but if not, it should in some sense be treated as stochastic, and that affects the assessment of error.

The point is similar in spirit to what is sometimes called the marginalization principle, namely that it rarely makes sense to consider models with nonzero interaction but exactly zero main effects.

1.5 Three versions of causality

In general terms, it is helpful to distinguish three types of statistical causality, all of importance in appropriate contexts (Cox and Wermuth, 2004). The first is that essentially a multiple-regression like analysis shows a dependence not explained away by other appropriate explanatory variables. In a time series context this forms Wiener–Granger causality; see, also, the more general formulation by Schweder (1970). The second definition is in the spirit of the previous section in terms of a real or notional intervention and implies a restriction on the nature of variables to be treated as possibly causal. This is the approach that has received most attention in recent years (Holland, 1986). The third notion requires some evidence-based understanding in terms of the underlying process.

The general principle that causality operates in time and rarely instantaneously has implications for model formulation. Thus if two variables measured as discrete-time series (X_t, Y_t) are such that each component in some sense causes the other a suitable formulation is that (X_{t+1}, Y_{t+1}) depends on (X_t, Y_t), with a corresponding differential equation form in continuous time. There are appreciable difficulties for subjects like macro-economics, in which data are typically quite heavily aggregated in time (Hoover, 2001).

1.6 Conclusion

It seems clear that the objective of many lines of research is to establish dependencies that are in some sense causal. Models and associated methods of statistical analysis that point towards causality are therefore very appealing and important. However, the circumstances under which causality in a meaningful sense can be inferred from a single study may be relatively restricted, mainly to randomized experiments with clear effects and no possibility of appreciable noncompliance and sometimes to observational studies in which large relative risks are encountered. Really secure establishment of causality is most likely to emerge from qualitative synthesis of the conclusions from different kinds of study. A fine example is the paper on smoking and lung cancer by Cornfield *et al.* (1959), reprinted in 2009. Here large-scale population data, the outcomes of longitudinal prospective and retrospective studies and the results of laboratory work were brought together to provide a powerful assertion of causal effect in the face of some scepticism at the time.

References

Cochran, W.G. (1965) The planning of human studies of human populations (with discussion). *Journal of the Royal Statistical Socity science* A, **128**, 234–266.

Cornfield, J., Haenszel, W., Hammond, E.C., Lilienfeld, A.M., Shimkin, M.B. and Wynder, E.L. (1959) Smoking and lung cancer: recent evidence and a discussion of some questions. *Journal of the National Cancer Institute* **22**, 173–203. Reprinted 2009 in *International Journal of Epidemiology*, **38**, 1175–1191.

Cox, D.R. (1958a) *Planning of Experiments*. New York: John Wiley & Sons, Inc.

Cox, D.R. (1958b) The interpretation of the effects of non-additivity in the Latin square. *Biometrika*, **45**, 69–73.

Cox, D.R. and Wermuth, N. (1996) *Multivariate Dependencies*. London: Chapman & Hall.

Cox, D.R. and Wermuth, N. (2004) Causality: a statistical view. *International Statistical Review*, **72**, 285–305.

Fisher, R.A. (1926) The arrangement of field experiments. *Journal of the Ministry of Agriculture Great Britain*, **33**, 503–513.

Fisher, R.A. (1935) *Design of Experiments*. Edinburgh: Oliver and Boyd, and subsequent editions.

Holland, P.W. (1986) Statistics and causal inference (with discussion). *Journal of the American Statistical Association*, **81**, 945–970.

Hoover, K.V. (2001) *Causality in Macroeconomics*. Cambridge: Cambridge University Press.

Kempthorne, O. and Wilk, M.B. (1957) Nonadditivity in a Latin square design. *Journal of the American Statistical Association*, **52**, 218–236.

Neyman, J. (1923) On the application of probability theory to agricultural experiments. Essay on principles. *Rocniki Nauk Rolniiczych*, **10**, 1–51. English translation of Section 9 by D.M. Dabrowska and T.P. Speed, *Statistical Science*, **9**, 465–480.

Neyman, J., Iwaszkiewicz, K. and Kolodziejczyk, St. (1935) Statistical problems in agricultural experimentation (with discussion). *Supplement of the Journal of the Royal Statistical Society*, **2**, 107–180.

Pearl, J. (2000) *Causality*. Cambridge: Cambridge University Press, 2nd edn, 2010.

Schweder, T. (1970) Composable Markov processes. *Journal Applied Probability*, **7**, 400–410.

Yates, F. (1938) *Design and Analysis of Factorial Experiments*. Harpenden: Imperial Bureau of Soil Science.

Yates, F. (1951) Bases logiques de la planification des experiences. *Annals of the Institute of H. Poincaré*, **12**, 97–112.

2

The language of potential outcomes

Arvid Sjölander
Department of Medical Epidemiology and Biostatistics, Karolinska Institutet, Stockholm, Sweden

2.1 Introduction

A common aim of empirical research is to study the association between a particular exposure and a particular outcome. In epidemiology, for instance, many studies have been conducted to explore the association between smoking and lung cancer. The ultimate goal, however, is often more ambitious; typically we want to establish a causal relationship, e.g. that smoking actually causes lung cancer. It is well known that association does not imply causation; an observed association between smoking and lung cancer may have several explanations, which do not necessarily involve a causal smoking effect. Thus, an important question for empirical researchers is this:

> Under what conditions can an observed association be interpreted as a causal effect?

This question is logically precluded by the following:

> What do we mean, more precisely, when we talk about causal effects?

Despite their obvious relevance, these questions were not treated formally in the statistical literature until the late 1970s. The dominating paradigm was that 'statistics can only tell us about association and not causation'. Thus, for most of the 20th century, causality remained an

Causality: Statistical Perspectives and Applications, First Edition. Edited by Carlo Berzuini, Philip Dawid and Luisa Bernardinelli.

ill-defined concept, and empirical researchers who wanted to draw causal conclusions from data had to resort to informal reasoning and justification. During the end of the century, however, a formal theory of causal inference emerged, based on *potential outcomes*. The foundation was laid out by Rubin (1974, 1978, 1980) for point exposures, that is exposures which are measured at a single point in time. Robins (1986, 1989) extended the potential outcome framework to time-varying exposures. Pearl (1993) demonstrated that important conceptual insights can be made by interpreting the potential outcomes as solutions to a nonparametric structural equation (NPSE) system. Galles and Pearl (1998) provided a complete axiomatization of the framework. In this chapter we provide a brief summary of the central ideas in the potential outcomes framework, as formulated by Rubin (1974, 1978, 1980). Specifically, we demonstrate in Sections 2.2 and 2.3 how the two questions above can be formally phrased and answered within this framework.

2.2 Definition of causal effects through potential outcomes

2.2.1 Subject-specific causal effects

Suppose that a particular subject dies after a surgical intervention. Did the intervention cause the subject's death? A natural way to answer this question would be to try to imagine what would have happened to this subject had he not been operated on. If the subject would have died anyway, we would not say that intervention was the cause. If, on the other hand, the subject would not have died had he not been operated on, then we would say that intervention caused the death. Thus, to determine whether an exposure causes an outcome we typically make a mental comparison between two scenarios; one where the exposure is present and one were the exposure is absent. If the outcome differs between the two scenarios we say that exposure has a causal effect on the outcome. The potential outcomes framework formalizes this intuitive approach to causality.

Specifically, suppose that we want to learn about the causal effect of a particular point exposure on a particular outcome. Let Y_x denote the outcome for a randomly selected subject in the study population, if the subject would hypothetically receive exposure level x. Depending on what exposure level the subject actually receives, Y_x may or may not be realized. For this reason, Y_x is referred to as a *potential* outcome. Let $\mathbb{X}$ be the set of all possible exposure levels. Each subject is coupled with a set of potential outcomes, $\{Y_x\}_{x\in\mathbb{X}}$, which completely describes how the outcome would look like for the particular subject, under each possible level of the exposure. For instance, when the exposure is binary we may define $\mathbb{X} = \{0, 1\}$, with 0 and 1 corresponding to 'unexposed' and 'exposed', respectively. In this case, each subject is coupled with a potential outcome vector $\{Y_0, Y_1\}$, where Y_0 is the outcome if the subject is unexposed and Y_1 is the outcome if the subject is exposed.

In practice, the outcome for a given subject may depend on the exposure and/or outcome for other subjects in the population. This would, for example, be the case in a vaccine intervention trial for an infectious disease, where the infection status of one subject depends on whether other subjects are infected as well. When subjects interfere in this way, a causal analysis based on potential outcomes becomes more involved. We will proceed by assuming that the outcome for a given subject does not depend on the exposure or outcome for other subjects. This assumption is a part of the 'stable-unit-treatment-value-assumption (SUTVA)' by Rubin

(1980). See Hudgens and Halloran (2008) and Tchetgen Tchetgen and VanderWeele (2010) for extensions to scenarios with subject interference.

Using potential outcomes, we define the *subject-specific causal effect* of exposure level x' versus x'' as some contrast of the potential outcomes $Y_{x'}$ and $Y_{x''}$, for example $Y_{x'} - Y_{x''}$. When the exposure is binary, the subject-specific causal effect is a contrast between Y_0 and Y_1. If, for a given subject, all potential outcomes are equal (i.e. Y_x does not depend on x), then, for this subject, the exposure has no causal effect on the outcome. If the exposure has no causal effect on the outcome for any subject in the study population, then we say that the *sharp causal null hypothesis* holds. A fundamental problem with subject-specific causal effects is that they are notoriously difficult to identify. This is because we cannot in general observe the same subject under several exposure levels simultaneously. Let X and Y denote the observed exposure and outcome, respectively, for a given subject. If a subject is exposed to level $X = x'$, then the potential outcome $Y_{x'}$ is assumed to be equal to the observed factual outcome Y for that subject. This link between the potential outcomes and the factual outcome is usually referred to as the 'consistency assumption', and is formally expressed as

$$X = x' \Rightarrow Y_{x'} = Y \tag{2.1}$$

We remain ignorant, however, about would have happened to the subject had it been exposed to some other level. Thus, for a subject who is exposed to level $X = x'$, all potential outcomes in $\{Y_x\}_{x \in \mathbb{X}}$, except $Y_{x'}$, are unobserved, or *counterfactual*. The word 'counterfactual' echoes the fact that the unobserved potential outcomes correspond to scenarios which are 'contrary to fact' – they did not happen. When the exposure is binary, Y_0 is unobserved if the subject is exposed and Y_1 is unobserved if the subject is unexposed. Thus, constrasts between Y_0 and Y_1 cannot be observed for any subject. Because subject-specific causal effects are in general not identifiable, they are of limited practical value.[1]

2.2.2 Population causal effects

A more useful concept is the *population causal effect*, which measures the aggregated impact of the exposure over the study population. Because the potential outcome Y_x may vary across subjects, we may treat it as a random variable, following a probability distribution $\Pr(Y_x)$ (we use $\Pr(\cdot)$ generically for both probabilities and densities). We interpret $\Pr(Y_x = y)$ as the population proportion of subjects with an outcome equal to y under the hypothetical scenario where everybody receives exposure level x. The population causal effect of exposure level x' versus x'' is defined as a contrast between the potential outcome distributions $\Pr(Y_{x'})$ and $\Pr(Y_{x''})$, for example the causal mean difference $E(Y_{x'}) - E(Y_{x''})$. When the outcome Y is binary, it would be natural to consider the causal risk ratio $\Pr(Y_x = 1)/\Pr(Y_{x'} = 1)$ or the causal odds ratio $[\Pr(Y_x = 1)/\Pr(Y_x = 0)]/[\Pr(Y_{x'} = 1)/\Pr(Y_{x'} = 0)]$. If $\Pr(Y_x)$ does not depend on x, then the exposure has no population causal effect on the outcome; we say that the causal null hypothesis holds. The sharp causal null hypothesis implies the causal null hypothesis. The converse is not true though; it is logically possible that the exposure has a causal effect for some subjects, but that these effects 'cancel out' in such a way that there is no aggregated effect over the population. In Section 2.3 we demonstrate that the population causal effect can

[1] A rare exception is when we are able to observe the same subject under several exposure levels subsequently, without any crossover effects. In these situations, subject-specific causal effects can be identified.

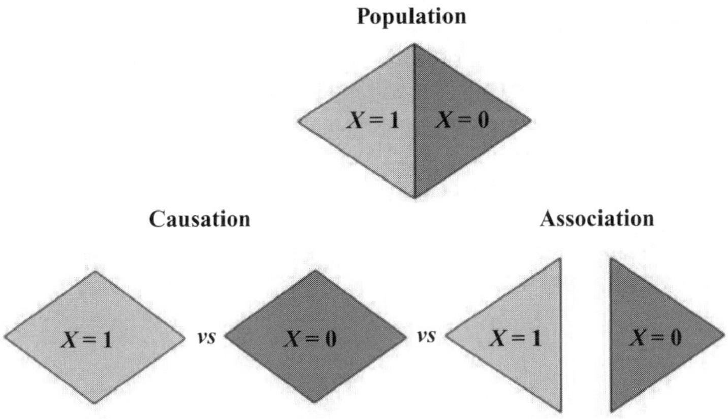

Figure 2.1 Association versus causation.

be identified even if we are not able to observe the same subject under several exposure levels simultaneously.

2.2.3 Association *versus* causation

Using potential potential outcomes, the fundamental difference between association and causation becomes clear. In the population of Figure 2.1 (adopted from Hernán and Robins forthcoming), some subjects are exposed ($X = 1$) and some subjects are unexposed ($X = 0$). We say that exposure and outcome are associated in the population if the outcome distribution differs between the exposed and unexposed. To quantify the association we may, for instance, use the mean difference $E(Y|A = 1) - E(Y|A = 0)$ or the risk ratio $\Pr(Y = 1|A = 1)/\Pr(Y = 1|A = 0)$. Thus, when we assess the exposure–outcome association we are by definition comparing two *different groups of subjects*: those who are factually exposed against those who are factually unexposed. In contrast, the population causal effect compares the potential outcomes for *the same* subjects (the whole population) under two hypothetical scenarios: everybody being exposed versus everybody being unexposed. This fundamental difference is the reason why association is in general not equal to causation. When we compare different subjects, there is always a risk that the subjects are different in other aspects than in the received exposure levels. If they are, then we may observe different outcome distributions for the exposed and the nonexposed, even if the exposure has no causal effect on the outcome.

2.3 Identification of population causal effects

2.3.1 Randomized experiments

As discussed in the previous section, population causal effects compare the outcome distributions for the *same* population under two different exposure levels. In practice, however, we cannot observe the same population under several exposure levels simultaneously. Nevertheless, intuition tells us that observed associations (i.e. comparisons between *different* subjects) can be interpreted as causal effects under certain conditions. Specifically, it is widely accepted that 'association equals causation' in well-designed randomized experiments. Using potential outcomes we can give formal support to this notion. If the exposure is properly randomized, then all pre-randomization variables (i.e. variables whose values are determined prior to randomization, e.g. age, sex, eye color) are equally distributed, asymptotically, across levels of the exposure. Now, note that the potential outcomes $\{Y_x\}_{x\in\mathbb{X}}$ are by definition pre-randomization variables; they describe how the given subject will react to different exposure levels, which depends on intrinsic subject characteristics determined prior to randomization. Thus, in a randomized experiment, the potential outcomes are statistically independent of the exposure. We express this formally as

$$\{Y_x\}_{x\in\mathbb{X}} \perp\!\!\!\perp X \tag{2.2}$$

When (2.2) holds, subjects are said to be *exchangeable* across exposure levels.[2] Under consistency (2.1) and exchangeability (2.2), the conditional probability of Y, among those who actually received exposure level x, is equal to the probability of Y, had everybody received level x:

$$\Pr(Y = y|X = x) = \Pr(Y_x = y|X = x) = \Pr(Y_x = y) \tag{2.3}$$

The first equality in (2.3) follows from (2.1) and the second equality follows from (2.2). Thus, under consistency and exchangeability, any measure of association between X and Y equals the 'corresponding' population causal effect of X on Y. For instance, the associational mean difference $E(Y|A = 1) - E(Y|A = 0)$ equals the causal mean difference $E(Y_1) - E(Y_0)$ and the associational relative risk $\Pr(Y = 1|X = 1)/\Pr(Y = 1|X = 0)$ equals the causal risk ratio $\Pr(Y_1 = 1)/\Pr(Y_0 = 1)$. Because randomization produces exchangeability it follows that population causal effects are identifiable in randomized experiments.

We make a technical remark. Exchangeability, as defined in (2.2), means that all potential outcomes in $\{Y_x\}_{x\in\mathbb{X}}$ are *jointly* independent of X. Although this is a sufficient criterion for identification of the population causal effects, it is slightly stronger than necessary. By inspecting (2.3) we observe that $\Pr(Y_x)$ is identified for all x if the potential outcomes in $\{Y_x\}_{x\in\mathbb{X}}$ are *separately* independent of X:

$$Y_x \perp\!\!\!\perp X \quad \forall x \tag{2.4}$$

In the literature, the word 'exchangeability' is sometimes used for the relation in (2.4). Rosenbaum and Rubin (1983) referred to (2.2) and (2.4) as 'strongly ignorable treatment assignment' and 'ignorable treatment assignment', respectively. Strong ignorability implies ignorability, but not the other way around. It is difficult, however, to imagine a practical scenario where the latter holds but not the former.

[2] We note that this definition of 'exchangeable' is different from the definition used in other branches of statistics, e.g. de Finetti (1974).

2.3.2 Observational studies

For ethical or practical reasons, randomization of the exposure is not always feasible. Indeed, many scientific 'truths' have been established through observational (i.e. nonrandomized) studies. When the exposure is not randomized, exchangeability does not necessarily hold, and an observed association cannot in general be interpreted as a causal effect. Violations of (2.2) typically occur when the exposure and the outcome have common causes. As an illustration, suppose that we wish to study the effect of a medical treatment (X) on the prognosis (Y) for patients with a certain disease. Suppose that a patients's general health status affects what treatment level the patient is assigned to (patients in a critical condition may, for example, receive higher doses than patients in a noncritical condition). Moreover, a patient's health status clearly affects his/her prognosis. That a patients's health status affects both treatment and prognosis implies that X and $\{Y_x\}_{x \in \mathbb{X}}$ are associated, which violates (2.2). When the exposure and outcome have common causes we say that the exposure–outcome association suffers from 'confounding'. The standard way to deal with confounding is to 'adjust for' (i.e. condition on) a selected set of potential confounders, for example by stratification or regression modeling. The rationale for this approach is that after adjustment, it may be reasonable to consider the exposure as being randomized 'by nature'. Formally, the aim of confounding adjustment is to produce conditional exchangeability:

$$\{Y_x\}_{x \in \mathbb{X}} \perp\!\!\!\perp X | C \tag{2.5}$$

Under consistency (2.1) and conditional exchangeability (2.5), $\Pr(Y = y|X = x, C) = \Pr(Y_x = y|C)$. It follows that any measure of the conditional association between X and Y, given C, equals the corresponding conditional population causal effect. For instance, $\Pr(Y = 1|X = 1, C)/\Pr(Y = 1|X = 0, C)$ equals $\Pr(Y_1 = 1|C)/\Pr(Y_0 = 1|C)$. The population (i.e. not C-specific) causal effect can be computed through 'standardization', i.e. by averaging over the marginal confounder distribution:

$$\Pr(Y_x = y) = E\{\Pr(Y = y|X = x, C)\}$$

2.4 Discussion

The potential outcome framework provides a natural definition of causal effects, based on the idea of comparing the same subjects under different exposure levels. Furthermore, it provides an elegant characterization of (conditional) exchangeability, which is the necessary requirement for drawing causal conclusions from observational data. During the last decades, the potential outcome framework has proven to be extremely useful, both in conceptualizing causal concepts (e.g. confounding and direct effects) and in the development of new methodology (e.g. propensity scores (Rosenbaum and Rubin, 1983), g-estimation (Robins, 1986), inverse probability weighting (Robins, 1997), and principal stratification (Frangakis and Rubin, 2002)).

One weakness of the potential outcome framework is that it is not so intuitive in the formulation and judgement of prior assumptions. Typically, subject matter experts understand well the causal structure of the problem (e.g. which the important confounders are and how the confounders are causally related to each other). In the potential outcome framework, however, such expert knowledge must be recoded as independence statements involving counterfactual variables, as in (2.5). This task may be very difficult, since counterfactual independencies are rather technical conditions, which are often awkward to interpret. Pearl (2010) argued that in the potential outcome framework 'it is hard to ascertain whether all relevant judgments have been articulated, whether the judgments articulated are redundant, or whether those judgments are self-consistent'. Major progress was made when Pearl (1993) demonstrated that potential outcomes can be emulated through nonparametric structural equation (NPSE) systems. In an NPSE system, subject matter knowledge about underlying causal structures can be modeled explicitly, without having to be recoded into counterfactual independencies. All relevant counterfactual independencies can then be derived from the postulated causal model through simple algorithms. Pearl (2009) provides an exhaustive review of NPSEs and their relation to potential outcomes (see also Chapter 3 in this volume).

One common criticism of counterfactuals is that their empirical content is often somewhat vague. For example, suppose that we carry out a study to evaluate the effect of obesity on the risk for cardiovascular disease (CVD). A commonly used proxy for obesity is body mass index (BMI). Using this proxy we may, for example, formulate the research question as 'What is the effect of an increase in BMI from 25 to 35, on the risk for CVD?' In the potential outcome framework, this question calls for a comparison between two hypothetical scenarios: one where everybody has BMI equal to 25 and one where everybody has BMI equal to 35. But what does this mean? For any given subject, we can easily imagine a wide range of body compositions, all yielding a BMI equal to 35, but crucially different in other outcome-related aspects. A subject with BMI equal to 35 may, for instance, have either an extreme excess of body fat or an extreme excess of muscles. Since the research question allows for these crucially different interpretations, we may consider the question as being ill-posed and the underlying counterfactuals as being vague. The example highlights an important difference between association and causation. We would typically consider the BMI–CVD association to be a well-defined concept, because we agree on what it means to factually have a certain BMI level. However, in order for the causal effect to be well defined, this is not enough; we must also agree on what it means to have a *counter*factual BMI level, e.g. to have BMI equal to 35 when the factual BMI is equal to 25. This latter requirement is much stronger, and typically a source vagueness in causal effects, which is not present in measures of pure associations. One way to reduce such vagueness is to make precise the definition of obesity, including, for instance, the ratio of muscles to body fat, placement of fat on the body, etc. Under a sufficiently detailed definition of obesity, the research question may no longer be interpretable in crucially different ways.

Another option, which has been promoted by several authors (e.g. Robins and Greenland, 2000, and Hernán, 2005), is to make precise a hypothetical intervention able to produce the counterfactual scenarios of interest. Possible interventions that could potentially yield a specific BMI level are, for example, a strict diet or training program, surgery, genetic modification, etc. It is important to recognize, though, that although the latter strategy may result in a well-defined research question, it combines any 'intrinsic' effect of obesity with potential side-effects of the intervention. For instance, a training program may reduce a subject's risk for CVD, even if it fails to reduce the subject's BMI levels. Whether this feature is desirable or not clearly depends on the ultimate aim of the study. In practical scenarios, training programs may be the most realistic way to modify peoples BMI levels. Thus, if the aim is to guide policy makers rather than learning about the 'ethiology of obesity', then the effect of a training program may be a more suitable target for analysis than the effect of obesity per se. Chapter 9 in this volume provides a discussion of counterfactuals for nonmanipulable exposures.

In most applications, counterfactuals are assumed to be deterministic quantities. That is, if we could somehow 'play back time' and observe the same subject repeatedly under the same exposure level, it is assumed that we would always observe the same outcome. Whether this assumption can ever be reasonable has been debated in the causal inference literature. Pearl (2000) argued that the assumption follows naturally if we view a 'subject' as 'the sum of all experimental conditions that might possibly affect the individual's reaction, including biological, physiological, and spiritual factors, operating both before and after the application of the treatment'. Dawid (2000) argued that '... any attempt to refine the description of [subject] u still further, so as to recover the desired [deterministic] structure, surely risks becoming utterly ridiculous, and of no relevance to the kind of causal questions we really wish to address'. Although this debate may appear somewhat esoteric it has important practical implications; any results derived from a deterministic counterfactual model may be meaningless if the world is truly stochastic. Fortunately, some of the results that have been derived under deterministic counterfactual models have also been reproduced under models utlilizing stochastic couterfactuals (e.g. Dawid, 2003).

References

Dawid, A.P. (2000) Causal inference without counterfactuals. *Journal of the American Statistical Association*, **95** (450), 407–448.

Dawid, A.P. (2003) Causal inference using influence diagrams: the problem of partial compliance (with discussion), in *Highly Structured Stochastic Systems* (eds P.J. Green, N.L. Hjort and S. Richardson). Oxford University Press, pp. 45–81.

de Finetti, B. (1974) *Theory of Probability*. New York: John Wiley & Sons, Inc.

Frangakis, C.E. and Rubin, D.B. (2002) Principal stratification in causal inference. *Biometrics*, **58**, 21–29.

Galles, D. and Pearl, J. (1998) An axiomatic characterization of causal counterfactuals. *Foundations of Science*, **3** (1), 151–182.

Hernán, M.A. (2005) Invited commentary: hypothetical interventions to define causal effects–afterthought or prerequisite? *The American Journal of Epidemiology*, **162** (7), 618–620.

Hernán, M.A. and Robins, J.M. (2006) Instruments for causal inference: an epidemiologists dream? *Epidemiology*, **17** (4), 360–372.

Hernán, M.A. and Robins, J.M. (forthcoming) *Causal Inference*. Chapman & Hall/CRC.

Hudgens, M.G. and Halloran, M.E. (2008) Toward causal inference with interference. *Journal of the American Statistical Association*, **103** (482), 832–842.

Pearl, J. (1993) Aspects of graphical models connected to causality, in *Proceedings of the 49th Session of the International Statistical Institute*, Florence, Italy, Tome LV, Book 1, pp. 399–401.

Pearl, J. (2000) The logic of counterfactuals in causal inference (Discussion of 'Causal inference without counterfactuals' by A.P. Dawid). *Journal of the American Statistical Association*, **95** (450), 428–431.

Pearl, J. (2001) Direct and indirect effects, in *Proceedings of the Seventeenth Conference on Uncertainty in Artificial Intelligence*. (eds D. Koller and J. Breese). San Francisco, CA: Morgan Kaufmann, pp. 411–420.

Pearl, J. (2009) *Causality: Models, Reasoning and Inference*, 2nd edn. Cambridge: Cambridge University Press.

Pearl, J. (2010) An introduction to causal inference. *The International Journal of Biostatistics*. **6** (2), Article 7.

Robins, J.M. (1986) A new approach to causal inference in mortality studies with sustained exposure periods–application to control of the healthy worker survivor effect. *Mathematical Modelling*, **7**, 1393–1512.

Robins, J.M. (1989) The analysis of randomized and non–randomized AIDS treatment trials using a new approach to causal inference in longitudinal studies, in *Health Service Research Methodology: a focus on AIDS* (eds L. Sechrest, H. Freeman and A. Mulley) pp. 113–159.

Robins, J.M. (1997) Marginal structural models. In *Proceedings of the American Statistical Association. Section on Bayesian Statistical Science*, pp. 1–10.

Robins, J.M. and Greenland, S. (2000) Discussion of 'Causal inference without counterfactuals' by A.P. Dawid. *Journal of the American Statistical Association*, **95** (450), 431–435.

Rosenbaum, P.R. and Rubin, D.B. (1983) The central role of propensity score in observational studies for causal effects. *Biometrika*, **70**, 41–55.

Rubin, D.B. (1974) Estimating causal effects of treatments in randomized and nonrandomized studies. *Journal of Educational Psychology*, **66** (5), 688–701.

Rubin, D.B. (1978) Bayesian inference for causal effects: the role of randomization. *The Annals of Statistics*, **6** (1), 34–58.

Rubin, D.B. (1980) Discussion of 'Randomization analysis of experimental data: the Fisher randomization test' by D. Basu. *Journal of the American Statistical Association*, **75**, 591–593.

Tchetgen Tchetgen. E.J. and VanderWeele, T.J. (2010) On causal inference in the presence of interference. *Statistical Methods in Medical Research*. DOI: 10.1177/0962280210386779.

3

Structural equations, graphs and interventions

Ilya Shpitser
Department of Epidemiology, Harvard School of Public Health, Boston, Massachusetts, USA

3.1 Introduction

Causality is fundamental to our understanding of the natural world. Causal statements are a part of everyday speech, as well as legal, scientific and philosophical vocabulary. Human beings reach an intuitive consensus on the meaning of many causal utterances and there have been numerous attempts to formalize causality in a way that is faithful to this consensus (Wright, 1921; Neyman, 1923; Tinbergen, 1937; Lewis, 1973; Rubin, 1974; Robins, 1987; Pearl, 2000).

The notion central to most of these formalizations is that of an intervention – a kind of idealized randomized experiment imposed on a 'set of units', which can represent patients in a medical trial, a culture of cells, and so on. Many scientific questions can be phrased in terms of effects of such experiments. Two influential formalizations of causation based on interventions are the framework of potential outcomes (Neyman, 1923; Rubin, 1974) and the framework of nonparametric structural equation models (NPSEMs) (Pearl, 2000).

The potential outcome framework, discussed by Sjölander in Chapter 2 of this volume, considers the effect of an intervention on a *treatment variable* A on an *outcome variable* Y. The resulting potential outcome, which we – in accord with the previous chapter – shall denote by Y_a, or $Y(a)$, is a random variable representing the behavior of Y in a hypothetical situation where the value of a random variable A was set to a, without regard for the normal behavior of A, a setting that may be implemented by randomization in practice. In the potential outcome framework, causal questions are encoded as statements about joint distributions made up

Causality: Statistical Perspectives and Applications, First Edition. Edited by Carlo Berzuini, Philip Dawid and Luisa Bernardinelli.

of potential outcome random variables, and causal assumptions are expressed as various restrictions on these distributions.

Nonparametric structural equation models define a very general data-generating mechanism suitable for encoding causation. In such models observable variables are functions of other observable variables, and possibly of noise terms, with the entire model behaving as a kind of stochastic circuit. An intervention on a variable A in an NPSEM replaces the function determining the value of A by a constant function, resulting in a new circuit. Causal relationships between variables in NPSEMs can be summarized by means of a graph called a causal diagram, with some causal assumptions entailed by the model represented by missing edges in the graph. NPSEMs induce joint distributions over pre- or post-intervention variables. These distributions provide a close link between the NPSEM formalism and the potential outcome formalism.

In this chapter we introduce the NPSEMs framework along with the relevant background from graph theory and probability theory, show the close connection between structural equations and potential outcomes, and argue for the conceptual complementarity of the two frameworks, with NPSEMs providing transparency and visual intuition of graphs, and potential outcomes giving a coherent way of thinking about causation in the "epistemically impoverished" situations frequently encountered in scientific practice.

3.2 Structural equations, graphs, and interventions

3.2.1 Graph terminology

We first introduce graph-theoretic terminology which is necessary for discussing NPSEMs in this chapter, and identifiability problems in Chapter 6 of this volume. In mathematics and computer science, graphs are used to model objects, represented by nodes, and their relationships, represented by edges. In this chapter, we will restrict our attention to graphs where the edges are either directed (contain a single arrowhead), or bidirected (contain two arrowheads). More specifically, we shall restrict our attention to directed and mixed graphs. A *directed graph* consists of a set of nodes and directed arrows connecting pairs of nodes. A *mixed graph* consists of a set of nodes and directed and/or bidirected arrows connecting pairs of nodes. A *path* is a sequence of distinct nodes where any two adjacent nodes in the sequence are connected by an edge. A *directed path* from a node X to a node Y is a path where all arrows connecting nodes on the path have an arrowhead pointing away from X and towards Y. If an arrow has a single arrowhead pointing from X to Y, then X is called a *parent* of Y and Y a *child* of X. If X and Y are connected by a bidirected arrow, they are called *spouses*. The set of spouses of a node X is denoted by $Sp(X)$. If X has a directed path to Y then X is an *ancestor* of Y, and Y a *descendant* of X. Non-descendants, descendants, ancestors, parents, and children of X are denoted, respectively, by $Nd(X)$, $De(X)$, $An(X)$, $Pa(X)$, and $Ch(X)$. By convention, X is both an ancestor and a descendant of X. A directed acyclic graph (DAG) is a directed graph where for any directed path from X to Y, Y is not a parent of X. An *acyclic directed mixed graph* (ADMG) is a mixed graph, which is a DAG if restricted to directed edges.

A consecutive triple of nodes W_i, W_j, W_k on a path is called a *collider* if the edge from W_i to W_j and the edge from W_k to W_j both have arrowheads pointing to W_j. Any other consecutive triple is called a *noncollider*. A path between two nodes X and Y is said to be *blocked* by a set Z if either for some noncollider on the path, the middle node is in Z, or for some collider

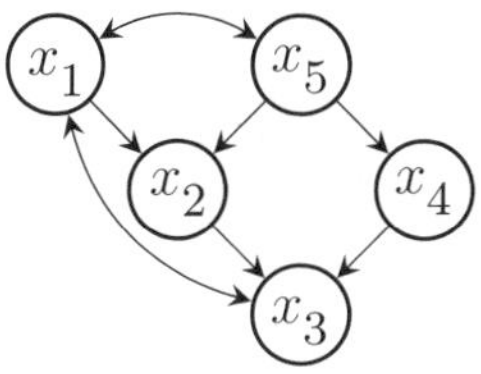

Figure 3.1 An acyclic directed mixed graph.

on the path, no descendant of the middle node is in Z. For disjoint sets X, Y, Z of nodes in an ADMG we say X is m-separated from Y given Z if every path from a node in X to a node in Y is blocked by Z. If the ADMG is also a DAG, then we say X is d-separated from Y given Z. If X is not m-separated (or d-separated) from Y given Z, we say X is m-connected (or d-connected) to Y given Z. See the graph in Figure 3.1 for an illustration of these concepts. In this graph:

- $X_1 \rightarrow X_2 \rightarrow X_3 \leftarrow X_4$ is a path from X_1 to X_4;
- $X_1 \rightarrow X_2 \rightarrow X_3$ is a directed path from X_1 to X_3;
- X_1 is a parent of X_2 and an ancestor of X_3;
- $X_2 \rightarrow X_3 \leftarrow X_4$ is a collider;
- X_1 is m-separated from X_4 given X_5;
- X_1 is m-connected to X_4 given X_5 and X_3.

3.2.2 Markovian models

A Markovian model, also known as a nonparametric structural equation model (NPSEM), is a tuple $\langle U, V, F, P(u) \rangle$, where the tuple elements are defined as follows. The symbol V denotes a set of observable random variables and the symbol U denotes a set of unobserved background random variables with one $U_i \in U$ assigned to each $V_i \in V$. The symbol F denotes a set of functions, one for each element V_i in V. Each $f_i \in F$ corresponding to $V_i \in V$ maps values of a subset of $V \setminus \{V_i\}$ and of the background variable $U_i \in U$ corresponding to V_i to values of V_i. Finally, $P(u) = \prod_i P(u_i)$ is a joint distribution over U. This distribution $P(u)$, along with F, induces a joint distribution $P(v)$ over V. A Markovian model may be represented by a directed graph, where there is a vertex for every random variable in V, and for two random variables $V_i, V_j \in V$, there is a directed edge $V_i \rightarrow V_j$ if and only if the subset of V whose values are mapped on to values of V_j by f_j contains V_i. This graph is said to be induced by the Markovian model and is called the *causal diagram* representation of the NPSEM. When the causal diagram is acyclic, the corresponding Markovian model is called *recursive*. In the remainder of this section, we will restrict our attention to recursive models.

As is the case in a standard Bayesian network (Pearl, 1988), the distribution $P(v)$ under a Markovian model is Markov relative to its induced graph $\mathcal{G}$. In other words, we have $P(v) = \prod_i P(v_i | pa(v_i)_{\mathcal{G}})$. This is equivalent to the local Markov property, which states that each V_i is independent of $V \setminus (De(V_i)_{\mathcal{G}} \cup Pa(V_i)_{\mathcal{G}})$ given $Pa(V_i)_{\mathcal{G}}$, and in turn equivalent to the global Markov property defined by d-separation (Pearl, 1988), which states that if X is d-separated by Y given Z in the model-induced graph, then X is independent of Y given Z in the joint distribution $P(v)$ of the model variables. The importance of directed graphs in both Bayesian networks and Markovian models stems from this relationship between d-separation, a path property defined on a graph, and probabilistic independence in an appropriate joint distribution.

Unlike Bayesian networks, Markovian models permit formalization of causality by means of the *intervention operation*, denoted by $do(v_i)$ in Pearl (2000), and defined as the action of setting the value of V_i to be equal to v_i by an external intervention. This operation, applied to a Markovian model M, constructs a *submodel* M_{v_i}, which is obtained from M by replacing the function f_i determining V_i in M by a new constant function f_i^* that always gives value v_i, more formally defined as $(\forall x) f_i^*(x) = v_i$. Interventions represent idealized experiments. The result of such an experiment, called the *causal effect* of $do(v_i)$, is the distribution induced by the intervention on the remaining variables $V \setminus \{V_i\}$ in M_{v_i}, this distribution often being called 'interventional', and denoted as $P(v \setminus \{v_i\}|do(v_i))$ or $P_{v_i}(v \setminus \{v_i\})$. The causal effect of $do(v_i)$ on a subset of interest $Y \subset V \setminus \{V_i\}$ is denoted by $P(y|do(v_i))$ or $P_{v_i}(y)$. Note that, in the causal inference literature, the so-called population causal effect of interest may actually be a contrast between two intervention distributions, induced by setting V_i at v_i and v_j, respectively. Examples of contrasts are the causal mean difference $E(y|do(v_i)) - E(y|do(v_j))$ or the causal relative risk $P(y|do(v_i))/P(y|do(v_j))$. However, these quantities can all be computed from $P(y|do(v_i))$ if this distribution is known for all exposure levels v_i. Thus, without loss of generality, we will refer to $P(y|do(v_i))$ as the causal effect of the intervention $do(v_i)$.

The causal diagram induced by M_{v_i} can be obtained from the causal diagram G induced by M by removing all arrows pointing to V_i in G. We will denote the resulting graph, called the *mutilated graph* in Pearl (2000), by $G_{\overline{v_i}}$. Since the variable V_i after the intervention $do(v_i)$ is a constant, it is sometimes omitted from the mutilated graph, in which case the graph representing M_{v_i} is simply obtained by removing V_i and all edges adjacent to V_i from G.

Many scientific questions can be formalized as being about causal effects in an appropriate Markovian model. In practice, interventions can be implemented by means of an actual randomized experiment, which allows the intervention distributions of interest to be estimated directly from the generated experimental data. The difficulty is that experimentation is often expensive, impractical, or illegal, which leads naturally to the question of whether causal effects can be expressed in terms of the observational distribution in a Markovian model, since this distribution can be estimated just by passive observation of a set of collected samples, rather than by performing experiments. This is a question of *identification* of causal effects from $P(v)$. In Markovian models, the effect $P(v \setminus \{v_i\}|do(v_i))$ is always identifiable and equal to

$$P(v \setminus \{v_i\}|do(v_i)) = \frac{P(v)}{P(v_i|pa(v_i))}$$

where it is assumed the value v_i of V_i is consistent with values v of V. This is known as the truncation formula (Spirtes *et al.*, 1993; Pearl, 2000), or the g-formula (Robins, 1986). This formula generalizes in an obvious way to interventions on multiple variables and subsets of $V \setminus \{V_i\}$. In particular, for any V_j, $P(v_j|do(pa(v_j))) = P(v_j|pa(v_j))$.

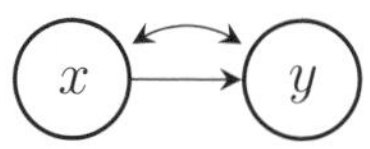

Figure 3.2 *A graph representing semi-Markovian models where $P(y|do(x))$ is not identifiable from $P(v)$.*

3.2.3 Latent projections and semi-Markovian models

Latent variables are very common in causal inference problems. For this reason it is useful to consider Markovian models where a subset of variables in V are unobserved. Such models can be represented by a directed acyclic graph with a subset of nodes labeled as latent. However, another alternative is to represent such models by acyclic directed mixed graphs (ADMGs) containing both directed arrows, representing "direct causation", and bidirected arrows, representing 'unspecified causally ancestral latent variables'. For example, in the ADMG shown in Figure 3.2, the X variable is a direct cause of Y and, in addition, these two variables share unspecified common causes. Every Markovian model with a set of nodes marked latent can be represented by an ADMG called a *latent projection*.

Definition 1. Latent projection *Let M be a Markovian model inducing $\mathcal{G}$, where V is partitioned into a set L of latents and a set O of observables.*

Then the latent projection $\mathcal{G}(O)$ is a graph containing a vertex for every element in O, where for every two nodes $O_i, O_j \in O$:

- *$\mathcal{G}(O)$ contains an edge $O_i \to O_j$ if there exists a d-connected path $O_i \to I_1 \to \cdots \to I_k \to O_j$, where $I_1, ... I_k \in L$.*
- *$\mathcal{G}(O)$ contains an edge $O_i \leftrightarrow O_j$ if there exists a d-connected path $O_i \leftarrow I_1 ... I_k \to O_j$, $O_i \leftarrow I_1 ... I_k \leftrightarrow O_j$, or $O_i \leftrightarrow I_1, ..., I_k \to O_j$, where $I_1, ..., I_k \in L$.*

Latent projections are a convenient way to represent latent variable Markovian models because they simplify the representation of multiple paths of latents while preserving many useful constraints over observable variables.

An alternative way to think about latent projections is to consider a generalization of Markovian models that induces ADMGs directly. This generalization is the semi-Markovian model (Pearl, 2000). Like Markovian models, semi-Markovian models are tuples $\langle U, V, F, P(u) \rangle$. The sole difference is we allow the values of a single $U_k \in U$ to influence either one or two functions in F, unlike the Markovian case where values of a single U_i influenced a single function $f_i \in F$. The induced graph of a semi-Markovian model contains a vertex for every element in V, with two elements V_i, V_j connected by a directed arrow as before, and connected by a bidirected arrow if f_i, f_j are both influenced by values of a single $U_k \in U$.

3.2.4 Interventions in semi-Markovian models

Since semi-Markovian models are causal, the definitions of intervention and causal effect carry over from Markovian models without change. However, the identification problem of causal effects in semi-Markovian models is significantly more difficult than in Markovian models.

In particular, there exist ADMGs such that in some semi-Markovian models inducing these ADMGs certain causal effects are not identifiable. The simplest such graph is known as the "bow arc graph" and is shown in Figure 3.2. It is possible to construct two models M_1 and M_2 that will both induce the shown graph and have the same distributions $P(v)$, but which will disagree on the value of the causal effect $P(y|do(x))$. Hence the causal effect is not a function of the observed data in the class of model inducing the graph shown in Figure 3.2, and thus cannot be estimated without additional assumptions.

The natural question is the characterization of ADMGs where a particular causal effect is identifiable. This problem has received much attention in literature, with complete solutions given in Tian and Pearl (2002), Huang and Valtorta (2006), Shpitser and Pearl (2006). Chapter 6 in this volume will review the identification problem of causal effects and its solutions in greater detail.

3.2.5 Counterfactual distributions in NPSEMs

Markovian and semi-Markovian models are very general data generating mechanisms that can be viewed as stochastic circuits, where nodes in the model represent input, output, and intermediate wires, and functions that determine the value of variables are logic gates. In such models, the joint distribution representing the effect of $do(x)$ on a singleton outcome variable Y, represented using the do notation as $P(y|do(x))$, can also be represented in the language of potential outcomes as a random variable Y_x or $Y(x)$. We will call such variables *counterfactual*.

Note that if we fix all U variables in a semi-Markovian model, the values of observable variables in V become deterministically fixed. This holds true even after interventions. In other words, if we fix the unobservable variable set U to have values u, then the counterfactual random variable Y_x becomes a constant denoted by $Y_x(u)$.[1]

This allows us to use the distribution $P(u)$ to define joint distributions over counterfactual variables, even if the interventions that determine these variables disagree with each other. The precise definition is as follows:

$$P(Y^1_{x^1} = y^1, ..., Y^k_{x^k} = y^k) = \sum_{\{u|Y^1_{x^1}(u)=y^1 \wedge ... \wedge Y^k_{x^k}(u)=y^k\}} P(u) \tag{3.1}$$

where U is the set of unobservable variables in the model. In words, this definition says that a joint probability of counterfactual variables $Y^1_{x^1}, ..., Y^k_{x^k}$ assuming values $y^1, ..., y^k$ is defined to be the sum of probabilities (according to $P(u)$) of values u of unobservable variables U, which result in constants $Y^1_{x^1}(u), ..., Y^k_{x^k}(u)$ being equal to $y^1, ..., y^k$.

The value of such distributions is that they allow us to formalize counterfactual reasoning. Consider the following question: "I have a headache. Would I have a headache had I taken an aspirin one hour ago?" Questions of this type are very natural[2] and frequently arise both in informal and scientific discourse. In Daurd (2007) they are called "questions about the causes of effects". We can formalize this question by considering a Markovian model M with two variables: A (representing the decision to take aspirin one hour ago) and Y (for headache now).

[1] Interpreting value assignments u as 'units', this implies that the stable unit treatment value assumption (SUTVA) holds in NPSEMs.

[2] One piece of evidence for the naturalness of such questions, at least in English, is that the English counterfactual connective 'had' is a short word.

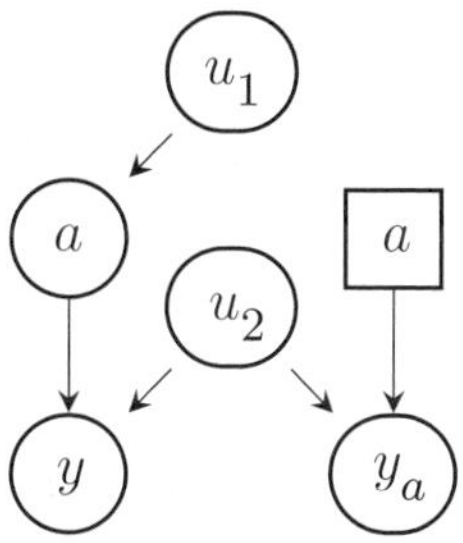

Figure 3.3 A counterfactual graph for the aspirin example. The square node represents an intervened-on node set to a constant.

Both variables are binary, with value 1 representing aspirin taken and headache present, and value 0 representing aspirin not taken and headache absent. It is reasonable to assume that the decision to take aspirin has some sort of effect on headache, but not vice versa, so our causal diagram will have A as the parent of Y. The question we are trying to formalize is referencing two possible worlds, the 'actual world,' where headache happened, and the hypothetical world, which is 'as close as possible' to the actual world, except for the fact that aspirin was taken one hour ago. The hypothetical world is represented by a submodel $M_{a=1}$. The expression "headache had I taken aspirin" refers to the variable Y in this submodel, in other words to $Y_{a=1}$. The original question also had a statement 'I have a headache,' referencing the variable Y in the actual world, represented by the model M.

Hypothetical worlds being 'as close as possible' to the actual world is modelled in NPSEMs by having all counterfactual worlds share unobservable variables. These variables represent the causal past, which lead to the currently observable state of affairs, with interventions representing minimal relevant differences between worlds. There is a graphical representation of these connected worlds known as the *counterfactual graph* (Shpitser and Pearl, 2007). Such a graph contains copies of the original causal diagram, one copy for each counterfactual world of interest. A copy corresponding to a hypothetical world where the intervention $do(x)$ took place is actually the mutilated graph $G_{\overline{X}}$. Each such copy shares the unobservable variables with other copies. In our aspirin example, the counterfactual graph is shown in Figure 3.3. See Shpitser and Pearl (2007) for technical details of constructing counterfactual graphs for arbitrary counterfactual queries. The counterfactual graph for the special case of two hypothetical worlds, as in our example, has been proposed in Balke and Pearl (1994a, 1994b) and is referred to as the *twin network graph*.

We can view the counterfactual graph in our example as a causal diagram representing another NPSEM, which contains both counterfactual worlds. In particular, d-separation in the counterfactual graph implies conditional independences in a joint distribution over counterfactual (potential outcome) variables $P(A, Y, Y_{a=1})$, defined as in Equation (3.1). Our counterfactual question: 'I have a headache. Would I have had a headache had I taken an aspirin one hour ago?' can now be represented as a conditional distribution derived from the above joint, specifically by the probability $P(Y_{a=1}|Y = 1)$. Identification of such distributions from either interventional or observational data, assuming a complete causal diagram is known, is sometimes possible. A treatment of this version of the identification problem appears in the literature (Shpitser and Pearl, 2007).

3.2.6 Causal diagrams and counterfactual independence

One of the advantages of using NPSEMs to talk about potential outcomes is that a fully specified causal diagram is an oracle that makes causal assumptions explicit as missing edges.

Consider an ADMG G representing a semi-Markovian model. For any node X in G, it is the case that $X_{pa(x)} = X_{nd(x)}$. In words, this assumption states, for all X, that any direct arrow missing from a nondescendant W of X to X implies fixing W to have no effect on X as long as all parents of X are already fixed. These assumptions are called exclusion restrictions in Pearl (2000). Similarly, for any missing bidirected arrow from W to X, it is the case that $W_{pa(w)}$ is independent of $X_{pa(x)}$. In words, two counterfactual variables (with their respective parents fixed to any value) cannot be dependent if the original causal diagram does not connect them by a bidirected arc. These assumptions are called independence restrictions in Pearl (2000).

These assumptions can be used to derive other counterfactual independences by means of an important axiom, which we call generalized consistency, but which is called composition in Pearl (2000). This axiom, which is generally assumed to hold in the potential outcome framework, is also assumed to hold in NPSEMs. Formally, it states that

$$(\forall u) W_x(u) = w \Rightarrow (Y_x(u) = Y_{x,w}(u))$$

In words, for any value assignment u to unobservable variables U, if the counterfactual variable $W_x(u)$ assumes value w, then the values assumed by counterfactual variables $Y_x(u)$ and $Y_{x,w}(u)$ will be the same. This assumption can be viewed as restating the convention of the sharing of unobserved variables between hypothetical worlds making up the counterfactual graph.

Coupled with generalized composition, exclusion and independence restrictions can be used logically to derive all independences in any joint distribution over counterfactual variables obtained from an NPSEM (Halpern, 2000). In some sense, these restriction form a kind of "logical basis" for counterfactual constraints implied by an NPSEM. Causal diagrams not only make this basis explicit via missing edges, but can also vastly simplify derivation of implications of this basis. Indeed, the counterfactual graphs of the kind shown in the last section can be used to show counterfactual independence directly by means of a graphical d-separation test. For example, we can conclude using d-separation that Y_a is independent of A, but dependent of A given Y. In complex causal models, this sort of counterfactual independence test is very difficult to perform without graphical aid.

3.2.7 Relation to potential outcomes

An NPSEM with a given causal diagram G implicitly defines counterfactual distributions over variables that can be viewed as potential outcomes. The standard assumptions made in the potential outcome framework also hold in the NPSEM framework. In particular, the SUTV assumption and the generalized consistency (composition) assumption hold in NPSEMs.

The advantage of NPSEMs is making assumptions explicit by means of graphs. Graphical language is more easily understood by the human mind, and counterfactual independence tests are much easier to verify via graphs than via unaided causal judgement. Furthermore, NPSEMs, as stochastic circuits, give a very reasonable data generating mechanism for a wide range of processes encountered in science. On the other hand, the reality of the scientific process is that unsettled areas of empirical science, by their very nature, suffer from lack of reliable knowledge of causality. This lack of knowledge is precisely what is motivating scientific investigation in the first place. When causal knowledge is scarce, it is not generally possible to justify causal assumptions needed to construct the full graph necessary for reasoning with NPSEMs. Instead, scientists may be justified in making a handful of assumptions on the data generating process, such as conditional ignorability (Rosenbaum and Rubin, 1983), which may be just enough to identify one causal quantity of interest, and little else.

One effective compromise between the "epistemically extravagant" NPSEMs and minimal assumptions made in the potential outcomes literature are the so-called minimal causal models (MCMs) (Robins and Richardson, 2011). These models are similar in spirit to NPSEMs in the sense of representing independences by means of graphs. At the same time, MCMs never make counterfactual independence claims that cannot be tested in principle, which is the cautious stance reminiscent of the practitioner willing to adopt potential outcomes, but not willing to commit to a detailed causal theory.

An alternative compromise is to use the detailed theory of identification developed in NPSEMs as a springboard to further results that relax graphical assumptions as much as possible while still yielding identification of causal quantities of interest. One recent result in this vein characterized assumptions necessary for covariate adjustment for identifying causal effects (Shpitser *et al.*, 2010).

References

Balke, A. and P. J. (1994a) Counterfactual probabilities: computational methods, bounds and applications, in *Proceedings of UAI-94*, pp. 46–54.

Balke, A. and Pearl, J. (1994b). Probabilistic evaluation of counterfactual queries, in *Proceedings of AAAI-94*, pp. 230–237.

Dawid, A. P. (2007) Countertactuals, hypotheticals and potential responses: a philosophical examination of statistical causality, in *Causality and Probability in the Sciences*, Texts in Philosophy, vol. 5 (eds F. Russo and J. Williamson). London: College Publications, pp. 503–532.

Halpern, J. (2000) Axiomatizing causal reasoning. *Journal of A.I. Research*, pp. 317–337.

Huang, Y. and Valtorta, M. (2006) Identifiability in causal Bayesian networks: a sound and complete algorithm, in *Twenty-First National Conference on Artificial Intelligence*, 2006.

Lewis, D. (1973) *Counterfactuals*. Cambridge, MA: Harvard University Press.

Neyman, J. (1923) Sur les applications de la thar des probabilities aux experiences agaricales: essay des principle. Excerpts reprinted (1990) in English. *Statistical Science*, **5**, 463–472.

Pearl, J. (1988) *Probabilistic Reasoning in Intelligent Systems*. San Mateo, CA: Morgan and Kaufmann.

Pearl, J. (2000) *Causality: Models, Reasoning, and Inference*. Cambridge: Cambridge University Press.

Robins, J.M. (1987) A graphical approach to the identification and estimation of causal parameters in mortality studies with sustained exposure periods. *Journal of Chronic Disease*, **2** 139–161.

Robins, J.M. and Richardson, T.S. (2011) Alternative graphical causal models and the identification of direct effects, in *Causality and Psychopathology: finding the determinants of disorders and their cures*.

Robins, J.M. (1986) A new approach to causal inference in mortality studies with sustained exposure periods – application to control of the healthy worker survivor effect. *Mathematical Modeling*, **7**, 1393–1512.

Rosenbaum, P.R. and Rubin, D.B. (1983) The central role of the propensity score in observational studies for causal effects. *Biometrika*, **70** (1), 41–55.

Rubin, D.B. (1974) Estimating causal effects of treatments in randomized and non-randomized studies. *Journal of Educational Psychology*, **66**, 688–701.

Shpitser, I. and Pearl, J. (2006) Identification of conditional interventional distributions. in *Uncertainty in Artificial Intelligence*, vol. 22.

Shpitser, I. and Pearl, J. (2007) Complete identification methods for the causal hierarchy. Technical Report R–336, UCLA Cognitive Systems Laboratory.

Shpitser, I. VanderWeele, T. and Robins, J.M. (2010) On the validity of covariate adjustment for estimating causal effects, in *International Joint Conference on Artificial Intelligence*, vol. 22.

Spirtes, P. Glymour, C. and Scheines, R. (1993) *Causation, Prediction, and Search.* New york: Springer Verlag.

Tian, J. and Pearl, J. (2002) A general identification condition for causal effects, in *Eighteenth National Conference on Artificial Intelligence*, pp. 567–573.

Tinbergen, J. (1937) *An Econometric Approach to Business Cycle Problems.* Paris: Hermann.

Wright, S. (1921) Correlation and causation. *Journal of Agricultural Research*, **20**, 557–585.

4

The decision-theoretic approach to causal inference

Philip Dawid
Statistical Laboratory, Centre for Mathematical Sciences, University of Cambridge, Cambridge, UK

4.1 Introduction

This chapter outlines the decision-theoretic (DT) approach to causal inference. In contrast to other approaches, as described elsewhere in this volume,[1] for DT we need no new philosophical concepts (such as counterfactuals), formal constructions (such as potential response variables) or arbitrary restrictions (such as functional relationships between variables, with all stochasticity being confined to 'error variables'): all the concepts and techniques required are readily available from the standard toolboxes of Probability and Statistics. These simple tools are adequate to analyse all meaningful problems of causal inference – at any rate so long as we confine ourselves to studying the 'effects of causes'.[2]

Section 4.2 introduces the DT approach, and in sections 4.3 and 4.4 we see how it can be used to address the important problem of confounding. The following Sections 4.5, 4.6 and 4.7 consider a variety of problems of causal inference – propensity analysis, instrumental variables and the effect of treatment on the treated – from the standpoint of DT. Finally, section 4.8

[1] See, in particular, Chapter 2, 'The language of potential outcomes', by Arvid Sjölander, and Chapter 3, 'Structural equations, graphs and interventions', by Ilya Shpitser.

[2] See Holland (1986) and Dawid (2000) for discussions of the important distinction between inferences about 'the effects of causes' and inferences about 'the causes of effects'. Studying the causes of effects is a more delicate enterprise, arguably needing some form of counterfactual reasoning and different tools (Dawid, 2000, 2011), which we shall not be considering here.

Causality: Statistical Perspectives and Applications, First Edition. Edited by Carlo Berzuini, Philip Dawid and Luisa Bernardinelli.

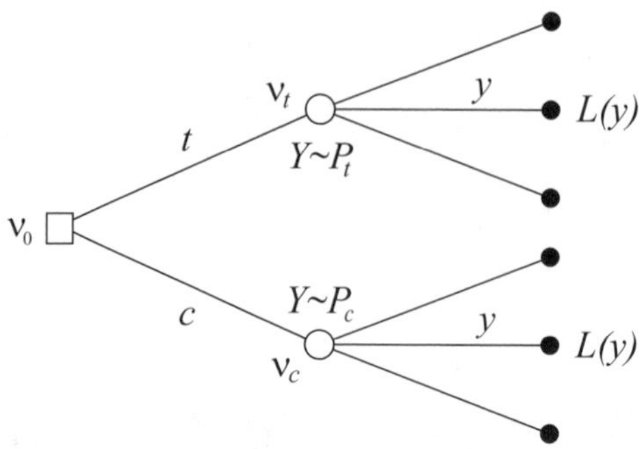

Figure 4.1 *Decision tree.*

compares and contrasts the DT approach with the alternative approaches of potential response modelling and causal graphs.

4.2 Decision theory and causality

4.2.1 A simple decision problem

I have a headache and wonder if it would help it go away if I were to take two aspirins. I can consider this as a simple statistical decision problem, involving a nonstochastic decision variable T (under my control) and a stochastic outcome variable Y (subject to the whim of Nature). The decision variable T acts as a parameter, indicating whether I decide to take the aspirins ($T = t$) or not ($T = c$). The outcome variable Y might be the length of time, measured in minutes beyond my decision point, that my headache will last.

For either decision that I might take, I will be uncertain as to what the ensuing value of Y will be and can represent this uncertainty by a probability distribution. Let P_i ($i = t$ or c) denote the distribution of Y when $T = i$ and let $L(y)$ be the loss I will suffer (measured on an expected utility scale) if in fact $Y = y$.[3] A decision tree (Raiffa, 1968) representation of this problem is shown in Figure 4.1.

The precepts of Bayesian decision theory now tell me I should take the aspirins if $E_t\{L(Y)\} < E_c\{L(Y)\}$, where E_i denotes expectation under the distribution P_i. Although it is important to be aware of the impact of nonlinear loss functions, for simplicity (and with no real loss of generality, since we could simply recode the outcome variable) we will henceforth take $L(y) \equiv y$. Then the aspirins should be taken if $E_t(Y) - E_c(Y) < 0$.

The uncertainty wrapped up in P_i might be a complex mix of the purely aleatory randomness of the way in which headaches respond to aspirin, combined with my personal epistemological uncertainty as to the features and relevance of this random process. The aleatory uncertainty can be regarded as an irreducible nub, but by conducting or examining suitable studies of the way in which headaches do typically react to treatment by aspirin, t, or by nothing, c, I might hope to reduce the penumbra of personal uncertainty surrounding it – though I will need to consider carefully the relevance of those studies to my personal case. However, in any event,

[3] For simplicity of exposition we suppose the loss depends only on y, but there would be no difficulty in principle in allowing further dependence of the loss on the treatment i applied.

what I need to focus on is the pair of distributions P_t and P_c – each relevant in just one of the two hypothetical circumstances I am considering – which (for any loss function) are all I need in order to solve my decision problem.

4.2.2 Causal inference

What has the above scenario to do with causal inference? While it admittedly does little to address deep and scientifically important questions such as the chemical, pharmacological, physical and psychological mechanisms by which aspirin works to cure headaches – the 'How?' and the 'Why?' of things – it does directly address the 'What?' – i.e. what is the *effect* (here described by the distribution P_t) of the *causal action* of taking aspirins ($T = t$) on the response variable Y? It further addresses the important question of how this compares with the effect (P_c) of the alternative 'action' of not taking aspirin ($T = c$).

The quantity needed to solve our decision problem is

$$\text{ACE} := E_t(Y) - E_c(Y). \tag{4.1}$$

This is the DT explication of the concept of *average causal effect* (ACE,) which features heavily in the causal literature. The DT approach allows us to make clear and intuitive distinctions between causal and associational concepts and quantities. Thus ACE is flagged as a causal, rather than associational, quantity by its dependence on the interventional distributions P_t and P_c. Indeed, as there is only one random variable, Y, involved, there is no question of an associational interpretation.

The study of purely empirical aspects of 'the effects of causes' has formed the focus of a very large part of the enterprise of statistics, at least since Fisher and Gosset (and, as seen above, it is equally relevant to a Bayesian decision theorist as to an inferential frequentist). It is my thesis that a correspondingly large part of the enterprise of causal inference can usefully be framed in the same way – indeed, that it is, at bottom, no more than a small and straightforward extension of the Fisherian/decision-theoretic approach.

The major novelty required in this extension is the need to address the crucial issue of the *relevance* of external data to the individual decision problem. For example, suppose we have performed a study in which the pair (Y, T) follows some bivariate distribution $P_\emptyset$. We might consider the 'face-value average causal effect':

$$\text{FACE} := E_\emptyset(Y \mid T = t) - E_\emptyset(Y \mid T = c) \tag{4.2}$$

where $E_\emptyset$ denotes expectation under the observational joint distribution $P_\emptyset$ for (Y, T). This is a purely associational quantity, which is estimable from the data of the study. Then we might be interested in what relationship (if any) there might be between FACE, given by (4.2), and the causal quantity of interest, ACE, given by (4.1).

Such concerns are traditionally considered informally, in terms of the twin attributes of *internal* and *external validity*. Internal validity can be achieved by means of a suitable randomised experiment, to ensure that when we compare the outcomes for the different treatment groups we are indeed comparing like with like. In that case, a new patient who could be regarded as exchangeable with those in the study (and thus – because of internal validity – equally so with either treatment group) could regard the results as directly informative about his or her (aleatory) P_c and P_t.[4] External validity addresses the harder question of whether, and if so how, the study results might be useful for a new patient *not* exchangeable with the study patients – which, in view of the restrictive criteria typically imposed for entry to a trial, will be the usual state of affairs. Even with the best randomised experiment in the world, this issue of external validity will necessarily involve a good deal of 'expert judgement', which will be open to debate and attack. Similar issues arise for the criterion of internal validity when the data arise from a nonrandomised study, with all the possibilities that raises for bias and confounding. Again, to address such issues one must resort to disputable 'expert judgements'.

From this perspective, the major focus of causal inference is understanding when and how we can transfer information gained from one 'regime' (typically, that operating in some experimental study, whether or not randomised) to another (e.g. that relevant to my individual decision problem). Since any justification for this will necessarily involve judgement, to allow informed discussion we need to be able to specify exactly what assumptions have been employed to support this information transfer and what the effects of these assumptions have been.

4.3 No confounding

Consider the assertion that a simple study of treated and untreated individuals has internal validity. We do not here ask how, why or when this assertion might be made (perhaps the experiment was randomised: see Section 4.4 below), or how reasonable it is, or how it might be defended. We merely confine ourselves to the task of expressing the assertion formally. For simplicity we suppose that all patients admitted to the study can be regarded as having independent identically distributed (T, Y) pairs, with bivariate distribution $P_\emptyset$. Note that T is here a stochastic variable, rather than a decision variable.

Consider now a new patient, satisfying the criteria for admission to the study (so considered exchangeable with those in it). As in Section 4.2.1, we can consider the distribution P_i of this patient's pair (T, Y) if we decide to give him treatment $T = i$ ($i = t$ or c). Clearly under P_i we must have $T = i$ with probability 1, and the only randomness remaining is in Y (it is the marginal distribution for Y alone that was denoted by P_i in Section 4.2.1).

We have now introduced three 'regimes', with three different joint distributions for (T, Y). These can be labelled by a 'regime indicator' F_T, a parameter variable with possible values $\emptyset$, t and c.

The property of internal validity can now be expressed as follows. For $i = t$ or c, the distribution of Y under P_i is the same as the conditional distribution of Y, given $T = i$, in $P_\emptyset$. When this holds, it follows that, for a new patient similar to those in the study, if we want the 'interventional' distribution of Y under applied cause i, we can use instead the 'observational' distribution of Y given $T = i$ in the study, without having to correct for any confounding or other biases. This property is also described as 'no confounding'. Another way of saying it is that we can regard the conditional distribution of Y given T as an invariant 'modular component', which can be transferred unchanged between the three regimes. In particular, we will have equality of the causal quantity ACE, given by (4.1), and the associational quantity FACE, given by (4.2).

Modularity judgements such as these lie at the heart of causal inference. An important and useful observation is that it is possible to re-express such modularity properties using the

[4] This is the case so long as the study used exactly the same treatments under the same circumstances.

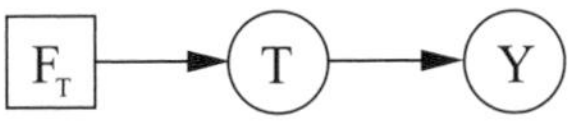

Figure 4.2 *No confounding.*

language and notation of conditional independence (CI), extended straightforwardly so as to accommodate nonrandom variables such as regime indicators (Dawid, 1979). We want to say that the conditional distribution of Y, given the values of (F_T, T), in fact depends only on the value T (which treatment was applied) and not further on the value of F_T (the regime in which it was applied). An equivalent statement is Y *is conditionally independent of* F_T, given T, expressed in notation as $Y \perp\!\!\!\perp F_T \mid T$. Consequently, as an optional but useful alternative to working directly with probability distributions and their relationships, we can use the algebraic theory of conditional independence to manipulate collections of causal assertions expressed as 'extended CI' (ECI) relations, so as to derive their implications.

There are also other mathematical tools – in particular, directed acyclic graph (DAG) representations (Cowell *et al.*, 1999) and their *influence diagram* (ID) extensions (Dawid, 2002), which also incorporate decision nodes – available to assist in performing such deductions. Thus the above ECI property $Y \perp\!\!\!\perp F_T \mid T$ can be represented graphically by the absence of an arrow from F_T to Y in the ID of Figure 4.2.[5]

At the mathematical level, at least, all the above considerations extend trivially beyond internal validity to encompass external validity also; we simply consider the regimes $F_T = t$ or c as relating to interventions on a new patient in the relevant new circumstances. However, the distributions P_c and P_t might well now differ from those appropriate to a patient fitting the study criteria and new substantive (and contestable) considerations will be required to justify the now bolder assertion $Y \perp\!\!\!\perp F_T \mid T$, which says that the conditional distribution of Y given $T = i$ extracted from the study correctly describes uncertainty about how the new patient's Y would respond to application of treatment i ($i = t$ or c).

4.4 Confounding

In many realistic contexts – in observational studies, almost invariably – the 'no confounding' property $Y \perp\!\!\!\perp F_T \mid T$ will be untenable. That is to say, we will have *confounding*. (Note that this definition of confounding does not depend on the existence of what are often called *confounding variables* or *confounders*, a usage that I do not consider helpful.)

4.4.1 Unconfounding

However, we might be able to tell an alternative, more convincing, story if we introduce a suitable further set U of variables and assert that we will have no residual confounding after conditioning on U. In order for this to be useful, it is necessary that such a variable U should be a *pre-treatment* variable, existent and measurable, at least in principle, before the point at which a treatment decision has to be made.

[5] In contrast to other uses of graphical models in causal inference, this representation of a causal property does not involve any probabilistic conditional independence properties within one regime; see Dawid (2010) for further discussion of such distinctions.

For example, if our data arise from an observational study on patients treated by a certain doctor, who might be allocating treatment according to his own observations of the general health of the patient (thus introducing confounding between treatment and outcome, and rendering the property $Y \perp\!\!\!\perp F_T \mid T$ inappropriate), it could be reasonable to suppose that, if we were to take U to comprise all the observations on the patient that the doctor might be taking into account in making his choice, then conditionally on U we would have no further confounding – in which case we might term U an *unconfounder*. We can then use the (associational) distribution of Y given $(U, T = i)$ operating in the study regime $F_T = \emptyset$ as the appropriate (causal) distribution of Y given U in the interventional regime $F_T = i$ ($i = t$ or c). In particular, if we can observe U in all cases we can use the study data to predict how the new patient will respond to either treatment. In ECI terms, the unconfounding property is expressed as

$$Y \perp\!\!\!\perp F_T \mid (U, T) \tag{4.3}$$

Also, the fact that U is a pre-treatment variable implies that it must have the same distribution in both the interventional regimes $F_T = 0$ and $F_T = 1$. We write this requirement as

$$U \perp\!\!\!\perp_I F_T \tag{4.4}$$

where the subscript I in the symbol $\perp\!\!\!\perp_I$ means that we are only allowing interventional values for F_T. However, (at any rate so long as we are restricting attention to internal validity), it will often be reasonable to strengthen (4.4) to include the observational regime $F_T = \emptyset$ also:

$$U \perp\!\!\!\perp F_T \tag{4.5}$$

which says that the way in which U arises for a new patient to be treated – the same under either of the interventional regimes – is also the same for a patient in the observational study. In this case we call the unconfounder U a *sufficient covariate*.

The two ECI properties (4.3) and (4.5) together are represented by the ID of Figure 4.3.

4.4.2 Nonconfounding

Two specialisations of the above structure can be obtained by removing either of the arrows marked a and b from Figure 4.3. It can be readily checked, using the semantics of DAG/ID representations (Lauritzen *et al.*, 1990, Verma and Pearl, 1990), that in either case both properties (4.3) and (4.5) will hold. Therefore, if we can make a case for leaving out either

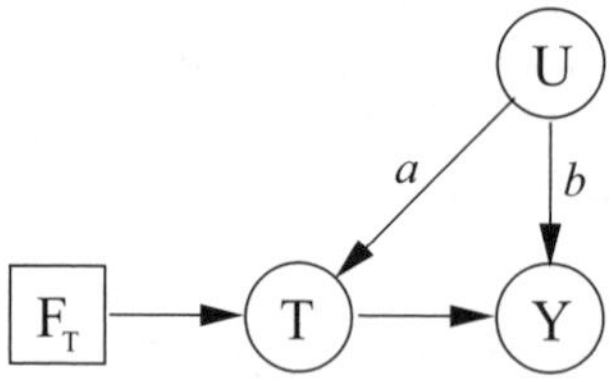

Figure 4.3 Unconfounding.

arrow, we will have justification for assuming no confounding. We might call such a variable U a *nonconfounder* and can safely forget that it ever existed.

The ID with a removed represents the additional[6] property $T \perp\!\!\!\perp U \mid F_T$: that in every regime T is independent of U. Since this condition holds trivially for the interventional regimes $F_T = t$ and $F_T = c$, where T is constant, it will hold if and only if T is independent of U in the study regime $F_T = \emptyset$; i.e. if the variables U that putatively might have affected the doctor's decision did not in fact do so (we can still allow dependence of Y on U in this case). If the study was genuinely randomised a pretty convincing case could be made for this assertion (thus showing why randomisation does indeed support internal validity); in the (rare) other situations that we can make a case for this independence property, we can treat the study *as if* it had been randomised. More generally, the structure of Figure 4.3 might be used to model the situation obtaining after already accounting for (i.e. conditioning on) additional pre-treatment variables, X. If it is considered that only the variables in X could affect the doctor's decision – which would thus not depend further on U – then we will have $T \perp\!\!\!\perp U \mid (X, F_T = \emptyset)$, and we could delete arrow a from the graphical representation of the conditional distribution given X. As an example, U might be a gene that affects outcome, but whose value is unknown to the decision maker.

As for the ID with b removed, this represents the additional property $Y \perp\!\!\!\perp U \mid T$: that the conditional distribution of Y given (T, U) (which, by (4.3), has already been supposed the same in all regimes) does not in fact depend on U. In that case, even if U is associated with treatment assignment in the study, this will not generate confounding. An example might be where treatment is allocated according to whether the subject's left or right foot is the larger. Again in real situations the condition $Y \perp\!\!\!\perp U \mid T$ for removing arrow b would typically be reasonable only when the whole structure is being modelled conditionally on further pre-treatment variables X (see Section 4.5 below for more on this).

4.4.3 Back-door formula

Suppose that U is a sufficient covariate (in particular, an unconfounder), but not a nonconfounder. We have seen that, if we can observe T, Y and U in all regimes, we can simply transfer the distribution of Y, given (T, U), from the study regime to the two interventional regimes under consideration, and so solve the new treatment decision problem. However, if we can only observe T and Y in all regimes we do not have access to any transferable modular distribution and remain confounded.

It may be that we can observe (T, Y, U) in the study, but for the new patient do not have access to U. In this case we can argue as follows (where for simplicity we suppose all variables discrete, though the argument is completely general[7]):

$$P_t(Y = y) = \sum_u P_t(Y = y \mid T = t, U = u)\, P_t(U = u) \tag{4.6}$$

$$= \sum_u P_\emptyset(Y = y \mid T = t, U = u)\, P_\emptyset(U = u) \tag{4.7}$$

[6] That is, over and above properties (4.3) and (4.5).

[7] This is subject to taking due care over events that might have probability 0 in one, but not all, regimes (Guo and Dawid, 2010).

In the first term of (4.6) extra conditioning on $T = t$ has been introduced, but that makes no difference, since under P_t, which corresponds to intervention to set $T = t$, the event $T = t$ is certain. Equation (4.7) now follows on after applying (4.3) to the first term and (4.5) to the second. A parallel expression can be found for $P_c(Y = y)$. Since we can estimate $P_\emptyset(Y = y \mid T = t, U = u)$, $P_\emptyset(Y = y \mid T = c, U = u)$ and $P_\emptyset(U = u)$ from the study data, we can compute $P_t(Y = y)$ and $P_c(Y = y)$ and thus solve our decision problem.

Alternatively, and more efficiently if we are interested only in ACE $:= E_t(Y) - E_c(Y)$, we can use the formula

$$\mathrm{ACE} = E_\emptyset\{\mathrm{SCE}(U)\} \tag{4.8}$$

where

$$\mathrm{SCE}(U) := E_t(Y \mid U) - E_c(Y \mid U) \tag{4.9}$$

$$= E_\emptyset(Y \mid U, T = t) - E_\emptyset(Y \mid U, T = c) \tag{4.10}$$

is the *specific causal effect* of treatment, given U. Here (4.10) follows by (4.3), and (4.8) by (4.5). Expressions (4.7) and (4.8) are versions of Pearl's *back-door formula* (Pearl, 1993).[8]

As a further variation, we may be able to extend the above argument to address external validity. Thus, suppose we can assume properties (4.3) and (4.4), but not property (4.5) (our new patient is not exchangeable with the study patients). If, nevertheless, we can assess a distribution for U relevant for the new patient (which by (4.4) is unaffected by our treatment decision), we can use this in place of $P_\emptyset(U = u)$ in (4.7) or for taking the expectation in (4.8).

The above general methodology can be extended to other situations in which credible causal assumptions, formulated as ECI properties, can be used to express a desired 'causal target' – a function of the interventional distributions P_t and P_c – in terms of the study distribution $P_\emptyset$. When the assumed ECI properties can be represented by an ID, this process can be effected by means of the 'do-calculus' (Pearl, 1995).[9]

This approach to causal inference is thus threefold: first, we use external subject matter considerations to suggest credible transferable modular patterns (this may require the introduction of further variables beyond those of immediate interest); next we express these formally in terms of ECI properties or (where possible) their ID representations; and finally, using do-calculus or the more general algebraic calculus of conditional independence, explore whether, and if so how, our assumptions allow us to express our causal target quantities, required for solving our new decision problem, in terms of distributions estimable from study data. The last of these stages, which is entirely technical, has been intensively researched in recent years: it is perhaps time to pay more attention to the very different requirements of the first stage. A good examplar of the application of such attention is the technique of *Mendelian randomisation* (Katan, 1986, 2004, Davey Smith and Ebrahim, 2003), where scientific understanding of the random way in which Nature assigns genes at birth is used to justify some of the assumptions needed to apply instrumental variable analysis (see Section 4.6 below). Even here care must be taken and possible modifications made to account for real-world disturbing factors (Didelez and Sheehan, 2007a).

[8] See also Chapter 6 of this volume, 'Graph-based criteria of identifiability of causal questions', by Ilya Shpitser.

[9] Pearl gives a set of rules in terms of the DAG remaining when intervention indicators are removed from the ID. However, these rules are more straightforwardly formulated and applied using the full ID (Dawid, 2007b, Chapter 8).

4.5 Propensity analysis

We have seen how the identification of a sufficient covariate makes it possible to transfer information from a study so as to address the new decision problem of interest. In general a sufficient covariate need not be unique, and if we have one, it might be possible to discard some of the information in it without sacrificing the sufficiency property. One way of doing this is by 'propensity analysis' (Rosenbaum and Rubin, 1983). This is generally defined in terms of potential responses, but that framework is inessential. Here we describe the decision-theoretic version (Guo and Dawid, 2010).

Suppose then that U is a sufficient covariate. A function V of U is a *sufficient reduction* of U if V is itself a sufficient covariate. Since property (4.5) for V follows immediately from the same property for U, we only need investigate whether the following property holds for V:

$$Y \perp\!\!\!\perp F_T \mid (V, T) \tag{4.11}$$

There are various additional conditions we can impose to ensure this. One is the following:

Condition 1. Treatment-sufficient reduction

$$T \perp\!\!\!\perp U \mid (V, F_T = \emptyset) \tag{4.12}$$

That is, in the observational regime, the choice of treatment depends on U only through the value of V.

It is notable that this condition does not involve the outcome variable Y, except for the essential requirement that the starting variable U be a sufficient covariate for the effect of T on Y.

Theorem 1 *Suppose U is a sufficient covariate and let be V be a function of U such that Condition 1 holds. Then V is a sufficient covariate.*

The following proof uses the algebraic calculus of conditional independence (Dawid, 1979).

Proof. Since, for $i = t$ or c, T is constant in regime $F_T = i$, we trivially have $U \perp\!\!\!\perp T \mid (V, F_T = i)$. Therefore (4.12) is equivalent to

$$U \perp\!\!\!\perp T \mid (V, F_T) \tag{4.13}$$

Now since V is a function of U, (4.5) implies $U \perp\!\!\!\perp F_T \mid V$, which together with (4.13) gives $U \perp\!\!\!\perp (T, F_T) \mid V$, whence $U \perp\!\!\!\perp F_T \mid (V, T)$. This now combines with (4.3) (equivalent, since V is a function of U, to $Y \perp\!\!\!\perp F_T \mid (U, V, T)$) to yield $(Y, U) \perp\!\!\!\perp F_T \mid (V, T)$, so that $Y \perp\!\!\!\perp F_T \mid (V, T)$, i.e. (4.11) holds. □

An alternative graphical proof can be based on the fact that the assumed conditional independence properties are represented by the ID of Figure 4.4, and then (4.11) can be

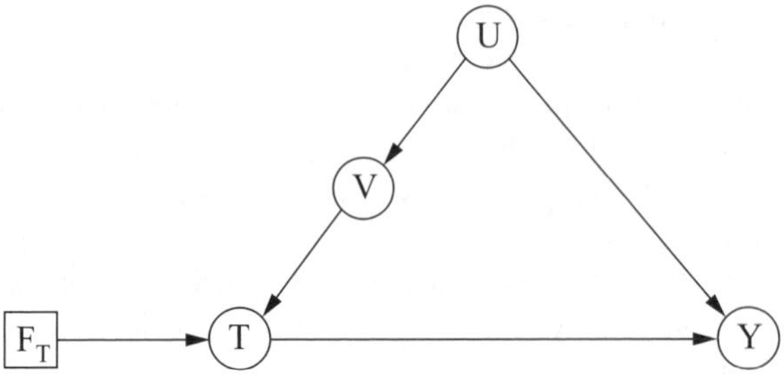

Figure 4.4 Treatment-sufficient reduction.

deduced by the moralisation criterion (Lauritzen *et al.*, 1990) or the equivalent *d*-separation criterion (Verma and Pearl, 1990).

Another, equivalent, description of treatment-sufficient reduction is as follows. Using the symmetry of conditional independence, the defining property (4.12) can be re-expressed as

$$U \perp\!\!\!\perp T \mid (V, F_T = \emptyset) \tag{4.14}$$

In this form it asserts that, in the observational regime, the conditional distribution of U given V is the same for both treatments (i.e. whether further conditioned on $T = t$ or on $T = c$): that is to say, V is a *balancing score* for U (Rosenbaum and Rubin, 1983). Property (4.14) can also be fruitfully interpreted as follows. Consider the family $\mathcal{Q} = \{Q_t, Q_c\}$ comprising the pair of observational conditional distibutions for U, given respectively $T = t$ and $T = c$. Then (4.14) asserts that V is a *sufficient statistic* (in the usual Fisherian sense) for this family. In particular, a *minimal* treatment-sufficient reduction is obtained as a minimal sufficient statistic for $\mathcal{Q}$: viz. any (1, 1)-function of the likelihood ratio statistic $\Lambda := q_t(X)/q_c(X)$. We might term such a minimal treatment-sufficient covariate a *propensity variable*, since one form for it is the treatment-assignment probability

$$\Pi := P_\emptyset(T = t \mid U) = \pi\,\Lambda/(1 - \pi + \pi\,\Lambda) \tag{4.15}$$

(where $\pi := P_\emptyset(T = t)$), which is known as the *propensity score* (Rosenbaum and Rubin, 1983). Either Λ or Π supplies a one-dimensional sufficient reduction of the orginal, perhaps highly multivariate, sufficient covariate U.[10]

4.6 Instrumental variable

When we have confounding, but cannot observe an unconfounder, we can sometimes make progress through the use of *instrumental variables* (or *instruments*). The property that a variable Z acts as an instrument for the effect of X on Y is typically described informally in terms such as the following (Martens *et al.*, 2006):

(i) Z has a causal effect on X.

[10] However, this property may not be as useful as may first appear (Guo and Dawid, 2010).

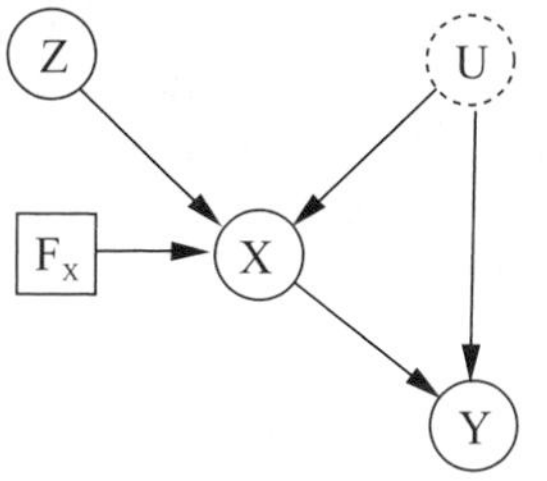

Figure 4.5 Instrumental variable.

(ii) Z affects the outcome Y only through X ('no direct effect of Z on Y').

(iii) Z does not share common causes with the outcome Y ('no confounding of the effect of Z on Y').

The following is an attempt to express these somewhat vague requirements more formally, in decision-theoretic terms. In addition to the observable variables X, Y, Z, we introduce a regime indicator F_X, where $F_X = \emptyset$ denotes the observational regime and $F_X = x$ the regime in which we intervene to set $X = x$, and an (unobservable) variable U, supposed to be a sufficient covariate for the effect of X on Y. Thus U satisfies

$$U \perp\!\!\!\perp F_X \tag{4.16}$$

$$Y \perp\!\!\!\perp F_X \mid (X, U). \tag{4.17}$$

As for Z, we assume:

$$X \not\perp\!\!\!\perp Z \mid F_X = \emptyset \tag{4.18}$$

$$Y \perp\!\!\!\perp Z \mid (X, U, F_X) \tag{4.19}$$

$$Z \perp\!\!\!\perp U \mid F_X \tag{4.20}$$

Property (4.18) replaces the causal requirement (i) with the purely associational requirement that X should not be independent of Z in the observational regime, which is sufficient for our purposes (Hernán and Robins, 2006, Dawid, 2010). Properties (4.19) and (4.20) are intended to capture the informal requirements (ii) and (iii). The ID of Figure 4.5 can be used to represent these ECI properties.[11]

Essentially the same figure, but with the regime indicator F_X omitted, is commonly used as the representation of an instrumental variable. This displays the following properties of the observational regime:

$$X \not\perp\!\!\!\perp Z \tag{4.21}$$

$$U \perp\!\!\!\perp Z \tag{4.22}$$

$$Y \perp\!\!\!\perp Z \mid (X, U) \tag{4.23}$$

[11] For (4.18), we need to assume that this is a faithful representation. Also, Figure 4.5 encodes the additional but inessential property $Z \perp\!\!\!\perp F_X$.

However, requirements relating only to the observational regime can never be enough to address the causal effect of X on Y, which is defined in terms of interventional regimes. We also need some assumptions relating the observational regime $F_X = \emptyset$ to the interventional regimes $F_X = x$. Such assumptions may be implicit, but it is safer to make them explicit, as is done in our (4.16)–(4.17) and Figure 4.5.

4.6.1 Linear model

Suppose all the observables are univariate and we can describe the dependence of Y on (X, U) (the same in all regimes, by (4.17)) by the linear model:

$$E(Y \mid X, U) = W + \beta X \tag{4.24}$$

for some function W of U.

Because (4.24) holds in the interventional regime $F_X = x$, we deduce

$$E(Y \mid F_X = x) = w_0 + \beta x$$

where $w_0 := E(W \mid F_X = x)$ is a constant independent of x, by (4.16). Thus β can be interpreted causally, as describing how the mean of Y responds to manipulation of X. Our aim is to identify β.

Now from (4.17) and (4.19) we have

$$Y \perp\!\!\!\perp (Z, F_X) \mid (X, U) \tag{4.25}$$

Therefore (4.24) is also $E(Y \mid X, Z, U, F_X = \emptyset)$. Then

$$E(Y \mid Z, F_X = \emptyset) = E(W \mid Z, F_X = \emptyset) + \beta\, E(X \mid Z, F_X = \emptyset)$$

However, by (4.20) the first term on the right-hand side is constant. Thus

$$E(Y \mid Z, F_X = \emptyset) = \text{constant} + \beta\, E(X \mid Z, F_X = \emptyset) \tag{4.26}$$

Equation (4.26) relates two functions of Z, each of which can be identified from observational data.[12] Consequently (so long as neither side is constant) we can identify the causal parameter β from such data. Indeed, it readily follows from (4.26) that (in the observational regime) $\beta = \text{Cov}(Y, Z)/\text{Cov}(X, Z)$, which can be estimated by the ratio of the coefficients of Z in the sample linear regressions of Y on Z and of X on Z.

4.6.2 Binary variables

When all the observable variables Z, X, Y are binary, without making further assumptions we cannot fully identify the 'causal probability' $P(Y = 1 \mid F_X = x)$ from observational data. However, we can develop inequalities it must satisfy. This approach was instigated by Manski

[12] Indeed, we do not even need to observe all the variables (Z, X, Y) simultaneously: it is enough to have observational data on the pair (Z, X) and perhaps quite separate observational data on the pair (Z, Y).

(1990). His inequalities were refined by Balke and Pearl (1997), under the strong additional condition of deterministic dependence of X on (Z, U) and of Y on (X, U).[13] This condition was, however, shown to be unnecessary by Dawid (2003), where a fully decision-theoretic approach was developed. In either approach, the analysis involves subtle convex duality arguments; see the above references for further details.

4.7 Effect of treatment of the treated

Suppose, in an observational setting, we confine attention to those who were selected to receive treatment t. We might ask: 'How much better have these individuals done than they would have done if given treatment c?' Since we can never know how they would have reacted to treatment c, this would seem to be an unanswerable question. Remarkably, it is not (Geneletti and Dawid, 2011).[14] In DT terms, this 'effect of treatment on the treated' (ETT) can be formulated as $E_{\emptyset}\{\text{SCE}(U) \mid T = t\}$, where U is a sufficient covariate and SCE(U) is the associated specific causal effect as given by (4.9). It turns out that this quantity does not depend on the choice of sufficient covariate U,[15] being expressible as

$$\text{ETT} = \frac{E_{\emptyset}(Y) - E_c(Y)}{P_{\emptyset}(T = 1)} \tag{4.27}$$

In particular, ETT can be estimated if we have experimental data on individuals taking the control treatment c, in addition to the data from our actual study. However, the estimability property in this case is something of an accident and does not readily extend to cases with more than two levels for the treatment variable.

4.8 Connections and contrasts

Other chapters in this volume describe various other frameworks for defining and manipulating causal concepts. The DT approach differs from these others in important ways, but of course there are relationships between them. Here we discuss these, noting similarities and differences.

4.8.1 Potential responses

In the DT approach we consider a single response variable Y, but with different probabilistic behaviours in different regimes. By contrast, in the popular *potential response* (PR) approach (Rubin, 1974, 1978; see also this volume, Chapter 2, 'The language of potential outcomes', by Arvid Sjölander) the response Y is itself multiplied into several versions, one for each value of the causal variable T, but we work with a single joint distribution $\mathbf{P}$. Thus in the simplest case we consider a pair of *potential responses* $\mathbf{Y} = (Y_t, Y_c)$, having bivariate probability distribution

[13] An alternative interpretation of this condition is in terms of potential outcomes; see Section 4.8.2 below and also Chapter 10 of this volume, 'Cross-classifications by joint potential outcomes', by Arvid Sjölander.

[14] We here regard the property of being chosen to receive treatment as identifying a particular subgroup of patients of interest. We are not, however, taking into account (conditioning on) any information about the responses of these patients, which would indeed render the question unanswerable without additional arbitrary assumptions.

[15] In particular (see footnote 17 below), within the PR framework of Section 4.8.1 we can take $U = \mathbf{Y}$, leading to $\text{ETT} = E(Y_t - Y_c \mid T = t)$, agreeing with the usual definition of ETT within PR.

P. The pair **Y** is conceived as existing prior to the assignment of treatment and the effect of applying treatment $T = i$ ($i = t$ or c) is merely to uncover the value Y_i: in particular the actual response is $Y := Y_T$. There is no explicit consideration of differing regimes, but since the pair **Y** is regarded as a fixed pre-existing attribute of the experimental subject, it is implicit that it has the same value in an observational as in an interventional regime. 'No confounding' is expressed as probabilistic independence between **Y** and T,[16] so that the distribution of the response $Y(= Y_T)$ given $T = i$ (implicitly referring to the observational regime), which is that of Y_i given $T = i$, will then be the same as the marginal distribution of Y_i (referring implicitly to the response in a regime in which we intervene with treatment i). In particular, in DT language we recover the property $Y \perp\!\!\!\perp F_T \mid T$.

A fundamental PR concept that has no immediate DT counterpart[17] is the *individual causal effect*, ICE $:= Y_t - Y_c$. It is aspects of ICE or its distribution that form the target of PR causal inferences. This enterprise is, however, complicated by what Holland (1986) termed 'the fundamental problem of causal inference' (more precisely, a fundamental problem of the PR approach to causal inference, since the issue simply does not arise in the DT approach): the fact that it is logically impossible ever to observe, simultaneously, both elements of the pair $\mathbf{Y} = (Y_t, Y_c)$. In particular, there are aspects of the bivariate distribution **P** of **Y** (such as its correlation coefficient) that it is simply impossible to learn from data. Nevertheless, some progress can be made. For example, it is possible to estimate ACE, now defined as **E**(ICE) (where **E** denotes expectation under **P**), since this can be expressed as $\mathbf{E}(Y_t) - \mathbf{E}(Y_c)$ (corresponding to the DT expression (4.1)), and the two individual terms can each be estimated from suitable experimental studies. However, no such let-out is available should we express interest in the ratio analogue of ACE, $\mathbf{E}(Y_t/Y_c)$, or in the variance **var**(ICE) of ICE; these cannot be identified from data unless we are willing to make further restrictive and essentially arbitrary assumptions (Dawid, 2000). (Since there is no DT counterpart of Y_t/Y_c or of **var**(ICE), DT analysis is spared this embarrassment.) It can be far from obvious just which PR quantities are empirically identifiable – so supporting meaningful causal inferences – and which are not.

A further distinction between the PR and DT approaches is that the former, being based on the single joint distribution **P**, necessarily imposes relationships between the derived distributions operating across distinct regimes. The DT approach is free to assume or not to assume such relationships, as may be considered appropriate to the circumstances. For example, we might asign different interventional distributions and make different ECI assumptions and graphical representations, depending on whether we are addressing internal or external validity.

Many inferences framed in the language of PR do have DT counterparts and the conclusions will then typically agree (though I consider that, once the bar of unfamiliarity has been overcome, the DT analysis can be seen as both more intuitive and more straightforward). Still other inferences, not derivable from DT, are apparently available from within the PR framework, but in my view these are often problematic, either because (as for **var**(ICE)) they exhibit a worrying dependence on arbitrary and empirically untestable additional assumptions or because they treat as real and meaningful certain concepts (such as principal strata: Frangakis

---

[16] Or by the weaker condition that T be independent of each Y_i separately.

[17] In the presence of a sufficient covariate U, the associated specific causal effect SCE(U) defined by (4.9) can be considered analogous to ICE, but the value of this quantity will depend on the choice of U. In the PR framework, it is implicit that **Y** has the formal properties of a sufficient covariate and we then have ICE $\equiv$ SCE(**Y**).

and Rubin, 2002; Pearl, 2011b; Dawid and Didelez, 2012) that depend on logically unknowable (and, for me at least, metaphysically mystifying) aspects of the bivariate quantity **Y**.

One apparent advantage of the PR approach over DT is its formal ability to represent counterfactual quantities, and thereby address issues of the causes of effects (CoE) (Dawid, 2007a). However, once again the ensuing inferences are – now in virtually every case – highly sensitive to arbitrary and empirically untestable assumptions (Dawid, 2000). In some cases, by considering all possible choices for these additional assumptions, we can extract empirically meaningful interval bounds (which can even reduce to a point) for causal quantities of interest (Shpitser and Pearl, 2007; Dawid, 2011), a form of inference that is admittedly not available from a DT analysis.

4.8.2 Causal graphs

Graphical representations of probabilistic and graphical processes have become increasingly popular, both as informal aids to causal conversation (with seemingly intuitive display of such concepts as 'direct cause', 'causal path', 'intermediate variable', etc.) and as more formal tools for expressing and processing causal properties. There can, however, be a dangerous – especially so since often unappreciated – mismatch between the informal and the formal uses (Dawid, 2010), which can lead to confusion and error. For formal use it is essential to have clearly defined and understood semantics, whereby we know what properties of causal and inferential processes are represented in graphical terms and exactly how this representation works (Didelez and Sheehan, 2007b). I have described above the specific semantics whereby (in suitable problems) we can represent DT causal properties, expressed as ECI relations, by means of an influence diagram. It cannot be too strongly emphasised that no causal meaning should be attributed to any aspects of this ID – for example the directions of its arrows – beyond the implied ECI relations: to do so would be to fall prey to the philosophical fallacy of 'reification' (Dawid, 2010).

The most popular formal graphical framework for causal reasoning is that developed by Pearl and expounded in his book (Pearl, 2009). In its basic form this uses DAGs to model collections of domain variables, but without our explicit introduction of regime indicators. However, the Pearlian semantics of such a DAG are identical to the DT semantics of the 'augmented DAG': the ID that elaborates the Pearlian DAG by including, for each domain node, an additional nonstochastic intervention indicator, equipped with an arrow feeding into its domain node. Just as for our regime indicator, this allows both for interventions setting a value of the domain node and for the 'hands-off' regime where Nature is allowed to run her course. This Pearlian framework can thus be regarded as specialisation of the DT framework, where the assumed ECI properties have a particular, DAG-related, structure.

In the later chapters of his book, Pearl introduces further structure: all stochasticity is now confined to a collection of introduced 'error variables', with the other variables having deterministic dependence on their graphical parents. This can be regarded as a generalisation of a linear structural equation model: Pearl terms it a 'structural causal model' (SCM) or 'non-parametric structural equation model' (NPSEM). Such models are formally isomorphic to potential response models (Pearl, 2011a); see also this volume, Chapter 3, 'Structural equations, graphs and interventions', by Ilya Shpitser.

For addressing problems of the effects of causes (EoC) this deterministic reformulation is simply of no consequence, since on marginalising out the error variables we recover a fully stochastic Pearlian DAG, having the identical implications for any EoC inferences (and agreeing with the DT analysis of the corresponding augmented DAG).[18] For addressing CoE issues, NPSEMs are subject to the same fundamental pros and cons as their equivalent PR models. In particular, distinct NPSEMs may be observationally indistinguishable, yet lead to differing causal 'inferences'.

Recently Pearl and his coworkers (Pearl and Bareinboim, 2011) have taken seriously the need to describe causal inference in terms of the relationships between different regimes, formalising these relationships in terms of selection variables (Bareinboim and Pearl, 2011; Didelez *et al.*, 2010) in stochastic causal diagrams. These can be considered as special cases of DT analyses.

4.9 Postscript

The existence of a variety of different formal explications of statistical causality is somewhat embarrassing – we can only pray for the arrival of a messianic figure who (just as Kolmogorov did for probability theory) will sweep away the confusion and produce a single theory that everyone can accept. Meanwhile, let us put a positive gloss on this babel of different languages: since different people seem to find different approaches naturally appealing, there may be something for everyone. In that understanding I suggest that DT deserves careful attention from those who currently choose to think about statistical causality in other terms.

Acknowledgements

The author has benefited from helpful comments by Carlo Berzuini and Vanessa Didelez.

References

Balke, A.A. and Pearl, J. (1997) Bounds on treatment effects from studies with imperfect compliance. *Journal of the American Statistical Association*, **92**, 1172–1176.

Bareinboim, E. and Pearl, J. (2011) Controlling selection bias in causal inference, in *Proceedings of the 25th AAAI Conference on Articial Intelligence* (ed W. Burgard and D. Roth), Menlo Park, CA: AAAI Press, pp. 1754–1755. http://www.aaai.org/ocs/index.php/AAAI/AAAI11/paper/view/3522.

Cowell, R.G., Dawid, A.P., Lauritzen, S.L. and Spiegelhalter, D.J. (1999) *Probabilistic Networks and Expert Systems*. New York: Springer.

Davey Smith, G. and Ebrahim, S. (2003) Mendelian randomization: Can genetic epidemiology contribute to understanding environmental determinants of disease? *International Journal of Epidemiology*, **32**, 1–22.

Dawid, A.P. (1979) Conditional independence in statistical theory (with Discussion). *Journal of the Royal Statistical Society, Series B*, **41**, 1–31.

Dawid, A.P. (2000) Causal inference without counterfactuals (with Discussion). *Journal of the American Statistical Association*, **95**, 407–448.

[18] When we confine attention to DAG models, any fully stochastic model can be generated by some (typically nonunique) NPSEM. This equivalence does not, however, extend to other graphical representations, such as chain graphs, where fully stochastic models can be strictly more general than their NPSEM counterparts that restrict stochasticity to error variables (Dawid, 2010, Example 11.2).

Dawid, A.P. (2002) Influence diagrams for causal modelling and inference. *International Statistical Review*, **70**, 161–189. Corrigenda, **70**, 437.

Dawid, A.P. (2003) Causal inference using influence diagrams: the problem of partial compliance (with discussion). In *Highly Structured Stochastic Systems* (eds. P.J. Green, N.L. Hjort and S. Richardson). Oxford University Press, pp. 45–81.

Dawid, A.P. (2007a) Counterfactuals, hypotheticals and potential responses: A philosophical examination of statistical causality, in *Causality and Probability in the Sciences*, Texts in Philosophy, vol. 5 (eds. F. Russo and J. Williamson). London: College Publications, pp. 503–532.

Dawid, A.P. (2007b) *Fundamentals of statistical causality*. Research Report 279, Department of Statistical Science, University College London, 94 pp. http://www.ucl.ac.uk/statistics/research/pdfs/rr279.pdf.

Dawid, A.P. (2010) Beware of the DAG!, in *Proceedings of the NIPS 2008 Workshop on Causality, Journal of Machine Learning Research Workshop and Conference Proceedings*, vol. 6 (eds. I. Guyon, D. Janzing and B. Schölkopf), pp. 59–86. http://tinyurl.com/33va7tm.

Dawid, A.P. (2011) The role of scientific and statistical evidence in assessing causality, in *Perspectives on Causation* (ed. R. Goldberg). Oxford: Hart Publishing, pp. 133–147.

Dawid, A.P. and Didelez, V. (2012) Imagine a can opener: the magic of principal stratum analysis. *International Journal of Biostatistics*. To appear.

Didelez, V. and Sheehan, N.A. (2007a) Mendelian randomisation as an instrumental variable approach to causal inference. *Statistical Methods in Medical Research*, **16**, 309–330.

Didelez, V. and Sheehan, N.A. (2007b) Mendelian randomisation: why epidemiology needs a formal language for causality, in *Causality and Probability in the Sciences*, Texts in Philosophy Series, vol. 5 (eds. F. Russo and J. Williamson). London: College Publications, pp. 263–292.

Didelez, V., Kreiner, S. and Keiding, N. (2010) Graphical models for inference under outcome-dependent sampling. *Statistical Science*, **25**, 368–387.

Frangakis, C.E. and Rubin, D.B. (2002) Principal stratification in causal inference. *Biometrics*, **58**, 21–29.

Geneletti, S. and Dawid, A.P. (2011) Defining and identifying the effect of treatment on the treated, in *Causality in the Sciences* (eds. P.M. Illari, F. Russo and J. Williamson). Oxford University Press, pp. 728–749.

Guo, H. and Dawid, A.P. (2010) Sufficient covariates and linear propensity analysis, in *Proceedings of the Thirteenth International Workshop on Artificial Intelligence and Statistics (AISTATS), 2010, Chia Laguna, Sardinia, Italy, May 13–15, 2010. Journal of Machine Learning Research Workshop and Conference Proceedings* (eds. Y.W. Teh and D.M. Titterington), pp. 281–288. http://tinyurl.com/33lmuj7.

Hernán, M.A. and Robins, J.M. (2006) Instruments for causal inference: an epidemiologist's dream? *Epidemiology*, **17**, 360–372.

Holland, P.W. (1986) Statistics and causal inference (with discussion). *Journal of the American Statistical Association*, **81**, 945–970.

Katan, M.B. (1986) Apolipoprotein E isoforms, serum cholesterol and cancer. *The Lancet*, **327** (8479), 507–508.

Katan, M.B. (2004) Commentary: Mendelian randomization, 18 years on. *International Journal of Epidemiology*, **33**, 10–11.

Lauritzen, S.L., Dawid, A.P., Larsen, B.N. and Leimer, H.G. (1990) Independence properties of directed Markov fields. *Networks*, **20**, 491–505.

Manski, C.F. (1990) Nonparametric bounds on treatment effects. *American Economic Review, Papers and Proceedings*, **80**, 319–323.

Martens, E.P., Pestman, W.R., de Boer, A., Belitser, S.V. and Klungel, O.H. (2006) Instrumental variables: applications and limitations. *Epidemiology*, **17**, 260–267.

Pearl, J. (1993) Comment: graphical models, causality and intervention. *Statistical Science*, **8**, 266–269.

Pearl, J. (1995) Causal diagrams for empirical research (with discussion). *Biometrika*, **82**, 669–710.

Pearl, J. (2009) *Causality: Models, Reasoning and Inference*, 2nd edn. Cambridge: Cambridge University Press.

Pearl, J. (2011a) Graphical models, potential outcomes and causal inference: comment on Linquist and Sobel. *NeuroImage*, **58**, 770–771.

Pearl, J. (2011b) Principal stratification – a goal or a tool? *International Journal of Biostatistics*, **7** (1), Article 20.

Pearl, J. and Bareinboim, E. (2011) Transportability of causal and statistical relations: a formal approach, in *Proceedings of the 25th AAAI Conference on Artificial Intelligence* (eds. W. Burgard and D. Roth). Menlo Park, CA: AAAI Press, pp. 247–254.

Raiffa, H. (1968) *Decision Analysis*. Reading, MA: Addison-Wesley.

Rosenbaum, P.R. and Rubin, D.B. (1983) The central rôle of the propensity score in observational studies for causal effects. *Biometrika*, **70**, 41–55.

Rubin, D.B. (1974) Estimating causal effects of treatments in randomized and nonrandomized studies. *Journal of Educational Psychology*, **66**, 688–701.

Rubin, D.B. (1978) Bayesian inference for causal effects: the rôle of randomization. *Annals of Statistics*, **6**, 34–68.

Shpitser, I. and Pearl, J. (2007) What counterfactuals can be tested, in *Proceedings of the Twenty-Third Conference Annual Conference on Uncertainty in Artificial Intelligence (UAI-07)* (eds. R. Parr and L. van der Gaag). Corvallis, OR: AUAI Press, pp. 352–359.

Verma, T. and Pearl, J. (1990) Causal networks: semantics and expressiveness, in *Uncertainty in Artificial Intelligence 4* (eds. R.D. Shachter, T.S. Levitt, L.N. Kanal and J.F. Lemmer). Amsterdam: North-Holland, pp. 69–76.

5

Causal inference as a prediction problem: Assumptions, identification and evidence synthesis

Sander Greenland
Department of Epidemiology and Department of Statistics, University of California, Los Angeles, California, USA

5.1 Introduction

I thank the editors for their kind and persistent invitation to contribute an essay. What follows is not a scholarly overview (in the sense of laying out history and development with accuracy and detailed citation) and provides no formal development. Instead it is only a brief, informal outline of my perspective on statistical treatments of causal inference that have emerged over the past several decades and discussion of problems that seem important in light of my applied experience. That experience and hence my comments are limited to health and medical research. Those who dislike philosophical treatments (which allow poorly constrained speculations, belaboring the obvious, and obvious errors for the sake of portraying analogies between models, language, and reality), as opposed to mathematical treatments (which allow only sound deduction within axiomatically constrained systems), are advised to read no further. Those who do read on are recommended to Greenland (2010) for more elaboration of several points herein.

I especially emphasize that, in applications, I have often found formal developments in causal inference could play only minor if valuable roles. Among the reasons for this limitation are that the simplifications required to make the formal machinery work are often too extreme

---

Causality: Statistical Perspectives and Applications, First Edition. Edited by Carlo Berzuini, Philip Dawid and Luisa Bernardinelli.

to yield credible results (Greenland, 2010). Furthermore, currently accepted machinery does not effectively incorporate the highly disparate evidence streams that have to be considered in risk assessment, except as externally imposed model constraints or prior distributions. While there have been many methodologies proposed for combining disparate evidence sources, as yet none has seen widespread adoption. I hope that (despite its incoherencies) my discussion and anecdotes will suggest reasons for that state of affairs, both in terms of the complexity of the problem and human limitations in facing that complexity. Throughout, I will assume readers are familiar with basic causal models.

5.2 A brief commentary on developments since 1970

It is remarkable to look back on the four decades I have been in health-science research and see the transformation of its theoretical foundation for causal inference. By the 1970s key formal elements in that foundation (chiefly potential outcomes, causal diagrams, structural equations) had been around in some form for a half-century and seen use in the social sciences. Yet they were largely unknown in training and applications in medical statistics, which instead focused on ascribing informal causal meaning to association measures (such as regression coefficients), subject to equally informal considerations of how well the surrounding evidence warranted that meaning (e.g., Hill, 1965).

Today these formalisms have merged into the vibrant topic area of causal inference and have become part of mainstream methods instruction (e.g., Morgan and Winship, 2007; Rothman *et al.*, 2008). There has in parallel been a subtle conceptual revolution that recognizes causal inference as a prediction problem, the task of predicting outcomes under alternative actions or decisions. The need to consider mutually exclusive actions imposes special structure and special identification issues that parallel problems in destructive testing in engineering, namely the fact that each unit can be observed to fail only under one treatment (e.g., operation at 220 volts rather than 120 volts). Thus we need a basis for inferring from several different groups observed under different treatments to one group that might be subject to any one of the treatments. In product testing this basis is ordinarily provided by random sampling and randomization, but these devices are often unavailable in health and medical studies.

The explication of the predictive structure of causal inference is most explicit in the decision-theoretic formalization of Dawid and colleagues (Dawid, 2004, 2007, and Chapter 4 in this volume). Nonetheless, all formalizations with mainstream acceptance (chiefly potential-outcome models and the parallel decision-theoretic and do-calculus formulations) can be expressed as containing an intervention, action or decision variable (or node) encoding a target 'cause' or treatment choice. Any temporally subsequent variable can then be structured as an 'effect' or outcome variable, whether it mediates or terminates the events of interest.

Within this framework, causal inference can be called 'active prediction', forecasting consequences of alternative action or decision sequences (as in Robins, 1986, 1987a, 1987b; Pearl and Robins, 1995; and Chapter 9 in this volume by VanderWeele and Hernán), albeit sometimes in retrospect, in which case all but one of the alternative actions is counterfactual. The basic idea is traceable at least back to Fisher and Neyman, with their strong emphasis on experimental design and the role of randomization in estimating effects. For nonexperimental inference, it manifests in the importance of *treatment* prediction as opposed or in addition to the outcome prediction of traditional regression analysis (Rubin, 1991), especially in high-dimensional problems (Robins and Wasserman, 2000, Section 5). That importance stands in contrast to 'passive prediction', which forecasts outcomes without formalized intervention or decision nodes, and thus with no formal role for treatment prediction or counterfactuals. It also stands in contrast to most causal inference in practice, where passive prediction methods dominate formal analysis and 'treatment' is only a specially labeled regressor that lacks any formal distinction from other regressors.

5.2.1 Potential outcomes and missing data

All statistical procedures rely on assumptions to identify quantities of interest, the most prominent including treatment randomization, random selection and no or random measurement error (usually conditional on certain observed variables, but these are strong assumptions nonetheless). Of these, treatment randomization stands out as an assumption distinguishing causal inference from passive prediction. This assumption can be reformulated in a number of ways, notably via the isomorphism between the potential-outcome and missing-data formalisms, in which counterfactual outcomes (potential outcomes corresponding to unassigned treatments) are treated as missing values to be imputed (Rubin, 1978, 1991).

Here, causal inference is transformed into a framework of inference under missing data. Given that, for any observation unit, at most one treatment assignment from a list $x_1, \ldots, x_m$ can be made, at most one component of the corresponding potential-outcome vector $\mathbf{Y} = (Y_1, \ldots, Y_m)$ can be observed on the unit. Letting A be the variable coding actual treatment, for a given unit with $A = a$, x_a is received treatment, and at best only Y_a is observed. All other treatment assignments $x_c \neq x_a$ are counterfactual for the unit, and their corresponding potential outcomes Y_c (or at least some summaries of their distribution) must be imputed for causal inference to proceed.

In ideal physics experiments, the potential outcomes are known functions of treatment determined from established laws or overwhelming evidence, and may be superfluous to list given the function (Chapter 9 in this volume by VanderWeele and Hernán). For example, if $X =$ temperature applied and $Y =$ liquification of a lead sample, knowing the melting point of lead renders superfluous listing all potential outcomes of the sample at different temperatures. At the other extreme, if no relevant law is known, imputation of missing potential outcomes can be enabled by a missing at random (MAR) assumption. The latter is an immediate consequence of treatment randomization plus random actual-outcome censoring given treatment and baseline covariates. It is an especially delicate imputation problem, however, insofar as no two potential outcomes Y_j and Y_k are ever seen on the same observation unit. Thus randomization/MAR cannot identify any aspect of the joint distribution of $Y_1, \ldots, Y_m$, which has led to debate on whether it is meaningful to even talk of such entities (at least in typical health and social science applications, where no law is available to determine the joint distribution) (e.g., see Dawid and discussants, 2000).

5.2.2 The prognostic view

A useful feature of the missing-data perspective on causation is that it is more oriented toward individual predictions, making it closer to clinical prediction than population-health prediction (epidemiology). It is of course not unique in that regard. The same idea is embodied in the development of passive prediction models for outcomes Y_a, and then re-computation of the fitted outcomes as prognostic scores by (possibly counterfactually) setting all units to the same treatment value x_j (e.g., Cornfield, 1971). This prognostic scoring constitutes imputation of the expected potential outcome $\mu_j = E(Y_j)$ under treatment x_j.[1]

The idea of prognostic prediction under different treatments has been reborn repeatedly since Cornfield's time, e.g. under the rubric of g-computation, in which the distribution of potential outcomes $p(y_j)$ is computed under different treatment regimes x_j (Robins, 1986, 1987a, 1987b), and again in the do-calculus in which $p(y_j) = p(y|\text{do}[x_j])$ (Pearl and Robins, 1995; Pearl, 2009). Arguably, a prognostic structure could be discerned in Neyman's original potential-outcomes model for randomized experiments (Neyman, 1923); the half-century delay in wide dissemination of this approach to formalizing causation may have been because it requires computers and regression software for serious application value.

5.3 Ambiguities of observational extensions

Formal causal models have produced many valuable insights, and are essential to make sense of mediation analysis and complex interventions (e.g., sequential treatment regimes). Nonetheless, for observational studies (as well as for severely compromised experiments) major challenges to causal inference derive from ambiguous mappings between models and data, and the nonidentification of model parameters from data after ambiguities are resolved.

A preliminary issue that is too often overlooked in observational applications is that potential outcomes (let alone their joint distribution) can be quite ambiguous in the absence of a well-defined *physical* mechanism that determines their assignment index (Greenland, 2005a; Hernán, 2005; Chapter 9 in this volume by VanderWeele and Hernán). Consider 'sex'. What does it mean for a female patient to have given treatment 'male' versus 'female'? The missing variable Y_{male} is undefined without some operational mechanism that transforms the given observation unit from 'female' to 'male'. To avoid this problem, some authors have maintained we should only assume potential outcomes exist when we can specify (perhaps we should add, with a straight face) an assignment mechanism that makes the purported treatment index correspond to values of an action or decision node. Others suggest that extension is possible in theory (again, for example, when potential outcomes can be determined from known laws) (Pearl, 2009; Chapter 3 in this volume by Shpitser), although the practical obstacles to extension may be insurmountable in typical health and social science settings (Chapter 9 in this volume by VanderWeele and Hernán).

In the absence of a well-defined treatment-assignment mechanism, it may be sensible and more relevant to switch to another type of observation unit for which such a mechanism can be specified (Greenland, 2005a). For example, to study hiring discrimination, one can change status shown on initial applications from 'male' instead of 'female' and using a masculine instead of feminine name, to see whether that matters to (say) being invited for an interview. Similar strategies can be applied to study effects of 'race', genotypes and other factors intrinsic to the individual (Kaufman, 2008). Here, the actual treated unit would be the person or group deciding on interviews (not the person applying) and the causal variable of interest would be their perception of gender, rather than the gender of the person applying (Chapter 9 in this volume by VanderWeele and Hernán).

[1] This imputation assumes implicitly a stochastic model for the actual outcomes Y_a, so that potential outcomes are unit-specific expectations $\mu_j = E(Y_a|\text{do}[x_j])$ or the corresponding distributions $p(y_j)$ of the Y_j (potential distributions of Y_a) under alternative treatments.

5.4 Causal diagrams and structural equations

In parallel with potential-outcome and decision-theoretic developments, there gradually emerged a fusion of graphical with causal concepts in the form of path analysis, structural equations and (eventually) modern causal diagrams (Spirtes *et al.*, 2001; Pearl, 2009). The diagrams themselves are unobjectionable to the extent they merely encode conditional independence assumptions (Dawid, 2010; Greenland, 2010). There are, however, several formal causal structures that can be mapped onto these diagrams, with ambiguities and controversies extending from problems with potential outcomes (Robins and Richardson, 2010).

The weakest formulations (which seem to underlie most health and social science uses) say only that the arrows in the graph represent causation, without much clarity about what that means beyond (a) temporal interpretation of arrows and (b) the conditional independencies (Markov structure) implied by the graph. The usual condition added for causality is that (c) all confounders are in the graph; nonetheless, it is difficult to understand what a confounder is without a definition of causation attached to the system portrayed. These vagaries may not hinder proper usage, but have compelled theoreticians to promote more precise formulations.

The strongest of these formulations treats structural-equation systems as response schedules giving (what look like, formally) potential outcomes for each endogenous node as a function of its parents. A causal graph then represents a nonparametric structural-equation model (NPSEM), in which a node Y with parents $\mathrm{pa}(Y)$ is taken to be determined by a functional relation $Y = f_Y(\mathrm{pa}(Y), \varepsilon_Y)$, where ε_Y is exogenous and an ancestor of to Y only (Pearl, 2009; Chapter 3 in this volume by Shpitser). When f_Y is a purely hypothetical function (as opposed to an experimentally tested law), there are serious objections to such formulations, at least if f_Y is considered to supply potential outcomes under different settings for $\mathrm{pa}(Y)$ and ε_Y.

One objection is that the NPSEM formulation entails a possibly enormous experimentally nonidentified joint distribution over all the potential-outcome vectors $\mathbf{Y} = \langle Y_j \rangle$, where j ranges over values of $\mathrm{pa}(Y)$ and Y ranges over endogenous graphical nodes. Of at least equal practical concern is the ambiguity of labeling arbitrary variables in $\mathrm{pa}(Y)$ as treatments or interventions (discussed above). This problem leads to questions about the realism and utility of such NPSEM formulations (Dawid, 2006; Robins and Richardson, 2010), unless great care is taken to define variables and observation units so that there are acceptably precise meanings for an operator like $\mathrm{do}[\mathrm{pa}(X)]$ for every variable X in the graph. In response to these objections, one can turn to weaker causal models to match the diagrams, which yield correspondingly less effect identification from a given diagram, or else turn to diagram expansion to match the identification given by the full NPSEM corresponding to the original diagram (Robins and Richardson, 2010).

5.5 Compelling versus plausible assumptions, models and inferences

If one has established acceptably precise and relevant treatment units and variables, precise causal models can be formulated. In practice, causal models typically encode numerous assumptions (constraints) beyond those implicit in their definitions of cause and effect. Because those further assumptions turn out to be essential for identification of effects, it is important to distinguish compelling assumptions from those that are merely plausible. This is an obvious point worth belaboring because it is so often forgotten in theory *and* practice.

Evidentially compelling assumptions are supported by a body of evidence taken to refute all of what might have otherwise been plausible alternatives. They are thus felt safe to treat as givens. Consequently, a model may be labeled as compelling if it entails only compelling assumptions or is flexible enough to approximate such a model. Similarly, an argument is compelling if all its assumptions are compelling and its deductions are sound.

Study-design features are often taken as compelling evidence for strict constraints (dimension-reducing assumptions). For example, baseline (pre-treatment) matching on an exogenous covariate (e.g., age) when constructing a study cohort can compel the strict (sharp-null) assumption that the covariate is unassociated with treatment and thus unconditionally not a confounder of any treatment effect on the cohort. On the other hand, background mechanistic information may only compel relatively vague constraints, although those may involve strict inequalities. As an example, monotone dose-response (e.g., monotonicity of the plot of μ_j against x_j) is often taken to be compelling, such as the relation of carbon monoxide concentration to death by asphyxiation.

An assumption is sometimes said to be warranted or credible or justified if there is some justification for use, such as supporting evidence that refutes explicit competitors. A plausible assumption may have no specifically supporting evidence or justification, and may simply be one among many unrefuted competitors. Plausibility often stems only from entailment of compelling consequences, or from some heuristic or esthetic advantages such as simplicity. In that case it is important to remember that many other assumptions will have the same source of plausibility. For example, a linear dose response (i.e., linearity of the X effect on Y, linearity of the plot of Y_j against x_j) is often plausible and so is common in everyday 'working models'. Linearity implies strict monotonicity, so when the latter is found compelling, the plausibility of linearity is enhanced; but the same can be said of many simple alternatives such as loglinearity.

To use plausible assumptions safely, the consequences of their violation under plausible alternatives need to be understood in some detail to see if those consequences are acceptable within our estimation task. Linearity may be relatively harmless for population inferences when (a) strict monotonicity holds and (b) one restricts inferences to what happens on average to Y as X varies (i.e., the marginal effect, which is the slope across the plots of Y_j against x_j averaged over the distribution of the potential-outcome vector $\mathbf{Y}$ in the source population of units). Nonetheless, if we assume linearity when responses are in fact nonlinear, this average slope can lead us into costly mistakes in predicting treatment effects (e.g., at high doses). This is so even if our average-slope estimate is from a perfect randomized trial.[2] Furthermore, even if individual responses are linear, heterogeneity of the individual slopes may also lead to costly mistakes in individual (clinical) predictions of treatment effects, especially if the slopes change sign across individuals (e.g., if the average slope is negative, indicating a benefit on average, but the treatment is harmful in some individuals).

A model may be labeled as plausible if it entails *only* plausible assumptions or is flexible enough to approximate such a model. It is a logical fallacy to claim plausibility of a model from its entailment of plausible or compelling consequences; nonetheless, if one quantifies plausibility with subjective probability, it is logical to increase the plausibility of a model upon

[2] For example, suppose in a cohort the death risks (expected potential outcomes) at doses $X = 0, 1, 2$ were 0.4, 0.1 and 0.1, and we randomize the cohort among these three doses. Then the expected fitted values from a linear model would make $X = 2$ appear the most beneficial dose even though full benefit is achieved at $X = 1$. If instead the death risks at $X = 0, 1, 2$ were 0.4, 0.4 and 0.1, the expected fitted values from a linear model would make $X = 1$ appear beneficial even though it is not.

observing one of its logical consequences or upon refutation of alternatives (e.g., see Adams, 1998). Unfortunately, purely logical relations (even probabilistic ones) do not seem to get us very far in statistical applications, which may explain (although not excuse) why logic is so neglected in statistical training.

As an example, suppose that 20 assumptions $A_1, \ldots, A_{20}$ entail a conclusion B. This means that we should assign B no less than the probability of A_1 and A_{20}, with B certain (probability 1) if we are certain of every A_j. It is then tempting to think that B must be probable if evidence leads us to assign each A_j a probability of 0.95, a common interpretation of 'very probable'. Nonetheless, if these are subjectively independent assumptions, this information alone does not imply that B should have probability greater than $0.95^{20} = 0.36$, with an even lower minimum if (as is usual) B is not perfectly entailed by $A_1, \ldots, A_{20}$. Thus, even if we assign 'very high' probability to each of 20 independent supporting pieces of evidence A_j (which sounds very impressive) and B can be deduced logically from the 20, it does not follow that B must be highly probable. For B to be highly probable, the *conjunction* of the assumptions must have high probability, which might not be the case if each assumption is subject to independent uncertainties or unknown errors.

Given that pure logic and probability may not take us far when given only independent bits of uncertain assumptions or evidence, causal inference often turns to highly interdependent and coordinated sets of assumptions in the form of mechanisms (mechanistic models). Although these models can be treated conveniently as singular hypotheses or assumptions, such treatment should not obscure their compound nature, especially when evaluating their credibility. In particular, a mechanistic model, by virtue of its elegance or other nonevidential properties, may induce a conjunction fallacy in which the model attains higher probability than any of its assumptions (Tversky and Kahneman, 1983).

Even if we agree that a particular mechanistic model is correct or compelling, the model's scope and impact may be far more narrow than believed, at least when used to argue against causation (which itself is a very large compound hypothesis embracing all causal mechanisms). Take the sharp null hypothesis that cell phones cannot cause brain cancer in humans. Clearly, this hypothesis has never been tested by a 'black box' experiment in which people are randomized to different exposure levels. Nonetheless, Shermer (2010) argued that this null hypothesis is compelling (in fact, certain), based on the mechanical argument that cell-phone microwaves (CPMs) are nonionizing and nonionizing radiation cannot alter chemical bonds including those in cellular DNA. The underlying physics is a mechanistic model, which is being offered as a compelling assumption that leads to a compelling conclusion that a particular etiologic pathway is not operating. This physics is uncontroversial and Shermer uses that fact as part of an appeal for the general causal null hypothesis.

Shermer's fallacy comes in applying the physics to the biology. To do so, he inserts the assumption that an exposure (such as CPM) can elevate cancer risk only by direct breakage of DNA. This assumption is not even plausible to some cancer biologists because (among other things) most carcinogenesis does not involve ionizing radiation and may not even require direct DNA damage; merely impeding DNA repair mechanisms or gene expression may suffice. Several experimental studies have reported that CPMs can in fact influence DNA (e.g., see the list in Slesin, 2008). Furthermore, the DNA argument applies only to tumor initiation; it does not apply to tumor promotion, which can be affected by many factors including metabolic processes that favor neoplastic over normal cells. In this regard it may be noted that CPMs have been experimentally reported to affect brain metabolism (Salford *et al.*, 2003; Volkow *et al.*, 2011). Consequently, the null hypothesis is not compelling to everyone – at best, the physics only argues against a certain narrow class of mechanisms (involving direct DNA damage) that are alternatives to the null. By ruling out this class, the physics thus should increase the plausibility of the null for those who accept the mechanistic argument, but cannot render the null certain until all other mechanisms are ruled out as well. Thus, there are far too many mechanisms left open by current evidence to even claim that the physics should render the null a warranted assumption.

Formal methods for incorporation of narrow or uncertain mechanistic information into causal inference as yet seem limited compared to methods for incorporating data (whose evidential value can be measured and added to the inferential process via loglikelihoods or penalties). The recent surge of interest in mechanism representations such as sufficient-cause models (Chapter 13 in this volume by VanderWeele) may, however, open an avenue for such incorporation (e.g., perhaps via introduction of prior distributions over sufficient causes). Conversely, these representations allow testing of certain mechanistic hypotheses (e.g., about 'interactions') with epidemiologic data (Rothman *et al.*, 2008, Chapters 5 and 16; Chapter 14 in this volume by Berzuini and Dawid; Chapter 13 in this volume by VanderWeele), but only under identifying assumptions that biases are negligible. Under compelling assumptions, epidemiologic data *alone* rarely identify the global causal-null hypothesis, let alone interactions or specific mechanisms (Greenland, 2005b, 2009, 2010).

5.6 Nonidentification and the curse of dimensionality

It seems natural to ask why we do not always use compelling models instead of merely plausible models. One reason is the controversy surrounding most certainty claims (as just exemplified for cell phones); another common reason is lack of data to justify such claims. The result is that a high degree of flexibility is needed to produce constraints that might be accepted by all involved as 'compelling' in a situation.

Consider approximation of multivariate monotonicity without use of additivity or linearity, as in nonparametric regression. Such flexibility will demand at least several terms for each variable alone and an explosion of products among them, along with complex constraints on the terms. With enough variables, conventional modeling rapidly hits the 'curse of dimensionality' in which small-sample artifacts begin to plague conventional fits, even with what appear to be large data sets (large numbers overall, but not enough to support maximum likelihood for the model). To some extent these artifacts can be minimized and performance greatly improved by use of priors or penalty functions (shrinkage) on model parameters (e.g., see Greenland, 2008, pp. 525–527, for citations to some examples and studies). Thus penalized estimation is arguably a better norm for practice (especially since conventional methods are a special case of penalized methods with zero penalty). Nonetheless, penalized estimation used alone retains the same 'effective' dimensional limits; it simply allows more flexible allocation of degrees of freedom.

The algorithmic-modeling (machine-learning) literature tackles the dimensionality problem directly via intensive simulation (Breiman, 2001; Hastie *et al.*, 2009) and enables consistent estimation of many properties of high-dimensional distributions, as long as those properties are identified by given constraints. It has been applied to good effect in causal-inference problems, both for treatment-probability estimation (McCaffrey *et al.*, 2004; Lee *et al.*, 2011) and more directly (van der Laan and Rose, 2011). Nonetheless, in common observational settings, compelling assumptions may no longer identify targeted causal quantities, at least if (as in classical frequentist inference) we are limited to strict constraints. Put another way, relaxation of assumptions to move from the plausible to the compelling can lead to interval estimates whose width increases without bound.

On the other side, re-introducing merely plausible assumptions (e.g., linearity, additivity) to impose identification can create inferences that are seriously biased and overconfident, especially in failing to incorporate the uncertainty surrounding those assumptions (Greenland, 2005b; Richardson *et al.*, 2010). This caution applies whether the assumptions are within a fully parametric, semiparametric, or nonparametric framework. For example, most causal-modeling methods assume randomization (often couched as nonconfounding or ignorability) conditional on certain observations, which nonparametrically identifies effects via conditional permutation tests. But, that assumption has no compelling support in typical epidemiologic contexts, and that lack often leads to considerable controversy about the meaning, implication and value of the test results.

It should also be appreciated that although an identifying assumption may look simple, it may entail an enormous number of important conditions and thus be far stronger than its form makes it seem. For example, conditional randomization/no-confounding/ignorability can be written simply as independence of treatment X and potential outcomes Y_j given confounders. Nonetheless, when there are many unmeasured or poorly measured potential confounders or there are many measured potential confounders, the assumption that any fitted propensity score or other empirical conditionalization produces the needed independency comes down to assuming that the many omitted terms are nonconfounding (unrelated to both X and the Y_j at once).

To deal with recognized nonidentification, most statistical treatments turn to sensitivity analyses in which identifying assumptions are varied. Hopefully these variations are over plausible assumptions, since without identification a complete sensitivity analysis will only display the range allowed for the target parameter by the fixed part of the model and any bounds on parameter exploration (Vansteelandt *et al.*, 2006). This type of analysis can be used to see whether plausible violations could be responsible for observed associations or could be masking important effects. From considerations of plausibility it is but one more step to impose prior distributions over the identifying assumptions (suitably parameterized) and proceed with Bayesian or similar analyses (e.g., Leamer, 1974; Eddy *et al.*, 1992; Phillips, 2003; Lash *et al.*, 2009; Greenland, 2005b, 2009; Gustafson, 2005; Turner *et al.*, 2009). Nonetheless, these variations or priors (or variations on priors) will again face a curse of dimensionality in large problems. Unless strict constraints are imposed, there will be too many sensitivity dimensions to explore let alone set priors on effectively (e.g., see the general confounding sensitivity formulas in VanderWeele and Arah, 2011).

5.7 Identification in practice

As a result of formal nonidentification, most analyses remain a pastiche of highly formalized and completely informal activities that induce an informal sort of identification. These activities often follow a pattern along the following lines:

(a) Assume a relatively simple, plausible identifying model (usually from a *very* short list that will go unquestioned in the context: linear, loglinear, etc.).

(b) Go through the mechanized and often highly elaborate fitting process.

(c) Add conditionals that amount to ‘if our model is correct...’ to the resulting formal statistical inferences.

(d) Pull back from those inferences with cautions that amount to ‘our model may not be correct’, along with speculations (often highly biased) about the likelihood and consequences of violations of single assumptions (which may not reflect accurately the likelihood and consequences of multiple violations).

(e) Conclude by reasserting the model-based inferences in step (c) as if they were observations, not inferences (‘we observed an/no effect of X on Y’).

Of course step (e) is not necessary and savvy researchers water it down to the extent referees and editors will allow. Nonetheless, in step (d), particular assumptions may be defended or criticized based on appeals to other studies or mechanisms.

Mechanistic hypotheses are difficult to calibrate or weight in the inference and as a result are often used overconfidently. At the extremes, authors may argue as if mere description of a plausible mechanism demonstrates causation or as if the failure to describe a plausible mechanism supports or even proves no causation. Thus the assumption that cell phones cannot produce direct DNA damage is based on ‘physics’, which is code for a plethora of studies establishing laws (theories accepted as facts) taken to prove that CPMs do not have sufficient energy for direct chemical effects. Again, such invocation does not exclude other causal pathways, and so does not justify adopting strictly the null hypothesis of no CPM effect, although it may shrink our prior distribution toward that null.

Analogous overconfidence in empirical evidence arises when studies are cited as if their results are more certain or demonstrative than they turn out to be when scrutinized closely. The latter problem is indicated by statements that studies ‘showed’ or ‘found’ some result (whether an association or no association). Inevitably, examination of the cited sources (whether experimental or epidemiological) reveal nothing to warrant such certainty upon considering interval estimates and study limitations. Yet the overconfidence placed in other reports often contributes to overconfident conclusions in the current report, which in turn are cited overconfidently in further reports. In addition, all the reports were overconfident from the use of arbitrary identifying constraints.

In a parallel fashion, there is often overconfident dismissal of still other reports, usually based on limitations inherent in the study design and execution. Typically, a limitation introduces the possibility of a bias mechanism (e.g., lack of randomization opens the door to confounding mechanisms), but rarely does it imply that the mechanism operated, and still more rarely would we know the extent of bias it produced. For example, a collection of data from interviews after the outcome opens the door to differential recall (dependence of recall on outcome, ‘recall bias’), but that fact alone does not imply that this phenomenon had any important impact and nor does it imply (as commonly assumed) that any impact it had was to push estimates away from the null (Drews and Greenland, 1990). A design limitation leaves open bias pathways or uncertainty sources, just as the corresponding design strength removes those pathways. Thus post-outcome recall leaves open bias pathways operating through outcome effects on response, while baseline data collection excludes that pathway, although the general nonidentitication problem remains.

One way of breaking the cycle of citation overconfidence (whether excessive credence or unjustified dismissal) is to enter all available data into a single analysis, treating design features as regressors (meta-regression). But, it is extremely difficult to obtain those data, and when they are obtained, conventional analyses can amplify another source of overconfidence – the computed statistical precision of the final estimate. With large enough numbers and forced identification, the standard errors will shrink to the point that they represent a minor component of warranted uncertainty.

Sensitivity analysis can provide an antidote to this statistical overconfidence by showing how apparently precise inferences may evaporate under certain model expansions. Nonetheless, it can be misused in the opposite extreme to claim that no inference is possible, even when diverse evidence points to a conclusion. One approach to the middle ground employs a greatly expanded model that incorporates all the known major sources of nonidentification and displays results from a system of plausible priors (Greenland, 2005b, 2009; Gustafson, 2005; Lash *et al.*, 2009). Unfortunately, like other approaches, these expansive analyses retain the difficulties of incorporating mechanistic information, and can be criticized for their inevitable failure to explore all imaginable dimensions or plausible priors for the problem.

5.8 Identification and bounded rationality

For better or worse, we make inferences and decisions when there are too many uncertain dimensions to be evaluated by data or deduction. It seems that some persons are much better at this than others, as evidenced by how well they do subsequent to their decisions (given their background). We thus might look to behavioral studies to discover heuristics for dealing with the dimensionality problem.

Those studies are justly celebrated for revealing the severe and often damaging biases that enter into common heuristics (Gilovich *et al.*, 2002; Kahnemann *et al.*, 1982). Notable among them is overconfidence, e.g., from overvaluation of evidence or opinions (one's own or those of trusted parties), which leads to adoption of unfounded strict constraints. Yet, as described in theories of bounded utility and rationality (Simon, 1991; Gigerenzer and Selten, 2002) and recognized in common cautions against 'overthinking', rational decisions require constraints on our deductive activity; i.e., we must rationally allocate or constrain analysis (called satisficing by Simon, 1956, and type-II rationality by Good, 1983). Indeed, the most common rational objections to new methods that I have heard from epidemiologists are of the form 'Is this technique really worth the effort?' Often (but by all means not always) the only honest and well-informed answer is that it depends on the cost of applying the method and the risk of error in applying the method versus the risk of error from failing to apply the method.

Adoption of an initial model represents a tentative commitment to constrain analysis within the confines of the model. Where do such initial constraints come from? Values (including values assigned to traditions as well as effort costs) are one source. It has long been recognized that values enter into methodologies, including those for scientific inference (Kuhn, 1977; Poole, 2001), and they are not hard to discern in modeling. Consider again linearity. Its adoption represents a commitment to not consider nonlinear models unless persuasive evidence of nonlinearity comes to our attention. In the absence of compelling empirical support (which would obviate the study under analysis), linear models are still rationalized based on interrelated values of simplicity, parsimony and transparency, plus conformity to tradition (in order to minimize criticism and questioning of methodology).

The role of values (as opposed to direct rationalization) becomes more obvious and dubious in medical-evidence ratings, which simply declare some studies stronger evidence than others based on gross classification schemes, without attention to finer details that could undermine the rational basis for the declaration in specific cases. A simple example is randomized trials versus nonrandomized cohort studies. Medical journals automatically classify trials as a better 'level of evidence' than observational cohorts, thus leading themselves and readers to value trials more than cohorts. Yet it is quite possible for a trial to be more misleading than an observational cohort. This can occur in many ways, even with neither study being biased in the narrow sense of statistical bias or internal invalidity.

As an example, typical drug-approval trials are too small and too short to reveal rare or long-term side effects and may be done in selected low-risk populations. As a result, they are prone to give an impression of safety that may be dispelled by studies of large health-care data bases. As another example, suppose the trial took place among low-risk volunteers but the cohort comprised high-risk patients and high-risk patients will receive the treatment in practice. The effect could be quite different in low-risk and high-risk populations; such a difference would constitute a bias present in the trial and absent in the cohort, and might even reverse their true evidential value. The lack of provision for such reversal in current medical evidence schemes illustrates the hazards of expedient simplifications for evidence evaluation, even when simplifications are needed on type-II rational grounds.

5.9 Conclusion

Many of the problems discussed here apply to scientific inference in general, whether labeled causal or not, especially problems of identification and evidence synthesis. There is a huge literature on methodologies for combination of evidence, so large that I cannot even begin to provide representative citations, let alone claim any expertise. Issues of combining highly disparate evidence arises in legal and management as well as scientific settings, and may provide valuable clues for improving causal-inference methodology (at least for those not averse to connecting inference and decision making). But, as yet none of those methodologies has become prominent in health-science research or the causal-inference literature, despite some overlap between evidence-synthesis and causal-inference research (e.g., Hepler *et al.*, 2007; Turner *et al.*, 2009).

The current lack of a received methodology for actual causal inference (with synthesis of diverse types of evidence) should be unsurprising, given the difficulty of formalizing such a complex (but essential) task and user limitations in applying complex formalisms. In light of this situation, it has been suggested that causal inferences might often just as well be left informal, at least whenever causal effects are far from identified in any realistic model (Greenland, 2010). After all, despite decades of formal developments, there is much wisdom to be had in informal guidelines for inference, such as the Hill (1965) considerations and later refinements (e.g., Susser, 1991), as long as they are not taken as codified dogmas or criteria (Phillips and Goodman, 2004).

None of this is to dismiss the utility of formal causal models. We need these models to give precise causal meaning to the associations we offer as estimated effects. That need becomes particularly acute with complex treatments, longitudinal interventions and mediation analysis. Nonetheless, the primary challenge in producing an accurate causal inference or effect estimate is most often that of integrating a diverse collection of ragged evidence into predictions to an as-yet unobserved target. This process does not fit into formal causal-inference methodologies currently in use, which by and large assume we have homogeneous observations from the target (embodied in phrases like 'imagine N independent identically distributed copies of X'). Thus, while theory and methods for causal modeling have come a long way in the past 40 years, they still have a very long way to go before they approach a complete system for causal inference.

Acknowledgments

The author would like to thank the editors for their kind and persistent invitation to contribute an essay. The author also wishes to thank Charles Poole, Katherine Hoggatt, Jay Kaufman and Tyler VanderWeele for helpful comments on this chapter.

References

Adams, E. W. (1998) *A Primer of Probability Logic*. Chicago: CSLI Publications/University of Chicago Press.

Breiman, L. (2001) Statistical modeling: the two cultures. *Statistical Science*, **16**, 199–215.

Cornfield, J. (1971). The University Group Diabetes Program: a further statistical analysis of the mortality findings. *Journal of the American Medical Association*, **217**, 1676–1687.

Dawid, A.P. (2000). Causal inference without counterfactuals (with discussion). *Journal of the American Statistical Association*, **95**, 407–448.

Dawid, A. P. (2004). Probability, causality and the empirical world: a Bayes–de Finetti–Popper–Borel synthesis. *Statistical Science*, **19**, 44–57.

Dawid, A. P. (2007) Counterfactuals, hypotheticals and potential responses: a philosophical examination of statistical causality, in *Causality and Probability in the Sciences, Texts in Philosophy*, vol. 5 (eds F. Russo and J. Williamson). London: College Publications, pp. 503–532. Available at http://www.ucl.ac.uk/statistics/research/pdfs/rr269.pdf.

Dawid, A. P. (2010) Beware of the DAG!, in *Proceedings of the NIPS 2008 Workshop on Causality* (eds I. Guyon, D. Janzing and B. Scholkopf). *Journal of Machine Learning Research, Workshop and Conference Proceedings*, **6**, 59–86. Available at http://jmlr.csail.mit.edu/proceedings/papers/v6/dawid10a/dawid10a.pdf.

Drews, C. and Greenland, S. (1990) The impact of differential recall on the results of case-control studies. *International Journal of Epidemiology*, **19**, 1107–1112.

Eddy, D.M., Hasselblad, V. and Shachter, R. (1992) *Meta-analysis by the Confidence Profile method.* New York: Academic Press.

Gigerenzer, G. and Selten, R. (2002) *Bounded Rationality: An Adaptive Toolbox*. Cambridge, MA: MIT Press.

Gilovich, T., Griffin, D. and Kahneman, D. (2002) *Heuristics and Biases: The Psychology of Intuitive Judgment*. New York: Cambridge University Press.

Good, I.J. (1983) *Good Thinking*. Minneapolis: University of Minnesota Press.

Greenland, S. (2005a) Epidemiologic measures and policy formulation: lessons from potential outcomes (with discussion). *Emerging Themes in Epidemiology* (online journal), **2**, 5. Available at http://www.ete-online.com/content/pdf/1742-7622-2-5.pdf.

Greenland, S. (2005b) Multiple-bias modeling for analysis of observational data (with discussion). *Journal of the Royal Statistical Society, Series A*, **168**, 267–308.

Greenland, S. (2008) Variable selection and shrinkage in the control of multiple confounders (invited commentary). *American Journal of Epidemiology*, **167**, 523–529; erratum: 1142.

Greenland, S. (2009) Relaxation penalties and priors for plausible modeling of nonidentified bias sources. *Statistical Science*, **24**, 195–210.

Greenland, S. (2010). Overthrowing the tyranny of null hypotheses hidden in causal diagrams, in *Heuristics, Probabilities, and Causality: A Tribute to Judea Pearl* (eds R. Dechter, H. Geffner and J.Y. Halpern). London: College Press, Chapter 22, pp. 365–382. Available at http://intersci.ss.uci.edu/wiki/pdf/Pearl/22_Greenland.pdf.

Gustafson, P. (2005) On model expansion, model contraction, identifiability, and prior information: two illustrative scenarios involving mismeasured variables (with discussion). *Statistical Science*, **20**, 111–140.

Hastie, T., Tibshirani, R. and Friedman, J. (2009) *The Elements of Statistical Learning: Data Mining, Inference, and Prediction*, 2nd edn. New York: Springer.

Hepler, A. B., Dawid, A. P. and Leucari, V. (2007) Object-oriented graphical representations of complex patterns of evidence. *Law, Probability and Risk*, **6**, 275–293.

Hernán, M.A. (2005) Hypothetical interventions to define causal effects – afterthought or prerequisite? *American Journal of Epidemiology*, **162**, 618–620.

Hill, A.B. (1965) The environment and disease: association or causation? *Proceedings of the Royal Society of Medicine*, **58**, 295–300.

Kahneman, D., Slovic, P. and Tversky, A. (1982) *Judgment under Uncertainty: heuristics and biases*. New York: Cambridge University Press.

Kaufman, J.S. (2008) Epidemiologic analysis of racial/ethnic disparities: some fundamental issues and a cautionary example (with discussion). *Social Science and Medicine*, **66**, 1659–1680.

Kuhn, T.S. (1977) Objectivity, value judgment, and theory choice, in *The Essential Tension* (ed. T.S. Kuhn). Chicago: University of Chicago Press, pp. 320–343.

Lash, T.L., Fox, M.P. and Fink, A.K. (2009) *Applying Quantitative Bias Analysis to Epidemiologic Data*. Boston, MA: Springer Publishing Company.

Leamer, E.E. (1974) False models and post-data model construction. *Journal of the American Statistical Association*, **69**, 122–131.

Lee, B.K., Lessler, J. and Stuart, E.A. (2011) Weight trimming and propensity score weighting. *PLoS One*, **6** (3), e18174. DOI: 10.1371/journal.pone.0018174. Available at http://www.plosone.org/article/info%3Adoi%2F10.1371%2Fjournal.pone.0018174.

McCaffrey, D.F., Ridgeway, G., Morral, A.R. (2004) Propensity score estimation with boosted regression for evaluating causal effects in observational studies. *Psychological Methods*, **9**, 403–425.

Morgan, S. and Winship, C. (2007) *Counterfactuals and Causal Inference: Methods and Principles for Social Research*. New York: Cambridge University Press.

Neyman, J. (1923) On the application of probability theory to agricultural experiments. Essay on principles, Section 9. English translation in *Statistical Science*, 1990, **5**, 465–480.

Pearl, J. (1995) Causal diagrams for empirical research (with discussion). *Biometrika*, **82**, 669–710.

Pearl, J. (2009) *Causality: Models, Reasoning, and Inference*, 2nd edn. New York: Cambridge University Press.

Pearl, J. and Robins, J.M. (1995) Probabilistic evaluation of sequential plans from causal models with hidden variables, in *Proceedings of the Eleventh Conference on Uncertainty in Artificial Intelligence* (eds P. Besnard and S. Hanks). San Mateo, CA: Morgan Kaufmann Publishers, pp. 444–453.

Phillips, C.V. (2003) Quantifying and reporting uncertainty from systematic errors. *Epidemiology*, **14**, 459–466.

Phillips, C.V. and Goodman, K.J. (2004) The missed lessons of Sir Austin Bradford Hill. *Epidemiol. Perspect. Innov.*, **1**, 3 (online journal). DOI: 10.1186/1742-5573-1-3.

Poole, C. (2001) Causal values. *Epidemiology*, **12**, 139–141.

Richardson, T.S., Evans, R.J. and Robins, J.M. (2010) Transparent parametrizations of models for potential outcomes, in *Bayesian Statistics*, vol. 9 (eds J.M. Bernardo, M.J. Bayarri, J.O. Berger, A.P. Dawid, D. Heckerman, A.F.M. Smith and M. West). New York: Oxford University Press.

Robins, J.M. (1986) A new approach to causal inference in mortality studies with a sustained exposure period: applications to control of the healthy worker survivor effect. *Mathematical Modeling*, **7**, 1393–1512.

Robins, J.M. (1987a) Addendum to: A new approach to causal inference in mortality studies with a sustained exposure period. *Computers and Mathematics with Applications*, **14**, 917–945; errata 1987, **18**, 477.

Robins, J.M. (1987b) A graphical approach to the identification and estimation of causal parameters in mortality studies with sustained exposure periods. *Journal of Chronic Disease* (40, Supplement), **2**, 139s–161s.

Robins, J.M. and Richardson, T.S. (2010) Alternative graphical causal models and the identification of direct effects, in *Causality and Psychopathology: finding the determinants of disorders and their cures* (ed. P. Shrout). New York: Oxford University Press.

Robins, J.M. and Wasserman L. (2000) Conditioning, likelihood, and coherence: a review of some foundational concepts. *Journal of the American Statistical Association*, **95**, 1340–1346.

Rothman, K.J., Greenland, S. and Lash, T.L. (2008) *Modern Epidemiology*, 3rd edn. Philadelphia: Lippincott-Wolters-Kluwer.

Rubin, D.B. (1978) Bayesian inference for causal effects: the role of randomization. *Annals of Statistics*, **6**, 34–58.

Rubin, D.B. (1991) Practical implications of modes of statistical inference for causal effects, and the critical role of the assignment mechanism. *Biometrics*, **47**, 1213–1234.

Salford, L.G., Brun, A.E., Eberhardt, J.L., Malmgren, L. and Persson, B.R.R. (2003) Nerve cell damage in mammalian brain after exposure to microwaves from GSM mobile phones. *Environmental Health Perspectives*, **111**, 881–883.

Shermer, M. (2010) Can you hear me now? *Scientific American*, October. Available at http://www.michaelshermer.com/2010/10/can-you-hear-me-now/.

Simon, H. A. (1956). Rational choice and the structure of the environment. *Psychological Review*, **63** (2), 129–138.

Simon, H. (1991) Bounded rationality and organizational learning. *Organization Science*, **2**, 125–134.

Slesin, L. (2008) Science magazine gets it wrong on DNA breaks. *Microwave News*, **28**, 1–3. Available at http://www.microwavenews.com/docs/mwn.11%2810%29-08.pdf.

Spirtes, P., Glymour, C. and Scheines, R. (2001) *Causation, Prediction, and Search*, 2nd edn. Cambridge, MA: MIT Press.

Susser, M. (1991) What is a cause and how do we know one? A grammar for pragmatic epidemiology. *American Journal of Epidemiology*, **133**, 635–648.

Turner, R.M., Spiegelhalter, D.J., Smith, G.C.S. and Thompson, S.G. (2009) Bias modeling in evidence synthesis. *Journal of the Royal Statistical Society, Series A*, **172**, 21–47.

Tversky, A. and Kahneman, D. (1983) Extension versus intuitive reasoning: The conjunction fallacy in probability judgment. *Psychological Review*, **90**, 293–315.

van der Laan, M.J. and Rose, S. (2011) *Targeted Learning: causal inference for observational and experimental data*. New York: Springer.

VanderWeele, T.J. and Arah, O.A. (2011) Bias formulas for sensitivity analysis of unmeasured confounding for general outcomes, treatments and confounders. *Epidemiology*, **22**, 42–52.

Vansteelandt, S., Goetghebeur, E., Kenward, M.G. and Molenberghs, G. (2006) Ignorance and uncertainty regions as inferential tools in a sensitivity analysis. *Statistica Sinica*, **16**, 953–980.

Volkow, N.D., Tomasi, D., Want, G., *et al.* (2011) Effects of cell phone radiofrequency signal exposure on the brain glucose metabolism. *Journal of the American Medical Association*, **305**, 808–814, 828–829.

6

Graph-based criteria of identifiability of causal questions

Ilya Shpitser
Department of Epidemiology, Harvard School of Public Health, Boston, Massachusetts, USA

6.1 Introduction

In Chapter 3 in this volume, we introduced nonparametric structural equation models (NPSEMs), which give a data-generating mechanism for describing causal inference problems. We have introduced an *intervention* operation, corresponding to a kind of idealized randomized experiment, and shown how the notion of a *causal effect* can be represented by post-intervention probability distributions. Many interventions of interest, such as disease or environmental exposure, cannot be implemented by randomization in humans. An important problem in causal inference is to characterize causal assumptions necessary for a particular post-intervention distribution to be expressible as a function of observational data. This is known as the problem of identification of causal effects. In this chapter, we review this problem, discuss known special cases of identification, and present a general identification algorithm, shown complete for NPSEMs. Finally, we discuss a relationship between causal effect identification and a class of constraints in latent variable graphical models.

6.2 Interventions from observations

Consider the graph shown in Figure 6.1, representing the simplest possible situation in causal inference: there is a single treatment, X, and a single outcome of interest, Y. X affects Y, but not vice versa, and we represent this by a directed arrow from X to Y. Since there is no bidirected arc (see Chapter 3) between X and Y, the graph imposes a counterfactual independence restriction,

---

Causality: Statistical Perspectives and Applications, First Edition. Edited by Carlo Berzuini, Philip Dawid and Luisa Bernardinelli.

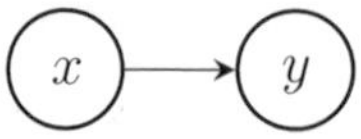

Figure 6.1 *The simplest causal inference problem: What is the causal effect of $do(x)$ on Y?*

namely that X is independent of Y_x (which we write as $Y_x \perp\!\!\!\perp X$, following Dawid, 1979). This assumption, also known as ignorability (Rosenbaum and Rubin, 1983), allows us to identify the causal effect of $do(x)$ on Y, which we write as either $P(y|do(x))$ or $P(Y_x = y)$, via the following derivation:

$$P(Y_x = y) = P(Y_x = y|X = x) = P(Y = y|X = x) \tag{6.1}$$

The first equality follows by the ignorability independence restriction imposed by the graph. The second equality follows by the consistency assumption, which states that if X is observed to attain the value x, then $Y = Y_x$. This assumption is a special case of the generalized consistency assumption introduced in Chapter 3. Consistency assumptions are crucial in causal inference, because they describe conditions under which it is reasonable to transfer information from observational to interventional worlds. There is no way, in general, to test consistency assumptions. However, consistency is often viewed as reasonable by human causal intuition.

Recall from Chapter 3 that interventions on X in an NPSEM inducing a mixed graph G are represented by a new graph $G_{\overline{X}}$, which is obtained from G by removing all arrows pointing to X. Recall also that the observable distribution $P(v)$ of an NPSEM inducing a DAG G obeys the global Markov property with respect to G. In other words, for any three disjoint sets of nodes X, Y, Z in G, if X is d-separated from Y given Z in G, then X is independent of Y given Z in $P(v)$. In some distributions the converse holds, that is whenever X is d-connected to Y given Z in G, X is also dependent on Y given Z in these distributions. Such distributions are said to be faithful to G, and for these distributions, d-separation characterizes conditional independence precisely.

In distributions faithful to a graph G, influence between two variables X and Y can be viewed as flow along d-connected paths connecting X and Y in the graph. The effect of $do(x)$ on Y in a post-intervention distribution can also be viewed as flow along d-connected paths from X to Y in the mutilated (see Chapter 3) graph $G_{\overline{X}}$. By definition of mutilated graphs, however, all d-connected paths from X to Y in $G_{\overline{X}}$ begin with arrows pointing away from X. This implies, in particular, that in any NPSEM DAG where there do not exist any d-connected paths from X to Y that start with arrows pointing to X, both the notions of influence of X on Y and effect of $do(x)$ on Y are represented graphically by the same bundle of paths – d-connected paths that start with arrows away from X. This gives a graphical, if somewhat informal, argument for the following claim.

Theorem 1 *Let* G *be a DAG where all d-connected paths from* X *to* Y *start with arrows pointing away from* X. *Then in any NPSEM inducing* G, $P(y|do(x)) = P(y|x)$.

The graph in Figure 6.1 is the simplest DAG where this theorem holds. We hasten to add that the above argument does not constitute a formal proof. Such a proof would have to rely on the consistency assumption in a crucial way. The value of graphs is in abstracting away formal logical reasoning with consistency in favor of more easily grasped visual arguments.

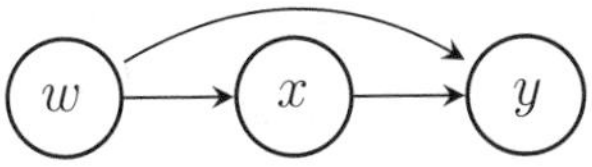

Figure 6.2 The simplest causal inference problem with confounding: What is the causal effect of $do(x)$ on Y?

The potential outcome analog of the above theorem would be to state that the post-intervention distribution $P(Y_x)$ and the conditional distribution $P(Y|X = x)$ are equal if the ignorability assumption $Y_x \perp\!\!\!\perp X$ is true.

6.3 The back-door criterion, conditional ignorability, and covariate adjustment

Practical causal inference problems are frequently plagued by the issue of confounding, that is the presence of variables that make the causal effect of $do(x)$ on the outcome Y different from the observed influence of X on Y. A common refrain "correlation does not imply causation" is common precisely because co-dependence of potential cause and potential effect may entirely be explained away by, for instance, a third unobserved cause of both X and Y.

As a standard example, consider the graph in Figure 6.2. We are interested in the causal effect of $do(x)$ on Y. However, in this graph there is a common cause of both X and Y. X being highly dependent on Y may be explained either by a strong causal effect of $do(x)$ on Y (via the path $X \rightarrow Y$), or via influence from X to Y via the path $X \leftarrow W \rightarrow Y$, or by some combination of the two. Translating the situation to the language of potential outcomes, in most reasonable distributions obtained from NPSEMs inducing the graph in Figure 6.2, it is not the case that Y_x is independent of X. In fact, constructing the counterfactual graph for $P(Y_x|X = x)$ as described in Chapter 3 results in Figure 6.3, where lack of independence between Y_x and X is shown by m-separation.[1]

However, assume we could show that Y_x is independent of X conditional on W. Then we could perform the following derivation similar in spirit to (6.1):

$$
\begin{aligned}
P(Y_x) &= \sum_w P(Y_x|W = w)P(W = w) \\
&= \sum_w P(Y_x|W = w, X = x)P(W = w) \\
&= \sum_w P(Y|W = w, X = x)P(W = w) \qquad (6.2)
\end{aligned}
$$

Here, the first equality is by rules of probability, the second is by our assumed independence, and the last by consistency. There are two ways of deriving this counterfactual independence from the NPSEM. The first is to list counterfactual assumptions implied by the graph explicitly and attempt to derive our assumption logically. Note that independence restrictions implied by the graph in Figure 6.2 imply that $Y_{x,w}$ is independent of both X_w and W. This results in

[1] Note that there are two hypothetical worlds, but only a single W node. This is because W and W_x variables are the same in this model, since X is not an ancestor of W and thus cannot causally influence W.

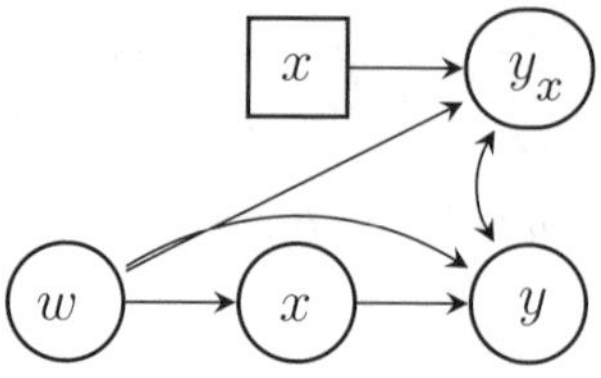

Figure 6.3 *The counterfactual graph for $P(Y_x|X)$ obtained from the graph in Figure 6.1.*

the following derivation:

$$P(Y_{x,w}|W) = P(Y_{x,w}) = P(Y_{x,w}|X_w, W = w) = P(Y_x|X = x, W = w) \qquad (6.3)$$

Here the first two equalities follow by the independence restrictions, and the last by generalized consistency (composition). Note that we also made use of an axiom called compositionality, which states that

$$(X \perp\!\!\!\perp Y|Z) \wedge (X \perp\!\!\!\perp W|Z) \Rightarrow (X \perp\!\!\!\perp Y \cup W|Z)$$

In words, this states that conditional independence on singletons implies conditional independence on sets. This axioms is not true for independences in arbitrary probability distributions. The reason this axiom holds for independences represented by d-separation is that d-separation is a path criterion and statements about d-separation of sets naturally reduce to statements about d-separation of singletons.

An alternative approach for deriving counterfactual conditional independences is to use the NPSEM graph to construct the appropriate counterfactual graph and read off the desired independence using m-separation (the counterfactual graphs obtained from DAGs are often drawn as mixed graphs due to the sharing of unobserved U variables creating bidirected arcs). In fact, it is easy to check that Y_x is m-separated from X given W in the counterfactual graph shown in Figure 6.3.

The expression obtained for $P(y|do(x))$ in (6.2) is known as the *adjustment formula* a version of the same formula has been derived in Chapter 4 in this volume, in the form of Equation (4.7), via decision-theoretic arguments. One can give a (somewhat informal) justification for the adjustment formula in our graph as follows. If we were interested in the *conditional* causal effect of $do(x)$ on Y in a subpopulation where W was observed to attain the value w, that is $P(y|w, do(x)) = P(y, w|do(x))/P(w|do(x))$, we would be justified in concluding this effect is equal to $P(y|w, x)$. This is because conditioned on W, the only d-connected paths from X to Y start with arrows pointing away from X in fact there is only one such path: $X \rightarrow Y$. This implies that there is no difference between the influence of X on Y conditioned on W and the causal effect of $do(x)$ on Y conditioned on W. However, we are not interested in a conditional causal effect given $W = w$, but the overall causal effect averaged across subpopulations. Thus we take an average of all such conditional causal effects, weighted by the prior probability of occurence of $W = w$, to obtain the adjustment formula $\sum_w P(y|x, w)P(w)$.

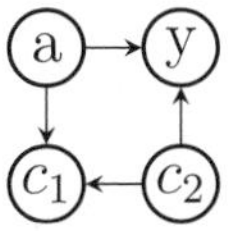

Figure 6.4 A graph where adjusting for $\{C_1, C_2\}$ correctly estimates the causal effect of A on Y, yet the back-door criterion fails.

This argument generalizes to more complex graphs, to give the following theorem

Theorem 2 ***Back-door criterion*** *Let* G *be an ADMG and let X, W, Y be disjoint sets of nodes in* G*. Then if W is a nondescendant of X and blocks all paths from X to Y that start with arrows pointing to X (back-door paths), then $P(y|do(x)) = \sum_w P(y|x, w)P(w)$ in any semi-Markovian model inducing G.*

The potential outcome analog of this theorem would be to invoke the conditional ignorability assumption, which states that $(Y_x \perp\!\!\!\perp X|W)$. Conditional ignorability always implies that the adjustment formula is valid. Conditional ignorability is also more general than the back-door criterion in the sense that there exist NPSEM graphs where conditional ignorability (and thus the adjustment formula) holds, but the back-door criterion does not. See, for instance, the graph in Figure 6.4. See Shpitser *et al.* (2010) for a further discussion of this issue. The back-door criterion was first proposed in Pearl (1993). See Pearl (2000) for a more detailed discussion of the back-door criterion.

6.4 The front-door criterion

Covariate adjustment is a technique capable of handling confounding in situations where sufficiently many potential confounders are observable. However, in practical causal inference problems it is often the case that a noncausal path between the treatment X and outcome Y exists that consists entirely of unobserved variables. The simplest such case is shown in Figure 6.5.

The relevant counterfactual restrictions implied by this graph are: $(Y_{w,x} = Y_w)$, $(W_x \perp\!\!\!\perp X)$, $(Y_{w,x} \perp\!\!\!\perp W_x)$. These restrictions can be used to produce the following derivation:

$$
\begin{aligned}
P(Y_x) &= \sum_w P(Y_x, W_x = w) \\
&= \sum_w P(Y_x, w, W_x = w) \\
&= \sum_w P(Y_x, w)P(W_x = w) \\
&= \sum_w P(Y_w)P(W_x = w) \\
&= \sum_w \sum_{x'} P(Y|w, x')P(x')P(W = w|x)
\end{aligned}
$$

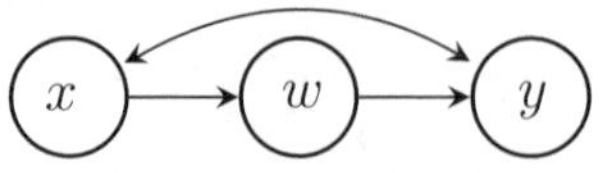

Figure 6.5 The simplest causal inference problem with unresolvable confounding: What is the causal effect of $do(x)$ on Y?

Here the first equality is by definition, the second is by generalized consistency, and third and fourth are by restrictions above, and the last by above restrictions used to repeat the derivations in the last section. In words, this derivation expresses the causal effect of interest $P(Y_x)$ as a product of effects $P(W_x)$ and $P(Y_w)$, and then identifies each effect in this product separately.

Just as before, rather than using an algebraic derivation using a combination of counterfactual restrictions implied by the graph and the generalized consistency assumption, it is possible to give an explicit graphical criterion for identifying causal effects by the functional $\sum_w \sum_{w'} P(Y|w, x')P(x')P(w|x)$.

Theorem 3 ***Front-door criterion*** *Let G be an ADMG and let X, Y be nodes and W a set of nodes in G. Then if W intercepts all directed paths from X to Y, there are no back-door paths from X to W, and X blocks all back-door paths from W to Y, then* $P(y|do(x)) = \sum_w \sum_{w'} P(Y|w, x')P(x')P(w|x)$ *in every semi-Markovian model inducing G.*

One difficulty with using the front-door criterion in practice is that a multitude of counterfactual assumptions must hold. In particular, there must exist observable variables that mediate *every* causal path from the effect variable *X* to the outcome variable *Y* and, moreover, those mediating variables must satisfy ignorability assumptions with respect to both effect and outcome variables. Nevertheless, one advantage of the front-door method of identification is that it gives an alternative way of handling confounding if covariate adjustment or instrumental variable methods are unreasonable. In some situations it may be possible to either structure data collection or randomize mediator variables in a way that would result in the front-door structure.

6.5 Do-calculus

Back-door and front-door criteria are capable of identifying causal effects if certain graphical pre-conditions (or alternatively certain counterfactual assumptions) are true. However, there may well exist situations where neither criterion applies, yet the causal effect of interest is still identified. One approach is to generalize algebraic derivations in the previous sections using a combination of generalized composition axiom and counterfactual restrictions imposed by the causal diagram (Halperm, 2000; Pearl, 2000).

An alternative approach are three rules of inference known as do-calculus (Pearl, 1993, 2000), which are as follows:

- Rule 1: $P_{\mathbf{X}}(\mathbf{y}|\mathbf{z}, \mathbf{w}) = P_{\mathbf{X}}(\mathbf{y}|\mathbf{w})$ if $(\mathbf{Y} \perp\!\!\!\perp \mathbf{Z}|\mathbf{X}, \mathbf{W})_{G_{\overline{\mathbf{X}}}}$.
- Rule 2: $P_{\mathbf{X},\mathbf{Z}}(\mathbf{y}|\mathbf{w}) = P_{\mathbf{X}}(\mathbf{y}|\mathbf{z}, \mathbf{w})$ if $(\mathbf{Y} \perp\!\!\!\perp \mathbf{Z}|\mathbf{X}, \mathbf{W})_{G_{\overline{\mathbf{X}},\underline{\mathbf{Z}}}}$.
- Rule 3: $P_{\mathbf{X},\mathbf{Z}}(\mathbf{y}|\mathbf{w}) = P_{\mathbf{X}}(\mathbf{y}|\mathbf{w})$ if $(\mathbf{Y} \perp\!\!\!\perp \mathbf{Z}|\mathbf{X}, \mathbf{W})_{G_{\overline{x},\overline{z(\mathbf{W})}}}$.

Here $Z(\mathbf{W}) = \mathbf{Z} \setminus An(\mathbf{W})_{G_{\overline{\mathbf{X}}}}$ and the graph $G_{\overline{\mathbf{X}}\underline{\mathbf{Z}}}$ is the graph where all arrows pointing into $\mathbf{X}$ are removed and all arrows pointing away from $\mathbf{Z}$ are removed.

Rule 1 states that d-separation in the mutilated graph implies conditional independence in the corresponding post-intervention distribution. Rule 2 states that absence of back-door paths from a node set $\mathbf{Z}$ to a node set $\mathbf{Y}$ implies that there is no difference between conditioning on $\mathbf{Z}$ and intervening on $\mathbf{Z}$ with respect to $\mathbf{Y}$. Rule 2 can be thought of as a kind of generalization of the back-door criterion. Finally, rule 3 states that any intervention that does not affect outcome or conditioning variables can be safely ignored.

To illustrate how do-calculus rules can be used to derive causal effect identification results, we rederive the front-door functional as a valid expression for $P(Y_x)$ in the graph shown in Figure 6.5. The derivation is as follows:

$$
\begin{aligned}
P(y|do(x)) &= \textstyle\sum_w P(y|w, do(x))P(w|do(x)) && (p)\\
&= \textstyle\sum_w P(y|w, do(x))P(w|x) && \text{(by rule 2)}\\
&= \textstyle\sum_w P(y|do(w, x))P(w|x) && \text{(by rule 2)}\\
&= \textstyle\sum_w P(y|do(w))P(w|x) && \text{(by rule 3)}\\
&= \textstyle\sum_w \sum_{x'} P(y|x', do(w))P(x'|do(w))P(w|x) && (p)\\
&= \textstyle\sum_w \sum_{x'} P(y|x', do(w))P(x')P(w|x) && \text{(by rule 3)}\\
&= \textstyle\sum_w \sum_{x'} P(y|x', w)P(x')P(w|x) && \text{(by rule 2)}
\end{aligned}
$$

Here (p) denotes using standard case analysis in probability, (by rule 2) denotes the application of rule 2, and (by rule 3) denotes the application of rule 3.

One advantage of using do-calculus over reasoning with counterfactual variables using generalized composition and graph restrictions is that the former automates reasoning involving independence and exclusion restrictions by explicitly using d-separation on graphs. This results in derivations that are easier to follow and verify.

Though the example derivation using do-calculus rules involves proving the validity of the front-door functional, do-calculus can be used to derive identification in situations where neither the back-door nor the front-door criteria hold. In fact, it has been shown that do-calculus derivations are capable of identifying *any* identifiable causal effect. We describe the way this result was derived when we consider the issue of identification of arbitrary causal effects in the next section.

6.6 General identification

In the previous sections we considered special cases of the causal effect identification problem. The back-door criterion gives conditions on the graph where the causal effect $P(y|do(x))$ can be identified even in the presence of confounding by means of covariate adjustment. The front-door criterion gives conditions where the causal effect $P(y|do(x))$ is identifiable even when there is confounding and covariate adjustment is impossible. Finally, do-calculus is a set of algebraic rules for manipulating post-intervention distributions that can be used to derive identification results.

In this section we consider the question of causal effect identifiation at its most general. Consider an ADMG G induced by some semi-Markovian model and disjoint sets X, Y of nodes in this graph. Is it possible to characterize situations when the causal effect of $do(x)$ on

function $\mathbf{ID}(\mathbf{y}, \mathbf{x}, P, \mathcal{G})$
INPUT: $\mathbf{x}$,$\mathbf{y}$ value assignments, P a probability distribution, $\mathcal{G}$ an ADMG.
OUTPUT: Expression for $P_{\mathbf{x}}(\mathbf{y})$ in terms of P or **FAIL**(F,F').

1 if $\mathbf{x} = \emptyset$ return $\sum_{\mathbf{v} \setminus \mathbf{y}} P(\mathbf{v})$.

2 if $\mathbf{V} \setminus An(\mathbf{Y})_{\mathcal{G}} \neq \emptyset$
return $\mathbf{ID}(\mathbf{y}, \mathbf{x} \cap An(\mathbf{Y})_{\mathcal{G}}, \sum_{\mathbf{v} \setminus An(\mathbf{Y})_{\mathcal{G}}} P, \mathcal{G}_{An(\mathbf{Y})})$.

3 let $\mathbf{W} = (\mathbf{V} \setminus \mathbf{X}) \setminus An(\mathbf{Y})_{\mathcal{G}[\mathbf{V} \setminus \mathbf{X}]}$.
if $\mathbf{W} \neq \emptyset$, return $\mathbf{ID}(\mathbf{y}, \mathbf{x} \cup \mathbf{w}, P, \mathcal{G})$.

4 if $\mathcal{D}(\mathcal{G}[\mathbf{V} \setminus \mathbf{X}]) = \{S_1, ..., S_k\}$
return $\sum_{\mathbf{v} \setminus (\mathbf{y} \cup \mathbf{x})} \prod_i \mathbf{ID}(s_i, \mathbf{v} \setminus s_i, P, \mathcal{G})$.

if $\mathcal{D}(\mathcal{G}[\mathbf{V} \setminus \mathbf{X}]) = \{S\}$

5 if $\mathcal{D}(\mathcal{G}) = \{\mathbf{V}\}$, throw $\mathbf{FAIL}(\mathbf{V}, \mathbf{V} \cap S)$.

6 if $S \in \mathcal{D}(\mathcal{G})$ return $\sum_{s \setminus \mathbf{y}} \prod_{\{i | V_i \in S\}} P(v_i | v_{\pi}^{(i-1)})$.

7 if $(\exists S')S \subset S' \in \mathcal{D}(\mathcal{G})$ return $\mathbf{ID}(\mathbf{y}, \mathbf{x} \cap S',$
$\prod_{\{i | V_i \in S'\}} P(V_i | V_{\pi}^{(i-1)} \cap S', v_{\pi}^{(i-1)} \setminus S'), \mathcal{G}[S'])$.

Figure 6.6 A complete identification algorithm. ***FAIL*** *propagates through recursive calls like an exception and returns the hedge, which witnesses nonidentifiability.* $V_{\pi}^{(i-1)}$ *is the set of nodes preceding* V_i *in some topological ordering* π *in* G.

Y is identifiable in any model inducing *G*? It turns out that the answer is "yes", and there is an explicit, closed-form algorithm for deriving the functional for such a causal effect whenever it is identifiable. This algorithm was first developed in Tian and Pearl (2002). A version of this algorithm is shown in Figure 6.6. This version first appeared in Shpitser and Pearl (2006b). In this algorithm, the notation $G[X]$ stands for $G_{\overline{v \setminus x}}$.

This algorithm is recursive and uses a divide-and-conquer strategy, using three possible simplification methods, based on districts and ancestral sets. A district (or c-component) is a maximal set of nodes pairwise connected by bidirected arcs. Districts uniquely partition the set of nodes in an ADMG. We will denote the set of districts in an ADMG *G* by $\mathcal{D}(G)$. An ancestral set is a set *S* of nodes with the property that if $X \in S$, then $An(X) \subseteq S$. First, the algorithm splits the problem into subproblems based on districts (line 4). Second, the algorithm marginalizes out some variables such that the margin that is left is an ancestral set (line 2). Finally, the algorithm 'truncates out' some variables in situations where the DAG truncation formula applies to a particular intervention (lines 6 and 7).

When the algorithm fails, it returns a witness for this failure. It turns out there is a particular graph structure called the *hedge* which characterizes cases when this algorithm fails.

Definition 1 C-forest *Let* R *be a subset of nodes in an ADMG* $\mathcal{G}$. *Then a set* F *is called an* R*-rooted C-forest if:*

$R \subseteq F$.

F *is a bidirected connected set.*

Every node in F *has a directed set to a node in* R *with every element on the path also in* F.

Definition 2 Hedge *Let* X, Y *be disjoint node sets in an ADMG* $\mathcal{G}$. *Then two* R*-rooted C-forests* F, F' *form a hedge for* (X, Y) *in* $\mathcal{G}$ *if:*

$F \subset F'$.

$R \subset An(Y)_{\mathcal{G}[V \setminus X]}$.

$X \subset F' \setminus F$.

Hedges characterize nonidentification of causal effects due to the following theorem.

Theorem 4 $P(y|do(x))$ *is identifiable in an ADMG* $\mathcal{G}$ *if and only if there is no hedge for any* (X', Y') *in* $\mathcal{G}$, *where* $X' \subseteq X, Y' \subseteq Y$.

The simplest graph where causal effect identification is not possible is shown in Figure 6.7. In this graph, the hedge witnessing nonidentifiability of $P(y|do(x))$ is two sets, $\{Y\}$ and $\{X, Y\}$. Mapping to the definition, $R = \{Y\}, F = \{Y\}, F' = \{X, Y\}$.

Frequently, the causal effect of interest may be *conditional*. In other words, we may be interested in the effect of $do(x)$ on Y in a subpopulation where an observable set of attributes Z attains values z; in other words $P(y|z, do(x)) = P(y, z|do(x))/P(z|do(x))$. An algorithm for identifying such causal effects is given in Figure 6.8, and first appeared in Shpitser and Pearl (2006a). The algorithm works by applying rule 2 of do-calculus to as many variables in Z as possible, to convert them from variables being conditioned on to variables being fixed, and then derives the functional for the resulting effect by calling the unconditional identification algorithm as a subroutine. All identifiable conditional causal effects can be identified in this way.

Theorem 5 *Let G be an ADMG,* X, Y, Z *disjoint sets of nodes in G. Then* $P(y|z, do(x))$ *is identifiable in all semi-Markovian models inducing G if and only if* ***IDC*** *successfully terminates if given the sets* X, Y, Z *and G as arguments.*

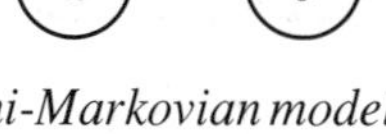

Figure 6.7 A graph representing semi-Markovian models where $P(y|do(x))$ *is not identifiable from* $P(v)$.

function **IDC**(**y**, **x**, **z**, P, G)
INPUT: **x**,**y**,**z** value assignments, P a probability distribution, G a causal diagram (an I-map of P).
OUTPUT: Expression for $P_{\mathbf{x}}(\mathbf{y}|\mathbf{z})$ in terms of P or **FAIL**(F,F').

1 if $(\exists Z \in \mathbf{Z})(\mathbf{Y} \perp\!\!\!\perp Z|\mathbf{X}, \mathbf{Z} \setminus \{Z\})_{G_{\overline{\mathbf{x}}, \underline{z}}}$,
return $\mathbf{IDC}(\mathbf{y}, \mathbf{x} \cup \{z\}, \mathbf{z} \setminus \{z\}, P, G)$.

2 else let $P' = \mathbf{ID}(\mathbf{y} \cup \mathbf{z}, \mathbf{x}, P, G)$.
return $P' / \sum_{\mathbf{y}} P'$.

Figure 6.8 A complete identification algorithm for conditional effects.

Furthermore, it turns out that every line of the **ID** and **IDC** algorithms can be represented as applications of sequences of do-calculus rules. Since **ID** and **IDC** are both complete for their respective identification problems, we have the following result.

Theorem 6 *Do-calculus is complete for causal effect identification. In other words, for every ADMG G and disjoint sets of nodes X, Y, Z in G, $P(y|z, do(x))$ is identifiable in G if and only if there exists a do-calculus derivation of this fact.*

See Shpitser and Pearl (2007) for a more detailed discussion of causal effect identification.

6.7 Dormant independences and post-truncation constraints

Consider the ADMG shown in Figure 6.9. This graph entails no m-separation statements, which implies that the statistical model consisting of all distributions obtained from semi-Markovian models inducing this graph has no conditional independence constraints. Is this model saturated? Surprisingly, the answer turns out to be "no"! In particular, this model contains the following constraint:

$$\frac{\partial}{\partial x_1} \sum_{x_2} p(x_4 \mid x_1, x_2, x_3) p(x_2 \mid x_1) = 0 \tag{6.4}$$

There are two interpretations for constraints of this type, which we call post-truncation independence constraints. The first is that these constraints generalize conditional independences in models with latent variables (recall that the interpretation of bidirected arcs in

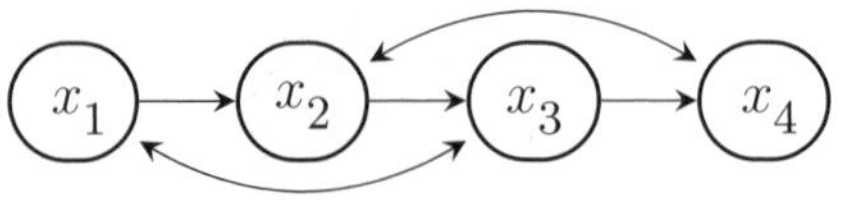

Figure 6.9 A mixed graph model not entailing any conditional independence constraints, but which has a smaller dimension than a saturated model.

ADMGs involve latent variables). In particular, a conditional independence ($X \perp\!\!\!\perp Y|Z$) in a distribution P means that X is independent of Y in $P^* = P(v)/P(z)$. In other words, conditional independence given Z in P is marginal independence in another distribution P^* obtained by dividing P by a marginal $P(z)$. A post-truncation independence can be viewed as a marginal independence in another distribution P^* obtained by dividing P by a *conditional* $P(z|w)$. In particular, the constraint in (6.3) can be viewed as stating that X_4 is marginally independent of X_1 in $P^* = P(x_1, x_2, x_3, x_4)/P(x_3|x_2, x_1)$.

The second interpretation of post-truncation constraints is due to a close connection between these constraints and identification of causal effects. In fact, it turns out that if we try to identify the causal effect $P(x_4|do(x_3, x_1))$ in the graph shown in Figure 6.9 using the **ID** algorithm in the previous section, we will find that this effect is identifiable by the functional $\sum_{x_2} p(x_4 \mid x_1, x_2, x_3)p(x_2 \mid x_1)$. This functional is precisely the one that appears in the constraint (6.3). Thus, a causal interpretation of this constraint is that it asserts the absence of direct effect of X_1 on X_4, in other words that $P(x_4|do(x_3, x_1)) = P(x_4|do(x_3))$. This causal constraint follows from exclusion restrictions imposed by the graph in Figure 6.9, namely that $X_{4_{x_3,x_1}} = X_{4_{x_3}}$. In Shpitser and Pearl (2008), these causal constraints were called (identifiable) dormant independences. See Shpitser and Pearl (2008) and Shpitser *et al.*, (2009) for a more detailed discussion of the causal interpretation of post-truncation constraints. The existence of these constraints was first pointed out by Robins (1986). These constraints were discussed in the context of graphs by Verma and Pearl (1990).

Post-truncation independences are also present in DAG models. However, in DAG models these independences are always logically implied by conditional independences, and thus are not particularly interesting. Intriguingly, as Figure 6.9 shows, in some latent variable models, post-truncation independences are not implied by any conditional independence. This implies that it is possible to use such constraints to distinguish certain latent variable models from saturated (and thus uninformative) models based on data, given appropriate assumptions such as generalizations of faithfulness (Shpitser *et al.*, 1993).

References

Dawid, A.P. (1979) Conditional independence in statistical theory. *Journal of the Royal Statistical Society*, **41**, 1–31.

Halpern, J. (2000) Axiomatizing causal reasoning. *Journal of A.I. Research*, 317–337.

Pearl, J. (1993) Graphical models, causality, and intervention. *Statistical Science*, **8**, 266–269.

Pearl, J. (1994) A probabilistic calculus of actions, in *Proceedings of the Tenth Annual Conference on Uncertainty in Artificial Intelligence (UAI-94)*, vol. 10, San Francisco, CA: Morgan Kaufmann, pp. 454–462.

Pearl, J. (2000) *Causality: Models, Reasoning, and Inference*. Cambridge: Cambridge University Press.

Robins, J.M. (1986) A new approach to causal inference in mortality studies with sustained exposure periods with application to control of the healthy worker survivor effect. *Mathematical Modeling*, **7**, 1393–1512.

Rosenbaum, P.R. and Rubin, D.B. (1983) The central role of the propensity score in observational studies for causal effects. *Biometrika*, **70** (1), 41–55.

Shpitser, I. and Pearl, J. (2006a) Identification of conditional interventional distributions, in *Proceedings of the Twenty-Second Conference on Uncertainty in Artificial Intelligence* (eds R. Dechter and T.S. Richardson), vol. 22. Corvallis, OR: AVAI Prress, pp. 437–444.

Shpitser, I. and Pearl, J. (2006b) Identification of joint interventional distributions in recursive semi-Markovian causal models, in *Proceedings of the Twenty-First National Conference on Artificial Intelligence*, vol. 2, Boston, MA: AVAI Press, pp. 1219–1226.

Shpitser, I. and Pearl, J. (2007) Complete identification methods for the causal hierarchy. Technical Report R–336, UCLA Cognitive Systems Laboratory.

Shpitser, I. and Pearl, J. (2008) Dormant independence. Technical Report R-340, Cognitive Systems Laboratory, University of California, Los Angeles, CA.

Shpitser, I., Richardson, T.S. and Robins, J.M. (2009) Testing edges by truncations, in *International Joint Conference on Artificial Intelligence*, vol. 21, pp. 1957–1963.

Shpitser, I., VanderWeele, T. and Robins, J.M. (2010) On the validity of covariate adjustment for estimating causal effects, in *International Joint Conference on Artificial Intelligence*, vol. 22.

Spirtes, P., Glymour, C. and Scheines, R. (1993) *Causation, Prediction, and Search*. New York: Springer Verlag.

Tian, J. and Pearl, J. (2002) A general identification condition for causal effects. in *Eighteenth National Conference on Artificial Intelligence*, pp 567–573.

Verma, T.S. and Pearl J. (1990) Equivalence and synthesis of causal models. Technical Report R-150, Department of Computer Science, University of California, Los Angeles.

7

Causal inference from observational data: A Bayesian predictive approach

Elja Arjas
Department of Mathematics and Statistics, University of Helsinki, Helsinki, Finland

7.1 Background

Causality is a challenging topic for anyone to consider in a formal way. Already the concept itself is problematic, and often people have sharply different opinions of its foundations. However, causality is perhaps a particularly challenging topic for a statistician. Rather than just trying to formulate views on some underlying philosophical issues, a statistician is often faced with the concrete problem of how to find empirical support in favour or against, or even trying to prove or disprove, a causal claim made in some substantive scientific or nonscientific context.

We shall here concentrate on three important aspects of causality. The first aspect is temporality, which manifests itself already in the fundamental requirement that a cause must precede the effect in time. Somewhat surprisingly, the temporal aspect of causality is only rarely explicitly accounted for in the major part of the literature on causal modeling. The second aspect to be emphasized here is confounding. When analyzing observational data with the aim of finding empirical support to a causal claim, there is always a possibility that the differences that are found may in fact be due to spurious associations. While unconfounded inference is ultimately always based on hypotheses that cannot be verified from data, it is important that these hypotheses are formulated in an intuitively understandable and, in the considered context, meaningful way. Third, our approach to causality is completely probabilistic, including aspects

Causality: Statistical Perspectives and Applications, First Edition. Edited by Carlo Berzuini, Philip Dawid and Luisa Bernardinelli.

of statistical inference. This leads us to using predictive distributions as the main summary measures for causal claims.

7.2 A model prototype

To start from a simple setting, suppose we are interested in studying the effect which a contemplated causal variable A might have on some measured response Y. Suppose further that the considered unit, individual or object being studied is described by an observed covariate X. Finally, let U be a generic notation for corresponding unobserved background characteristics and parameters that we think could be relevant to the causal question. When setting up a probability model p for these variables, we consider them in the natural time order $U \to X \to A \to Y$ (meaning U temporally precedes X, which temporally precedes A, and so on), in which case the chain multiplication rule then leads to the joint distribution $p(U, X, A, Y) = p(U)p(X \mid U)p(A \mid U, X)p(Y \mid U, X, A)$. For a graphical representation of this, in the form of a directed acyclic graph (DAG), see Figure 7.1(a).

As the variable U is not observed, predictions concerning Y would be based on only knowing the values of X and A, that is on $p(Y \mid X, A)$. This conditional distribution can be obtained from the joint distribution of all four variables in the obvious way by first forming the marginals $p(X, A, Y)$ and $p(X, A)$ by integration, and then computing $p(X, A, Y)/p(X, A)$.

Here we have particularly in mind the situation arising in purely observational studies, where the investigator has had no real physical control of the value of A even after the value of X was observed. Following a notation introduced by Lindley (2002), this aspect can be emphasized by writing $p(Y \mid X, \text{see}(A))$ for such a prediction. However, for a causal interpretation, where A would be viewed as a cause of Y, it seems essential that one can think of there being a parallel, perhaps only hypothetical, situation in which the value of A would be determined by some external mechanism. This would be the case, for example, in a randomized clinical trial, where A could be the indicator of assigning a patient either to receiving a treatment or to placebo and the external mechanism would be the device used for the randomization. Such a situation is similarly emphasized in the notation, following Pearl (2009), by writing $p(Y \mid X, do(A))$. In what follows, the basis for presenting probabilistic evidence

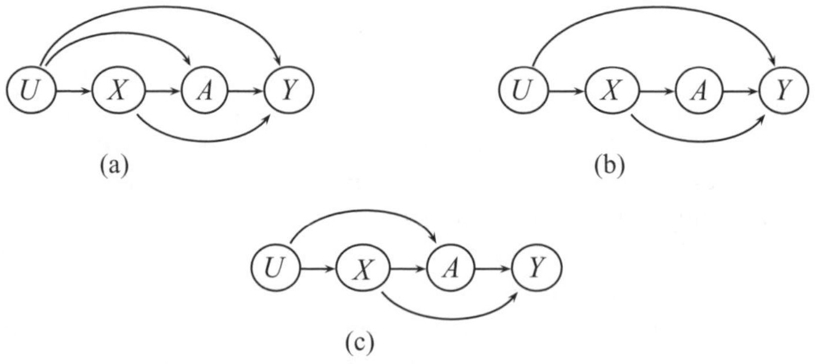

Figure 7.1 A graphical representation of the dependencies considered in the model prototype. In (a) all dependencies may be present and therefore U is a potential confounder. (b) represents the situation of Definition 1, where A is unconfounded relative to U and in (c) the model is unconfounded since U is not a potential confounder.

for the claim 'A causes Y' is by considering contrasts between predictive probabilities of the form $p(Y \mid X, do(A = a))$ and $p(Y \mid X, do(A = a'))$, where a and a' are two different values of A that are being compared.

From the perspective of statistical inference, the key question is now the following: 'How can such *do*-probabilities be evaluated empirically when the supporting data come from an observational *see*-study?' The *see*- and *do*-probabilities must obviously be related to each other in some way. However, they are not the same! In particular, the *do*-experiment does not in itself give rise to a specification of a conditional distribution, which would correspond to $p(A \mid U, X)$ in a *see*-study. Indeed, in a *do*-experiment it would be irrelevant for a causal analysis of what random or other mechanism was used to deliver the value of A as long as it was somehow exogenous from the perspective of the observational *see*-study. In this sense one could say that in the *do*-situation the joint distribution of (U, X, A, Y) is only partly specified. In an observational *see*-study, on the other hand, one generally views the variable A as being endogenous, with values being determined in the same fashion as those of the covariate X.

In contrast to this, in a causal context it seems both natural and necessary to assume that otherwise the probabilistic structure and description of the *see*-study is lifted to the, perhaps only hypothetical, *do*-experiment without change (cf. Lindley, 2002). In particular, one should then assume that:

(i) the joint (prior) distribution of the variables U and X is the same in the hypothetical *do*-experiment as in the real *see*-study, and also that

(ii) the response Y, given X, A and U, behaves in the same way regardless of whether the event $\{A = a\}$ was *done* or merely *seen*. Stated more explicitly, we assume that $p(Y \mid U, X, \text{see}(A)) = p(Y \mid U, X, \text{do}(A))$.

The problems therefore seem to centre around the conditional distribution $p(A \mid U, X)$, the potential fallacy for a causal analysis being that of *confounding*: in an observational study it could happen that differences between the predictions $p(Y \mid X, see(A = a))$ and $p(Y \mid X, see(A = a'))$ given to two items or individuals, or to the same item or individual in two different circumstances, with different values a and a' of the contemplated causal variable A, would actually be due, not to these differences in A but to differences in the unobserved variable U. In an extreme situation, A could be only a marker of some important background variable U such that its value alone would completely determine the value of Y. A manipulation that would change only the value of the marker A, and not that of U, would then leave the value of Y unchanged.

This idea can be formalized as follows. Consider the causal problem where X is an observed covariate, A is a contemplated cause and Y the response of interest. Then we call the variable U a *potential confounder* in this causal problem if the prediction of Y based on knowing the values of X, A and U, as expressed by the distribution $p(Y \mid X, A, U)$, actually depends on U. If the considered problem formulation does not contain such potential confounders, then we can say that this causal model is *unconfounded*.

Consider then the more interesting situation where the model contains an unobserved potential confounder U. (If there are more than one, we include them all in U and consider U to be a vector.) Then, as we cannot ignore this variable in our causal analysis, we have to find some condition under which the predictions based on a *see*-study could be utilized also in a situation in which A was thought to arise from a manipulation, that is from a *do*-study.

This becomes possible under the following conditional independence postulate (cf. Arjas and Parner (2004), Definition 1):

Definition 1 *We say that A is unconfounded relative to a potential confounder U if A and U are conditionally independent given X, that is if*

$$p(A \mid U, X) = p(A \mid X)$$

This postulate, which is expressed in the form of a DAG in Figure 7.1(b), has an important consequence. The (posterior) distribution of U based on the observations X and A does not actually depend on A, that is we have:

Lemma 1 *If A is unconfounded in the sense of Definition 1, then the two posterior distributions $p(U \mid X, see(A))$ and $p(U \mid X)$ are the same.*

This result justifies why it is natural to call A satisfying the condition of Definition 1 unconfounded. The result itself is an obvious consequence of Bayes' formula. For deriving the posterior $p(U \mid X, see(A))$ we would start from the prior $p(U)$ and consider the expression $p(X, A \mid U) = p(X \mid U)p(A \mid U, X) = p(X \mid U)p(A \mid X)$ as the likelihood. Similarly, for deriving $p(U \mid X)$, we would start from the prior $p(U)$ and use $p(X \mid U)$ as the likelihood. However, as the priors are the same and the likelihood expressions are proportional (in U), the posteriors will also be the same. This is because the proportionality constant $p(A \mid X)$ appears in Bayes' formula both in the numerator and the denominator, and therefore can be cancelled.

Now, if A in a *see*-study is unconfounded in the sense of Definition 1, we can replace the rule $p(A \mid X)$ by which the selection of the value of A was modeled, by an arbitrary distribution that does not depend on U. As an extreme case, we could imagine that the values of A observed in the data had in fact been chosen by us in advance, and then fixed. Alternatively, we could think that they had been determined by some rule that had been chosen in advance, as functions of the corresponding covariate values X. The rule could even involve some external information such as independent randomization, as long as it does not depend on the unobserved background variable U. In all such situations, the posterior inference concerning U, when based on X, will remain the same as in the original *see*-study because it does not depend on A.

These considerations become perhaps more clear if we introduce separate notations for the probability distributions describing these two situations, using subscript 'obs' for the observational *see*-study that gave rise to the data and subscript 'ex' for the hypothetical *do*-experiment, this notation referrring to the idea that the selection of the value of the contemplated causal variable A is *exogenous* when viewed from the perspective of a causal model used for describing the observational data. The subscript 'ex' could also be thought of as referring to 'external' or to 'experimental'. (Note that in Arjas and Parner, 2004 the subscript 'opt' has been used in place of 'ex', as an abbreviation of 'optional'. However, such a notation can lead to a confusion between 'optional' and 'optimal'. In some instances we are actually interested in finding an optimal regime for assigning values to A, so we will reserve the subscript 'opt' to denote such a rule; see, for example, Arjas and Saarela, 2010.)

Thus we write, by again applying the chain rule,

$$p_{\text{obs}}(U, X, A, Y) = p_{\text{obs}}(U)p_{\text{obs}}(X \mid U)p_{\text{obs}}(A \mid U, X)p_{\text{obs}}(Y \mid U, X, A) \qquad (7.1)$$

for the joint distribution of the variables (U, X, A, Y) in the data, and then assuming that $p_{\text{obs}}(A \mid U, X) = p_{\text{obs}}(A \mid X)$ if A is unconfounded. In the hypothetical case where the value of A is assigned by some external mechanism, we write similarly

$$p_{\text{ex}}(U, X, A, Y) = p_{\text{obs}}(U)p_{\text{obs}}(X \mid U)p_{\text{ex}}(A \mid U, X)p_{\text{obs}}(Y \mid U, X, A) \tag{7.2}$$

Thus, corresponding to the assumed links (i) and (ii) above, the other parts of the model p_{obs}, apart from the one describing assignment of A, have been retained when switching from 'obs' to 'ex'. Using this notation we can then conclude the following.

Lemma 2 *If* A *is unconfounded in the sense of Definition 1, then:*

(i) The posterior distributions of U based on the observed data (X, A) are the same in both schemes, that is

$$p_{\text{obs}}(U \mid X, A) = p_{\text{ex}}(U \mid X, A) \tag{7.3}$$

Here neither of these posterior distributions depends on A.

(ii) The posterior distributions of U based on the observed data (X, A, Y) are the same in both schemes, that is

$$p_{\text{obs}}(U \mid X, A, Y) = p_{\text{ex}}(U \mid X, A, Y) \tag{7.4}$$

(iii) The predictive distributions of Y based on observed data (X, A) are the same in both schemes, that is

$$p_{\text{obs}}(Y \mid X, A) = p_{\text{ex}}(Y \mid X, A) \tag{7.5}$$

The first claim is a restatement of Lemma 1. The second claim follows directly from the first, by an application of Bayes' formula, when interpreting $p_{\text{obs}}(U \mid X, A)$ and $p_{\text{ex}}(U \mid X, A)$ in (7.3) as prior distributions and then using the assumption that the respective conditional distributions (or likelihood expressions) $p_{\text{obs}}(Y \mid U, X, A)$ and $p_{\text{ex}}(Y \mid U, X, A)$ of Y are the same in both schemes. Proving the third conclusion is similar; it follows at once from (7.3) when multiplying the left-hand side by $p_{\text{obs}}(Y \mid U, X, A)$ and the right-hand side by $p_{\text{ex}}(Y \mid U, X, A)$, which factors were assumed to be identical, and then integrating U out on both sides. Note that the same conclusion (7.5) holds trivially if U is not a potential confounder, that is, $p_{\text{obs}}(Y \mid U, X, A) = p_{\text{obs}}(Y \mid X, A)$, a situation displayed graphically in Figure 7.1(c).

Stated briefly, the first two results say that under the unconfoundedness condition the posterior inferences concerning U when based on X and A, and similarly when based on X, A and Y, are the same regardless of whether the value of A has been *done* or merely *seen*. As a consequence, under that condition, we can perform Bayesian estimation of the unknown variable U in the context of an observational study as if treatment A had been assigned, in a designed experiment, some pre-specified fixed value.

Essentially the same conclusion as in (7.4) holds if, instead of applying Bayesian methods and terminology, one prefers using direct likelihood inference and a corresponding formulation of the result. Then, fixing X and considering $p_{\text{obs}}(A, Y \mid U, X) = p_{\text{obs}}(A \mid U, X)p_{\text{obs}}(Y \mid U, X, A)$, as a function of U, to be the likelihood arising from observing A and Y, we see readily from the unconfoundedness postulate that its first factor does not depend on U and that it is therefore proportional (in U) to its second factor $p_{\text{obs}}(Y \mid U, X, A)$. Thus only this second factor, which is common to both schemes, needs to be considered when inferring on U.

Remark 1 The above simple model is in many ways similar to the well-known Rubin causal model (Rubin, 1974; Holland, 1986). Both models can be viewed as providing descriptions of a situation in which, on a given individual i, covariates X_i are measured at time 0 and this is immediately followed by an assignment A_i of i to either treatment or placebo. At a later point in time τ, a response Y_i is measured. In the Rubin causal model, however, it is assumed that for each individual i there exist, already before the experiment is performed, two fixed *potential outcomes*, $Y_0(i)$ under placebo and $Y_1(i)$ under treatment (for a detailed discussion of the concept of potential outcomes, see Chapter 2 in this volume by Sjölander). The treatment assignment A_i then determines which potential outcome will actually be realized and observed (and only one will). The randomness in the Rubin model is thus contained in the assignment variables A_i for different individuals i. In the terminology of Rubin, and now suppressing the index i in the notation, A is called *ignorable* if it is conditionally independent of (Y_0, Y_1) given X, that is $(Y_0, Y_1) \perp A \mid X$. This corresponds closely to our Definition 1 of unconfounded inference, and actually formally coincides with it if we set $U = (Y_0, Y_1)$. Note also that if this connection between the two models is made, U becomes a trivial potential confounder in our setting because the value of Y is completely determined by U and A. The main difference between the Rubin model and ours is in the interpretation of U and, more generally, in how randomness is understood. A technical difference is that in the Rubin model separate variables and notations are introduced for different potential outcomes, whereas in our approach there is a single response variable Y and we consider its conditional distributions given A.

7.3 Extension to sequential regimes

These simple considerations extend readily to a longitudinal setting in which we have in the data a sequence of event times and corresponding descriptions of the events that have taken place. Such a setting is of particular interest in clinical studies, where one attempts to establish an optimal regime for the treatments and where the individual treatment decisions are allowed to depend on how a patient has responded to the treatments that were received earlier. Here is a general formulation of such a situation.

Suppose that a random number N_τ events occur over the considered time interval $(0, \tau]$. At each event time T_k, covariates X_k are measured and an action, or treatment, A_k follows immediately upon this. Hence, the recorded data consist of $(T_k, (X_k, A_k))$, $k = 0, 1, \ldots, N_\tau$, with $0 \equiv T_0 < T_1 < T_2 < \cdots < T_{N_\tau}$, and finally, of a measured response Y. Such a sampling scheme can be formulated naturally by using the general framework of marked point processes (see, for example, Brémaud, 1981; Arjas, 1989; Karr, 1991; Andersen *et al.*, 1993), with T_k being the event times and $Z_k = (X_k, A_k)$ the corresponding marks. As a convention, we can also treat the considered response Y as a marked point, identifying it with 'the last observed covariate value' X_{N_τ}. This can be done without any restriction to the generality and allows, for example, situations in which the final response is some summary measure determined on the basis of the entire observed sample path of the process.

The marked point process formalism is also able to accommodate latent variables and developments, which are potentially relevant for describing the considered causal problem, but which were not observed. As before, we use the generic notation U for such variables and then make the convention that they can be imbedded, as a sequence of additional random marks, into the marked point process. Having denoted by T_k the time of the kth event, we denote by $\tilde{Z}_k = (U_k, X_k, A_k)$ the corresponding event, or mark, where (X_k, A_k) is observed in the data and U_k is unobserved.

Before moving on, we make some further conventions on how the marks can be interpreted. In a situation in which we have data on several individuals, say n individuals indexed by $i = 1, 2, \ldots, n$, we can form the natural superposition of the marked point processes describing the considered individuals, then arriving at a formulation in which the components of $\tilde{Z}_k = (U_k, X_k, A_k)$ are vectors with coordinates indexed according to the individuals. In some designs, e.g. randomized clinical trials, there may not be a covariate measurement X_k preceding a corresponding assignment A_k to a treatment, say a, in which case we could write $(\varnothing, a)$ as the value of (X_k, A_k). On the other hand, in some sampling schemes a number of repeated covariate measurements are made before there is an actual assignment A_k to a treatment, and then we can similarly write $(x, \varnothing)$. A more general but also notationally more elaborate description and analysis of such situations is considered in Parner and Arjas (1999).

In problems of this type it seems natural to think that there is some structural interest parameter θ, which is common to all these individuals and with which the inferential problem is primarily concerned. On a more abstract level, the existence of such a parameter would be justified by an exchangeability assumption and the de Finetti (1974) representation theorem. We use θ as a generic notation for model parameters; its exact meaning will then depend on the particular application and model that are considered. We now make the convention that θ is imbedded into the latent mark U_0 as a coordinate. Thus our inferences concerning the latent marks $U_k, k = 0, 1, \ldots, N_\tau$, will also cover inferences on θ.

Setting up a probability for the canonical sample paths of such a marked point process turns out to be very simple because the marked point process framework allows us to proceed by induction, always moving from a time point T_k to the next point at T_{k+1} and then considering it jointly with the corresponding mark $(U_{k+1}, X_{k+1}, A_{k+1})$. Setting up a probability model for the sample paths of the marked point process can then be done by a sequential application of the chain rule of multiplication. In its kth step we consider conditional probabilities of the form $p_{\text{obs}}(T_{k+1}, U_{k+1}, X_{k+1}, A_{k+1} \mid \mathcal{F}_k)$, where $\mathcal{F}_k = \{(T_i, U_i, X_i, A_i); i = 0, 1, \ldots, k\}$ is the 'full' history of the marked point process up to time T_k. In this way it will be sufficient to consider a single generic step in such an induction. Denote similarly by $\mathcal{H}_k = \{(T_i, X_i, A_i), i = 0, 1, \ldots, k\}$ the observed history up to time T_k.

Consider then the issue of potential confounding in this framework. We start with the following definition, which extends our earlier Definition 1 to the present sequential setting and corresponds to Definition 2 in Arjas and Parner (2004) (see also Section 8.8 in Chapter 8 in this volume by Berzuini, Dawid and Didelez and Definition 2 in Dawid and Didelez 2010).

Definition 2 Unconfounded inference *We say that a sequence of contemplated causal variables (A_k) in an observational study described by sample path $\mathcal{F}_{N_\tau} = \{(T_i, U_i, X_i, A_i), i = 0, 1, \ldots, N_\tau\}$ and probability p_{obs} is unconfounded relative to latent variables (U_k) if, for each* k, *A_k and $\{U_i, i = 0, 1, \ldots, k\}$ are conditionally independent given $(\mathcal{H}_{k-1}, T_k, X_k)$,* *that is*

$$p_{\text{obs}}(A_k|\mathcal{F}_{k-1}, T_k, U_k, X_k) = p_{\text{obs}}(A_k|\mathcal{H}_{k-1}, T_k, X_k), k = 1, 2, \ldots, N_\tau \tag{7.6}$$

Remark 2 The above setting bears a close resemblance to the sequential trial design studied by Robins (1986). In particular, our condition of unconfounded inference corresponds to his 'no unmeasured confounders' condition. Robins' model can be viewed as a extension of the Rubin causal model to sequential designs, and its main technical – and perhaps also conceptual – difference to ours is its reliance on the potential outcomes framework. More specifically, Robins considers potential outcomes of the form $Y_{\mathbf{a}}$, where $\mathbf{a} = (a_0, \ldots, a_K)$ is a regime of a fixed sequence of K treatments at predetermined time points. Its interpretation is similar to the Rubin causal model, in the sense that the potential outcomes $Y_{\mathbf{a}}$ are then assumed to exist, in some sense, already before the experiment has been performed, for all admissible regimes $\mathbf{a}$.

Let us now see how this postulate can be used in a context where we would like to relate an observed sequence of events to a causal problem, viewing the observed variables (A_k) as 'causes'. Following the same idea as in Section 7.2, we connect the inferences, which can be drawn from the observational data and which are described in terms of a probability denoted by p_{obs}, to corresponding statements relative to another probability denoted by p_{ex}. To do so, we link these two probabilities to each other by the following requirements:

$$\begin{aligned} p_{\text{ex}}(U_0, X_0) &= p_{\text{obs}}(U_0, X_0) \\ p_{\text{ex}}(T_{k+1}, U_{k+1}, X_{k+1} \mid \mathcal{F}_k) &= p_{\text{obs}}(T_{k+1}, U_{k+1}, X_{k+1} \mid \mathcal{F}_k),\ k = 0, 1, \ldots, N_\tau \end{aligned} \tag{7.7}$$

Note that again in here, we can view the probability p_{ex} as being only partially defined in the sense that the conclusions of Theorem 1 below remain valid regardless of the way in which the treatment assignment probabilities for (A_k) are specified, as long as they are exogenous in the sense that they do not depend on the potential confounder variables (U_k).

Our main conclusion can now be stated as follows.

Theorem 1 *Suppose that contemplated causal variables (A_k) are unconfounded in the sense of Definition 2. Then, for each $k = 0, 1, \ldots, N_\tau$:*

(i) The posterior distributions of the complete history $\mathcal{F}_k = \{T_i, U_i, X_i, A_i; i = 0, 1, \ldots, k\}$, given the corresponding observed history $\mathcal{H}_k = \{T_i, X_i, A_i; i = 0, 1, \ldots, k\}$, are the same in both schemes, that is

$$p_{\text{obs}}(\mathcal{F}_k|\mathcal{H}_k) = p_{\text{ex}}(\mathcal{F}_k|\mathcal{H}_k) \tag{7.8}$$

Here neither of these posterior distributions depends on the latest assignment A_k.

(ii) The predictive distributions of the next marked point $(T_{k+1}, U_{k+1}, X_{k+1})$ given the corresponding observed history $\mathcal{H}_k = \{T_i, X_i, A_i, i = 0, 1, \ldots, k\}$ are the same in both schemes, that is

$$p_{\text{ex}}(T_{k+1}, U_{k+1}, X_{k+1} \mid \mathcal{H}_k) = p_{\text{obs}}(T_{k+1}, U_{k+1}, X_{k+1} \mid \mathcal{H}_k) \tag{7.9}$$

For a proof of (7.8), we only need to go through the steps of Lemma 2, replacing (U, X, A) with $(U_i, (T_i, X_i), A_i)_{0 \leqslant i \leqslant k}$, and use induction in k. The crucial argument in such an induction is that, when moving from T_k to T_{k+1}, the likelihood contribution corresponding to the 'new data' $(T_{k+1}, X_{k+1}, A_{k+1})$ is, possibly up to a proportionality factor not depending on $(U_i)_{0 \leqslant i \leqslant k+1}$, the same as what would correspond to only (T_{k+1}, X_{k+1}). To prove (7.9), note first that it states the same relationship between the conditional distributions of $(T_{k+1}, U_{k+1}, X_{k+1})$ as (7.7), except that now the conditioning is with respect to the observed history $\mathcal{H}_k$. However, this follows from (7.8) when the transition probabilities (7.7), considered as functions of U_k, are integrated out with respect to the corresponding (identical) posterior distributions p_{ex} and p_{obs}.

Remark 3 This same result could have been derived and presented by using the framework of counting processes and the associated stochastic intensities (more generally, compensators) based on a continuous time parameter $t \geqslant 0$. The postulate of unconfounded inference, corresponding to our Definition 2, would then be phrased in a natural way as a *local independence* condition (Schweder, 1970; Aalen, 1987; Didelez, 2008), technically stating that the local characteristics corresponding to treatment assignments in the compensators would be the same, regardless of whether the compensators were considered to be relative to the observed histories $(\mathcal{H}_t)_{t \geqslant 0}$ or relative to the larger histories $(\mathcal{F}_t)_{t \geqslant 0}$ generated, in addition, by the unobserved process $(U_t)_{t \geqslant 0}$. (Note that, in the somewhat different context of time series, the concept of *noncausality* of Granger (1969) expresses a very similar idea as local independence.)

These two ways of formulating the key condition, one based on considering a sequence of marked points and the associated conditional distributions as above, and the other on the corresponding counting processes in continuous time and their compensators with respect to alternative histories, are easily seen to be equivalent. This is so because, in a marked point process, the compensators between any two consecutive event time points T_k and T_{k+1} are actually completely determined by the conditional distributions of $(T_{k+1}, U_{k+1}, X_{k+1}, A_{k+1})$ given the respective histories $\mathcal{H}_k$ and $\mathcal{F}_k$ at T_k, and conversely. More generally, both these approaches can be viewed to be special cases of *filtering* of a marked point process (see, for example, Arjas *et al.*, 1992; Arjas and Haara, 1992).

The approach based on continuous time and a local independence condition has been considered in complete detail in Parner and Arjas (1999). However, instead of considering posterior and predictive distributions as in Theorem 1, the main result there is formulated in terms of likelihood processes. It says that, under the local independence condition, the contributions from the treatment assignments A_k to the likelihood expression do not depend on the parameter of interest. These contributions can therefore be ignored in direct likelihood inference concerning that parameter. Thus the comments on likelihood inference that were made earlier, after Lemma 2, remain valid also in this more general context of sequential assignments.

Remark 4 There is an interesting difference in how knowledge of A_k influences conditional probability distributions describing the *past* and the *future*. While, as stated in part (i) of Theorem 1, the posterior distribution of the latent history $\{U_i; i = 0, 1, \ldots, k\}$, based on $\mathcal{H}_k = \{T_i, X_i, A_i; i = 0, 1, \ldots, k\}$, does not depend on A_k, all future variables in the marked point process after time T_k, including the final response, may well depend on it. Indeed, such dependence is precisely what a statistician anticipating a causal effect would expect to see.

Note also that a similar statement does not hold for covariates X_k; a new covariate reading will generally have both inferential value backwards in time and predictive value forwards in time.

7.4 Providing a causal interpretation: Predictive inference from data

Looking now at the contents of Theorem 1 one may wonder what bearing it has for considering concrete problems in causality. While it provides a condition under which unconfounded statistical inferences can be drawn from observational data, it does not at first sight appear to say how such inferences could be used to arrive at useful conclusions relating to causality. In the following we try to straighten this, before following the same basic ideas that were, in a simple case, presented in Section 7.2.

Suppose that we, after having carried out data analysis of the above kind, would like to express our conclusions by using 'causal language'. For such a purpose it would be natural to try to predict how a generic (real or hypothetical) item or individual would respond to some specific treatment, or a sequence of such treatments. Using from now on the word 'individual' in this case, we then assume that (i) the contemplated causal variables (A_k) considered in the data are unconfounded in the sense of Definition 2 and (ii) this generic individual is exchangeable with those considered in the data in the sense that their behavior can be modeled by the same probability model, parametrized with a common parameter θ. In other words, we assume that the inferences drawn from the observational data are valid information about how a similar individual would respond to treatments assigned by some exogenous rule or mechanism.

The inferences that were drawn from the data $\mathcal{H}_{N_\tau} = \{T_k, (X_k, A_k), k = 0, 1, \ldots, N_\tau\}$ (henceforth denoted simply by 'data'), on the common model parameter θ can then be utilized also for predicting what will happen to this generic individual if he/she is to be given some specific sequence of treatments. Adding a star (*) to the notation to signify the considered generic individual, possibly described by some covariate values $X_0^* = x_0^*$ at the baseline, we would, in a simple case, be interested in predicting the response Y^* under a given fixed sequence of 'forced' treatment assignments, say $A_i^* = a_i^*, i = 0, 1, \ldots, k$ (cf. Dawid, 2002). Such a prediction can then be expressed in terms of the predictive distribution $p_{\text{ex}}(Y^* \mid x_0^*, \mathbf{a}^*, \text{data})$, where we have denoted $\mathbf{a}^* = (a_i{}^*)_{0 \leqslant i \leqslant k}$.

More generally, there could be a *dynamic treatment regime*, say $\mathcal{A}$, such that each A_k^* could be allowed to be a function of the past observed history of that individual, consisting of past event times, covariate readings and possible earlier treatment assignments. In fact, the treatment assignments of an individual could in principle depend also on the past histories of the other individuals in the sample. Thus we are not, generally speaking, postulating here a 'no interference between units' property. More generally still, such a regime could be randomized, as long as the randomization mechanism does not depend on the past potential confounder variables U_k^*. In order to make the role of the regime $\mathcal{A}$ explicit in the notation we add it to the subscript of p_{ex} by writing $p_{\text{ex}(\mathcal{A})}$. The considered predictive distribution will then be denoted by $p_{\text{ex}(\mathcal{A})}(Y^* \mid x_0^*, \text{data})$. The exact specification of this probability will then depend on the considered assignment mechanism $\mathcal{A}$.

Note that before finally obtaining a response Y^*, the considered generic individual is thought to experience a realization of the marked point process; that is there may be a sequence of event times T_k^*, covariate readings X_k^* and assignments A_k^* according to the chosen regime $\mathcal{A}$. Apart from possible fixed attributes (such as gender), which have been initially chosen to characterize the considered individual, the covariates will generally evolve over time in ways that cannot be known precisely at the time at which the prediction is made. As a consequence, the computation of these predictive distributions involves an integration in the assumed probability model with respect to such possibly time-dependent random variables.

It may be useful to distinguish between three different ingredients, or sources of randomness, which need to be accounted for when computing the predictive distributions $p_{\mathrm{ex}(\mathcal{A})}(Y^* \mid x_0^*, \text{data})$. The first is the intrinsic or endogenous randomness in the behavior of the considered generic individual, that is in the event times $T_k{}^*$, covariate readings $X_k{}^*$ and ultimately response Y^*, as specified by our probability model p_{ex} in a situation in which the values of θ, X_0^* and $\mathbf{A}^*$ would be fixed at some known values. The second potential source is the treatment regime $\mathcal{A}$, which may be randomized by some mechanism. Finally, the model parameters θ are unknown and need to be estimated from data. Here, using our earlier convention that the parameters θ are already contained in the common latent variable U_0, which belongs to $\mathcal{F}_0$, the necessary results are summarized by the corresponding marginal of the posterior distribution $p_{\mathrm{obs}}(\mathcal{F}_{N_\tau} \mid \text{data})$, which was provided in Theorem 1 (i). We can then compute, by an integration with respect to this posterior, the desired predictive distribution $p_{\mathrm{ex}(\mathcal{A})}(Y^* \mid x_0^*, \text{data})$.

Having now described how to derive, for a given regime $\mathcal{A}$, the predictive distribution $p_{\mathrm{ex}(\mathcal{A})}(Y^* \mid x_0^*, \text{data})$ of the generic response Y^*, given the values of the baseline covariates, we can consider any two such regimes of interest, say $\mathcal{A}_1$ and $\mathcal{A}_2$, and compare the corresponding predictive distributions $p_{\mathrm{ex}(\mathcal{A}_1)}(Y^* \mid x_0^*, \text{data})$ and $p_{\mathrm{ex}(\mathcal{A}_2)}(Y^* \mid x_0^*, \text{data})$, or the corresponding expectations of some suitable test functions, to each other. In practice, the necessary numerical integration can be carried out efficiently by Monte Carlo simulation, by applying data augmentation alongside the computations that are needed for statistical inference (see also Section 8.10 in Chapter 8 in this volume by Berzuini, Dawid and Didelez).

Practical illustrations of this general method can be found, for example, in Arjas and Liu (1995, 1996), Arjas and Haastrup (1996), Arjas and Andreev (2000), Arjas and Saarela (2010) and Härkänen *et al.* (2000).

Remark 5 The above predictive distributions bear a close resemblance to the G-computation algorithm of Robins (1986) (see also Dawid and Didelez, 2010 and Parner and Arjas, 1999, for some comments, the latter being available on the web page http://wiki.helsinki.fi/display/biometry/Elja+Arjas). Technically, there seems to be the difference that our predictive distributions involve additional integrations, not only of time points (T_k^*) and (potentially time dependent) future covariates (X_k^*) but also of the model parameter θ (with respect to the posterior) and of future treatment assignments (with respect to the considered regime $\mathcal{A}$). The expression obtained by the G-computation algorithm thus, generally speaking, depends on the model parameter and a chosen sequence of treatment assignments, where the latter could be said to represent a sequence of 'forced conditionings' (instead of standard conditioning of probabilities). Our use of predictive distributions, on the other hand, integrates both the results from statistical inference (represented by the posterior) and adherence to a considered dynamic regime into the same computation. Another difference is more conceptual, stemming from the fact that we are here not making use of formulations based on the idea of potential or counterfactual outcomes. This aspect seems to separate us from a large body of the causality literature.

7.5 Discussion

The main issue in this paper has been our attempt to clarify how, and under what circumstances, unconfounded statistical inferences can be drawn from observational data by applying Bayesian methods and modeling based on marked point processes. It was argued that the causal conclusions from such study, in the sense of effects of causes (Dawid, 1995), can be formulated in a natural way in terms of posterior predictive distributions or the corresponding expectations. Such results can then be seen as ingredients in the lengthy process of arriving at a causal understanding and interpretation of the empirical results obtained in some substantive context of interest (Cox, Chapter 1 in this volume). However, since our approach is entirely probabilistic, it becomes important how the concept of probability is understood in such a context. Our point of view has been that probabilities should be viewed primarily, not as objects or characteristics of the physical world but as expressions of what is known about it on the basis of the information that is available. Consistent with this, we are here not claiming to be providing tools for proving the existence, in a literal sense, of some causal relationships in Nature, nor for identifying the magnitudes of corresponding effects.

Considering briefly two well-known popular alternatives to the approach presented here, graphical models provide a convenient tool for exploring and visualizing the dependence relations between random variables that are involved (e.g. Pearl, 2009; Didelez, 2008). This is especially true for conditional independencies since they can be read off directly from the graph. Although directed acyclic graphs (DAGs) often reflect the time order in which sampling on a single individual was performed, they do not fully incorporate the time aspect. More specifically, they do not specify *when* in time measurements and actions were taken. This can be crucial if inference is based on a sample of individuals and sampling does not follow a sequential trial design with predetermined time points. Moreover, graphical representations of such situations tend to become impracticable enough to lose their intuitive appeal and usefulness for visualization. (Some additional aspects relating to the role of the time in causal considerations are provided in Arjas and Eerola, 1993.) Finally, one sees only rarely explicit suggestions on how statistical inferences drawn from existing empirical data should be used for assigning probabilities to a DAG.

The second popular framework for considering causal modeling and inference, based on the concept of potential or counterfactual outcomes, may be useful because of the intuitive interpretations they offer. However, their drawback, particularly in the context of sequential regimes, is the large number of random variables representing such potential outcomes 'that might have occurred' and the consequent tedious notation. One may also wonder how much sense the modeling convention, according to which all these variables are 'fixed', makes as a description of reality. Marked point processes, the approach advocated here, seem to be more flexible in these respects while also allowing for natural extensions.

It is generally accepted that the randomization device, at least in principle, will lead to valid causal inferences, whereas results from observational studies normally are given the lower status of statistical associations. Even then, such inferences are in practice based on an observed finite sample and will therefore depend on how well the realization of the randomization managed to distribute the unobserved characteristics over treatment groups. This can be a problem, particularly in small samples. Thus, while the conditions for unconfounded inference, given in Definitions 1 and 2, are guaranteed to hold in designs in which treatment assignment is through a genuine randomization device, this will only protect us against biasing our causal conclusions systematically by unwarranted confounding. In concrete applications, randomization is therefore a valid argument for claiming that the analysis is not confounded, but in no way an absolute safeguard.

In practice, suggesting candidates for the marks (U_k, X_k, A_k) is limited to the *causal field* under consideration, that is to observed and unobserved variables and processes that the investigator is conjecturing, or knows, to be relevant to the considered causal problem. Any statement about causal dependence would then be relative to that setting. In practice, however, one must limit both the number of different covariates being measured and the amount of unobserved variables included in the causal field in order to justify the 'unconfounded inference' assumption of Definition 2.

Controlling a potential confounder process by observation, whenever possible, is of course a direct way in which one can try to assure that the design is unconfounded. This may, however, be a dangerous policy because conditioning on additional random variables may destroy an already existing conditional independence property. This is also the reason why, when computing predictive distributions of the form $p_{\text{ex}(\mathcal{A})}(Y^* \mid x_0^*, \text{data})$, we are not controlling the values of the covariates X_k^*, except for $X_0^* = x_0^*$ determined already at the baseline, or those of the later assignments A_k^*, but integrate them out, the latter according to the chosen regime $\mathcal{A}$. This is so for two reasons: first, such variables are often intermediate in the sense of being in time between a contemplated causal variable and the considered response, and can sometimes be viewed as mediators of the causal effect of the former on the latter, and, second, because such conditioning can influence the predicted response also indirectly, by feeding back to the distribution of the unobservables in the past. This is also reflected in our notation of the predictive distributions when we write $p_{\text{ex}(\mathcal{A})}(Y^* \mid x_0^*, \text{data})$, instead of using standard conditioning of the form $p_{\text{ex}}(Y^* \mid x_0^*, \mathbf{A}^*, \text{data})$, where $\mathbf{A}^* = (A_i{}^*)_{0 \leqslant i \leqslant k}$ would be some sequence of treatment assignments.

Acknowledgement

Parts of this article have been taken, after some modification, from Parner and Arjas (1999) and Arjas and Parner (2004).

References

Aalen, O.O. (1987) Dynamic modelling and causality. *Scandinavian Actuarial Journal*, pp. 177–190.

Andersen, P.K., Borgan, Ø., Gill, R.D. and Keiding, N. (1993) *Statistical Models Based on Counting Processes*. New York: Springer.

Arjas, E. (1989) Survival models and martingale dynamics. *Scandinavian Journal of Statistics*, **16**, 177–225.

Arjas, E. and Andreev, A. (2000) Predictive inference, causal reasoning, and model assessment in nonparametric Bayesian analysis: a case study. *Lifetime Data Analysis*, **6**, 187–205.

Arjas, E. and Eerola, M. (1993) On predictive causality in longitudinal studies. *J. Statist. Plan. Inf.*, **34**, 361–386.

Arjas, E. and Haara, P. (1992) Observation scheme and likelihood. *Scandinavian Journal of Statistics*, **19**, 111–132.

Arjas, E. and Haastrup, S. (1996) Claims reserving in continuous time; a nonparametric bayesian approach. *A.S.T.I.N. Bulletin*, **26**.

Arjas, E. and Liu, L. (1995) Assessing the losses caused by an industrial intervention: a hierarchical bayesian approach. *Applied Statistics*, **44**, 357–368.

Arjas, E. and Liu, L. (1996) Nonparametric Bayesian approach to hazard regression: a case study involving a large number of missing covariate values. *Statistics in Medicine*, **15**, 1757–1770.

Arjas, E. and Parner, J. (2004) Causal reasoning from longitudinal data. *Scandinavian Journal of Statistics*, **31**, 171–187.

Arjas, E. and Saarela, O. (2010) Optimal dynamic regimes: presenting a case for predictive inference. *The International Journal of Biostatistics*, **6**, Article 10 (electronic). DOI: 10.2202/1557-4679.1204.

Arjas, E., Haara, P. and Norros, I. (1992) Filtering the histories of a partially observed marked point process. *Stochastic Processes and their Applications*, **40**, 225–250.

Brémaud, P. (1981) *Point Processes and Queues*. New York: Springer.

Dawid, A.P. (1995) Discussion of 'Causal diagrams for empirical research' by J. Pearl. *Biometrika*, **82**, 689–690.

Dawid, A.P. (2002) Influence diagrams for causal models and inference. *International Statistical Review*, **70**, 161–189.

Dawid, A.P. and Didelez, V. (2010) Identifying the consequences of dynamic treatment regimes: a decision theoretic overview. *Statistics Surveys*, **4**, 184–231.

de Finetti, B. (1974) Bayesianism: its unifying role for both the foundations and applications of statistics. *International Statistical Review*, **42**, 117–130.

Didelez, V. (2008) Graphical models for marked point processes based on local independence. *Journal of the Royal Statistical Society, Series B*, **70**, 245–264.

Granger, C.W.J. (1969) Investigating causal relations by econometric models and cross-spectral methods. *Econometrica*, **37**, 424–438.

Härkänen, T., Virtanen, J. and Arjas, E. (2000) Caries on permanent teeth: a nonparametric bayesian analysis. *Scandinavian Journal of Statistics*, **27**, 577–588.

Holland, P.W. (1986) Statistics and causal inference, with discussions. *Journal of the American Statistical Association*, **81**, 945–970.

Karr, A. (1991) *Point Processes and Their Statistical Inference*, 2nd edition. Marcel Dekker.

Lindley, D. (2002) Seeing and doing: the concept of causation. *International Statistical Review*, **70**, 191–214.

Parner, J. and Arjas, E. (1999) Causal reasoning from longitudinal data. Research Report A27, Rolf Nevanlinna Institute, Helsinki.

Pearl, J. (2009) *Causality: Models, Reasoning and Inference*. Cambridge: Cambridge University Press.

Robins, J. (1986) A new approach to causal inference in mortality studies with a sustained exposure period – application to control of the healthy worker survivor effect. *Mathematical Modelling*, **7**, 1393–1512.

Rubin, D.B. (1974) Estimating causal effects of treatments in randomized and nonrandomized studies. *Journal of Educational Psychology*, **66**, 688–701.

Schweder, T. (1970) Composable Markov processes. *Journal of Applied Probability*, **7**, 400–410.

8

Assessing dynamic treatment strategies

Carlo Berzuini[1], Philip Dawid[1], and Vanessa Didelez[2]
[1]*Statistical Laboratory, Centre for Mathematical Sciences, University of Cambridge, Cambridge, UK*
[2]*Department of Mathematics, University of Bristol, Bristol, UK*

8.1 Introduction

We continue the discussion of sequential data-gathering and decision-making processes, started in the preceding chapter in this volume. The archetypical context is that of a sequence of medical decisions, taken at different time points during the follow-up of the patient, each decision involving choice of a treatment in the light of any interim responses or adverse reactions to earlier treatments. The problem consists of predicting the consequences that a (possibly new and possibly hypothetical) treatment plan will have on a future patient, on the basis of what we have learnt from the performances of past medical decision makers on past patients. While we make constant reference to a medical application context, the scope of the method is of much broader relevance.

Our approach to this problem owes immensely to the seminal work of James Robins. In a rich series of papers, Robins (1986, 1987, 1988, 1989, 1992, 1997, 2000, 2004), Robins and Wasserman (1997) and Robins *et al.* (1999) introduce the idea of different treatment strategies applied in different hypothetical studies (these studies being analogous to our notion of 'regimes'). These papers examine conditions on the relationships between these studies, under which those treatment strategies can be compared. Among these, the *sequential randomization* condition is closely related to the 'stability' assumption used in this chapter. Furthermore, Robins (1986) introduced the *G*-computation algorithm (a special version of

Causality: Statistical Perspectives and Applications, First Edition. Edited by Carlo Berzuini, Philip Dawid and Luisa Bernardinelli.

dynamic programming) to evaluate a sequential treatment strategy, which works where standard regression – even under the sequential randomization assumption – fails. Most of the concepts expounded in this chapter have a counterpart in Robins' work, although, in the interest of readability, we shall frequently abstain from making the relationships with his work explicit.

We shall assume that we have an idea of which variables informed past decisions, without assuming that these variables have all been observed in the past. Nor shall we assume that the rules that governed past decisions are discernible or similar to those that will guide future actions.

In contrast with the previous chapter, we do use (fully stochastic) causal diagrams (more precisely, influence diagrams). In the context of sequential plan identification, graphical representations of causality were first advocated by Pearl and Robins (1995). In accord with these authors, we assume that the relevant causal knowledge is encoded in the form of a diagram with a completely specified topology, whose associated numerical conditional probabilities are given only for a *subset* of its variables, so-called *observed* variables. The remaining variables in the diagram are 'unobserved' or 'latent'. We shall, however, introduce a form of graphical representation of causality that differs from Pearl and Robins's diagrams in some respects. Part of the problem will be to characterize situations in which the estimation of the causal effects of the treatment plan of interest is not invalidated by unobserved confounding introduced by the latent variables. Throughout this chapter we use previous results by Dawid and Didelez (2010).

Elja Arjas, both in his chapter (Chapter 7) in this volume and in the joint paper Arjas and Parner (2004), considers two probability models. One (called the 'obs' model) models the observational study that has generated the data. The other (called the 'ex' model) models the consequences of the future (perhaps hypothetical) application of the treatment plan of interest. In our approach, this distinction is embodied in a decision parameter, called the 'regime indicator', which indicates the particular regime, 'obs' or 'ex', in operation. Supplementing the set of domain variables with the mentioned regime indicator results in an 'augmented' set of variables. In our approach, conditions for the equivalence of Arjas's 'obs' and 'ex' distributions are phrased in terms of conditional independence conditions on this augmented set. Both chapters, ours and that of Arjas, avoid formulations based on ideas of potential outcomes or counterfactuals. We illustrate the methods and their motivations with the aid of two examples. Following Robins, we choose the first example in the area of the treatment of HIV infection. The second example will be in the treatment of abdominal aortic aneurysm.

This chapter is about the identifiability of treatment plan effects, rather than about the methods of estimation and computation one is supposed to use if the problem turns out to be identifiable. The latter problems are discussed in greater detail in Chapter 17 in this volume, by Daniel, De Stavola and Cousens, who illustrate their method with the aid of the same HIV example we use here. An introduction to this example is given in the next section.

8.2 Motivating example

We start with an example in human immunodeficiency virus (HIV) infection. The standard treatment of this disease involves an *initiation therapy* based on a combination of three antiretroviral agents, chosen from a larger pool of candidate drugs. In many patients, this initial therapy will eventually fail. Failure criteria include virologic evidence, for example failure to achieve HIV RNA levels lower than 200 copies/mL, or repeated observation of HIV RNA levels above that threshold, or development of life-threatening toxic effects. Therapy failure will prompt consideration of a possible replacement of one or more components of the initial drug cocktail. The decision whether to do the replacement, and – in case of a positive decision – the choice of the drug components to be replaced or used as replacement, will be performed in the light of the patient's past history, in accord with a collection of rules that define, jointly with the therapy failure criteria, the *treatment strategy* (or plan) in this particular application.

One possible strategy consists of applying a fixed, pre-defined, sequence of drug cocktails and dosages, where the switches from one cocktail to the next occur at pre-established times, irrespective of any incoming information about the patient. This is called a *static* strategy or plan. More interesting are *dynamic* treatment plans, where the treatment decisions adapt to, and depend on, the accruing series of patient events. Treatment plans are usually compared in terms of their impact on the patient's disease course, for example in terms of the length of the AIDS-free survival time, although cost-effectiveness considerations may easily be brought into the framework (we shall not say much about this). This chapter addresses the problem of predicting the consequences on a hypothetical future patient of a treatment plan of interest, without requiring this plan to have ever been considered or implemented in the past. Application of the method to alternative treatment plans provides a basis for *comparing* them, as well as a basis for determining the *best* plan under given constraints, the so-called *optimization* problem. A discussion of methodological and computational issues in optimization is outside the scope of the present chapter.

8.3 Descriptive versus causal inference

Standard statistical software offers tools to describe the endpoint experience of groups of patients subjected to different treatment strategies. Application of these tools to the data might, for example, reveal that 'patients that have been treated according to strategy 1 have fared better than those that have been treated according to strategy 2'.

The problem with this 'descriptive' approach is that – while possibly useful and correct in a sense – it may be misleading both scientifically and in terms of practical (medical) implications. What the clinicians and the scientists want is to predict the effect of assigning (by intervention) a specific treatment to a *future* case, at any point of the treatment sequence, conditional on the accumulating information. Can this knowledge be extracted from observational data, collected from past medical records?

The general answer is no, because future cases may differ from those in the data in various important ways, for example because, in the observed data, the doctor's choices of actions made use of additional – unrecorded – private information. In Arjas's terminology, this would correspond to the causal variables, $\{A_k\}$, *not* being unconfounded relative to the latent variables, in the sense of his Definition 2. Moving in a similar direction to Arjas, in this chapter we formalize the requirements under which past empirical evidence *can* be validly carried into future decisions. Whenever these requirements appear unreasonable in a real application, we must be prepared to conclude that the available data have little practical implication with respect to future (medical) practice. While often useful, descriptive analysis may be misleading if used to corroborate causal claims in those situations where the formal requirements discussed in this chapter are not carefully negotiated.

Of course, if we were able to carry out a large experiment in which sample patients are randomized over the set of treatment strategies of interest, these difficulties would, at least in part, disappear. Confounding would be eliminated by randomization. However, in very many circumstances, we shall need (or wish) to exploit data gathered under purely observational circumstances, and this motivates our effort.

8.4 Notation and problem definition

Consider the ordered sequence of time points (T_k: $k = 0, 1, \ldots, N+1$), with $0 \equiv T_0 < T_1 < T_2 < \cdots < T_{N+1}$. Unlike in the preceding chapter, we take this sequence to be fixed. Moreover, for simplicity, we shall take it to be the same in all patients. In a medical treatment decision context, these time points will usually be defined on a patient-specific time scale, such as, for example, time since diagnosis, rather than, say, calendar year. For $k = 1, \ldots, N+1$, we consider the following patient-specific variables:

X_k: the values of a set of covariates observed and recorded in the sample individuals at T_k,

A_k: an action or treatment decision, performed and recorded in the sample individuals right after T_k (undefined for $k = N+1$). In certain applications, this variable could denote the level of a certain exposure at time T_k.

Our assumption that the domain variables are completely ordered in time is not essential, but it will make our discussion simpler. We let the symbol $\overline{A}_k$ denote the partial sequence $(A_0, A_1, \ldots, A_k)$ of actions performed up to time T_k, and a similar notation is used for other variables in the problem. The (X_k) sequence contains a distinguished variable, $Y \equiv X_{N+1}$, called the *endpoint* variable. This could be the state of health of the patient at the end of a given time period or, for example, a summary of the entire realized history, $(\overline{X}_N, \overline{A}_N)$, of the recorded process. This will cover situations in which alternative plans are compared in terms of the treatment costs involved.

Throughout this chapter, we are assuming our data to have been collected under an *observational* regime, which we denote by o. Such a regime will generally involve the passive observation of the performance of past decision makers, such as when information about the treatments and their outcomes are drawn from medical records. In our notation, these data constitute sample realizations (or 'paths') of the $(\overline{X}_{N+1}, \overline{A}_N)$ process. The method we are going to describe does not require us to discern the decision criteria in operation under the observational regime, nor the source of these criteria, be it the doctor's personal experience, or public guidelines, imposed protocols, etc. In contrast with the observational regime, we have an *experimental* regime, denoted by e, in which a hypothetical future patient is treated according to a specific (static or dynamic) plan, which we wish to evaluate. Such a plan will consist of a specific set of rules for determining the value of each A_k, at time T_k, possibly on the basis of information about $(\overline{A}_{k-1}, \overline{X}_{k-1})$. Clearly, the observational and the experimental regimes differ in that, in the former, the actions A_k are passively observed, whereas in the latter they are imposed on the patient. The notion of 'experimental regime' can include the modalities whereby each treatment action will be carried out and, more generally, the environmental and technological conditions, the stage of evolution of the pathogens and the medical expertise that will surround the application of these actions. The aim of the method is to predict the effect of the experimental regime e on Y, on the basis of the data collected under o. We do not need to assume that the decision criteria that inform the plan under assessment have ever been adopted in the past.

We introduce a nonrandom *regime indicator*, σ, taking values in (o, e). We think of the value of σ as being determined externally, before any observations on $(\overline{X}_{N+1}, \overline{A}_N)$ are made. All probability statements about $(\overline{X}_{N+1}, \overline{A}_N)$ must then be explicitly or implicitly conditional on the value of σ. We use the symbol $p(Y \mid \sigma = e)$ to denote the marginal density of Y for a future patient under treatment regime e. If we are able to estimate this density, we shall be able to say something about how effective (in terms of whatever loss function we adopt) is the treatment plan under investigation. Of interest then are the conditions under which the past data, collected under regime o, can be used to determine $p(Y \mid \sigma = e)$. These we shall refer to as *plan identifiability conditions*. For simplicity, in the following, we suppose that the time grid, (T_k), and the $(\overline{X}_{N+1}, \overline{A}_N)$ variables are the same for both regimes, o and e. For the time being, we shall assume that treatment decisions under e are influenced by the same variables that have influenced the corresponding decisions under o.

8.5 HIV example continued

With reference to our HIV example, the sequence $(T_k\colon k = 1, \ldots, N)$, might represent a fixed set of times at which the patient is visited, with T_0 representing some clinically meaningful temporal origin. The X_k covariate might represent the observed values, at T_k, of viral RNA levels, CD4 levels, AIDS status and – more broadly – safety-related variables or indicators of possible unfavourable side effects of the therapy. Immediately after information X_k has been collected, a therapeutic action, A_k, will be performed on the patient. This might consist of switching from the current drug cocktail to a new one, by replacement of one or more components in response to the past clinical evolution of the patient.

8.6 Latent variables

In most situations, we shall need to introduce into the problem additional variables that are essential to describe the causal structure of the problem, but are not directly observable or have not been recorded in the data. These variables we call 'unobserved', or unrecorded, or *latent*, and collectively denote with the symbol $\mathcal{L}$. We shall let elements of this set be denoted by the generic symbols U and W. We shall have these variables time-subscripted, with U_k, say, denoting the value of U right before time T_k and after time T_{k-1}. The symbol $\overline{L}_k$ will denote the set of all $\mathcal{L}$ variables with a subscript equal or lower than k.

In the HIV example, we may use latent variables to represent, for example, patient-specific, possibly inheritable, patterns of response to drugs that cannot be directly measured and have not been recorded in the data. In a similar example in Chapter 17, Daniel and colleagues introduce latent variables that categorize the patients into 'types', *e.g.*, indicating lower-than-average responsiveness to specific therapies and/or vulnerability to the side effects of a specific therapy (see Chapter 17 for details). Our formulation allows these variables to be organized in a time process, which allows the model of HIV progression to acknowledge the tendency of some patients to switch type during the course of therapy. The introduction of latent variables will result in a more detailed causal structure, and a correspondingly more complex causal diagram representation of the problem. This has advantages. For one, it will often help justify the necessary assumptions. The stability assumption of the next section, for example, is often easier to justify in terms of local properties of a detailed causal diagram that includes latent variables than in terms of relationships between the observed variables of direct interest.

8.7 Conditions for sequential plan identifiability

We are now going to explore conditions for sequential plan identifiability, that is conditions under which our data, which have been collected under regime o, can be used to infer the distribution $p(Y \mid \sigma = e)$ of the endpoint variable when we apply the treatment strategy of interest under regime e. In general, this inference will *not* be possible, one reason being that, in general, the probability distribution of the variables will differ from one regime to the other. We have identifiability if certain properties of equivalence between the distributions of the variables under the two regimes are satisfied. These properties will now be discussed under the assumption that (X, A) variables are the same under the two regimes, with the treatment strategy under study specified in terms of a probability distribution of A_k given $\overline{A}_{k-1}$ and $\overline{X}_k$.

Under the experimental regime, by $\sigma = e$, the joint distribution of the observed variables factorizes into the product of the conditional distributions of each of these variables given the earlier observed variables:

$$p(y, \overline{x}_N, \overline{a}_N \mid \sigma = e) = \left\{ \prod_{k=0}^{N+1} p(x_k \mid \overline{x}_{k-1}, \overline{a}_{k-1}\, ; \sigma = e) \right\} \times \left\{ \prod_{k=0}^{N} p(a_k \mid \overline{x}_k, \overline{a}_{k-1}\, ; \sigma = e) \right\} \tag{8.1}$$

where $X_{N+1} \equiv Y$. The second factor on the right-hand side of Equation (8.1) represents the pre-established treatment strategy we want to assess and, as a consequence, is to be considered known *a priori*. Our inferential problem is solved if we are able, on the basis of the data collected under o, to estimate the probability distribution (8.1). We shall now see that we may indeed be able to do so if the properties of stability and positivity, discussed in the following, are satisfied.

8.7.1 Stability

In general, for a generic k, the conditional distribution $p(X_k \mid \overline{X}_{k-1}, \overline{A}_{k-1})$ will vary across regimes; that is, conditional on a particular *observed* history of the individual, the effect of a given sequence of actions s will, in general, differ systematically from one regime to the other. In our HIV example, this could be due, for example, to the fact that the doctors of one regime have a better visibility of the underlying patient's 'type' than those of the other regime. As a consequence, those patients of one regime who underwent the sequence s are not comparable (despite the conditioning on past observed history) with those patients of the other regime who underwent the same sequence. In those situations where, by contrast, the distribution $p(X_k \mid \overline{X}_{k-1}, \overline{A}_{k-1})$ does *not* differ from one regime to the other, we say we have 'stability'. More formally, we say that the *stability* condition is satisfied if, for $k = 0, \ldots, N+1$, the following identity holds:

$$p(x_k \mid \overline{x}_{k-1}, \overline{a}_{k-1}\, ; \sigma = e) = p(x_k \mid \overline{x}_{k-1}, \overline{a}_{k-1}\, ; \sigma = o) \tag{8.2}$$

whenever the conditioning event, $\overline{x}_{k-1}, \overline{a}_{k-1}$, has positive probability under either of regimes o and e. Our definition of stability is closely related to Robins' concept (1986) of sequential randomization (1986). Having previously defined σ to take values in (o, e), the stability condition is equivalent to the following conditional independence property:

$$X_k \perp\!\!\!\perp \sigma \mid (\overline{X}_{k-1}, \overline{A}_{k-1}) \quad (k = 0, \ldots, N+1). \tag{8.3}$$

Here and throughout, we use the notation and theory of conditional independence introduced by Dawid (1979) as generalized in Dawid (2002) to apply also to problems involving decision or parameter variables.

A major *caveat* is the fact that, because stability is a property of the relationship between the observational and the experimental regime, it cannot be empirically verified in those (considerably many) situations where no data have been collected under the latter. In these situations, the justification of the stability assumption will require genuine insight into the subject matter. In the next section, with reference to our HIV example, we shall see that a detailed causal diagram of the problem, complete with the relevant latent variables, can help in this task.

Intuitively, under the stability condition (8.4), or under the equivalent condition (8.3), one can formally replace the distribution $p(x_k \mid \overline{x}_{k-1}, \overline{a}_{k-1}\,; \sigma = e)$ in Equation (8.1) by $p(x_k \mid \overline{x}_{k-1}, \overline{a}_{k-1}\,; \sigma = o)$, and thus hope to be able to estimate the first factor on the right-hand side of Equation (8.1) from the available observational data. In this case, because the second factor on the right-hand side of Equation (8.1) is known a priori, one would hope to be able to estimate the distribution $p(y, \overline{x}_N, \overline{a}_N \mid \sigma = e)$ from the data. However, some extra care is needed here. We shall – in general – have to invoke a further property, *positivity*, which is discussed in the following.

8.7.2 Positivity

If, for example, we want to assess a treatment strategy, under which a particular action sequence s can arise, we will be unable to do so if, under the observational regime that has generated our data, that particular sequence of actions arises with probability zero. Define the *positivity* condition to be satisfied when, for any event E defined in terms of $(\overline{X}_N, \overline{A}_N, Y)$, that has a positive probability under the strategy to be evaluated, we have $p(E \mid o) > 0$. In our HIV example, positivity implies that for any combined sequence of patient events and medical actions that can arise under the treatment strategy that we wish to evaluate, there is a non-null probability of that particular sequence arising in the observational data. An extreme example of violated positivity occurs when the strategy of interest contemplates the possible administration of a drug that was not available in the observational study that generated the data. More subtle examples of violated positivity can arise.

When both stability and positivity are satisfied, we should be able to estimate the joint distribution (8.1) on the basis of the observational data. Then, on the basis of this distribution, we should – at least in principle – be able to obtain $p(Y \mid \sigma = e)$ by marginalization with respect to all variables but $X_{N+1} \equiv Y$. Note, however, that this procedure is cumbersome from a computational point of view. A computationally more efficient procedure is discussed later in this chapter.

8.8 Graphical representations of dynamic plans

As stated in the introduction section, we consider the use of causal diagrams to be a very helpful ingredient of our method. We shall here restrict ourselves to diagrams that have the form of a directed acyclic graph (DAG).

A 'causal DAG' is supposed to model causal relations. Unlike conditional independence, which is an unambiguous property of a probability distribution, causal relations lack a clear, mathematical, definition. A DAG representing causal properties should, in principle, be something totally different from a DAG representing conditional independence properties. In the former case, the arrows are supposed to have a direct interpretation in terms of cause and effect (whatever this means) whereas for conditional independence the arrows are nothing but incidental construction features supporting the d-separation semantics described by Shpitser in Chapter 3. In spite of all this, the idea of using DAGs that simultaneously represent conditional independence and causal properties (the latter understood as describing the effects of interventions), introduced by Judea Pearl and described in his book (2009), has been extremely fruitful.

Pearl's graphical representation applies to a collection of variables measured on some system, such that we can intervene (or at least can conceive of the possibility of intervening) on any one variable (possibly simultaneously on different variables), so as to *set* its value in a way that is determined entirely externally. From a probabilistic point of view, any instance of a Pearl causal DAG represents the assumption that the joint distribution of the variables satisfied, under anyone of the (possibly many) possible interventional regimes, the conditional independence properties one can read off the graph, for example via d-separation. From a causal point of view, the same graph represents the 'modularity' assumption that, for any node i of the graph, its conditional distribution, given its DAG parents, is the same, no matter which variables in the system (other than i itself) are intervened upon, and that only the DAG parents of i have a direct causal effect on i, relative to the other variables in the DAG. The above assumptions impose strong (and far from self-evident) relationships between the behaviours of the system under different regimes of intervention. These assumptions enable a single Pearl DAG to fulfil two interpretative functions simultaneously: the probabilistic and the causal one.

An important idea, previously discussed in Chapter 4, is to enrich the causal DAG in such a way that reflects more explicitly the necessary assumptions about the probabilistic behaviours of the domain variables across the regimes (Dawid, 2010). One way of doing so, which has been inspired by influence diagrams (Dawid, 2002, 2003), is to supplement the DAG with one or more nonrandom *regime indicators*, which we shall draw as squares. One example of a regime indicator, denoted by σ, has been previously introduced in this chapter to distinguish between regimes e and o. By incorporating σ in the causal diagram, we are able to express the identity of the conditional distribution of certain sets of domain variables, given certain other sets of domain variables, across different regimes.

This idea is illustrated in Figure 8.1 by the ADAG representation of our HIV example, for the special case of $N = 1$ to preserve visual simplicity. The diagram uses the variable symbols introduced in Section 8.5, plus the latent process (U_0, U_1). In line with the way this example is dealt with in Chapter 17 in this volume, take U_k to characterize the patient just before time T_k as belonging to one of a number of 'types', in the sense of Section 8.6. Now, the diagram obtained satisfies the stability condition (8.4), as can be checked by applying the usual

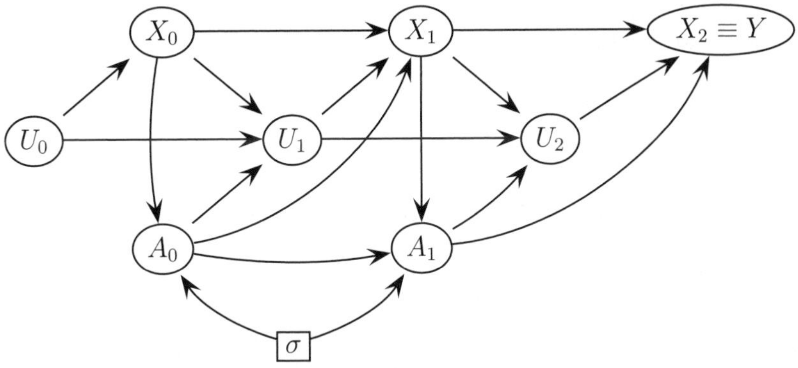

Figure 8.1 *Graphical representation of the HIV problem example, with* $N = 1$.

d-separation semantics of DAGs to it. In the graph of Figure 8.1, stability is a consequence of the following two topological properties:

1. $\sigma \to (X, U)$ arrows are missing. This expresses the assumption that, at any point of the treatment process, conditional on past history, the consequences of any specific action on the patient will be the same, regardless of the particular environment (hospital, doctors, evolution of the relevant pathogens, etc.) that characterizes the regime in operation.
2. $U \to A$ arrows are missing; that is, under each regime, and at any stage of the decision process, treatment assignments are determined by some deterministic, or randomizing, device, which only has the values of earlier *observed* (under both regimes) and recorded variables as inputs.

Property 1 is the graphical counterpart of what Dawid and Didelez (2010) call the *extended stability* condition. Property 2 is directly related to Robin's concept (1986) of sequential randomization. The conditional independence version of property 2, for $k = 1, \ldots, N$ and for $\sigma \in (o, e)$, is $A_k \perp\!\!\!\perp \overline{U}_k \mid \overline{X}_k, \overline{A}_{k-1}, \sigma$. This appears to be a discrete-time version of Definition 2 (ii) of Arjas and Parner (2004). We note that property 2, if not accompanied by property 1, does not *per se* guarantee stability. We also note that validity of the two properties above is a sufficient, but not necessary condition for stability. Further combinations of properties that result in stability are discussed in Dawid and Didelez (2010).

As noted by Daniel and colleagues in Chapter 17, property 1 appears to be plausible in our HIV example. The reason is that the effect of taking a particular anti-retroviral cocktail is likely be the same irrespective of whether the therapy is assigned by a doctor within the observational study o or assigned by some other doctor (assuming, of course, that no major changes in the selective resistance of viral agents have occurred between regimes o and e). Property 2 is realistic if no information about the latent patient's type was or will be used by the doctors under both regimes, except possibly indirectly through the observed patient's covariate process. We conclude that, under the assumptions represented by the graph of Figure 8.1, and if the positivity condition is also valid, the data collected from medical records of past HIV patients can be used to predict the effectiveness of alternative treatment strategies on future patients. In the next section we introduce a second example of study, in the field of abdominal aortic aneurysm.

8.9 Abdominal aortic aneurysm surveillance

An abdominal aortic aneurysm (AAA) is a serious vascular disease. An AAA occurs when the large blood vessel that supplies blood to the abdomen, pelvis and legs becomes abnormally large or balloons outward. A surgical intervention may be needed to avoid rupture, which is associated with a mortality rate of 80 %. While surgery can fix the AAA, it is a serious undertaking; mortality rates during surgery (in the United States) are in the range of 2 % to 6 % for repair under elective, nonemergency, circumstances. Possible complications from surgery include bleeding, infection and kidney or bowel damage. In addition, because coronary artery disease is so common among patients with AAA, a major worry is the risk of post-operative heart trouble. Surgery places a significant strain on the heart and can cause problems such as angina or heart attack.

Hence the difficult decision: should we immediately fix the AAA or should we wait watchfully until the AAA has grown to a size that represents a greater threat to the patient's life? This type of decision is typically made by trained experts, using both statistical data and memories of previous experiences.

A possible set of causal assumptions about the problem is represented in the graph of Figure 8.2. The graph is, once again for simplicity, confined to a set of three measurement times: $T_0 < T_1 < T_2$. Let variable X_k (for $k = 0, \ldots, 2$) represent relevant patient information, including the size of the aneurysm, at T_k. Let variable A_k indicate whether, at time T_k, the doctors have decided to surgically intervene or to 'wait watchfully'. Let variable W_0 represent unrecorded factors that vary with the 'doctor in charge' and the treatment centre, whereas U_0 represents pre-treatment unmeasured risk factors, whether of a genetic nature or dependent on exposure history.

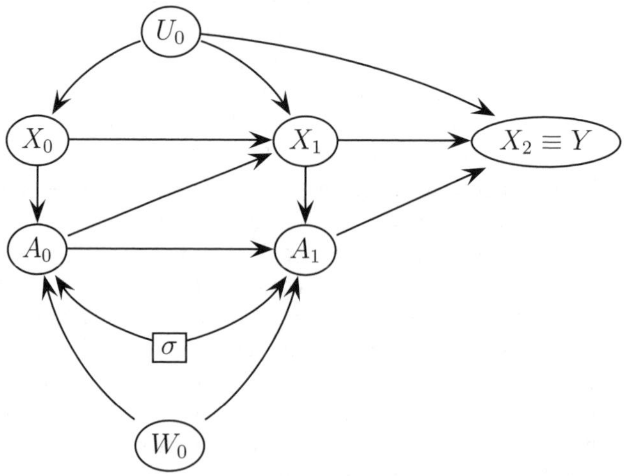

Figure 8.2 Graphical representation of the AAA surveillance example.

Two important assumptions are represented in Figure 8.2. The first assumption, corresponding to the missing $U_0 \to A_k$ arrows, states that no information about the risk factors U_0, unmeasured in the data, has been or will be consulted for the therapeutic interventions $(A_0, A_1, \ldots)$ performed on the study patients, under whichever regime. Again, this assumption must be checked by carefully interviewing the doctors in charge. The second assumption, corresponding to the missing $W_0 \to X_k$ and $\sigma \to X_k$ arrows, asserts that the impact of a therapy plan is the same, whoever the doctor, or Health Centre, in charge of the treatment, and whichever the regime under operation.

It is straightforward to verify, via the usual conditional independence semantics of directed acyclic graphs, that the stability condition (8.5) is satisfied in this particular example. This means that, under the assumptions discussed, observational data on patients under follow-up for AAA can be used as a basis for future therapeutic decisions.

8.10 Statistical inference and computation

Once we have made sure that the stability and positivity conditions are valid in the application of interest, the task will be to estimate the causal parameters from the data. One possibility is to impose some form of smoothness across the distributions $p(x_k \mid \overline{x}_{k-1}, \overline{a}_{k-1})$ or give them a parametric form (Berzuini and Allemani, 2004). The unknown parameters can be formally included in the U_k variables. The preceding chapter suggests a Bayesian approach to the problem. In a similar vein, we may put a prior distribution on the parameters and then update it into a posterior by using Markov chain Monte Carlo (MCMC) methods to integrate with respect to uncertainty about the estimated parameters, the stochastic component of the treatment strategy and the outcomes.

Then the Markov chain is run on a huge graphical model. To visualize this model, imagine that each sample individual is represented by a corresponding graph, of the type in Figure 8.1 or 8.2, all the resulting graphs being then interconnected through sharing the same model parameters. Upon convergence, the chain will generate samples of the parameters, as if obtained from the correct posterior. Under stability, each $p(x_k \mid \overline{x}_{k-1}, \overline{a}_{k-1})$ distribution is the same under the two regimes, observational and experimental. It follows that the same model, and what we have learned about its parameters, can be used to determine a predictive distribution for Y in a future (hypothetical) patient subject to the treatment strategy of interest. If we come up with a loss function, $L(Y)$, we can then use the generated MCMC samples to estimate the expectation $E\{L(Y); e\}$, that is the *expected loss* associated with the treatment strategy of interest. This will allow us to compare alternative strategies.

We have previously discussed the positivity-related requirements for the method. Are they *always* required? Once we make parametric or smoothness assumptions for the $p(x_k \mid \overline{x}_{k-1}, \overline{a}_{k-1})$ distributions, positivity could be relaxed. In a situation of data scarcity, even under positivity, some of the probabilities might be so small that we are unable to estimate them well on the basis of the available observational data, and this can create problems of convergence of the Markov chain. Smoothing may attenuate this problem.

Another possibility is to adopt a method consisting of two stages (Dawid and Didelez, 2010). In the first stage, given enough data collected under o, we estimate the relevant distributions $p(x_k \mid \overline{x}_{k-1}, \overline{a}_{k-1}\, ;\, o)$ $(k = 0, \ldots, N + 1)$. Under stability and positivity, this will also give us the distributions $p(x_k \mid \overline{x}_{k-1}, \overline{a}_{k-1}\, ;\, e)$, for $(k = 0, \ldots, N + 1)$. These distributions, together with the *a priori* known distributions $p(a_k \mid \overline{x}_k, \overline{a}_{k-1}\,;\, e)$, constitute the main ingredient of the second stage of the procedure, described in the following.

Let h denote a partial history of the form $(\overline{x}_k, \overline{a}_{k-1})$ or $(\overline{x}_k, \overline{a}_k)$ $(0 \leq k \leq N)$. We also include the 'null' history $\emptyset$ and 'full' histories $(\overline{x}_N, \overline{a}_N, y)$. We denote the set of all partial histories by $\mathcal{H}$. Under the experimental regime, e, define a function f on $\mathcal{H}$ by

$$f(h) := E\{L(Y) \mid h\,;\, e\} \tag{8.4}$$

In a medical treatment context, the quantity $f(h)$ represents the expected loss (a negative measure of the overall therapeutic success) for a patient with current history h, under the treatment strategy of interest. The term 'history' must here be interpreted to denote the past evolution of both the disease and the treatment.

Simple application of the laws of probability yields

$$f(\bar{l}_k, \overline{a}_{k-1}) = \sum_{a_k} p(a_k \mid \bar{l}_k, \overline{a}_{k-1}\,;\, e) \times f(\bar{l}_k, \overline{a}_k) \tag{8.5}$$

$$f(\bar{l}_{k-1}, \overline{a}_{k-1}) = \sum_{l_k} p(l_k \mid \bar{l}_{k-1}, \overline{a}_{k-1}\,;\, e) \times f(\bar{l}_k, \overline{a}_{k-1}) \tag{8.6}$$

For h a full history $(\bar{l}_N, \overline{a}_N, y)$, we have $f(h) = L(y)$. Using these as starting values, by successively implementing (8.6) and (8.7) in turn, starting with (8.6) for $k = N + 1$ and ending with (8.7) for $k = 0$, we step down through ever shorter histories until we have computed $f(\emptyset) = E\{L(Y) \mid e\}$, the expected loss for the treatment strategy under evaluation. More generally, we could consider loss function Y^* of $(\overline{X}_N, \overline{A}_N, Y)$. Starting now with $f(\overline{x}_N, \overline{a}_N, y) := Y^*(\overline{x}_N, \overline{a}_N, y)$, we can apply the identical steps to arrive at $f(\emptyset) = E\{Y^* \mid e\}$, which yields the desired expected loss associated with the treatment strategy under evaluation. Under suitable further conditions we can combine this recursive method with the selection of an optimal strategy from among a set of candidate plans, when it becomes dynamic programming. Although the method is applied when we have data from all possible *static* strategies, it is nevertheless vital that the stability assumption includes all possible dynamic regimes.

When e is a nonrandomized strategy, the distribution of A_k given $\overline{X}_k = \bar{l}_k$, when $\sigma = e$, is degenerate, at $a_k = g_k = g_k(\bar{l}_k\,;\, e)$, say, and the only randomness left is for the variables $(X_0, \ldots, X_N, Y)$. We can now consider $f(h)$ as a function of only the (x_k) appearing in h, since, under e, these then determine the (a_k). Then (8.6) holds automatically, while (8.7) becomes

$$f(\overline{x}_{k-1}) = \sum_{x_k} p(l_k \mid \overline{x}_{k-1}, \overline{g}_{k-1}; e) \times f(\overline{x}_k) \tag{8.7}$$

When, further, the regime e is static, each g_k in the above expressions reduces to the fixed action a_k^* specified by e.

The conditional distributions in (8.6), (8.7) and (8.5) are undefined when the conditioning event has probability 0 under s. One possibility is to define $f(h)$ in (8.6) to be equal to 0 whenever $p(h\,;\, \emptyset,\, e) = 0$.

Robins and Wasserman (1997) warn about the uncritical use of parametric models for the conditional distributions involved, on the grounds that they can lead to a so-called *null paradox*, which prevents discovering that different plans have the same effect. In the light of these considerations, these authors propose the use of *marginal* or *nested structural models* (Robins, 1998, 2004) that avoid the null paradox.

Computational aspects of the theory and their complex relationship with substantive considerations are illustrated in Chapter 17 in this volume.

8.11 Transparent actions

Let the term *transparent actions* denote the situation where the actions performed under e are only influenced by earlier variables that are observed under both regimes and have been recorded in the data. This is what occurs, for example, in Figure 8.1. Under this assumption, the distribution of Y under the experimental regime can be written as

$$P(Y \mid \sigma = e) = \sum_{\overline{a}_N, \overline{x}_N} \left[\prod_{k=0}^{N} P(a_k \mid pa_{A_k}; \sigma = e) \times \left(\sum_{\overline{u}_N} P(Y \mid pa_Y) \prod_{k=0}^{N} P(x_k \mid pa_{X_k}) \prod_{k=0}^{N} P(u_k) \right) \right] \tag{8.8}$$

where the outer summation is over all possible multivariate configurations of $\overline{a}_N$ and $\overline{x}_N$, and where the symbol pa_Z represents the set of the parents of a node Z of interest, in the graph. The expression on the right-hand side is obtained by factorizing the joint distribution over the graph into a product of conditional distributions of each variable given its parents in the graph, and by then averaging with respect to $(\overline{u}_N, \overline{a}_N, \overline{x}_N)$. A consequence of the transparent action assumption is that the parental sets for the action variables, pa_{A_k}, do not contain U variables, and this justifies the fact that, in the above expression, the $P(a_k \mid pa_{A_k}; \sigma = e)$ factors are not averaged over $\overline{u}_N$.

Importantly, Tian (2008) notes that the expression within round brackets corresponds to the distribution of $(\overline{X}_N, Y)$ under an atomic intervention on $\overline{A}_N$. Let this distribution be denoted by $P(Y, \overline{X}_N \mid \sigma = \overline{a}_N)$. Equation (8.9) can then be rewritten as

$$P(Y \mid \sigma = e) = \sum_{\overline{a}_N, \overline{x}_N} \left[P(Y, \overline{x}_N \mid \sigma = \overline{a}_N) \prod_{k=0}^{N} P(a_k \mid pa_{A_k}; \sigma = e) \right] \tag{8.9}$$

Note that, because the $P(a_k \mid pa_{A_k}; \sigma = e)$ terms are *given*, the only unknown quantities in the above expression are the $P(Y, \overline{x}_N \mid \sigma = \overline{a}_N)$, from which the following criterion follows (Shpitser and Pearl, 2006).

Criterion
Under the transparent actions assumption, the distribution of interest, $P(Y \mid \sigma = e)$, is estimable if we are able to identify the distribution of $(\overline{X}_N, Y)$ under a generic atomic intervention on $\overline{A}_N$ from the data.

A complete algorithm for checking identifiability under atomic interventions is the *do*-calculus (Tian and Pearl, 2003; Shpitser and Pearl, 2006), reviewed in Chapter 6 in this volume. We conclude that, under the transparent actions assumption, the atomic (static) semantics of the *do*-calculus can be employed to analyse the identifiability of the effects of the dynamic plans.

8.12 Refinements

The above theory can be refined in a number of directions. For example, Tian (2008) considers situations where some X variables are *not* ancestors of Y and refines the theory correspondingly. However, situations in scientific practice where such refinement is crucial have not often been shown.

Another direction is towards problems in which the experimental and the observational distributions are characterized by different graphs, specifically, where the former graph is a subgraph of the latter. Consider the example of Figure 8.3, which is an elaboration of the HIV graph of Figure 8.1. The elaboration consistes of adding dotted arrows from U_k to A_k, for $k = 0, \ldots, N$. These arrows are intended to be present in the observational graph, but missing in the experimental graph. In other words, the graph of Figure 8.3 assumes that, while the data were generated by actions taken in the light of U-information, inferential interest focuses on treatment strategies that do *not* use that information. A possible justification is that the treatments are intended for application in routine medical contexts where the expensive tests and the medical experience necessary to probe the patient's "type" cannot be afforded. In this case, the distribution of inferential interest, $P(Y \mid \sigma = e)$, is still given by Equation (8.9). This suggests that application of the criterion of the preceding section to the experimental graph (that is to the graph without the dotted arrows) will provide correct guidance in assessing identifiability. In the particular example of Figure 8.3, this will lead to the conclusion that the strategies of interest are identifiable from the data.

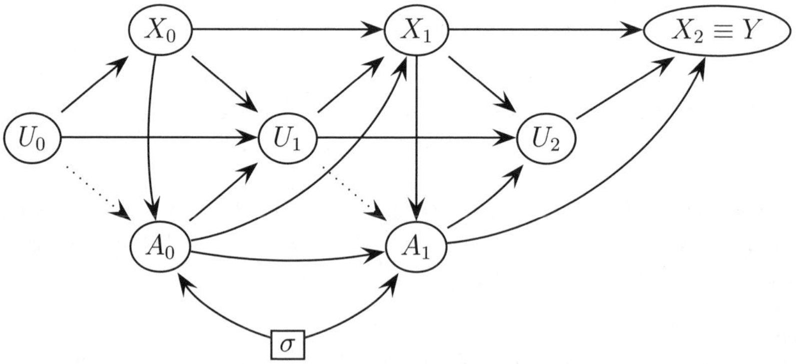

Figure 8.3 Elaboration of the HIV problem of Figure 8.1. Some of the arrows are dotted to indicate that information about the U variables influenced the action decisions in the observational distribution, but not in the experimental one.

8.13 Discussion

Stability and positivity are sufficient conditions for the estimability of the effect of treatment strategies in the sense of the present chapter. However, it turns out they are not necessary. The stability condition, in particular, requires that, for each time T_i, the conditional distribution of X_i given the earlier observed variables should be the same under both regimes, observational and experimental. This is, indeed, a strong assumption. In many applicative situations, one might not be willing to accept it. Work has been done to establish more general conditions for identifiability (see Dawid and Didelez, 2010, for example), mainly by imposing restrictions on the kinds of information made available for the decisions under the strategy e.

Finally, an important development will be to extend the methods for use when event times, on a continuous support, are explicitly incorporated in the analysis.

Acknowledgements

We thank Dr Silvia Corona for providing insight into the HIV example.

References

Arjas, E. and Parner, J. (2004) Causal reasoning from longitudinal data. *Scandinavian Journal of Statistics*, **31** (2), 171–187.

Berzuini, C. and Allemani, C. (2004) Effectiveness of potent antiretroviral therapy on progression of human immunodeficiency virus: Bayesian modelling and model checking via counterfactual replicates. *Journal of the Royal Statistical Society. Series C (Applied Statistics)*, **53** (4), 633–650.

Dawid, A.P. (1979) Conditional independence in statistical theory (with discussion). **41**, 1–31.

Dawid, A.P. (2002) Influence diagrams for causal modelling and inference. *International Statistical Review*, **70**, 161–189. Corrigenda, **70**, 437.

Dawid, A.P. (2003) Causal inference using influence diagrams: the problem of partial compliance (with discussion), in *Highly Structured Stochastic Systems* (eds P.J. Green, N.L. Hjort and S. Richardson), Oxford University Press, pp. 45–81.

Dawid, A.P. (2010) Beware of the DAG!, in *Proceedings of the NIPS 2008 Workshop on Causality. Journal of Machine Learning Research Workshop and Conference Proceedings* (eds D. Janzing, I. Guyon and B. Schoelkopf), vol. 6, pp. 59–86.

Dawid, A.P. and Didelez, V. (2010) Identifying the consequences of dynamic treatment strategies: a decision-theoretic overview. *Statistics Surveys*, **4**, 184–231.

Pearl, J. (2009) *Causality*, 2nd edn. Cambridge: Cambridge University Press.

Pearl, J. and Robins, J. (1995) Probabilistic evaluation of sequential plans from causal models with hidden variables. In *Uncertainty in Artificial Intelligence* **11**, (eds. P. Besnard and S. Hanks), Morgan Kaufmann, San Francisco, CA, pp. 444–453.

Robins, J.M. (1986) A new approach to causal inference in mortality studies with sustained exposure period: application to control of the healthy worker survivor effect. *Mathematical Modelling*, **7**, 1393–1512.

Robins, J.M. (1987) Addendum to a new approach to causal inference in mortality studies with sustained exposure period: application to control of the healthy worker survivor effect. *Computers and Mathematics with Applications*, **14**, 923–945.

Robins, J.M. (1989) The analysis of randomized and nonrandomized aids treatment trials using a new approach to causal inference in longitudinal studies, in *Health Service Research Methodology: A Focus on AIDS* (eds H. Freeman L. Sechrest and A. Mulley), US Public Health Service, pp. 113–159.

Robins, J.M. (1992) Estimation of the time-dependent accelerated failure time model in the presence of confounding factors. *Biometrika*, **79**, 321–324.

Robins, J.M. (1997) Causal inference from complex longitudinal data, in *Latent Variable Modeling and Applications to Causality* (ed. M. Berkane), vol. 120 of Lecture Notes in Statistics, New York: Springer-Verlag, pp. 69–117.

Robins, J.M. (1998) Structural nested failure time models, in *Survival Analysis*, vol. 6 of *Encyclopedia of Biostatistics* (eds P.K. Andersen and N. Keiding). Chichester, UK: John Wiley & Sons, Ltd. pp. 4372–4389.

Robins, J.M. (2000) Robust estimation in sequentially ignorable missing data and causal inference models, in *Proceedings of the American Statistical Association Section on Bayesian Statistical Science 1999*, vol. 79, pp. 6–10.

Robins, J.M. (2004) Optimal structural nested models for optimal sequential decisions, in *Proceedings of the Second Seattle Symposium on Biostatistics* (eds D.Y. Lin and P. Heagerty). Springer, New York: Springer, pp. 189–326.

Robins, J.M. and Wasserman, L.A. (1997) Estimation of effects of sequential treatments by reparameterizing directed acyclic graphs, in *Proceedings of the 13th Annual Conference on Uncertainty in Artificial Intelligence* (eds Gerger and P. Shenoy). San Francisco, CA: Morgan Kaufmann Publishers, Francisco, pp. 409–420.

Robins, J.M. Greenland, S. and Hu, F.C. (1999) Estimation of the causal effect of a time-varying exposure on the marginal mean of a repeated binary outcome. *Journal of the American Statistical Association*, **94**, 687–700.

Shpitser, I. and Pearl, J. (2006) Identification of joint interventional distributions in recursive semi-Markovian causal models, in *Proceedings of the Twenty-First National Conference on Artificial Intelligence*, Menlo Park, CA, July 2006.

Tian, J. (2008) Identifying dynamic sequential plans, in *Proceedings of the Twenty-fourth Uncertainty in Artificial Intelligence Conference* (eds D. McAllester and A. Nicholson). AUAI Press, Corvallis, Oregon, 554–561.

Tian, J. and Pearl, J. (2003) *On the Identification of Causal Effects*. Technical Report R-290-L, Department of Computer Science, University of California, Los Angeles.

9

Causal effects and natural laws: Towards a conceptualization of causal counterfactuals for nonmanipulable exposures, with application to the effects of race and sex

Tyler J. VanderWeele[1] and Miguel A. Hernán[2]
[1]Departments of Epidemiology and Biostatistics, Harvard School of Public Health, Boston, Massachusetts, USA
[2]Department of Epidemiology, Harvard School of Public Health, Boston, Massachusetts, USA

9.1 Introduction

In this chapter we offer reflections on the extent to which the laws of nature in the physical sciences might be used as a framework for thinking about causation in observational research in the social and biomedical sciences. The motivation for these reflections arises from the tension between (i) in the causal inference literature in statistics and the social sciences, most notions of causation and counterfactuals are grounded in hypothetical interventions or manipulations but (ii) many of the exposures of interest do not easily lend themselves to manipulation. As will be clear below, the reflections offered here are heavily dependent

Causality: Statistical Perspectives and Applications, First Edition. Edited by Carlo Berzuini, Philip Dawid and Luisa Bernardinelli.

on a good deal of existing writing in causal inference (e.g. Cartwright, 1997; Heckman, 2005; Greenland, 2005; Pearl, 2009; Commenges and Gégout-Petit, 2009; among others). In this paper, we discuss some of the relationships between natural laws and contrary to fact statements and the extent to which natural laws in the physical sciences might be used as a framework for understanding causal effects of a nonmanipulable exposure at a single point in time in the social and biomedical sciences. Although we believe that counterfactuals related to manipulable quantities are of primary interest for policy purposes, causation related to nonmanipulable quantities can be of scientific interest and arguably constitute a substantial portion of instances of causation in science. We consider what this approach to conceptualizing causal effects might contribute to discussion of the effects of sex and race. We conclude by offering some further observations about parallels between the research methods and the conceptualizations of causality found in the physical sciences, on the one hand, and the social and biomedical sciences, on the other. As will be evident below, the account we provide here is not intended to constitute a formal system for reasoning but we do nevertheless hope that these reflections will stimulate further thinking on conceptualizing causal effects for nonmanipulable exposures.

9.2 Laws of nature and contrary to fact statements

The physical sciences do not ordinarily make explicit reference to notions of causation. Instead, the physical sciences describe natural laws thought to govern a particular system at all times and places. By virtue of their universality, these natural laws then also entail various contrary to fact statements. Under classical physics, the positions of various balls on a billiard table are deterministically governed by the initial position of each ball, their velocity and acceleration, the friction of the table surface, etc. Not only do the laws of nature allow for prediction of what will actually occur given information on all variables within the system at a particular point in time but they also allow for prediction under contrary to fact scenarios if some particular ball had been in a different place, moving at a different velocity, etc. All of these predictions, actual and counterfactual, are governed by the laws of nature themselves.

Because the laws of nature are to govern all systems at all times, hypotheses about what the laws of nature actually are can be empirically tested by manipulations. Hypotheses about the laws of nature give rise to predictions concerning scenarios that can be brought about by manipulations in controlled settings. If the predictions are contradicted by what actually occurs in a controlled experimental setting the hypothesis is rejected or at least revised. The 'scientific method' generally proceeds by repeated experiment.

Our own everyday causal language is arguably often a shorthand for statements concerning the laws of nature. We would ordinarily say 'X caused Y' if '(i) X, along with the state of the universe and the laws of nature entail Y and (ii) not X, along with the state of the universe and the laws of nature do not entail Y.' Counterfactual accounts of causation (Lewis, 1973a, 1973b) proceed along similar lines but introduce greater formality. Instances of 'X, along with the state of the universe and the laws of nature entailed Y...' are generally not the only circumstances in which we make statements like 'X caused Y' but the basic point here is that the physical sciences usually manage to do away with explicit causal language because they focus on the laws of nature themselves, which are thought to govern all manner of contrary to fact statements.

9.3 Association and causation in the social and biomedical sciences

Empirical associations uncovered by statistical analysis in observational epidemiology and in the social sciences also allow for prediction. As we observe associations we can sometimes predict what might happen to a particular individual given certain covariates or given the past. However, the associations that are discovered in such observational research do not in general allow for prediction under contrary to fact scenarios, e.g. under certain manipulations to set things to other than they were. The causal inference literature in statistics, epidemiology, the social sciences, etc., attempts to clarify when predictions of contrary to fact scenarios are warranted. We describe associations as 'causal' when the associations are such that they allow for accurate prediction of what would occur under some intervention or manipulation.

The literature often proceeds by articulating various 'ignorability' or 'unconfoundedness' or 'exogeneity' assumptions under which association would reflect causal relationships and by then evaluating the plausibility of such assumptions, controlling for other variables that might make these assumptions more plausible and assessing the sensitivity of the causal conclusions to the assumptions being made.

Most of the causal inference literature in epidemiology, statistics and the social sciences is focused on issues of 'internal validity', whether associations allow for accurate prediction under contrary to fact scenarios for the particular population under study. There is an acknowledgement that the effect of a particular intervention or manipulation may differ in other populations with different characteristics. There is an acknowledgement that not all of the dynamics of the exposure or phenomenon under study have been captured. Thus, in the social sciences, we often are seeking causal relationships or laws that govern systems only 'locally', for a particular population; even attaining this is difficult. Questions of generalizability or 'external validity' require much stronger assumptions and are difficult to assess (Shadish *et al.*, 2002; Heckman, 2005) because not all aspects of the dynamics governing social or biomedical systems are understood or easily characterized. In contrast, however, the physical sciences often seek to describe laws that have universal or 'global' governance; the laws are to govern all systems at all times and places.

9.4 Manipulation and counterfactuals

The definitions of a causal effect in most of the causal inference literature in statistics is based on the idea of hypothetical manipulations of the exposure of interest (Holland, 1986). In contrast, in the physical sciences a causal effect is generally a secondary or derivative concept from the laws governing a system. Furthermore, predictions made in the physical sciences extend to contrary to fact cases that could not come about by any conceivable manipulation. We can use the laws of nature to predict what would have occurred to an object if, at a particular point in time, it had been somewhere other than where it was and we can make such predictions even if it is not in fact possible to move it from where it in fact is to where we are speculating about by the time we are speculating about. The laws of nature are sufficiently general to allow for predictions under scenarios that could not come about given the present state of the world. The laws govern systems in which one thing is considered to be different from what it in fact was and all other things are considered to be the same. Statements about systems in which one thing is considered to be different from what in fact was and all other things are considered to be the same are sometimes referred to as 'ceteris paribus' statements.

In the biomedical and social sciences we generally do not have a sufficient understanding of the dynamics of systems to be able to make ceteris paribus statements concerning most exposures of interest. We do not have a clear sense of what we mean if, for example, we were to talk about 'everything being the same as it was except that an individual's body mass index (BMI) were other than it in fact is.' We generally find it easier to make ceteris paribus statements about particular manipulations. We can more easily conceive of ceteris paribus statements about changing an individual's exercise without doing anything else or changing an individual's diet without doing anything else. Both of these interventions might change an individual's BMI, perhaps even change it to the same level. However, even if both of these interventions were to result in the same BMI, the effects of these interventions on some other outcome, cardiovascular disease say, may be very different. Thus, although we can perhaps meaningfully speak of the effect of an exercise regime or the effect of a diet it becomes difficult to conceptualize what we might mean when we try to speak of the effect of BMI or of obesity.

The difficulty in conceiving of causal effects for which we cannot conceive of interventions gives rise to the notion in the causal inference literature that specifying a manipulation or an intervention gives rise to a meaningful causal counterfactual and that the counterfactual is ill-defined to the extent that we fail to precisely specify what the intervention is (Robins and Greenland, 2000). The causal inference literature has therefore ordinarily conceived of counterfactual or potential outcomes (Rubin, 1974, 1978) in terms of hypothetical interventions on the exposure of interest. Manipulability becomes an important criterion for the meaningfulness of speaking about causal effects. Some authors have gone so far as to put forward the slogan 'no causation without manipulation' (Holland, 1986).

In the remainder of this paper we will consider the extent to which notions of natural laws from the physical sciences might be taken as a framework for conceptualizing causal effect in observational research in epidemiology and the social sciences.

9.5 Natural laws and causal effects

One consequence of the laws of nature, at least as conceived of under classical physics, is that for a particular system, given complete information on the state of a system at a particular point in time, it is possible to predict what will happen to a particular object within the system at any future time. Let Y denote the position of an object at time t and suppose that the initial state of the system at time 0 may be characterized by variables $(x_0, \ldots, x_n)$, which include the object's initial position, velocity and acceleration, the position, velocity and acceleration of other objects, etc. For any set of inputs $(x_0, \ldots, x_n)$ the laws of nature under classical physics would entail that Y is a deterministic function of $(x_0, \ldots, x_n)$ and we could write

$$Y = f(x_0, \ldots, x_n)$$

We could define the causal effect of X_i comparing levels x_i and x_i' with respect to the initial state $(x_0, \ldots, x_{i-1}, x_{i+1}, \ldots, x_n)$ by

$$f(x_0, \ldots, x_{i-1}, x_i, x_{i+1}, \ldots, x_n) - f(x_0, \ldots, x_{i-1}, x_i', x_{i+1}, \ldots, x_n)$$

In quantum physics, the outcome of the measurement of some particle at a future time is no longer a deterministic function of the initial state of the system but rather a random function of the initial state. The laws governing the system again imply

$$Y = f(x_0, \ldots, x_n)$$

but where $f(x_0, \ldots, x_n)$ is now a random variable rather than a fixed value. Conceived of another way, we have that the expected value of the measurement of the particle, $E[Y]$, is a deterministic function $g(x_0, \ldots, x_n)$ of the initial state:

$$E[Y] = g(x_0, \ldots, x_n)$$

If we consider settings that are typical in the social or biomedical sciences we generally cannot hope to characterize the laws of nature in the same way as can be done in the physical sciences. Nevertheless, we could perhaps hypothetically conceive of a set of laws governing such more complex systems even if we cannot articulate what those laws in fact are. Suppose that Y now denotes whether an individual has cardiovascular disease at age 50. If we thought that the individual's physiology was completely governed by deterministic laws we could conceive a set of inputs, $(x_0, \ldots, x_n)$, with n very large, all measured at a particular point in time denoted arbitrarily as 0, and with some of the x_i perhaps denoting also aspects of the history of the individual. We would then have

$$Y \sim f(x_0, \ldots, x_n)$$

In essence this is a nonparametric structural equation (Pearl, 2009) in which all inputs are measured at the same time 0 but in which some inputs may also denote the history, up until time 0, of certain variables. What is important about this equation is that if we have truly captured all of the dynamics of dependence of Y on the initial state of the system then the equation will give the value of Y for any initial state $(x_0, \ldots, x_n)$ irrespective of how the state $(x_0, \ldots, x_n)$ comes about. Perhaps somewhat more realistically, the function $f(x_0, \ldots, x_n)$ may once again be a random variable rather than giving a single value. However, we could still make ceteris paribus type statements provided that the function f sufficiently captured the dynamics of the system so that the distribution of Y would be given by $Y = f(x_0, \ldots, x_n)$ irrespective of how the initial state $(x_0, \ldots, x_n)$ comes about. If $(x_0, \ldots, x_n)$ are all measured at some reference time 0 and it is the case that $Y = f(x_0, \ldots, x_n)$ irrespective of how the initial state $(x_0, \ldots, x_n)$ comes about, then we will say that f is a law for Y with respect to $(x_0, \ldots, x_n)$ at time 0. If it is not the case that $Y = f(x_0, \ldots, x_n)$ irrespective of how the initial state $(x_0, \ldots, x_n)$ comes about, then the function f has not been sufficiently elaborated so as to capture all the dynamics of the system; other variables at the initial time 0 may need to be included so that f fully captures the system dynamics.

Analogous with natural laws, the function f will give the value of Y for contrary to fact values of $(x_0, \ldots, x_n)$, even ones that could not come about through ordinary manipulations. If the function f captures dynamics in this way we can make *ceteris paribus* type statements. Suppose that there were a law f governing cardiovascular disease Y at age 50, with the initial state (*BMI, diet, exercise, sex, genotype*, $x_5, \ldots, x_n$) at age 25, where $x_5, \ldots, x_n$ denote a potentially large number of variables beyond BMI, diet, exercise, sex and genotype. We would have that

$$Y = f(BMI, diet, exercise, sex, genotype, x_5, \ldots, x_n)$$

We could then define the causal effect of BMI comparing a BMI of 30 and a BMI of 20 with the initial state (diet, exercise, sex, genotype, $x_5, \ldots, x_n$) as

$$f(30, diet, exercise, sex, genotype, x_5, \ldots, x_n) \\ -f(20, diet, exercise, sex, genotype, x_5, \ldots, x_n)$$

This is arguably a different sort of causal effect than any that could possibly be defined by reference to interventions. Suppose we are again interested in attempting to conceive of what might be meant by the effect of a BMI of 30 versus a BMI of 20 at age 25 on cardiovascular disease at age 50 but wanted to conceptualize this in terms of interventions. If we had access to the subject of interest at age 24.5 we could conceive of different sorts of interventions to ensure a BMI of 20 or of 30 by age 25. A surgical or liposuction intervention could be used to bring down BMI, but this might itself have effects on exercise at age 25; an exercise regime intervention may be used to change BMI, but this may have an effect on the individual's diet. Any conceivable intervention to change BMI will also have effects on other factors in the function $f(BMI, diet, exercise, sex, genotype, x_5, \ldots, x_n)$ and thus we would be capturing the effect of the intervention not just the effect of BMI itself. Causal statements about interventions are still meaningful (and in this context arguably more useful and of greater public health importance) but they are conceptually quite different from the sorts of ceteris paribus statements that can be made in the natural sciences.

The notion of a function that captures the entirety of the dynamics of a system has some history in structural models in econometrics (see Heckman, 2008, for a discussion) and is sometimes referred to as an 'all causes model'. The approach is of course controversial as it will generally be infeasible to collect data on a sufficient number of variables to even hope to approximate such a model. Trying to formulate an 'all causes' model would require controlling for covariates not simply so as to avoid traditional problems of confounding and selection in the identification of causal effects but to attempt to capture the dynamics of the system better. Although this 'all causes' model approach may not generally be feasible in practice, it does provide an alternative manner in which to at least conceive of causal effects. See also Commenges and Gé gout-Petit (2009) for a more elaborate dynamic approach attempting to embed causal statements in social and biomedical research within the context of laws governing dynamic systems.

Although we generally cannot realistically ever hope to describe a function f that captures the dynamics of social or biomedical systems in this manner, the discussion above does potentially suggest that manipulation may not be entirely necessary in conceiving of causal effects. In some cases it may be possible to formulate causal statements for counterfactual scenarios that are not or cannot be achieved by manipulation. The discussion suggests that manipulation may in fact be secondary for our understanding of causal counterfactuals. It is nevertheless the case that for purposes of policy and decision-making, counterfactuals based on well-defined manipulable quantities will be of primary interest (Holland, 1986; Greenland, 2005; Hernán and VanderWeele, 2011).

9.6 Consequences of randomization

Our discussion thus far has concerned conceptualizing causal effects for counterfactual scenarios that are not or cannot be achieved by manipulation, rather than on the estimation of such effects. In most cases, because we generally cannot realistically ever hope to describe a function f that captures the dynamics of social or biomedical systems, we will likely only be able to conceive of causal effects in this manner, not estimate them. Exceptions may arise when some exposure is effectively randomized and in this section we discuss such randomization within the context of causal laws governing systems.

As noted above, in an 'all causes' model, covariates are included not simply to avoid traditional problems of confounding and selection in the identification of causal effects but to attempt to capture the dynamics of the system better. Thus, even if, in a particular city under study, say air pollution were completely randomly distributed so that control for it was not necessary if one were interested in the association between BMI and subsequent cardiovascular disease, if air pollution were related to cardiovascular disease, the 'all causes' model approach would suggest including it in the model in order to capture the dynamics of the system.

Suppose that for some outcome Y at time t there were a law f for Y with respect to the initial state $(x_0, \ldots, x_n)$ at time 0:

$$Y = f(x_0, \ldots, x_n)$$

We will now let ω denote particular individuals so that $Y(\omega)$ is the outcome for individual ω and the initial state for individual ω is $X_0(\omega), \ldots, X_n(\omega)$. We then have that

$$Y(\omega) = f(X_0(\omega), \ldots, X_n(\omega))$$

Suppose we are interested in the effects of some particular variable $A = X_0$. We might then define the counterfactual variables for individual ω by

$$Y_a(\omega) = f(a, X_1(\omega), \ldots, X_n(\omega))$$

and the causal effect for individual ω comparing exposure levels a_1 and a_0 would be

$$Y_{a_1}(\omega) - Y_{a_0}(\omega)$$

If we knew the function f we could calculate average causal effects by

$$E[Y_{a_1} - Y_{a_0}] = E[f(a_1, X_1, \ldots, X_n)] - E[f(a_0, X_1, \ldots, X_n)].$$

With f unknown we could still potentially apply the ordinary principles of causal inference to estimate causal effects. We would say that the effect of A on Y is unconfounded given C (or that treatment assignment for A is ignorable conditional on C) if, for all a,

$$Y_a \perp\!\!\!\perp A | C$$

which denotes that Y_a is independent of A conditional on C. Following standard arguments, if the effect of A on Y is unconfounded given C then we have that

$$E[Y_{a_1} - Y_{a_0}] = E_C[E[Y|A = a_1, C]] - E_C[E[Y|A = a_0, C]]$$

In particular, if the effect of A on Y is unconfounded given $X_1, \ldots, X_j$ then we have

$$E[Y_{a_1} - Y_{a_0}] = E_{X_1,\ldots,X_j}[E[Y|A = a_1, X_1, \ldots, X_j]] - E_{X_1,\ldots,X_j}[E[Y|A = a_0, X_1, \ldots, X_j]]$$

and we do not need data on $X_{j+1}, \ldots, X_n$ in order to estimate causal effects. If A is randomized then $Y_a \perp\!\!\!\perp A$ and we do not need data on any variables except A and Y in order to estimate the causal effect of A, comparing exposure levels a_1 and a_0, on Y. We can estimate $E[Y_{a_1} - Y_{a_0}]$ by $E[Y|A = a_1] - E[Y|A = a_0]$. If A is not randomized, one might in principle consider attempting to evaluate $Y_a \perp\!\!\!\perp A|C$ by constructing a causal diagram (Pearl, 2009) in which $X_1, \ldots, X_n$ all have direct edges into Y but with a potentially large number of other variables that are ancestors of $X_1, \ldots, X_n$ and one could, again, at least in principle, use the backdoor path criterion (Pearl, 2009) to evaluate whether some set C (which may or may not be a subset of $X_1, \ldots, X_n$) blocks all backdoor paths from A to Y.[1]

9.7 On the causal effects of sex and race

We illustrate the concepts in the previous section by considering what might be defined as the causal effect of sex and race. Let Y be income at age 30 (or any other post-conception outcome at some fixed time). For simplicity here we will consider 'sex' as indicating XX versus XY sex chromosomes rather than self-perceived identity ('gender'). We might then hypothesize a law for Y of the form

$$Y = f(sex, parental\_income, x_2, \ldots, x_n)$$

where X_0 denotes sex, X_1 denotes parental income and $x_2, \ldots, x_n$ denote a potentially very large number of variables (such as parental education, parental age and numerous others) all measured at the time of conception. We could define the average causal effect of sex on Y as

$$E[f(1, X_1, \ldots, X_n)] - E[f(0, X_1, \ldots, X_n)]$$

Again, using more traditional counterfactual notation, we let $A = X_0$ denote sex and we could define for each individual ω, $Y_a(\omega) = f(a, X_1(\omega), \ldots, X_n(\omega))$. Irrespective of the form

[1] We can also consider the consequences of randomization with regard to preserving laws when omitting components of the initial state $(x_0, \ldots, x_n)$. If some variable X_k, comprising a component of the initial state, is randomized with some probability p at time 0 then by randomization for all $j \neq k$ we have

$$Y_{x_j} \perp\!\!\!\perp X_j | \{X_0, \ldots, X_n\} \backslash \{X_j, X_k\}$$

Furthermore, for all $(x_0, x_1, \ldots, x_{k-1}, x_{k+1}, \ldots, x_n)$ we would have that $f(x_0, x_1, \ldots, x_{k-1}, x_{k+1}, \ldots, x_n) = f(x_0, x_1, \ldots, x_{k-1}, 1, x_{k+1}, \ldots, x_n)$ with probability p and $f(x_0, x_1, \ldots, x_{k-1}, 0, x_{k+1}, \ldots, x_n)$ with probability $1 - p$ irrespective of how $(x_0, x_1, \ldots, x_{k-1}, x_{k+1}, \ldots, x_n)$ are set and thus f would constitute a law for Y with respect to $(x_0, x_1, \ldots, x_{k-1}, x_{k+1}, \ldots, x_n)$ at time 0. Randomization allows for the estimation of causal effects but it also permits the obtaining of a new law for Y even when omitting the variable that was randomized from the initial state.

that the law f takes because, sex is randomized at conception, we have $Y_a \perp\!\!\!\perp A$ and thus we have

$$E[f(1, X_1, \ldots, X_n)] - E[f(0, X_1, \ldots, X_n)] = E[Y_1 - Y_0] \\ = E[Y|A = 1] - E[Y|A = 0]$$

The total causal effect of sex can be estimated simply by taking the observed differences in Y between individuals who are male and female. We can define the total effect of sex even though we do not have in view manipulations to change it and we can estimate this effect because sex itself is randomized at conception. If sex is indeed randomized at conception then it will be independent of parental income, of parental education, of race and of all other components of the initial state at time 0 in the law f for Y.[2] The analysis given here would likewise arguably be applicable to particular genetic variants if effectively randomized at conception, even if these genetic variants are not (at least at present) considered to be manipulable.

The analysis above employing the law $Y = f(sex, parental\_income, x_2, \ldots, x_n)$ presupposes a deep form of determinism. However, a similar analysis would be applicable under the stochastic equivalent described in the previous section. Nothing in the argument given presupposes that the law governing the outcome Y and the initial state is deterministic rather than stochastic.

In a number of settings (e.g. cases of discrimination) it is not the total effect that is of interest but rather something along the lines of 'the direct effect of sex on employment or on receiving an interview controlling for education, skill, abilities' (Carson v. Bethlehem Steel Corp., 1996, cited in Pearl 2009). In such cases this simple approach for the total effect of sex will no longer suffice; the direct effect of sex is of interest and is potentially more difficult conceptually to define and estimate. However, in such cases where discrimination is in view, often the exposure of interest (sex) can be defined so as to not be conceived of as the actual effect of sex, but rather the effect of the employer's seeing what sex is listed on the application. Defining the exposure in this manner (as whatever is seen by the employer on the application) allows for one more easily to conceive of manipulations to change that variable and defining counterfactual and causal effects when the exposure is manipulable is then less problematic.

A distinction should perhaps be drawn between what one might call a disparity and what one might call discrimination. A disparity in, say, a particular health outcome between men and women might be conceived as simply any difference between the outcome for men and for women. Such a disparity may arise because of discrimination; it might also arise because of biological differences (e.g. breast cancer or prostate cancer being extreme examples) or because of sociological differences (e.g. smoking and subsequently lung cancer rates are higher for men than for women). The disparity would in general be calculated by simply comparing means; this would effectively just be the total effect described above. Discrimination, on the

[2] Note that this argument concerning randomization of sex presupposes a somewhat simplified model of fertilization. Sperm in fact compete to penetrate the ovum wall and whether a particular sperm penetrates the ovum is likely to be related to the possible eventual phenotypic characteristics of the offspring. The analysis above might still be considered reasonable if in spite of this process of selection it were the case that genetic variants on the sex chromosome were independent of variants on other chromosomes. It is generally assumed in the genetic literature that variants on different chromosomes are independent. However, if it is the case that independence fails to hold, either because variants on the sex chromosome and those on other chromosomes are unconditionally associated across sperm or because they are associated conditional on fertilization or birth, then the analysis that we have given may not be adequate.

other hand, is essentially the direct effect of sex controlling for all other variables at the time in which sex is perceived.

Let us now contrast our analysis of the causal effect of sex with a similar analysis that might be considered for race. Once again let Y be income at age 30. We might again hypothesize a law for Y of the form:

$$Y = f(skin\_color, parental\_skin\_color, genotype, neighborhood\_income, x_4, \ldots, x_n)$$

where X_0 denotes the individual's skin color, X_1 parental skin color, X_2 genotype, X_3 neighborhood income and $x_4, \ldots, x_n$ again denote a potentially very large number of variables, all measured at time 0. We could then define the causal effect of skin color with respect to the initial state ($parental\_skin\_color, genotype, neighborhood\_income, x_4, \ldots, x_n$) by

$$f(skin\_color^{\dagger}, parental\_skin\_color, genotype, neighborhood\_income, x_4, \ldots, x_n)$$
$$- f(skin\_color^{*}, parental\_skin\_color, genotype, neighborhood\_income, x_4, \ldots, x_n)$$

where $skin\_color^{\dagger}$ indicated black and $skin\_color^{*}$ indicated white or we could even define the average causal effect

$$E[f(skin\_color^{\dagger}, X_1, \ldots, X_n)] - E[f(skin\_color^{*}, X_1, \ldots, X_n)]$$

Alternatively, we might attempt to capture a more general effect of what might be considered 'race' by considering skin color, parental skin color and genotype jointly:

$$f(skin\_color^{\dagger}, parental\_skin\_color^{\dagger}, genotype^{\dagger}, neighborhood\_income, x_4, \ldots, x_n)$$
$$- f\ (skin\_color^{*}, parental\_skin\_color^{*}, genotype^{*}, neighborhood\_income, x_4, \ldots, x_n)$$

or simply skin color and genotype jointly:

$$E[f(skin\_color^{\dagger}, parental\_skin\_color, genotype^{\dagger}, neighborhood\_income, x_4, \ldots, x_n)$$
$$- f(skin\_color^{*}, parental\_skin\_color, genotype^{*}, neighborhood\_income, x_4, \ldots, x_n)$$

One point of interest here is that each of the 'race effects' defined above may well be different; the 'effect of race' is ambiguous and may vary depending on whether what is in view is skin color, or skin color and genotype considered jointly, or skin color, parental skin color and genotype all considered jointly. When the effect of race is in view it will generally be important to clarify what precisely is under discussion.

Although we could potentially conceptualize causal effects in this way with a hypothetical law f for Y, identifying and estimating these causal effects is not straightforward. Unlike the case for sex, race is not randomized. Irrespective of whether we consider skin color, or parental skin color or genotype, singly or jointly, all of these are likely to be correlated with neighborhood income, say, at the time of conception. We can no longer simply take averages of the outcome Y among different racial groups (defined by skin color or genotype, say) to obtain the effects defined above. Because the randomization that held with sex does not hold with race, although we can define effects that may be of interest, we cannot identify them.

In certain studies we may be able to identify aspects of 'the effect of race.' In family-based studies, aspects of genotype are effectively randomized so as to allow one to estimate the effects of a single nucleotide polymorphism for a particular sample of individuals. Also, once again, in the context of discrimination, we can redefine the exposure of interest to be, for example, the employer's seeing a particular race listed on the application. Again, the exposure defined in this manner is subject to conceivable manipulations and thus defining causal effects for this exposure is relatively unproblematic and randomized trials can even be conducted to assess this effect and evaluate discrimination (e.g. Bertrand and Mullainathan, 2004). However, we cannot in general hope to be able to conduct a randomized trial that would identify the 'effect of race' as conceived of above as the overall effect of skin color, or parental skin color, or genotype, considered singly or jointly.

Note that, at least conceptually, in the case of race, we might distinguish between discrimination, a disparity and particular aspects of the 'effect of race.' Discrimination could essentially be the direct effect of perceived race controlling for all other variables at the time in which race is perceived (which again could potentially be estimated using randomized experiments). A disparity could simply be conceived of as differences in, say, a health outcome between black and white individuals. Such disparities might arise through discrimination (e.g. employment decisions) or through different biologies (e.g. melanoma) or through different neighborhood characteristics at birth (e.g. educational achievement). The disparity is itself in general calculated by simply comparing means. However, unlike in the sex example, because of lack of randomization, such a disparity comparison no longer corresponds with anything that could be conceived of as the total effect of race. As in the account given above we can perhaps define aspects of the effect of race, e.g. the overall effect of skin color, or parental skin color, or genotype, considered singly or jointly, but we cannot estimate these by simply taking averages.

The discussion above does, however, provide a way in which to meaningfully, and hopefully more precisely, speak about the 'effect of race', even if we cannot in practice estimate such effects. Indeed, it seems ridiculous to suggest that race does not have a causal effect on a variety of outcomes. Outside of the context of specific discrimination cases, when it comes to trying to estimate what that 'effect of race' is on a variety of outcomes, interpretation becomes difficult. If 'race/ethnicity' is put in a regression this is likely to capture the effects of skin color, parental skin color and genotype along with various other factors such as neighborhood income, quality of schools, etc., which are correlated with skin color, parental skin color, and genotype, these correlations themselves arising from a complex historical process (Kaufman, 2008). When considering race it is important to clarify whether discrimination, a disparity or some aspect of what might be considered as an effect of race is in view. Moreover, if what is of principal interest is eliminating a disparity, then it will be important to think carefully about counterfactuals related to manipulable quantities.

9.8 Discussion

In this paper we have explored whether the conceptualization of contrary to fact statements as arising from natural laws as found in the physical sciences might be taken as a model for conceptualizing causal effects in the social and biomedical sciences. We have been concerned principally with conceptualizing causal effects for nonmanipulable exposures rather than with policy questions for which counterfactuals related to manipulable exposures are arguably of greater relevance. We have throughout taken a rather idealized view of the actual laws governing physical systems and of our knowledge of those laws; we contrasted this with the seemingly intractable complexity that we encounter in the social and biomedical sciences. We conclude this chapter by noting some of the parallels between research methods and reasoning in the physical versus the biomedical and social sciences that suggest that the contrast between the disciplines is perhaps more of a continuum rather than a dichotomy.

We noted that, in the physical sciences, often what is sought are universal or global laws that govern systems at all times and at all places whereas in the social and biomedical sciences the causal relationships that we are after are often at best restricted to particular populations and contexts. However, even in the physical sciences, many of laws that have been described are context specific. Ohm's law, for instance, which governs current, voltage and resistance, assumes constant temperature. Similarly, we discussed how in the physical sciences, systems are often sufficiently tractable so as to be captured by an articulable mathematical law whereas this is generally not possible with social and biomedical systems, such as those concerned with the effects of BMI. However, in the physical sciences, complexity also quickly becomes intractable, at least in practice, when we consider, say, systems governing the weather and we seek to make precise inferences weeks or months out. We noted that in the physical sciences we can often make inferences about individual contrary to fact scenarios whereas in the biomedical and social sciences we generally at best hope for average causal effects. However, the developments in quantum physics make clear that even with physical systems our inferences really concern averages over various quantum states. We noted that in the biomedical and social sciences, our knowledge generally arises from observations but our most reliable knowledge comes from manipulation and experiments. The physical sciences are ultimately not so very different: although mathematical hypotheses, theories and generalization play an important role, our knowledge still ultimately depends upon observations and principally upon controlled manipulable experiments. The differences between the physical and biomedical sciences in terms of methodology and goals are thus perhaps best conceived as laying on a continuum of complexity, specificity, generality and rigor.

Although the differences between the disciplines in terms of methodology and goals are perhaps only relative, we still believe that causation historically has been conceived of in a different manner in the physical sciences, on the one hand (as derivative on laws governing systems), and in the social and biomedical sciences, on the other. However, because these differences between these disciplines are arguably only relative, the conceptualization of causation in the physical sciences may prove useful in at least conceptualizing causal effects, for nonmanipulable exposures, in the social and biomedical sciences as well.

Acknowledgements

The authors thank Sander Greenland, Jay Kaufman, Edward Laird and David Chan for helpful comments.

References

Bertrand, M. and Mullainathan, S. (2004) Are Emily and Greg more employable than Lakisha and Jamal? A field experiment on labor market discrimination. *American Economic Review* **94**, 991–1013.

Carson v. Bethlehem Steel Corp. (1996) 70 FEP cases 921 (7th Cir. 1996).

Cartwright, N. (1997) What is a causal structure?, in *Causality in Crisis* (eds V.R. McKim and S.P. Turner). Notre Dame, IN: University of Notre Dame Press, University of Notre Dame, pp. 343–357.

Commenges, D. and Gégout-Petit, A. (2009) A general dynamical statistical model with possible causal interpretation. *Journal of the Royal Statistical Society, Series B*, **71**, 719–736.

Greenland, S. (2005) Epidemiologic measures and policy formulation: lessons from potential outcomes. *Emerging Themes in Epidemiology*, **2**, 5.

Heckman, J.J. (2005) The scientific model of causality. *Sociological Methodology*, 1–98.

Heckman, J.J. (2008) Econometric causality. *International Statistical Review*, **76**, 1–27.

Hernán, M.A. and VanderWeele, T.J. (2011) Compound treatments and transportability of causal inference. *Epidemiology*, May, **22**, **(3)**, 368–377.

Holland, P. (1986). Statistics and causal inference. *Journal of the American Statistical Association*, **81**, 945–960.

Kaufman, J.S. (2008) Epidemiologic analysis of racial/ethnic disparities: some fundamental issues and a cautionary example. *Social Science and Medicine*, **66**, 1659–1669.

Lewis, D. (1973a) Causation. *Journal of Philosophy*, **70**, 556–567.

Lewis, D. (1973b) *Counterfactuals*. Cambridge, MA: Harvard University Press.

Pearl, J. (2009) *Causality: Models, Reasoning, and Inference*. Cambridge: Cambridge University Press.

Robins, J.M. and Greenland, S. (2000) Comment on 'Causal inference without counterfactuals' by A.P. Dawid. *Journal of the American Statistical Association*, **95**, 477–482.

Rubin, D. (1974) Estimating causal effects of treatments in randomized and non-randomized studies. *Journal of Educational Psychology*, **66**, 688–701.

Rubin, D.B. (1978) Bayesian inference for causal effects: the role of randomization. *Annals of Statistics*, **6**, 34–58.

Shadish, W.R., Cook, T.D. and Campbell, D.T. (2002) *Experimental and Quasi-Experimental Designs for Generalized Causal Inference*. Boston, MA: Houghton Mifflin.

10

Cross-classifications by joint potential outcomes

Arvid Sjölander
Department of Medical Epidemiology and Biostatistics, Karolinska Institutet, Stockholm, Sweden

10.1 Introduction

Continuing the discussion of Chapter 2 in this volume on the use of potential outcomes in causal inference, suppose we wish to estimate the causal effect of a binary exposure X on a binary outcome Y, coded as 0 or 1. Then we think of each subject as being characterized by two potential outcomes, Y_0 and Y_1. The causal effect of X on Y is typically defined as some comparison between the marginal probabilities $\Pr(Y_1 = 1)$ and $\Pr(Y_0 = 1)$, e.g. $\Pr(Y_1 = 1) - \Pr(Y_0 = 1)$. Thus, the standard definition of causal effects does not exploit the joint structure of Y_0 and Y_1. Nevertheless, in some scenarios it may be useful to consider a cross-classification of subjects by joint potential outcomes. A classification by the joint potential outcomes $\{Y_0, Y_1\}$ yields four strata. The first stratum contains those subjects for which $(Y_0 = Y_1 = 0)$, the second contains those for which $(Y_0 = 0, Y_1 = 1)$, the third contains those for which $(Y_0 = 1, Y_1 = 0)$, and the fourth contains those for which $(Y_0 = Y_1 = 1)$. As discussed in Chapter 2, potential outcomes cannot in general be jointly observed. Thus, although we can envisage the existence of these four strata, we cannot in general tell to which of these four strata each subject belongs. For instance, if we observe that $(X = 0, Y = 0)$ for a particular subject, we can immediately tell that $Y_0 = 0$. The value of Y_1 remains unknown, however, so we cannot tell whether the subject belongs to the stratum $(Y_0 = Y_1 = 0)$ or the stratum $(Y_0 = 0, Y_1 = 1)$. Despite this limitation, cross-classifications by joint potential outcome have been used to derive several important results in the causal inference literature. In this chapter we review three applications.

Causality: Statistical Perspectives and Applications, First Edition. Edited by Carlo Berzuini, Philip Dawid and Luisa Bernardinelli.

1. ***Bounding nonidentifiable causal effects***. In observational (i.e. nonrandomized) studies, the possibility of unmeasured confounding can never be completely ruled out. Thus, in observational studies, causal effects are in general not identifiable. However, this does not mean that observational data contain no information about causal effects. In many scenarios, observational data can be used to construct bounds, i.e. a range of values that can be proved to include the causal effect of interest (Manski, 2007). Balke and Pearl (1994) developed a generic method for deriving such bounds, based on joint potential outcomes and linear programming techniques. They used this method to derive bounds for the causal treatment effect in randomized trials with noncompliance. Later, the method has been used to derive bounds for controlled direct effects and natural direct effects under various assumptions. In Section 10.2 we review the Balke-Pearl method within the context of randomized trials with non-compliance.
2. ***Identifying subpopulation causal effects with instrumental variables***. One way to deal with unmeasured confounding is to use instrumental variable (IV) techniques (Hernán, 2006). An IV is a variable that satisfies certain restrictions in relation to the exposure and outcome of interest. When an IV can be found, the so-called IV formula can be used to compute the causal effect for a subset of the population – a stratum defined by joint potential outcomes – even in the presence of unmeasured confounding. One particular example is the randomized trial with noncompliance. In this scenario, the 'treatment assigment' can be viewed as an IV and the IV formula gives, under certain assumptions, the treatment effect for the compliers – the subgroup of people who would comply regardless of which treatment level they are assigned to. In Section 10.3 we review the IV formula within the context of randomized trials with non-compliance.
3. ***Defining proper causal effects in studies suffering from 'truncation by death'***. In many epidemiological studies, the goal is to measure the long-term effect of an exposure on a certain outcome. A common problem in these studies is that some subjects may die before the outcome can be measured. One way to analyze such 'truncated' data is to compare the outcome distribution across exposure levels for those who did survive until the outcome could be measured. However, it has been long recognized that this contrast may suffer from so-called 'post-treatment selection bias' and cannot in general be interpreted as a causal effect (Kalbfleish and Prentice, 1980). Robins (1986 remark, 12.2 on page 1496) suggested an alternative target parameter based on joint potential outcomes. Although this parameter has a proper causal interpretation, it is in general not identifiable, and a lot of research has been devoted to the development of bounds (Zhang and Rubin, 2003; Jemiai, 2005) and sensitivity analyses (Gilbert *et al.*, 2003; Jemiai, 2005; Shepherd *et al.*, 2006; Jemiai *et al.*, 2007). In Section 10.4 we review the target parameter suggested by Robins (1986) and the bounds developed by Zhang and Rubin (2003) and Jemiai (2005).

10.2 Bounds for the causal treatment effect in randomized trials with imperfect compliance

To fix ideas, consider a randomized trial to evaluate the performance of a certain treatment on a certain outcome. Each subject is randomly assigned to either receive the treatment or

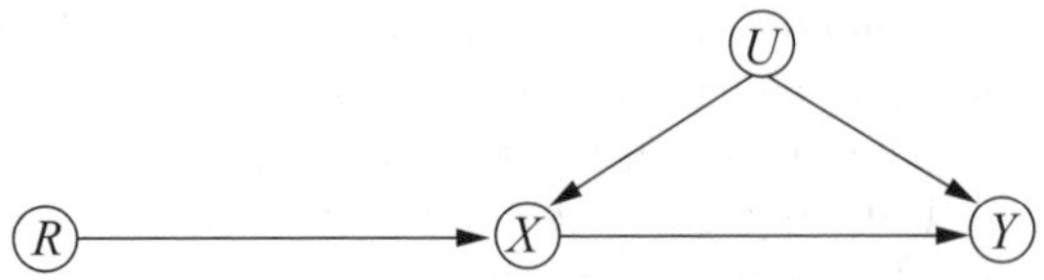

Figure 10.1 *R: assigned treatment, X: received treatment, Y: outcome, U: common causes of X and Y.*

not. We use R for assigned treatment level, with $R = 1$ for 'assigned to receive the treatment' and $R = 0$ for 'assigned not to receive the treatment'. Not all subjects comply with their assignment. Thus, some subjects who are assigned to receive the treatment may refrain, and vice versa. We use X to denote the received treatment level, with $X = 1$ for 'received the treatment' and $X = 0$ for 'did not receive the treatment'. We use Y to denote the outcome and we assume that Y is binary, with $Y = 1$ for 'favorable outcome' and $Y = 0$ for 'not favorable outcome'. The observed data consist of a random sample from the distribution $\Pr(R, X, Y)$. The scenario is displayed in the directed acyclic graph (DAG) of Figure 10.1. The arrow from R to X represents the influence of the assigned treatment on the received treatment. Ideally, all subjects comply with their assigned treatment, so that $X = R$. In the worst case, treatment assignment has no influence on whether the subjects under study choose to take the treatment or not, implying that the arrow from R to X is absent. The arrow from X to Y represents the causal effect of the received treatment on the outcome, which is the target of the analysis. U represents all potential common causes for X and Y, e.g. age, health status, etc. The absence of an arrow from U to R follows by randomization of the assignment. The absence of an arrow from R to Y encodes the assumption that all the influence of the assigned treatment on the outcome is mediated through the received treatment. This assumption was labeled 'the exclusion restriction' by Angrist and Imbens (1991)[1] and is commonplace in the compliance literature.

There are two standard analyses for randomized trials with imperfect compliance: the intention-to-treat (ITT) analysis and the as-treated (AT) analysis. The ITT analysis adresses the association between R and Y, ignoring X. The main motivation for the ITT analysis is that it produces a valid test for the null hypothesis of no causal effect of X on Y. Specifically, when the arrow from X to Y is absent in Figure 10.1, then R and Y are independent. Thus, an association between R and Y implies the presence of a causal effect of X on Y. A drawback of the ITT analysis is that when the influence of R on X is weak, then the power of this test will be low. In the extreme case, when the arrow from R to X is absent, then R and Y are independent, regardless of what effect X has on Y. The AT analysis addresses the association between X and Y, ignoring R. The motivation for the AT analysis is that in the absense of U (no commmon causes of X and Y), the AT analysis produces the causal effect of X on Y. In reality, however, the presence of U is often very likely. If U is present and fully observed, we may 'adjust for' (condition on) U in an AT analysis (e.g. through stratification or regression modeling) to obtain the conditional causal effect of X on Y, given U.

If U is present, and is partly or fully unmeasured, the causal effect of X on Y is not identifiable. To review the Balke–Pearl method for deriving bounds on the effect, we introduce

[1] Angrist and Imbens (1991) formulated the exclusion restriction in terms of potential outcomes, as in Equation (2.3) of Chapter 2 in this volume, not in terms of missing arrows on a DAG.

some further notation. We use potential outcomes to define the causal effect of X on Y. Let X_r denote the potentially received treatment for a randomly selected subject, if that subject would be assigned to treatment r, and let Y_{rx} denote the potential outcome for a randomly selected subject, if that subject would be assigned to treatment r and would receive treatment x. The potential outcomes are related to the observed outcomes through the following *consistency assumption* (according to Equation (2.1) of Chapter 2 in this volume)

$$\begin{aligned} R = r &\Rightarrow X_r = X \\ (R = r, X = x) &\Rightarrow Y_{rx} = Y \end{aligned} \tag{10.1}$$

Randomization of the assignment implies that subjects are *exchangeable* across levels of assignment (according to Equation (2.2) of Chapter 2 in this volume). Exchangeability means that the potential outcomes X_r and Y_{rx} are jointly independent of R, i.e.

$$(X_0, X_1, Y_{00}, Y_{01}, Y_{10}, Y_{11}) \perp\!\!\!\perp R \tag{10.2}$$

The exclusion restriction states that the potential outcome Y_{rx} does not depend on r, so we can write

$$Y_{rx} = Y_x, \qquad \forall (r, x) \tag{10.3}$$

In general, the (population) causal effect of X on Y is defined as some contrast between the potential outcome distributions $\Pr(Y_0)$ and $\Pr(Y_1)$. The linear programming technique developed by Balke and Pearl (1994) can be used to construct bounds for the population causal risk difference

$$CRD \equiv \Pr(Y_1 = 1) - \Pr(Y_0 = 1) \tag{10.4}$$

The causal null hypothesis ($CRD = 0$) can be represented in the DAG by deletion of the arrow from X to Y.[2] To construct bounds for CRD we classify subjects into $2^4 = 16$ strata by the joint potential outcomes (X_0, X_1, Y_0, Y_1). We enumerate these strata by two four-valued variables V_x and V_y. V_x determines the compliance behavior of a subject through the relation

$$V_x = \begin{cases} 0 & \text{if } X_0 = 0, X_1 = 0 \\ 1 & \text{if } X_0 = 0, X_1 = 1 \\ 2 & \text{if } X_0 = 1, X_1 = 0 \\ 3 & \text{if } X_0 = 1, X_1 = 1 \end{cases} \tag{10.5}$$

Following Imbens and Rubin (1997), we refer to a subject with compliance behavior $V_x = 0$, 1, 2, 3 (respectively) as a never-taker, a complier, a defier, and an always-taker. V_y determines

[2] Deletion of the arrow from X to Y actually represents the sharp causal null hypothesis $Y_0 = Y_1$; see Chapter 2 in this volume.

the outcome behavior of a subject through the relation

$$V_y = \begin{cases} 0 & \text{if } Y_0 = 0, Y_1 = 0 \\ 1 & \text{if } Y_0 = 0, Y_1 = 1 \\ 2 & \text{if } Y_0 = 1, Y_1 = 0 \\ 3 & \text{if } Y_0 = 1, Y_1 = 1 \end{cases} \tag{10.6}$$

Let q_{jk} be the proportion of subjects who belong to the stratum labeled $(V_x = j, V_y = k)$, i.e. $q_{jk} \equiv \Pr(V_x = j, V_y = k)$. The parameter of interest, *CRD*, is a linear function of $(q_{00}, \ldots, q_{33})$:

$$\begin{aligned} CRD &= \Pr(Y_1 = 1) - \Pr(Y_0 = 1) \\ &= \{\Pr(V_y = 1) + \Pr(V_y = 3)\} - \{\Pr(V_y = 2) + \Pr(V_y = 3)\} \\ &= \sum_{j=0}^{3} q_{j1} - \sum_{j=0}^{3} q_{j2} \end{aligned} \tag{10.7}$$

$(q_{00}, \ldots, q_{33})$ are constrained by

$$\sum_{j=0}^{3}\sum_{k=0}^{3} q_{jk} = 1 \tag{10.8}$$

and

$$q_{jk} \geq 0 \quad \text{for } j, k \in \{0, 1, 2, 3\} \tag{10.9}$$

Because potential outcomes cannot be observed jointly, $(q_{00}, \ldots, q_{33})$ are not directly observable either. Let $p_{yx \cdot r}$ be the observed probability $\Pr(Y = y, X = x | R = r)$. Under (10.1) and (10.2), $(p_{00 \cdot 0}, \ldots, p_{11 \cdot 1})$ are linear functions of $(q_{00}, \ldots, q_{33})$:

$$p_{yx \cdot r} = \sum_{\substack{j : X_r = x \\ k : Y_x = y}} q_{jk} \tag{10.10}$$

Minimizing and maximizing (10.7) under the constraints (10.8), (10.9), and (10.10) is a linear programming problem. There is plenty of off-the-shelf software for numerical solutions of linear programming problems. Balke and Pearl (1994) developed a computer program for analytic solutions, i.e. solutions expressed as functions of $(p_{00 \cdot 0}, \ldots, p_{11 \cdot 1})$. The analytic solution to the problem in (10.7) is given by

$$CRD \geq \max \left\{ \begin{array}{c} p_{11\cdot1} + p_{00\cdot0} - 1 \\ p_{11\cdot0} + p_{00\cdot1} - 1 \\ p_{11\cdot0} - p_{11\cdot1} - p_{10\cdot1} - p_{01\cdot0} - p_{10\cdot0} \\ p_{11\cdot1} - p_{11\cdot0} - p_{10\cdot0} - p_{01\cdot1} - p_{10\cdot1} \\ -p_{01\cdot1} - p_{10\cdot1} \\ -p_{01\cdot0} - p_{10\cdot0} \\ p_{00\cdot1} - p_{01\cdot1} - p_{10\cdot1} - p_{01\cdot0} - p_{00\cdot0} \\ p_{00\cdot0} - p_{01\cdot0} - p_{10\cdot0} - p_{01\cdot1} - p_{00\cdot1} \end{array} \right\} \quad (10.11)$$

$$CRD \leq \min \left\{ \begin{array}{c} 1 - p_{01\cdot1} + p_{10\cdot0} \\ 1 - p_{01\cdot0} + p_{10\cdot1} \\ -p_{01\cdot0} + p_{01\cdot1} + p_{00\cdot1} + p_{11\cdot0} + p_{00\cdot0} \\ -p_{01\cdot1} + p_{11\cdot1} + p_{00\cdot1} + p_{01\cdot0} + p_{00\cdot0} \\ p_{11\cdot1} + p_{00\cdot1} \\ p_{11\cdot0} + p_{00\cdot0} \\ -p_{10\cdot1} + p_{11\cdot1} + p_{00\cdot1} + p_{11\cdot0} + p_{10\cdot0} \\ -p_{10\cdot0} + p_{11\cdot0} + p_{00\cdot0} + p_{11\cdot1} + p_{10\cdot1} \end{array} \right\} \quad (10.12)$$

The derivation of the bounds in (10.11) and (10.12) relies on the consistency assumption (10.1), exchangeability (10.2), and the exclusion restriction (10.3). An important question is whether these assumptions are testable. The answer is 'yes'; by requiring that the upper bound is not lower than the lower bound, one obtains (Pearl, 1995a) the following testable restriction on the observed data distribution:

$$\max_x \sum_y \{\max_r \Pr(Y = y, X = x|R = r)\} \leq 1 \quad (10.13)$$

The relation in (10.13) was called the 'instrumental inequality' by Pearl (2000). We note that although a violation of (10.13) implies a violation of either (10.1), (10.2), or (10.3), the converse is not true. Thus, any statistical test based on (10.13) is inconsistent.[3]

10.3 Identifying the complier causal effect in randomized trials with imperfect compliance

Imbens and Angrist (1994) attacked the problem of imperfect compliance from a different angle. They demonstrated that the causal effect of X on Y is identifiable for a certain subset of the population – the compliers (see also Angrist *et al.*, 1996). Identification of the complier causal effect relies on an additional assumption of *monotonicity*, but does not require the outcome, Y, to be binary, and is applicable to causal effects on any scale (e.g. mean differences, risk ratios, odds ratios, etc).

[3] A statistical test is said to be consistent if, for any fixed significance level and for any fixed alternative hypothesis, its power approaches 1 as the sample size approaches infinity (Cox and Hinkley, 1974).

The complier causal effect is defined as some contrast between the conditional distributions $\Pr(Y_0|X_0 = 0, X_1 = 1)$ and $\Pr(Y_1|X_0 = 0, X_1 = 1)$. The monotonicity assumption rules out the existence of defiers, i.e.

$$X_1 \geq X_0 \tag{10.14}$$

Under assumptions (10.1), (10.2), (10.3), and (10.14), we have that

$$\begin{aligned}
&\Pr(Y = y, X = 1|R = 1) - \Pr(Y = y, X = 1|R = 0) && \text{(10.15a)}\\
&\quad = \Pr(Y_{11} = y, X_1 = 1|R = 1) - \Pr(Y_{01} = y, X_0 = 1|R = 0) && \text{(10.15)}\\
&\quad = \Pr(Y_{11} = y, X_1 = 1) - \Pr(Y_{01} = y, X_0 = 1) && \text{(10.16)}\\
&\quad = \Pr(Y_1 = y, X_1 = 1) - \Pr(Y_1 = y, X_0 = 1) && \text{(10.17)}\\
&\quad = \Pr(Y_1 = y, X_0 = 0, X_1 = 1) + \Pr(Y_1 = y, X_0 = 1, X_1 = 1)\\
&\qquad - \Pr(Y_1 = y, X_0 = 1, X_1 = 1) && \text{(10.18)}\\
&\quad = \Pr(Y_1 = y, X_0 = 0, X_1 = 1) && \text{(10.19)}
\end{aligned}$$

where (10.15) follows from (10.1), (10.16) follows from (10.2), (10.17) follows from (10.3), and (10.18) follows from (10.14). From (10.15a) to (10.19) we have that $\Pr(Y_1 = y|X_0 = 0, X_1 = 1)$ is identifiable and given by

$$\Pr(Y_1 = y|X_0 = 0, X_1 = 1) = \frac{\Pr(Y = y, X = 1|R = 1) - \Pr(Y = y, X = 1|R = 0)}{\Pr(X = 1|R = 1) - \Pr(X = 1|R = 0)} \tag{10.20}$$

By a similar argument we have that $\Pr(Y_0 = y|X_0 = 0, X_1 = 1)$ is identifiable and given by

$$\Pr(Y_0 = y|X_0 = 0, X_1 = 1) = \frac{\Pr(Y = y, X = 0|R = 0) - \Pr(Y = y, X = 0|R = 1)}{\Pr(X = 1|R = 1) - \Pr(X = 1|R = 0)} \tag{10.21}$$

Thus, both $\Pr(Y_1 = y|X_0 = 0, X_1 = 1)$ and $\Pr(Y_0 = y|X_0 = 0, X_1 = 1)$ are identifiable, which implies that the complier causal effect of X on Y is identifiable on any scale. For instance, the causal mean difference is given by

$$E(Y_1|X_0 = 0, X_1 = 1) - E(Y_0|X_0 = 0, X_1 = 1) = \frac{E(Y|R = 1) - E(Y|R = 0)}{E(X|R = 1) - E(X|R = 0)} \tag{10.22}$$

The right-hand side of (10.22) is often refered to as the instrumental variable (IV) formula and has been used in various disciplines of science (Hernán, 2006). We note that Bloom (1984) and Angrist and Imbens (1991) both derived the IV formula under an assumption of 'no intrusion'

$$R = 0 \Rightarrow X = 0 \tag{10.23}$$

and without using joint potential outcomes. Under no intrusion (10.23), consistency (10.1), exchangeability (10.2), and the exclusion restriction (10.3), the IV formula gives the the causal treatment effect for the treated:

$$E(Y_1|X = 1) - E(Y_0|X = 1) \tag{10.24}$$

The good news so far is that we are able to identify the causal effect of X on Y for a subset of the population, i.e. the compliers. The bad news is that we cannot in general tell who belongs to this subset and who does not. Identifying a subject as a complier requires that we can observe the joint potential outcomes $\{X_0, X_1\}$, which is in general not possible. We can, however, determine the proportion of subjects that belong to this subset. From the equality between (10.15a) and (10.19) it follows that the proportion of compliers is identifiable and equal to

$$\Pr(X_0 = 0, X_1 = 1) = \Pr(X = 1|R = 1) - \Pr(X = 1|R = 0) \tag{10.25}$$

Identifiability of the complier causal effect relies on the consistency assumption (10.1), exchangeability (10.2), the exclusion restriction (10.3), and the monotonicity assumption (10.14). An important question is whether these assumptions can be tested. The equality between (10.15a) and (10.19) implies that

$$\Pr(Y = y, X = 1|R = 1) - \Pr(Y = y, X = 1|R = 0) > 0 \quad \forall y \tag{10.26}$$

By a similar argument we have that

$$\Pr(Y = y, X = 0|R = 0) - \Pr(Y = y, X = 0|R = 1) > 0 \quad \forall y \tag{10.27}$$

Thus, if either (10.26) or (10.27) is violated, then at least one of the assumptions (10.1), (10.2), (10.3), and (10.14) is false. The testable restrictions in (10.26) and (10.27) were derived by Balke and Pearl (1997) for binary Y.

10.4 Defining the appropriate causal effect in studies suffering from truncation by death

To fix ideas, consider a randomized trial to evaluate the effect of a certain medical treatment on a certain continuous outcome measured at one year after treatment assignment. Each subject in a cohort is randomly assigned to either have the treatment or not. We assume that the compliance is 100 %. We use X to denote the received treatment level, e.g. $X = 1$ for 'received treatment' and $X = 0$ for 'received no treatment'. The cohort is followed for 1 year. During follow-up, some subjects die. We let Z be the 'survival indicator' i.e. $Z = 1$ for 'survived during follow-up' and $Z = 0$ for 'did not survive'. For a surviving subject, the outcome Y is measured at the end of follow-up. Note that Y is only well defined for those subjects who survive ($Z = 1$). The observed data consists of a random sample from $\Pr(X, Z, Y)$, where by notational convenience we define $Y = *$ if $Z = 0$. The scenario is displayed in the DAG of Figure 10.2. The arrow from X to Z represents the influence of treatment on survival. The arrow from X to Y represents the causal treatment effect on the outcome, which is the target of the analysis. U represents all common causes for Z and Y, e.g. genetics, lifestyle factors, etc. The absence of an arrow from U to X follows by randomization of the treatment.

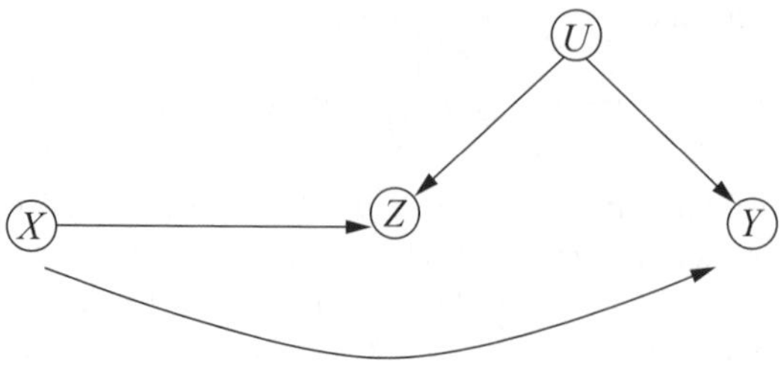

Figure 10.2 *X: treatment, Z: survival, Y: outcome, U: common causes of Z and Y.*

From the DAG in Figure 10.2 it is obvious that an analysis restricted to the survivors ($Z = 1$) does not produce a causal effect; conditioning on $Z = 1$ opens the path $X \rightarrow Z \leftarrow U \rightarrow Y$. In particular, even if there is no effect of the treatment on the outcome (no arrow from X to Y) there will be a conditional association between X and Y, given $Z = 1$.

It is not so obvious, though, how to define the causal effect of interest. A treatment 'effect' is by definition a comparison of two scenarios, one where the treatment is present and one where it is absent. However, if the outcome of interest is not well defined under either scenario, then the treatment effect is not well defined either. Robins (1986) proposed to solve this conceptual problem by restricting the analysis to those subjects who would survive, regardless of whether they are treated or not; we may call them 'healthy'. For the healthy subjects, the outcome is well defined under both scenarios and a causal comparison across treatment levels can be made. To formalize, let Z_x denote the potential survival status for a given under treatment level $X = x$. A cross-classification by the joint potential outcomes (Z_0, Z_1) yields four strata. The stratum $(Z_0 = 0, Z_1 = 0)$ contains those who die regardless of whether they are treated or not; we call them 'doomed'. The stratum $(Z_0 = 0, Z_1 = 1)$ contains those who survive if and only if they are treated; we call them 'helped'. The stratum $(Z_0 = 1, Z_1 = 0)$ contains those who survive if and only if they are *not* treated; we call them 'harmed'. Finally, the stratum $(Z_0 = 1, Z_1 = 1)$ contains the healthy subjects. In general, we cannot observe the two potential outcomes Z_0 and Z_1 jointly. This implies that the stratum of healthy subjects is in general not observable either and that the treatment effect for this stratum is in general not identifiable.

Let μ_x denote the mean potential outcome Y_x for the stratum of healthy subjects:

$$\mu_x \equiv E(Y_x | Z_0 = 1, Z_1 = 1) \qquad (10.28)$$

Zhang and Rubin (2003) and Jemiai (2005) derived bounds for the causal mean difference $\mu_1 - \mu_0$. They did not use the linear programming technique developed by Balke and Pearl (1994). In fact, linear programming techniques cannot be used for bounds on $\mu_1 - \mu_0$ since this parameter cannot be expressed as a linear function of joint potential outcome probabilities. In what follows, we review the derivation in Zhang and Rubin (2003) and Jemiai (2005). Let π be the proportion of healthy subjects, i.e. $\pi \equiv \Pr(Z_0 = 1, Z_1 = 1)$. π is not identifiable. However, Zhang and Rubin (2003) and Jemiai (2005) showed that π is bounded by

$$\min(\pi) = \max\{0, \Pr(Z = 1|X = 0) - \Pr(Z = 0|X = 1)\} \qquad (10.29)$$

$$\max(\pi) = \min\{\Pr(Z = 1|X = 0), \Pr(Z = 1|X = 1)\} \qquad (10.30)$$

Define $\gamma_x(\pi) \equiv \pi/\Pr(Z = 1|X = x)$. To proceed, we note that the group defined by $(X = 0, Z = 1)$ is a mixture of healthy and harmed, with mixing proportions $\gamma_0(\pi)$ and $1 - \gamma_0(\pi)$, respectively. This implies that, for any fixed value of π, the minimal value of μ_0 is the $\gamma_0(\pi)$-fraction with the smallest values of Y, among the subjects with $(X = 0, Z = 1)$, i.e.

$$\min(\mu_0; \pi) = E[Y|X = 0, Z = 1, Y \leq q_0\{\gamma_0(\pi)\}] \tag{10.31}$$

where $q_x(p)$ is the p-quantile satisfying $q_x(p) = y$ if $\Pr(Y \leq y|Z = 1, X = x) = p$. Similarly, the maximal value of μ_0 is the $\gamma_0(\pi)$-fraction with the largest values of Y, among the subjects with $(X = 0, Z = 1)$, i.e.

$$\max(\mu_0; \pi) = E[Y|X = 0, Z = 1, Y > q_0\{1 - \gamma_0(\pi)\}] \tag{10.32}$$

By a similar argument we have that

$$\min(\mu_1; \pi) = E[Y|X = 1, Z = 1, Y \leq q_1\{\gamma_1(\pi)\}] \tag{10.33}$$

and

$$\max(\mu_1; \pi) = E[Y|X = 1, Z = 1, Y > q_1\{1 - \gamma_1(\pi)\}] \tag{10.34}$$

We now note that $\min(\mu_x; \pi)$ is a nondecreasing function of π and $\max(\mu_x; \pi)$ is a nonincreasing function of π. Thus, we arrive at the following bounds for $\mu_1 - \mu_0$:

$$\begin{aligned} \min(\mu_1 - \mu_0) &= \min\{\mu_1; \min(\pi)\} - \max\{\mu_0; \min(\pi)\} \\ \max(\mu_1 - \mu_0) &= \max\{\mu_1; \min(\pi)\} - \min\{\mu_0; \min(\pi)\} \end{aligned} \tag{10.35}$$

Jemiai (2005) proved that the bounds in (10.35) are tight, i.e. they do not include any value of $\mu_1 - \mu_0$ that is not compatible with the observed distribution $\Pr(X, Z, Y)$. Jemiai (2005) also derived bounds for $\mu_1 - \mu_0$ when the outcome is binary.

10.5 Discussion

Cross-classifications by joint potential outcomes have been used to derive many important results in the causal inference literature. An implicit assumption is that potential outcomes are jointly well defined, even though they can never – even in principle – be jointly observed. Whether this assumption is reasonable or not has been subject to some debate (e.g. Dawid, 2000). Some of the results that have been derived using joint potential outcomes have also been reproduced with methods not relying on the existence of joint potential outcomes (e.g. Dawid, 2003).

One limitation of methods utilizing joint potential outcomes is that these methods are often intractable in scenarios with nonbinary variables. The complier causal effect, for instance, was defined in Section 10.3 for 'all or nothing' compliance. This definition can be generalized to allow for more complex patterns of compliance, but then the effect is no longer (nonparametrically) identifiable (Vansteelandt and Goethgebeur, 2005). Nonidentifiability of parameters, however, is not necessarily an argument against methods utilizing joint potential outcomes.

In the context of 'truncation by death', for instance, it can be argued that the estimand of interest is only meaningful within a stratum of joint potential outcomes. That this estimand is inherently difficult to identify simply follows from the complex nature of the problem.

We have used directed acyclic graphs (DAGs) to visualize the problems that we have considered. In the literature, however, these problems have mostly been formulated in terms of potential outcomes and not in terms of DAGs (e.g. Angrist and Imbens, 1991; Frangakis and Rubin, 2002; Zhang and Rubin, 2003; Gilbert *et al.*, 2003). There may be several reasons for this (see the discussion of Pearl, 1995b), including pure tradition. However, it is important to recognize that there is no logical contradiction between potential outcomes and DAGs. Pearl (1993) demonstrated that both these frameworks can be mathematically unified under the umbrella of nonparametric structural equations (NPSEs).

A special cross-classification by joint potential outcomes is the so-called 'principal stratification'. The term was coined by Frangakis and Rubin (2002). They argued that when the exposure of interest (X) is controlled by design (e.g. randomized or assigned through a well-defined protocol), then we have a clear understanding of what it means to receive hypothetically a counterfactual exposure level. However, when the exposure is not controlled, this may be less clear. Thus, they argued, when the exposure is not controlled the counterfactual outcome Y_x is not well defined.[4] Frangakis and Rubin (2002) referred to counterfactual variables under uncontrolled exposures as a priori counterfactual. They appeared to reserve the term 'principal stratification' for cross-classifications by joint potential outcomes under controlled exposures. With this definition, the strata defined by joint compliance behavior $(X_{r=0}, X_{r=1})$ in Section 10.2 are 'principal strata', since treatment assignment (R) is indeed randomized. However, the strata defined by joint outcome behavior $(Y_{x=0}, Y_{x=1})$ are not 'principal strata', since treatment taken (X) is uncontrolled. Refusing to rely on a priori counterfactuals typically means that population causal effects, as in Section 10.2, cannot be defined, and one has to resort to 'principal stratum' effects, as in Section 10.3.

References

Angrist, J.D. and Imbens G.W. (1991) Source of identifying information in evaluation models. Discussion Paper 1568, Department of Economics, Harvard University, Cambridge, MA.

Angrist, J.D., Imbens, G.W. and Rubin, D.B. (1996) Identification of causal effects using instrumental variables. *Journal of the American Statistical Asssociation*, **91** (434), 444–455.

Balke, A. and Pearl, J. (1994) Counterfactual probabilities: computational methods, bounds, and applications. Technical Report R-213-B, UCLA Cognitive Systems Laboratory.

Balke, A. and Pearl, J. (1997) Bounds on treatment effects from studies with imperfect compliance. *Journal of the American Statistical Association*, **92**, 1171–1176.

Bloom, H.S. (1984) Accounting for no-shows in experimental evaluation designs. *Evaluation Review*, **8**, 225–246.

Cox D.R. and Hinkley, D.V. (1974) *Theoretical Statistics*. London: Chapman and Hall.

Dawid, A.P. (2000) Causal inference without counterfactuals (with discussion). *Journal of the American Statistical Association*, **95** (450), 407–448.

[4] See Chapter 2 in this volume for a further discussion on well-defined versus ill-defined counterfactuals.

Dawid, A.P. (2003) Causal inference using influence diagrams: the problem of partial compliance (with discussion), in *Highly Structured Stochastic Systems* (eds P.J. Green, N.L. Hjort and S. Richardson). Oxford University Press, pp. 45–81.

Frangakis, C.E. and Rubin, D.B. (2002) Principal stratification in causal inference. *Biometrics*, **58**, 21–29.

Gilbert, P.B. and Bosch, J.B. and Hudgens, M.G. (2003) Sensitivity analysis for the assessment of causal vaccine effects on viral load in HIV vaccine trials. *Biometrics*, **59**, 531–541.

Hernán, M.A. and Robins, J.M. (2006) Instruments for causal inference: an epidemiologists dream? *Epidemiology*, **17** (4), 360–372.

Imbens, G.W. and Angrist, J.D. (1994) Identification and estimation of local average treatement effects. *Econometrica*, **62** (2), 467–475.

Imbens, G.W. and Rubin, D.B. (1997) Bayesian inference for causal effects in randomized experiments with noncompliance. *Annals of Statistics*, **25**, 305–327.

Jemiai, Y. (2005) *Semiparametric methods for the effect of treatment on an outcome existing only in a post-randomization selected subpopulation.* PhD Thesis, Harvard University, Cambridge, MA.

Jemiai, Y., Rotnitzky, A., Shepherd, B.E. and Gilbert, P.B. (2007) Semiparametric estimation of treatment effects on an outcome measured after a post-randomization event occurs. *Journal of the Royal Statistical Society*, **69**, 879–901.

Kalbfleish, J.D. and Prentice, R.L. (1980) *The Statistical Analysis of Failure Time Data*. New York: John Wiley & Sons, Inc.

Manski, C.F. (1990) Nonparametric bounds on treatment effects. *American Economic Review, Papers and Proceedings*, **80**, 319–323.

Manski, C.F. (2007) *Identification for Prediction and Decision*. Cambridge, MA: Harvard University Press.

Pearl, J. (1993) Aspects of graphical models connected to causality, in *Proceedings of the 49th Session of the International Statistical Institute*, Italy, Tome LV, Book 1, Florence, pp. 399–401.

Pearl, J. (1995a) Causal inference from indirect experiments. *Artificial Intelligence in Medicine*, **7**, 561–582.

Pearl, J. (1995b) Causal diagrams for empirical research. *Biometrika*, **82** (4), 669–710.

Pearl, J. (2000) *Causality: Models, Reasoning and Inference*. Cambridge: Cambridge University Press.

Robins, J.M. (1986) A new approach to causal inference in mortality studies with sustained exposure periods – application to control of the healthy worker survivor effect. *Mathematical Modelling*, **7**, 1393–1512.

Robins, J.M. (1989) The analysis of randomized and non-randomized AIDS treatment trials using a new approach to causal inference in longitudinal studies, in *Health Service Research Methodology: A focus on AIDS* (eds L. Sechrest, H. Freeman and A. Mulley). Washington, DC: US Public Health Service, National Center for Health Services Research, pp. 113–159.

Robins, J.M. and Greenland, S. (1992) Identifiability and exchangeability for direct and indirect effects. *Epidemiology*, **3**, 143–155.

Shepherd, B.E., Gilbert, P.B., Jemiai, Y. and Rotnitzky, A. (2006) Sensitivity analysis comparing outcomes only existing in a subset selected post-randomization, conditional on covariates, with application to HIV vaccine trials. *Biometrics*, **62**, 332–342.

Sjölander, A. (2009) Bounds on natural direct effects in the presence of confounded intermediate variables. *Statistics in Medicine*, **28** (4), 558–571.

Vansteelandt, S. and Goethgebeur, E. (2005) Sense and sensitivity when correcting for observed exposures in randomized clinical trials. *Statistics in Medicine*, **24**, 191–210.

Zhang, J.L. and Rubin, D.B. (2003) Estimation of causal effects via principal stratification when some outcomes are truncated by 'death'. *Journal of Educational and Behavioral Statistics*, **28**(4), 353–368.

11

Estimation of direct and indirect effects

Stijn Vansteelandt

Ghent University, Ghent, Belgium, and London School of Hygiene and Tropical Medicine, London, UK

11.1 Introduction

For many decades, scientists in most scientific fields – most notably psychology, sociology and epidemiology – have been occupied with questions as to whether an exposure affects an outcome through pathways other than those involving a given mediator/intermediate variable. The answer to such questions is of interest because it brings insight into the mechanisms that explain the effect of exposure on outcome (VanderWeele, 2009a) and because the presence of intermediate variables may sometimes complicate the interpretation of the targeted exposure effect (Joffe *et al.*, 2001; Rosenblum *et al.*, 2009). Mediation analyses are used for this purpose. They attempt to separate so-called 'indirect effects' from 'direct effects'. The former term is typically used in a loose sense to designate that part of an exposure effect that arises indirectly by affecting a (given) set of intermediate variables; the latter then refers to the remaining exposure effect.

In statistical mediation analysis, the direct effect is commonly connected with the residual association between outcome and exposure after regression adjustment for the mediator(s); the indirect effect is then obtained through a clever combination of the exposure's effect on the mediator and the mediator's effect on the outcome (Baron and Kenny, 1986; MacKinnon, 2008; Fosen *et al.*, 2006). For instance, when the associations between exposure A and mediator M and outcome Y can be modelled through linear regressions as

$$E(Y|A, M) = \beta_0 + \beta_a A + \beta_m M$$
$$E(M|A) = \alpha_0 + \alpha_a A,$$

Causality: Statistical Perspectives and Applications, First Edition. Edited by Carlo Berzuini, Philip Dawid and Luisa Bernardinelli.

then β_a is commonly interpreted as a direct effect and $\beta_m \alpha_a$ as an indirect effect (Baron and Kenny, 1986). Such decomposition is often transported to nonlinear models (MacKinnon, 2008; Fosen *et al.*, 2006), though not always justified for the following two reasons.

First, the precise interpretations of association parameters like β_a and $\beta_m \alpha_a$ as direct and indirect effects are ambiguous and sometimes flawed. One of the major contributions of the causal inference literature (Robins and Greenland, 1992; Pearl, 2001) has been to make clear that different definitions – and thus different interpretations – of the direct effect exist, not all of which enable a meaningful concept of the indirect effect. We refer the interested reader to Chapter 12 by Pearl in this volume, which shows that decomposition of a total effect into a direct and indirect effect becomes subtle when nonlinear associations exist between the mediator and outcome.

Second, the causal diagram (see Section 8.8 of Chapter 8 in this volume for a discussion of this representation tool) of Figure 11.1 displays a setting where some prognostic factors L of the mediator (other than the exposure) are also associated with the outcome, so that the association between the mediator and outcome is confounded. This situation is representative of most empirical studies, including randomized experiments, because the fact that the exposure is randomly assigned does not prevent confounding of the mediator–outcome association. In the presence of such confounding, the residual association between outcome and exposure after adjusting for the mediator(s) (cf. β_a in the above model) does not encode a direct exposure effect. This is technically seen because adjustment for a collider M (i.e. a node in which two edges converge; see a more detailed definition in Section 3.2.1 of Chapter 3 in this volume) along the path $A \rightarrow M \leftarrow L \leftarrow U \rightarrow Y$ may render A and Y dependent along that path, and may thus induce a noncausal association (Robins, 1999c; Cole and Hernán, 2002).

One of the major contributions of the causal inference literature has been to point this out and to make clear that specialized estimation techniques are often needed to be able to adjust for such confounders, as these may themselves be affected by the exposure. The goal of this chapter is to review various of these techniques, which will also be exemplary for more generic techniques to deal with time-dependent confounding. In this chapter we predominately use a potential outcomes (see Chapter 2 in this volume) terminology. For a more decision-theoretic formulation of problems of analysis of mediation, see Berzuini and colleagues (submitted 2011).

11.2 Identification of the direct and indirect effect

In this section, we briefly recap definitions of controlled and natural direct effects, referring the reader to Chapter 12 by Pearl for details. In addition, we review conditions under which these effects can be identified.

11.2.1 Definitions

Let for each subject $Y(a,m)$ denote the counterfactual outcome under exposure level a and mediator level m, and $M(a)$ be the counterfactual mediator under exposure level a. The *controlled direct effect* of exposure level a versus reference exposure level 0, controlling for M, can then be defined as the expected contrast $E\{Y(a, m) - Y(0, m)\}$ (Robins and Greenland, 1992; Pearl, 2001). It expresses the exposure effect that would be realized if the mediator were controlled at level m uniformly in the population. The so-called natural or pure direct effect

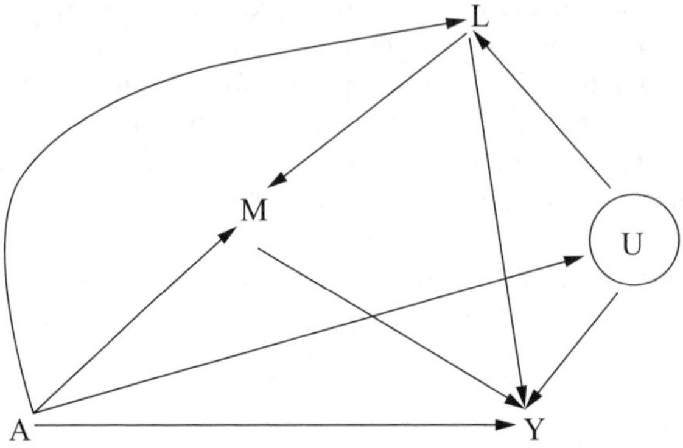

Figure 11.1 Causal diagram with measured variables A, L, M *and* Y, *and with* U *an unmeasured confounder of the* L-Y *relationship.*

(Robins and Greenland, 1992; Pearl, 2001) is defined as the expected contrast $E\{Y(a, M(0)) - Y(0, M(0))\}$ and the total direct effect (Robins and Greenland, 1992) as $E\{Y(a, M(a)) - Y(0, M(a))\}$. Note that these definitions naturally extend to scales other than the additive scale. For instance, on the odds ratio scale, VanderWeele and Vansteelandt (2010) define the controlled direct effect as

$$\frac{\text{odds}\,\{Y(a, m) = 1\}}{\text{odds}\,\{Y(0, m) = 1\}} \tag{11.1}$$

and the natural direct effect as

$$\frac{\text{odds}\,\{Y(a, M(0)) = 1\}}{\text{odds}\,\{Y(0, M(0)) = 1\}}$$

where for any two random variables A and B, $\text{odds}(A = 1|B) = P(A = 1|B)/P(A = 0|B)$.

Under the generalized consistency or composition assumption (Pearl, 2009; VanderWeele and Vansteelandt, 2009) that for each a, $Y(a, M(a)) = Y(a)$ with probability 1, the difference between the total causal effect and a pure natural direct effect

$$E\{Y(a) - Y(0)\} - E\{Y(a, M(0)) - Y(0)\} = E\{Y(a, M(a)) - Y(a, M(0))\}$$

measures an indirect effect. It expresses how much the outcome would change on average if the exposure were controlled at level a but the mediator were changed from level $M(0)$ to $M(a)$. It is termed the total indirect effect (Robins and Greenland, 1992). Likewise, the difference between the total effect and the total direct effect gives the natural indirect effect

$$E\{Y(a) - Y(0)\} - E\{Y(a) - Y(0, M(a))\} = E\{Y(0, M(a)) - Y(0, M(0))\}$$

This expresses how much the outcome would change on average if the exposure were controlled at level 0, but the mediator were changed from its natural level $M(0)$ to the level $M(a)$, which it would have taken at exposure level a.

Direct effects are sometimes defined using concepts of principal stratification (Frangakis and Rubin, 2002), namely as the exposure effect within the principal stratum of individuals whose mediator level was not affected by the exposure, i.e. $E\{Y(a) - Y(0)|M(0) = M(a)\}$. The appeal of principal stratum direct effects relative to the foregoing definitions of direct effect is that they do not require conceptualizing manipulation of the mediator. They may henceforth be meaningful even when interventions on the mediator are ill defined or hard to conceive. However, this comes at a cost: principal stratum direct effects have a more limited utility because of the inability to identify which individuals fall into which principal strata, because the principal strata are sparsely populated in many realistic applications and because they do not correspond to a natural definition of the indirect effect (Robins *et al.*, 2007; VanderWeele, 2008); see also Pearl (2011) and the following discussions. In view of these limitations, we will not discuss the identification and estimation of principal stratum direct effects in this chapter and refer the interested reader to Elliott *et al.* (2010), Emsley *et al.* (2010), Gallop *et al.* (2009) and Rubin (2004) and to VanderWeele (2008), who discusses the relations between controlled, natural and principal stratification direct effects.

11.2.2 Identification

All foregoing direct and indirect effect definitions involve counterfactuals. To link these to the observed outcomes, current estimation techniques work under the consistency assumption. Following this assumption, interventions that set A to the value a have no effect among those for whom the exposure level $A = a$ was naturally observed and likewise for interventions on the mediator, in the sense that $Y(a, m) = Y$ if $A = a$ and $M = m$. The practical implication is that the results obtained from mediation analysis express the effect of 'noninvasive' interventions or manipulations (VanderWeele and Vansteelandt, 2009), which involve changing the exposure while holding the mediator at a fixed level. The consistency assumption enables one to learn about the counterfactual $Y(a,m)$ in subjects with $A = a$ and $M = m$. To infer the distribution of the counterfactual $Y(a,m)$ in the entire study population, additional untestable assumptions are needed. Most procedures invoke assumptions about the exchangeability of subjects under different exposure levels (e.g. assumptions that all confounders of the association between exposure and outcome have been measured). Exchangeability conditions that are sufficient for identification are briefly reviewed below.

Controlled direct effects conceptualize atomic interventions on both the exposure and the mediator, and can therefore be identified when all confounders C of the exposure–outcome association have been measured in the sense that

$$Y(a, m) \perp\!\!\!\perp A|C, \quad \forall a, m \tag{11.2}$$

and, additionally, all confounders L of the mediator–outcome association have been measured in the sense that

$$Y(a, m) \perp\!\!\!\perp M|A = a, L, \quad \forall a, m \tag{11.3}$$

Here, L must include C, and may additionally contain components some of which are affected by the exposure, in which case we say there is intermediate confounding. In particular, with these assumptions, the controlled direct effect can be identified by G-computation (Robins,

1986; Pearl, 2001) as

$$E\{Y(a,m) - Y(0,m)|C\} = \int E(Y|A=a, M=m, L)f(L|A=a, C)\mathrm{d}L \\ - \int E(Y|A=0, M=m, L)f(L|A=0, C)\mathrm{d}L \quad (11.4)$$

For the identification of natural direct and indirect effects, assumptions are needed in addition to the above conditions with $L = C$. First, because natural direct and indirect effects involve counterfactuals $M(0)$, assumptions are needed regarding the availability of all confounders of the exposure–mediator association, e.g. that

$$M(a) \perp\!\!\!\perp A|C, \quad \forall a$$

Second, because natural direct and indirect effects involve composite counterfactuals $Y(a, M(0))$, further assumptions are needed. Pearl (2001) assumes that

$$Y(a,m) \perp\!\!\!\perp M(0)|C, \quad \forall a, m \quad (11.5)$$

which Petersen *et al.* (2006) relax to

$$E\{Y(a,m) - Y(0,m)|M(0), C\} = E\{Y(a,m) - Y(0,m)|C\}, \quad \forall a, m \quad (11.6)$$

Because these assumptions are difficult to interpret, Imai *et al.* (2010b) substitute all the above conditions with the following more easily interpretable conditions:

$$\{Y(a,m), M(0)\} \perp\!\!\!\perp A|C, \quad \forall a, m$$
$$Y(a,m) \perp\!\!\!\perp M(0)|A=0, C, \quad \forall a, m$$

All of the above sets of assumptions are satisfied when A and M are randomized within levels of C and lead one to identify the natural direct effect as (Pearl, 2001; Petersen *et al.*, 2006; Imai *et al.*, 2010b)

$$\begin{aligned} &E\{Y(a, M(0)) - Y(0, M(0))|C\} \\ &\quad = \int E\{Y(a,m) - Y(0,m)|C\} f(M=m|A=0, C)\mathrm{d}m \\ &\quad = \int \{E(Y|M=m, A=a, C) - E(Y|M=m, A=0, C)\} \\ &\quad\quad \times f(M=m|A=0, C)\mathrm{d}m \end{aligned} \quad (11.7)$$

and the total indirect effect as

$$E\{Y(a, M(a)) - Y(a, M(0))|C\} = \int \{f(M=m|A=a, C) - f(M=m|A=0, C)\} \\ \times E(Y|M=m, A=a, C)\,\mathrm{d}m \quad (11.8)$$

In the case that the mediator is dichotomous, Hafeman and VanderWeele (2010) show that identification of natural direct and indirect effects is sometimes possible under weaker assumptions.

The above identification results for natural direct and indirect effects essentially[1] exclude the possibility of intermediate confounding. This is an important limitation because it is often likely to believe that some prognostic factors of the mediator may themselves be affected by the exposure, especially when the mediator – and hence some of its prognostic factors – comes much later in time than the exposure (Robins, 2003); see VanderWeele and Vansteelandt (2009) for exceptions. Robins and Greenland (1992) therefore work under the assumption that, at the *individual level*, there is no interaction between exposure and the mediator in the effect that they produce on the outcome in the sense that $Y(a, m) - Y(0, m)$ is a random variable not depending on m. This no-interaction assumption enables one to equate natural and controlled direct effects, thus making natural direct effects identifiable in the presence of intermediate confounding. However, it has a more limited utility as it is often unrealistic (Petersen *et al.*, 2006), may be contradicted by the observed data and because exposure–mediator interactions are often of prime interest (MacKinnon, 2008). Vansteelandt and VanderWeele (2011) show that total direct and natural indirect effects in the exposed, i.e. $E\{Y(a, M(a)) - Y(0, M(a))|A = a\}$ and $E\{Y(0, M(a)) - Y(0, M(0))|A = a\}$ respectively, can essentially be identified under the same identification conditions as for controlled direct effects, and thus in particular in the presence of intermediate confounding.

A minimal requirement implied by all of the foregoing identification results is that the data analyst disposes of all confounders of the mediator–outcome association. This is sometimes perceived to be a strong limitation because such data are often lacking in practice. A root cause of this is that many applied researchers are ignorant of the precise identification conditions for mediation analysis, and thus in particular are ignorant of the fact that mediation analyses require measurements on confounders of the mediator–outcome relationship. To make progress in the absence of such confounder measurements, some authors have developed statistical methods that enable identification of direct and/or indirect effects in the presence of unmeasured confounders of the mediator–outcome relationship. Some of these are described in excellent review papers by Emsley *et al.* (2010) and Ten Have and Joffe (2011). For instance, Sobel (2008) assumes the absence of a direct effect and makes progress through instrumental variable methods. Robins and Greenland (1994) and Ten Have *et al.* (2007) obtain identification of controlled direct effects within a class of rank-reserving structural models by solely relying on the assumption that there are no unmeasured confounders of the exposure–outcome association (i.e. assumption (11.2)). A limitation is that identification is now partly obtained by imposing unverifiable parametric restrictions so that the considered models are no longer nonparametrically identifiable. For instance, Ten Have *et al.* (2007) include baseline covariates C in their model, but exclude interactions of these covariates with exposure and the mediator, as well as exposure–mediator interactions. They find that identification is possible when the exposure interacts with these covariates in the effect it produces on the mediator. As with related instrumental variable methods, identification under these assumptions is typically weak and inferences can be rather sensitive to the modelling assumptions. Our personal viewpoint is therefore that, as in other observational studies, progress in applied mediation

[1] The condition of Petersen *et al.* (2006) in principle allows for this, but Vansteelandt and VanderWeele (2011) argue that data-generating mechanisms that satisfy this condition are rather implausible in the presence of intermediate confounding.

analysis must ideally be made through efforts in collecting data on confounders of both the exposure–outcome and the mediator–outcome association. Such efforts are also key to the success of well-designed observational studies for total effects. In the remainder of this chapter, we will therefore restrict our discussion on estimation methods to those that work under the aforementioned exchangeability assumptions. Whether estimates of the considered effect measures can eventually be interpreted as 'direct' and 'indirect' effects in a specific application depends on whether the required untestable exchangeability assumptions are met. With concern over the validity of these assumptions, sensitivity analyses are recommended (see, for example, Imai *et al.* 2010b, VanderWeele 2010, Sjolander 2009).

11.3 Estimation of controlled direct effects

11.3.1 G-computation

Under assumptions (11.2) to (11.3), the controlled direct effect can be identified using G-computation as (11.4). In the absence of intermediate confounding (i.e. when L is empty), this reduces to

$$E\{Y(a,m) - Y(0,m)|C\} = E(Y|A = a, M = m, C) - E(Y|A = 0, M = m, C)$$

Controlled direct effects are then obtainable from a standard regression of outcome on exposure, mediator and confounders. For example, when

$$E(Y|A, M, C) = \gamma_0 + \gamma_m M + \gamma_a A + \gamma_{am} AM + \gamma_c C$$

then the controlled direct effect $E\{Y(a,m) - Y(0,m)|C\}$ equals $(\gamma_a + \gamma_{am} m)a$. In the more likely case that intermediate confounding is present, Equation (11.4) can still be used, but requires integration and an additional model for the joint distribution of all intermediate confounders, given the exposure and baseline confounders. For example, when

$$E(Y|A, M, C, L) = \gamma_0 + \gamma_m M + \gamma_a A + \gamma_{am} AM + \gamma_c C + \gamma_l L$$

then the controlled direct effect $E\{Y(a,m) - Y(0,m)|C\}$ equals

$$(\gamma_a + \gamma_{am} m)a + \gamma_l \{E(L|A = a, C) - E(L|A = 0, C)\}$$

When, furthermore, $E(L|A, C) = \alpha_0 + \alpha_a A + \alpha_c C + \alpha_{ac} AC$, this reduces to

$$\{\gamma_a + \gamma_{am} m + \gamma_l (\alpha_a + \alpha_{ac} C)\} a$$

The integration and modelling of intermediate confounders in (11.4) can make G-computation computationally cumbersome and greedy in demanding parametric modelling assumptions, especially when L is high-dimensional and nonlinearly related with the outcome. Further subtleties arise because Equation (11.4) is difficult to combine with pre-specified models for the controlled direct effect, in the sense that parsimonious models for the outcome mean and confounder distribution need not translate into parsimonious models for the controlled direct effect. This can make the results unattractive for reporting. Furthermore, as demonstrated by Robins (1997), it is essentially impossible to choose nonlinear models for $E(Y|A, M, C, L)$ and $f(L|A, C)$ that allow for the possibility of intermediate confounding while, at the same time, combining to controlled direct effects $E\{Y(a, m) - Y(0, m)|C\}$ equalling zero for all a, m and C. Adopting such models in the G-computation formula (11.4) may therefore imply guaranteed rejection of the null hypothesis of no effect in large samples (Robins, 1997), a phenomenon that is known as the null paradox. For example, suppose that C is empty and L is dichotomous and that the data analyst chooses the following models:

$$\begin{aligned} E(Y|A, M, L) &= \gamma_0 + \gamma_m M + \gamma_a A + \gamma_l L \\ E(L|A) &= \text{expit}(\alpha_0 + \alpha_a A) \end{aligned}$$

with $\text{expit}(x) = \exp(x)/\{1 + \exp(x)\}$. Then

$$E\{Y(a, m)\} = \gamma_0 + \gamma_m m + \gamma_a a + \gamma_l \text{expit}(\alpha_0 + \alpha_a a)$$

For continuous exposure A, one would need $\gamma_a = 0$ and either $\alpha_a = 0$ or $\gamma_l = 0$ for the null hypothesis of no direct effect to hold. However, the hypothesis that $\gamma_a = 0$ and either $\alpha_a = 0$ or $\gamma_l = 0$ could be rejected even if there is no direct effect because it requires that, in addition, either L is not affected by A or L is not associated with Y (conditional on A and M). In view of the above difficulties, recourse to alternative estimation strategies will be sought.

11.3.2 Inverse probability of treatment weighting

A large class of estimation methods for controlled direct effects relies on inverse probability of treatment weighting by the conditional density/probability of the observed mediator, given intermediate confounders, exposure and baseline confounders. Intuitively, this has the effect of removing the exposure's influence on the mediator and thereby also an indirect effect (Vansteelandt, 2009b). The remaining direct effect is then obtained from a (weighted) regression of outcome on exposure, mediator and baseline confounders. More formally, inverse probability of treatment weighting can be justified by rewriting the G-computation formula as

$$\begin{aligned} E\{Y(a, m)|C\} &= \int E(Y|A = a, M = m, L) f(L|A = a, C) \mathrm{d}L \\ &= \int E(Y|A = a, M, L) \frac{I(M = m)}{f(M = m|L, A = a, C)} f(M = m|L, A = a, C) \\ &\quad \times f(L|A = a, C) \mathrm{d}L \, \mathrm{d}m \\ &= E\left\{Y \frac{I(M = m)}{f(M = m|L, A = a, C)} |A = a, C\right\} \end{aligned}$$

VanderWeele (2009) focuses on so-called marginal structural models (Robins *et al.*, 2000) for the marginal expectation of the counterfactuals $Y(a, m)$. For example, in the marginal structural model

$$E\{Y(a, m)\} = \beta_0 + \beta_a a + \beta_m m + \beta_{am} am$$

the controlled direct effect $E\{Y(a,m) - Y(0,m)\}$ is given by $(\beta_a + \beta_{am}m)a$. Estimates for the unknown parameters indexing this model may be obtained via a weighted regression of the corresponding traditional regression model

$$E(Y|A, M) = \beta_0 + \beta_a A + \beta_m M + \beta_{am} AM$$

using the weights

$$\frac{f(M|A)f(A)}{f(M|L, A, C)f(A|C)}$$

Vansteelandt (2009) avoids inverse probability weighting by the exposure A by focusing on marginal structural models (Robins *et al.*, 2000) for the conditional expectation of the counterfactuals $Y(a,m)$, given baseline confounders C; for example,

$$E\{Y(a,m)|C\} = \beta'_0 + \beta'_a a + \beta'_m m + \beta'_{am} am + \beta'_c C \tag{11.9}$$

Estimates for the unknown parameters indexing this model may be obtained via a weighted regression of the corresponding traditional regression model

$$E(Y|A, M, C) = \beta'_0 + \beta'_a A + \beta'_m M + \beta'_{am} AM + \beta'_c C$$

using the weights

$$\frac{f(M|A, C)}{f(M|L, A, C)} \tag{11.10}$$

which tends to result in more stable and efficient inferences (but is also more prone to model extrapolation when A and C are highly correlated).

Application of these estimation techniques in practice requires models for the densities/probabilities appearing in the weights. Here, the models for $f(M|L, A, C)$ (and $f(A|C)$) must be correctly specified to ensure consistent estimation; the models for the numerator densities/probabilities are merely included to stabilize the weights. While ideally standard errors of the controlled direct effect estimators must acknowledge estimation of the weights, conservative standard errors are easily obtained from standard software using sandwich estimators that ignore the models relating to the weights (Robins *et al.*, 2000).

As an illustration, we reanalyse data from the Job Search Intervention Study (JOBS II), which was a randomized field experiment that investigated the efficacy of a job training intervention on unemployed workers. The programme was designed not only to increase reemployment among the unemployed but also to enhance the mental health of the job seekers; 1801 unemployed workers received a pre-screening questionnaire and were then randomly assigned to treatment and control groups. Those in the treatment group participated in job-skills workshops. Those in the control condition received a booklet describing job-search tips. We refer the reader to Imai *et al.* (2010b) for details and caution that, for illustrative purposes, we will ignore the complications imposed by missing data and analyse a single imputed data set that is available in the `mediation` package in R. Our interest lies in the controlled direct effect of the intervention on mental health (as measured based on the Hopkins Symptom

Checklist), other than through job finding. We consider baseline depression, age, gender and socioeconomic status as baseline confounders and job-search self-efficacy (a measure of participants' confidence in their ability to search for a job) as an intermediate confounder. Using the R software, we first fitted a model for the mediator `work` on the exposure `treat`, the intermediate confounder `jobseek` and any remaining baseline confounders as

```
> medmod <- glm(work ~ treat + jobseek + depress1
    + econhard + sex + age, family = binomial)
```

Next, we constructed a weight for each subject and fitted a weighted linear regression model of the depression outcome `depress2` on exposure, mediator and baseline confounders using software for generalized estimating equations to obtain sandwich standard errors:

```
> library(geepack)
> weight <- work/fitted(medmod)
    + (1-work)/(1-fitted(medmod))
> m <- geese(depress2 ~ treat + work + depress1
    + econhard + sex + age, id = 1:length(treat), weights
    = weight)
```

We conclude from the final output below that assignment to the intervention would still reduce the depression outcome with 0.029 (95% confidence interval -0.056 to 0.11) on average if all participants were to find work:

```
> summary(m)


                 estimate      san.se           p
(Intercept)  1.047770402 0.121640212  74.1955602 0.000000e+00
treat       -0.028767866 0.043443319   0.4384997 5.078475e-01
work        -0.230975392 0.040342238  32.7801889 1.031912e-08
depress1     0.425264961 0.039533235 115.7162870 0.000000e+00
econhard    0.032176907 0.021970620   2.1448848 1.430458e-01
sex          0.024368324 0.039535236   0.3799117 5.376505e-01
age         -0.003326291 0.002042503   2.6521330 1.034111e-01
```

The foregoing procedures are very appealing for applied mediation analysis because of their simplicity and straightforward extensibility to arbitrary marginal structural models, e.g. marginal hazard models of the form

$$\lim_{\Delta y \to 0} \frac{1}{\Delta y} P\left\{Y(a,m) \in [y, y+\Delta y[| Y(a,m) \geq y, C\right\} = \lambda(y) g(a,m,C;\beta)$$

for an unknown function $\lambda(.)$ and a known smooth function $g(.)$ of an unknown finite-dimensional parameter β or marginal mean models of the form

$$E\left\{Y(a,m)|C\right\} = g(a,m,C;\beta)$$

where, for instance,

$$g(a, m, C; \beta) = \text{expit}(\beta_0 + \beta_a a + \beta_m m + \beta_{am} am)$$

for logistic regression.

In the literature, several extensions have been devised to improve efficiency as well as robustness against model misspecification, but these give up on simplicity and generalizability. Robins (1999c) relaxes assumptions by postulating a model directly for the conditional controlled direct effect $E\{Y(a, m) - Y(0, m)|C\}$; for example,

$$E\{Y(a, m) - Y(0, m)|C\} = (\beta'_a + \beta'_{am} m)a \tag{11.11}$$

This model is less restrictive than the marginal structural model (11.9) because it imposes no restrictions on the expected counterfactual $E\{Y(0, m)|C\}$. Intuitively, the principle behind the estimation in model (11.11) is to remove the direct effect from the outcome by transforming it as $Y - \beta'_a A - \beta'_{am} AM$ and additionally to remove the indirect effect via inverse probability weighting by (11.10). Upon using the true effect sizes β'_a and β'_{am} in this transformation, no effect should remain so that, by assumption (11.2), the transformed outcome becomes conditionally independent of A_i, given C. This inspires estimating β'_a and β'_{am} as the solution to the estimating equation

$$0 = \sum_{i=1}^{n} \begin{pmatrix} 1 \\ M_i \end{pmatrix} \frac{f(M_i|A_i, C_i)}{f(M_i|L_i, A_i, C_i)} \{Y_i - \beta'_a A_i - \beta'_{am} A_i M_i - E\left(Y_i - \beta'_a A_i - \beta'_{am} A_i M_i | C_i\right)\}$$
$$\times \{A_i - E(A_i|C_i)\}$$

which, roughly, sets a weighted conditional covariance between $Y - \beta'_a A - \beta'_{am} AM$ and A, given C, equal to zero. Note that these estimating equations involve the conditional expectations $E\left(Y - \beta'_a A - \beta'_{am} AM|C\right)$ and $E(A|C)$, which are typically unknown; an exception is when A is a randomized exposure or when – in the context of a genetic association study – A is an offspring genotype, with C the corresponding parental genotype, in which case $E(A|C)$ may be a priori known (Vansteelandt *et al.*, 2009). When working models are used for these expectations, consistent estimation of the controlled direct effect requires correct specification of at least one of these two models, along with the model for the mediator density $f(M|L, A, C)$.

This previous estimation principle extends to multiplicative controlled direct effect models of the form

$$\frac{E\{Y(a, m)|C\}}{E\{Y(0, m)|C\}} = \exp\left\{(\beta'_a + \beta'_{am} m)a\right\} \tag{11.12}$$

upon substituting $Y_i - \beta'_a A_i - \beta'_{am} A_i M_i$ with $Y_i \exp(-\beta'_a A_i - \beta'_{am} A_i M_i)$ in the above equation. However, it is not generically applicable; e.g. it does not extend to logistic controlled direct effect models (Robins, 1999c). Goetgeluk *et al.* (2008) consider various extensions aimed at improving the efficiency by making use of an auxiliary model for $E(Y|M, L, A, C)$. Their development at the same time ensures robustness of these controlled direct effect estimators against misspecification affecting either the model for $f(M|L, A, C)$ or the model for $E(Y|M, L, A, C)$, but not both.

In spite of these extensions, simple or doubly robust inverse probability weighting is at present not ideally tailored to mediators that are continuous or have strong measured predictors, because of the ensuing large variability of the weights (11.10) and thus of the resulting direct effect estimators (Goetgeluk *et al.*, 2008; Vansteelandt, 2009). Under model (11.11), Goetgeluk *et al.* (2008) overcome this by deliberately misspecifying $f(M|L, A, C)$ to equal $f(M|A, C)$, and thereby the weights (10) to equal one, in their doubly robust estimator. The resulting estimator is consistent under correct specification of a model for $E(Y|M, L, A, C)$ and can equivalently be obtained using G-estimation in structural nested models, which we review in the following sections.

11.3.3 G-estimation for additive and multiplicative models

We will loosely refer to G-estimation of controlled direct effects as a procedure that works by first removing the mediator's effect (and thereby the indirect effect) from the outcome by transforming it, and subsequently estimating the remaining direct effect from the association between the exposure and that residual outcome. We give a specific illustration before a more general theory.

Suppose first that the following standard association model holds:

$$E(Y|M, L, A, C) = \gamma_0 + \gamma_m M + \gamma_l L + \gamma_a A + \gamma_c C + \beta'_{am} AM \tag{11.13}$$

which can be fitted by ordinary least squares regression. It then follows from (11.3) that $E(Y|M = m, L, A = a, C) = E\{Y(a, m)|L, A = a, C\}$ and hence that $\beta'_{am} AM + \gamma_m M$ encodes the mediator's effect on the outcome. Upon transforming the outcome as $Y - \beta'_{am} AM - \gamma_m M$ to remove the mediator's effect, only a direct effect remains. Assuming in particular that the controlled direct effect model (11.11) holds for $m = 0$, we have that

$$\begin{aligned} E(Y - \beta'_{am} AM - \gamma_m M|A, C) &= E\{Y(A, 0)|A, C\} && \text{by (11.3)} \\ &= E\{Y(0, 0)|A, C\} + \beta'_a A && \text{by (11.11)} \\ &= E\{Y(0, 0)|C\} + \beta'_a A && \text{by (11.2)} \end{aligned}$$

Thus regressing $Y - \beta'_{am} AM - \gamma_m M$ on A and C, for instance, by fitting the following model:

$$E(Y - \beta'_{am} AM - \gamma_m M|A, C) = \alpha_0 + \alpha_c C + \beta'_a A \tag{11.14}$$

one obtains a consistent estimator of the direct main effect β'_a (provided that all models are correctly specified and thus, in particular, $E\{Y(0, 0)|C\}$ is indeed linear in C). It finally follows from $E(Y|M = m, L, A = a, C) = E\{Y(a, m)|L, A = a, C\}$ and model (11.13) that

$$\begin{aligned} &E\{Y(a, m) - Y(0, m)|C\} - E\{Y(a, 0) - Y(0, 0)|C\} \\ &\quad = E\{Y(a, m) - Y(a, 0)|C\} - E\{Y(0, m) - Y(0, 0)|C\} \\ &\quad = E(\gamma_m m + \beta'_{am} am|C) - E(\gamma_m m|C) \\ &\quad = \beta'_{am} am \end{aligned}$$

so that model (11.11) holds for all m, where β'_{am} indexes model (11.13) and thus can be obtained by standard ordinary least squares regression. This approach has been independently advocated by Joffe and Greene (2009) and Vansteelandt (2009).

For illustration, we reconsider the Job Search Intervention Study (JOBS II). First, we fit a linear regression of the depression outcome on the exposure, mediator and confounders (but without exposure–mediator interaction) to assess the mediator's effect on the outcome:

```
> my <- lm(depress2 ~ treat + work + jobseek + depress1
  + econhard + sex + age)
> summary(my)

                Estimate Std. Error t value Pr(>|t|)
(Intercept)    1.5231877  0.1524466   9.992  < 2e-16 ***
treat         -0.0239943  0.0401265  -0.598  0.55001
work          -0.2166093  0.0410801  -5.273 1.69e-07 ***
jobseek       -0.1701083  0.0265846  -6.399 2.53e-10 ***
depress1       0.4280667  0.0361312  11.848  < 2e-16 ***
econhard       0.0626834  0.0204496   3.065  0.00224 **
sex            0.0376477  0.0383122   0.983  0.32604
age           -0.0004689  0.0018320  -0.256  0.79803
```

Note that the exposure coefficient in this model may not reflect the direct effect of interest as a result of collider-stratification bias (i.e. because the adjustment induces a spurious association along the path $A \rightarrow L \leftarrow U \rightarrow Y$). Next, we regress the outcome on exposure and baseline confounders after taking away the mediator effect:

```
> myt <- lm(I(depress2 + 0.2166093*work) ~ treat
          + depress1 + econhard + sex + age)
> summary(myt)

               Estimate Std. Error t value Pr(>|t|)
(Intercept)    0.839833   0.110260   7.617 6.62e-14 ***
treat         -0.034222   0.040905  -0.837   0.4030
depress1       0.472859   0.036166  13.075  < 2e-16 ***
econhard       0.045229   0.020691   2.186   0.0291 *
sex            0.033554   0.038959   0.861   0.3893
age           -0.001172   0.001844  -0.636   0.5251
```

We conclude from the final output that assignment to the intervention would still reduce the depression outcome with 0.034 on average if all participants were to find work; however, we cannot trust the standard error and p-value from the above output since it ignores the estimation uncertainty in the mediator effect. Valid standard errors can be obtained from the R-function `seqg`, which can be downloaded from `users.ugent.be/~svsteela`:

```
>  seqg(Y=depress2, X=treat, M=work, L=jobseek,
       C=cbind(depress1,econhard,sex,age))
```

```
$betax

   Coefficient Standard Error  95% Lower  95% Upper   P value
X -0.03422164      0.04107112 -0.1147196 0.04627628 0.4047157
```

Exposure-mediator interactions can be incorporated via:

```
> seqg(Y=depress2, X=treat, M=work, XM=treat*work,
     L=jobseek, C=cbind(depress1,econhard,sex,age))

$betax

   Coefficient Standard Error  95% Lower 95% Upper   P value
X  -0.042640818      0.1196375 -0.2771260 0.1918444 0.7215285
XM  0.006445017      0.0825603 -0.1553702 0.1682602 0.9377769
```

However, this analysis suggests no evidence of interactions.

In using the above approach, care has to be exercised to ensure that the second stage model (11.14) is congenial with the controlled direct effects model (11.11). Specifically, suppose that the association model is of the following general form:

$$E(Y|M, L, A, C) = \mu(A, C, L; \gamma) + \mu(A, C, L, M; \gamma) \tag{11.15}$$

where the mediator effect $\mu(A, C, L, M; \gamma)$ is a known function, smooth in γ, and satisfying $\mu(A, C, L, 0; \gamma) = 0$, and where the remainder term $\mu(A, C, L; \gamma)$ is also a known function, smooth in γ with γ an unknown finite-dimensional parameter. This model can be fitted using standard regression methods for conditional mean models. Consider, furthermore, the following model for the controlled direct effect at $m = 0$:

$$E\{Y(a, 0) - Y(0, 0)|C\} = g(a, C; \beta) \tag{11.16}$$

where $g(A, C; \beta)$ is a known function, smooth in β and satisfying $g(0, C; \beta) = 0$ with β an unknown finite-dimensional parameter. After fitting model (11.15), the suggested approach involves regressing $Y - \mu(A, C, L, M; \gamma)$ on A and C. Importantly, this second stage model must be of the form

$$\begin{aligned} E\{Y - \mu(A, C, L, M; \gamma)|A, C\} &= E\{Y(A, 0)|A, C\} \\ &= E\{Y(0, 0)|C\} + g(A, C; \beta) \end{aligned} \tag{11.17}$$

When a parametric model for $E\{Y(0, 0)|C\}$ is chosen, then the implied model (11.17) can again be fitted using standard regression methods for conditional mean models. In this way, an estimate of the controlled direct effect at $m = 0$ is obtained. To obtain an estimate of the controlled direct effect at other levels m, note from $E(Y|M = m, L, A = a, C) = E\{Y(a, m)|L, A = a, C\}$ that

$$\begin{aligned} &E\{Y(a, m) - Y(0, m)|C\} - E\{Y(a, 0) - Y(0, 0)|C\} \\ &= E\{Y(a, m) - Y(a, 0)|C\} - E\{Y(0, m) - Y(0, 0)|C\} \\ &= E\{\mu(a, C, L, m; \gamma) - \mu(0, C, L, m; \gamma)|C\} \end{aligned}$$

It follows that the controlled direct effect at arbitrary mediator level m,

$$E\{Y(a,m)-Y(0,m)|C\}=g(a,C;\beta)+E\{\mu(a,C,L,m;\gamma)-\mu(0,C,L,m;\gamma)|C\}$$

must be obtained from a combination of the controlled direct effect at $m=0$ and the mediator effect as obtained from the association model. Here, the term $\mu(a,C,L,m;\gamma)-\mu(0,C,L,m;\gamma)$ depends on L only when triple interactions between a, L and m exist. In the unlikely event that such interactions are considered, an additional model for the distribution of L given C is needed to evaluate the above conditional expectation.

Care must also be exercised to ensure that the second stage model for $E\{Y-\mu(A,C,L,M;\gamma)|A,C\}$ is congenial with the association model (11.15) for $E(Y|M,L,A,C)$ because of the restriction

$$E\{Y(0,0)|C\}=E\{\mu(0,C,L;\gamma)|A=0,C\}$$

In particular, when for the given the association model there is no observed data law $f(L|A=0,C)$ and parameter value in the model for $E\{Y(0,0)|C\}$ such that the above identity holds, then the model is guaranteed to be misspecified. Such problems of incongeniality are not likely to occur in additive models; however, with concern, one can avoid parametric restrictions on the main term $\mu(A,C,L;\gamma)$ in the association model by redefining it to be

$$E(Y|M,L,A,C)=\mu(A,C,L)+\mu(A,C,L,M;\gamma)$$

with $\mu(A,C,L,M;\gamma)$ defined as before, but $\mu(A,C,L)$ an unknown function. Given a parametric model for $f(M|L,A,C)$, the parameter γ can be estimated via G-estimation as the solution to an estimating equation of the form

$$\begin{aligned}0=\sum_i&[d(M_i,L_i,A_i,C_i)-E\{d(M_i,L_i,A_i,C_i)|L_i,A_i,C_i\}]\\&\times[Y_i-\mu(A_i,C_i,L_i,M_i;\gamma)-E\{Y_i-\mu(A_i,C_i,L_i,M_i;\gamma)|L_i,A_i,C_i\}]\end{aligned}$$

where $d(M_i,L_i,A_i,C_i)$ is an arbitrary vector function of the dimension of γ, e.g. $d(M_i,L_i,A_i,C_i)=\partial\mu(A_i,C_i,L_i,M_i;\gamma)/\partial\gamma$. The resulting estimator of γ can subsequently be used in the aforementioned G-estimation procedure.

All of the above ideas extend to multiplicative models. Specifically, suppose that the association model is of the form

$$E(Y|M,L,A,C)=\exp\{\mu(A,C,L;\gamma)+\mu(A,C,L,M;\gamma)\}$$

with $\mu(A,C,L,M;\gamma)$ and $\mu(A,C,L;\gamma)$ defined as before. Consider, furthermore, the following model for the controlled direct effect at $m=0$:

$$\frac{E\{Y(a,0)|C\}}{E\{Y(0,0)|C\}}=\exp\{g(a,C;\beta)\}$$

where $g(A, C; \beta)$ is defined as before. The suggested approach then involves regressing $Y \exp\{-\mu(A, C, L, M; \gamma)\}$ on A and C. This second stage model must now be of the form

$$\begin{aligned} E\left[Y \exp\{-\mu(A, C, L, M; \gamma)\} \mid A, C\right\} &= E\{Y(A, 0) \mid A, C\} \\ &= E\{Y(0, 0) \mid C\} \exp\{g(A, C; \beta)\} \end{aligned}$$

and can again be fitted using standard regression methods for conditional mean models. With concern for incongeniality of the association model and the model for $E\{Y(0, 0) \mid C\}$, the association model must be relaxed to

$$E(Y \mid M, L, A, C) = \exp\{\mu(A, C, L) + \mu(A, C, L, M; \gamma)\}$$

with $\mu(A, C, L)$ an unknown function. Given a parametric model for $f(M \mid L, A, C)$, the parameter γ can then be estimated via G-estimation as the solution to an estimating equation of the form

$$\begin{aligned} 0 = \sum_i & [d(M_i, L_i, A_i, C_i) - E\{d(M_i, L_i, A_i, C_i) \mid L_i, A_i, C_i\}] \\ & \times \left[Y_i \exp\{-\mu(A_i, C_i, L_i, M_i; \gamma)\} - E\{Y_i \exp\{-\mu(A_i, C_i, L_i, M_i; \gamma)\} \mid L_i, A_i, C_i\}\right] \end{aligned}$$

11.3.4 G-estimation for logistic models

The principle behind G-estimation does not extend to logistic controlled direct effect models (i.e. models for the controlled direct effect odds ratio (11.1)), because the logistic link function does not enable transformation of the outcome in order to remove the mediator's effect. This is unfortunate because alternative strategies rely on inverse probability weighting and are henceforth not easily applicable to continuous mediators. In this section, we discuss recent attempts to extend the G-estimation principle to logistic models for dichotomous outcomes.

Vansteelandt (2010) accommodates the difficulties of estimation under logistic structural direct effect models (Robins, 1999b) by parameterizing the controlled direct effect conditional on the exposure and *both* baseline and intermediate confounders

$$\frac{\text{odds}\{Y(a, 0) = 1 \mid L, A = a, C\}}{\text{odds}\{Y(0, 0) = 1 \mid L, A = a, C\}} = \exp\{g(a, C; \beta)\} \tag{11.18}$$

where $g(A, C; \beta)$ is defined as before. Model (11.18) is defined like a logistic structural direct effect model (Robins, 1999a, 1999b), except that it conditions additionally on the intermediate confounder L. This additional conditioning does not compromise the validity of (11.18) as a direct causal effect parameter. This can be seen because the estimand in the left-hand side of (11.18) involves comparing the same subset of the population, $\{A = a, L, C\}$, twice under different interventions. However, a disadvantage of the additional conditioning is that a dependence of the controlled direct effect (11.18) on the intermediate confounder L is not (nonparametrically) identified (Vansteelandt, 2010). While such dependence is not scientifically of interest, ignoring a real dependence could induce a bias. This is, however, not considered to be a major point of concern because the impact of ignoring interactions is typically to result in approximately 'averaged effects', as confirmed in simulation studies (Vansteelandt, 2010). A remaining disadvantage of the additional conditioning is that the controlled direct effect (11.18) cannot be used for making treatment decisions because the subset of the population, $\{A, L, C\}$, is unknown at the onset of a study when L is measured later in time. Vansteelandt (2010) argues that the aim of mediation analyses is typically not to make treatment decisions, but rather to infer mechanism. Furthermore, he shows how estimators of the controlled direct effect (11.18) can be used to infer the unconditional direct effect

$$\frac{\text{odds}\{Y(A, 0) = 1\}}{\text{odds}\{Y(0, 0) = 1\}}$$

whose interpretation does not require knowledge of L.

Estimation of the parameter β indexing the controlled direct effect model (11.18) largely follows the G-estimation principle explained in the previous paragraphs. For simplicity of exposition, let $g(A, C; \beta) = \exp(\beta A)$. First, a standard association model is postulated from which the mediator effect can be inferred, e.g.

$$\text{logit} P(Y = 1|M, L, A, C) = \gamma_0 + \gamma_m M + \gamma_l L + \gamma_a A + \gamma_c C + \beta'_{am} AM \qquad (11.19)$$

Subsequently, both the direct exposure effect and the mediator effect (and hence the indirect exposure effect) are deleted from the outcome via

$$\text{expit}\,(\gamma_0 + \gamma_l L + \gamma_a A + \gamma_c C - \beta A)$$

which equals $Y(0, 0)$ in expectation. Estimation of β_a then happens to ensure that this residual outcome is independent of the exposure, conditional on C, e.g. by solving the following estimating equation:

$$0 = \sum_{i=1}^{n} \{A_i - E(A_i|C_i)\}\,\text{expit}\,(\gamma_0 + \gamma_l L + \gamma_a A + \gamma_c C - \beta_a A)$$

for β_a. There are instances where the direct effect model and the association are incongenial, which may in particular result in estimating equations with no solution. In the previous section, this was remedied by means of association models that merely restrict the mediator effect, but similar semi-parametric regression models do not exist for dichotomous outcomes. When incongeniality forms a potential concern and adjustment for baseline covariates is not necessary, Vansteelandt (2010) recommends nonparametric modelling of the main exposure effect in the association model, e.g. by means of smoothing methods. Concerns for model incongeniality also make extension of this approach to controlled direct effect models for all mediator levels m (rather than just $m = 0$) difficult.

11.3.5 Case-control studies

In the absence of intermediate confounding, the estimation of controlled direct effects from (unmatched) case-control data is possible via standard logistic regression, as detailed in Section 11.3.1 of the present chapter, as noted by VanderWeele and Vansteelandt (2010). We here discuss two approaches that yield approximate results in the presence of intermediate confounding. Both these approaches use the standard logistic association model

$$E(Y|M, L, A, C) = \text{expit}\{\mu(A, C, L; \gamma) + \mu(A, C, L, M; \gamma)\}$$

with $\mu(A, C, L, M; \gamma)$ and $\mu(A, C, L; \gamma)$ defined as before, to estimate the mediator's effect on the outcome.

The first approach, which is proposed in Vansteelandt (2009) and works under the multiplicative model (11.16) for risks, involves approximately removing the direct and indirect effect from the outcome via $Y \exp\{-\mu(A, C, L, M; \gamma) - g(A, C; \beta)\}$. This approximation is accurate when the outcome mean is close to zero, as is often the case in case-control studies, but not otherwise. The unknown parameter β is then chosen so as to make this residual outcome independent of A conditional on C. This is realized by solving an estimating equation of the form

$$\begin{aligned} 0 = \sum_i &[d(A_i, C_i) - E\{d(A_i, C_i)|C_i, Y_i = 0\}]\, Y_i \exp\{-\mu(A_i, C_i, L_i, M_i; \gamma) \\ &- g(A_i, C_i; \beta)\}\,(1 - Y_i) \end{aligned}$$

for an arbitrary, conformable function $d(A_i, C_i)$. The above estimating equation is unbiased at the population level and equals zero in controls (with $Y = 0$), from which its unbiasedness in cases (with $Y = 1$) follows. This is the key to the validity of this approach. However, the conditional expectation $E\{d(A_i, C_i)|C_i\}$ refers to a population expectation, which is rarely known (an exception is in family-based genetic association studies when A_i and C_i refer to offspring and parental genotypes, respectively). Under a rare disease assumption, it can be approximated by averaging within the subgroup of controls.

The second approach, which is proposed in Vansteelandt (2010) and works under the logistic model (11.18) for risks, involves exactly removing the direct and indirect effect from the outcome via $\text{expit}\{\mu(A, C, L; \gamma) - g(A, C; \beta)\}$. The unknown parameter β is then estimated by solving an estimating equation of the form

$$0 = \sum_i [d(A_i, C_i) - E\{d(A_i, C_i)|C_i\}]\, \text{expit}\{\mu(A_i, C_i, L_i; \hat{\gamma}) - g(A_i, C_i; \beta)\}$$

for an arbitrary, conformable function $d(A_i, C_i)$ and $\hat{\gamma}$ a consistent estimator of γ as obtained via standard regression methods. While $\text{expit}\{\mu(A, C, L; \hat{\gamma}) - g(A, C; \beta)\}$ with β the population direct effect does not consistently estimate $E\{Y_i(0, 0)|L_i, A_i, C_i\}$ due to biased estimation of the logistic regression intercept under case-control sampling, Vansteelandt (2010) motivates and observes the resulting bias on the controlled direct effect estimator to be negligible when the population mean is relatively close to zero (i.e. below 10 %). Berzuini *et al.* (2011) investigate the G-estimation of controlled direct genetic effects from matched case-control data and present a decision-theoretic (in the sense of Chapter 4 in this volume) discussion of conditions for the identifiability of such effects; they apply the method to assess whether the FTO gene is associated with myocardial infarction other than via an effect on obesity.

11.3.6 G-estimation for additive hazard models

Structural accelerated failure time models (Robins and Tsiatis, 1991) adopt the structure of multiplicative structural models and thereby enable a G-estimation approach. This is worked out in Robins and Greenland (1994) under the sole assumption (11.2), but has to the best of my knowledge not been considered under assumptions (11.2) and (11.3). Censoring complicates the analysis under these models relative to the analysis of multiplicative structural models for uncensored outcomes. In particular, even when censoring is noninformative w.r.t. the observed survival time, it usually does not stay that way after removing the direct effects of mediator and exposure, because censoring may depend on the exposure itself. In view of this, a re-censoring approach is typically adopted whereby subjects who were originally uncensored may become censored to maintain unbiased estimating equations. This typically implies an important loss of information and can cause nonsmoothness of the estimating functions (Joffe *et al.*, 2011). In this section, we review an alternative approach due to Martinussen *et al.* (2011), which is based on the Aalen additive hazard model and avoids the use of re-censoring.

We denote the survival time by T, the counting process by $N(t) = I(\tilde{T} \leq t)$ and the history spanned by $N(t)$ by $\mathcal{F}_t^N$. We use $T(a, m)$, $N_{(a,m)}(t)$ and $\mathcal{F}_{(a,m),t}^N$ to denote the corresponding counterfactual survival time, counting process and history, respectively, at time t if the exposure were uniformly set to a and the intermediate variable to m. Considering a dichotomous exposure and no baseline covariates C, our initial focus is on the difference in hazard functions

$$\gamma_{A,m}(t)\mathrm{d}t = E\left\{\mathrm{d}N_{(1,m)}(t)|\mathcal{F}_{(1,m),t}\right\} - E\left\{\mathrm{d}N_{(0,m)}(t)|\mathcal{F}_{(0,m),t}\right\}$$

This cannot be interpreted as a direct (causal) effect because the two subgroups $\mathcal{F}_{(1,m),t}$ and $\mathcal{F}_{(0,m),t}$ may not be exchangeable at certain times t. We will therefore define the controlled (cumulated) direct effect of exposure on the survival outcome other than through M as

$$\Gamma_{A,m}(t) = \int_0^t \gamma_{A,m}(s)\,\mathrm{d}s$$

This indeed encodes a controlled direct effect because

$$\exp\left\{-\Gamma_{A,m}(t)\right\} = \frac{P\{T(1,m) > t\}}{P\{T(0,m) > t\}}$$

measures the effect of a unit change in exposure on the survival probability, controlling the mediator at some chosen value m.

Adopting the G-estimation principle requires the following three steps. First, the mediator's effect on the survival time is estimated by fitting the Aalen additive hazard model

$$E\left\{\mathrm{d}N(t)|\mathcal{F}_t, A, M, L\right\} = \left\{\psi_0(t) + \psi_A(t)A + \psi_M(t)M + \psi_L(t)L\right\} R(t)\mathrm{d}t$$

with $R(t)$ the at-risk indicator at time t, using the Aalen least squares estimator (Aalen, 1989). Next, the event time is corrected by removing the mediator effect. This requires correcting the increment $\mathrm{d}N(t)$ as well as the risk set $R(t)$ at each time t. The correction in the increment is achieved by modifying the counting process $\mathrm{d}N(t)$ into

$$\mathrm{d}N(t) - \psi_M(t)M\mathrm{d}t.$$

In expectation, this can be interpreted as the counterfactual intensity at time t if the exposure were as observed and the mediator set back to zero. Indeed, it follows from the Aalen additive hazard model under assumption (11.3) that

$$P\{T(A,0)>t|A,M,L\}=P(T>t|A,M,L)\exp\left\{\int_0^t \psi_M(s)M\mathrm{d}s\right\} \tag{11.20}$$

and henceforth that

$$\begin{aligned}E\left\{\mathrm{d}N_{(A,0)}(t)|\mathcal{F}_{(A,0),t},A,M,L\right\}&=-\frac{\mathrm{d}P\{T(A,0)>t|A,M,L\}/\mathrm{d}t}{P\{T(A,0)>t|A,M,L\}}\\&=E\{\mathrm{d}N(t)-\psi_M(t)M\mathrm{d}t|\mathcal{F}_t,A,M,L\}\end{aligned}$$

The correction in the risk set is achieved by substituting $R(t)$ with

$$R(t)\exp\left\{M\int_0^t \psi_M(s)\mathrm{d}s\right\}$$

which, from (11.20) measures the size of the risk set in expectation if the exposure were as observed and the mediator set back to zero. Using this, we now show that

$$U=\begin{pmatrix}1\\A\end{pmatrix}R(t)\exp\left\{M\int_0^t \psi_M(s)\mathrm{d}s\right\}\{\mathrm{d}N(t)-M\psi_M(t)\mathrm{d}t-\gamma_0(t)\mathrm{d}t-\gamma_A(t)A\mathrm{d}t\}$$

is an unbiased estimating function when

$$E\left\{\mathrm{d}N_{(A,0)}(t)|\mathcal{F}_{(A,0),t},A\right\}=\{\gamma_0(t)+\gamma_A(t)A\}R_{(A,0)}(t)\mathrm{d}t$$

with $R_{(A,0)}(t)$ the counterfactual at-risk indicator corresponding to the filtration $\mathcal{F}_{(A,0),t}$. Indeed, it follows from the previous arguments that $E(U|\mathcal{F}_t,A,M,L)$ equals

$$\begin{pmatrix}1\\A\end{pmatrix}R(t)\exp\left\{M\int_0^t \psi_M(s)\mathrm{d}s\right\}[\psi_0(t)+\psi_A(t)A+\psi_L(t)L-\gamma_0(t)\mathrm{d}t-\gamma_A(t)A\mathrm{d}t]$$

Taking expectations conditional on A, M and L yields (using (11.3))

$$\begin{pmatrix}1\\A\end{pmatrix}P\{T(A,0)>t|A,L\}[\psi_0(t)+\psi_A(t)A+\psi_L(t)L-\gamma_0(t)\mathrm{d}t-\gamma_A(t)A\mathrm{d}t]$$

which, in expectation, is identical to

$$\begin{pmatrix}1\\A\end{pmatrix}R_{(A,0)}(t)\left\{\mathrm{d}N_{(A,0)}(t)-\gamma_0(t)\mathrm{d}t-\gamma_A(t)A\mathrm{d}t\right\}.$$

The unbiasedness is now immediate from the definition of $\gamma_0(t)$ and $\gamma_A(t)$. Martinussen *et al.* (2011) show that the solution to this unbiased estimating equation yields a closed-form estimator. One of its attractions is that censoring is naturally dealt with whenever censoring is independent of the event time conditional on exposure, intermediate variable and confounders, as well as conditional on the exposure alone. Although the latter assumption would fail if censoring were dependent upon the intermediate variable or confounders, it can be easily remedied via inverse probability of censoring weighting.

11.4 Estimation of natural direct and indirect effects

Under the identification conditions (11.5) of Pearl (2001) or (11.6) of Petersen *et al.* (2006), natural direct effects can be identified as an average of controlled direct effects (see (11.7)). Total indirect effects can be identified as the difference[2] between the total effect and the natural direct effect, or directly via (11.8). Based on this, VanderWeele and Vansteelandt (2009, 2010) derive (approximate) closed-form expressions for natural/total direct and indirect effects when the mediator is normally distributed with mean that is linear in the exposure and baseline covariates, and constant variance, and when the outcome is linear in the exposure, mediator and baseline covariates on the identity or logit scale, possibly including exposure–mediator interactions. Their expressions extend the famous Baron–Kenny expressions (Baron and Kenny, 1986) and simplify those given in (Chapter 12 in this volume by Pearl). VanderWeele (2009) proposes a similar approach based on marginal structural models for the outcome and mediator, which we do not consider here because the identification condition (11.5) rules out the presence of intermediate confounders, in which case the usefulness of marginal structural models becomes more limiting; e.g. it cannot easily handle continuous mediators. Imai *et al.* (2010a) develop a Monte Carlo simulation approach to make the integration in (11.7) and (11.8) feasible and make it available in the R-package `mediation`. Their approach can handle more general models for the outcome and mediator, but has the disadvantage that, for continuous exposures, it may result in natural/total direct and indirect effects, which are complicated functions of the exposure a and reference mediator level m, which can make it less attractive for reporting.

All foregoing approaches infer natural/total direct and indirect effects indirectly by combining standard models for the outcome and mediator. A disadvantage is that parsimonious models for the outcome and mediator may not translate into parsimonious models of natural/total direct and indirect effects. This would not only make the results difficult to communicate, as in Chapter 12 by Pearl, where these effects are displayed as functions of the exposure, but also makes it challenging to test uniformly for the presence of a direct or indirect effect. Van der Laan and Petersen (2008) therefore propose to model the natural direct effect directly. Assume, for instance, that

$$E\,\{Y(a, M(0)) - Y(0, M(0))|C\} = g(a, C; \beta)$$

for a known smooth function $g(.)$ of an unknown finite-dimensional parameter β, which satisfies $g(0, C; \beta) = 0$, e.g. $g(a, C) = (\beta_a + \beta_c C)a$. Then van der Laan and Petersen (2008)

[2] In Pearl (Chapter 12 in this volume) it is suggested that the indirect effect cannot always be calculated as the difference between the total effect and the natural direct effect. Note that this is because Pearl does not use the definition of total indirect effect.

show how inverse probability weighting can be used to estimate β. In particular, they use inverse probability weighting by

$$\frac{f(M|A=0,C)}{f(M|A,C)}$$

to reconstruct the composite counterfactual $Y(a, M(0))$ and further subtract the natural direct effect $g(a, C; \beta)$ to arrive at $Y(0, M(0))$, which must be independent of A, given C, under assumption (11.2). The following estimating equation, from which β can be solved, is designed to make that conditional independence happen:

$$0 = \sum_{i=1}^{n} [d(A_i, C_i) - E\{d(A_i, C_i)|C_i\}] \frac{f(M_i|A_i=0, C_i)}{f(M_i|A_i, C_i)} \times [Y_i - g(A_i, C_i; \beta) - E\{Y_i - g(A_i, C_i; \beta)|C_i\}]$$

where $d(A_i, C_i)$ is an arbitrary, conformable index function. The total indirect effect can then be estimated as the difference between the total effect and the natural direct effect.

11.5 Discussion

In this chapter, we have mainly reviewed approaches for dealing with intermediate (time-dependent) confounding in the estimation of controlled direct effects. In addition, we have briefly reviewed the estimation of natural direct and indirect effects. In both cases, the use of inverse probability weighting entails relatively simple and generic approaches, whose practical implementation may nevertheless be problematic when the inverse weights are highly variable, as in the case of continuous mediators. We have great hopes of doubly robust estimators, and in particular of recent developments that have focused on improving the performance of doubly robust estimators in other settings (see, for example, van der Laan and Petersen, 2008, and Cao *et al.*, 2009).

A remaining limitation with all current approaches for estimating natural/total direct and indirect effects is that the identification condition (11.5) of Petersen *et al.* (2006) essentially rules out the presence of intermediate confounders. While the identification condition (11.6) of Petersen *et al.* (2006) in principle enables natural direct effects to be calculated by averaging controlled direct effects, even in the presence of intermediate confounding, Vansteelandt and VanderWeele (2011) argue that it is rather implausible for their condition to be satisfied in such settings. They propose alternative definitions of the direct and indirect effect on the exposed, which allow for effect decomposition even in the presence of intermediate confounding. Estimation relies on G-computation and remains to be extended to enable a more direct estimation approach.

Acknowledgements

The author is grateful to Maarten Bekaert (Ghent University, Belgium) for valuable help when writing the R-function `seqg` and acknowledges support from IAP Research Network Grant P06/03 from the Belgian government (Belgian Science Policy).

References

Aalen, O. (1989) A linear-regression model for the analysis of life times. *Statistics in Medicine*, **8**, 907–925.

Baron, R. and Kenny, D. (1986) The moderator–mediator variable distinction in social psychological research: conceptual, strategic, and statistical considerations. *Journal of Personality and Social Psychology*, **51**, 1173–1182.

Berzuini, C., Vansteelandt, S., Foco, L., Pastorino, R. and Bernardinelli, L. (2011) Direct genetic effects and their estimation from matched case-control data. arxiv: 1112.5130v1 [stat.ME] (submitted 21 December 2011).

Cao, W., Tsiatis, A. and Davidian, M. (2009) Improving efficiency and robustness of the doubly robust estimator for a population mean with incomplete data. *Biometrika*, **96**, 723–734.

Cole, S. and Hernán, M. (2002) Fallibility in estimating direct effects. *International Journal of Epidemiology*, **31**, 163–165.

Elliott, M.R., Raghunathan, T.E. and Li, Y. (2010) Bayesian inference for causal mediation effects using principal stratification with dichotomous mediators and outcomes. *Biostatistics*, **11**, 353–372.

Emsley, R., Dunn, G. and White, I.R. (2010) Mediation and moderation of treatment effects in randomised controlled trials of complex interventions. *Statistical Methods in Medical Research*, **19**, 237–270.

Fosen, J., Ferkingstad, E., Borgan, O. and Aalen, O.O. (2006) Dynamic path analysis – a new approach to analyzing time-dependent covariates. *Lifetime Data Analysis*, **12**, 143–167.

Frangakis, C. and Rubin, D. (2002) Principal stratification in causal inference. *Biometrics*, **58**, 21–29.

Gallop, R., Small, D.S., Lin, J.Y., Elliott, M.R., Joffe, M. and Ten Have, T.R. (2009) Mediation analysis with principal stratification. *Statistics in Medicine*, **28**, 1108–1130.

Goetgeluk, S., Vansteelandt, S. and Goetghebeur, E. (2008) Estimation of controlled direct effects. *Journal of the Royal Statistical Society, Series B*, **70**, 1049–1066.

Hafeman, D.M. and VanderWeele, T.J. (2011) Alternative assumptions for the identification of direct and indirect effects. *Epidemiology*, **22**, 753–764.

Imai, K., Keele, L. and Tingley, D. (2010a) A general approach to causal mediation analysis. *Psychological Methods*, **15**, 309–334.

Imai, K., Keele, L. and Yamamoto, T. (2010b) Identification, inference and sensitivity analysis for causal mediation effects. *Statistical Science*, **25**, 51–71.

Joffe, M. and Greene, T. (2009) Related causal frameworks for surrogate outcomes. *Biometrics*, **65**, 530–538.

Joffe, M., Byrne, C. and Colditz, G. (2001) Postmenopausal hormone use, screening, and breast cancer: characterization and control of a bias. *Epidemiology*, **12**, 429–438.

Joffe, M., Yang, W. and Feldman, H. (2011) G-estimation and artificial censoring: problems, challenges, and applications, in *Biometrics*. Blackwell Publishing Inc. DOI: 10.1111/j.1541-0420.2011.01656.x

MacKinnon, D. (2008) *An Introduction to Statistical Mediation Analysis*. New York: Lawrence Erlbaum Associates.

Martinussen, T., Vansteelandt, S., Gerster, M. and Hjelmborg, J. (2011) Estimation of direct effects for survival data using the Aalen additive hazards model. *Journal of the Royal Statistical Society, Series B*, **73**, 773–788.

Pearl, J. (2001) Direct and indirect effects, in *Proceedings of the Seventeenth Conference on Uncertainty and Artificial Intelligence*. San Francisco: Morgan Kaufmann, 411–420.

Pearl, J. (2009) *Causality: Models, Reasoning and Inference*. Cambridge: Cambridge University Press.

Pearl, J. (2011) Principal stratification – a goal or a tool? *The International Journal of Biostatistics*, **7**, Article 20.

Petersen, M., Sinisi, S. and van der Laan, M. (2006) Estimation of direct causal effects. *Epidemiology*, **17**, 276–284.

Robins, J. (1986) A new approach to causal inference in mortality studies with a sustained exposure period – application to control of the healthy worker survivor effect. *Mathematical Modelling*, **7**, 1393–1512.

Robins, J.M. (1997) Causal inference from complex longitudinal data, in *Latent Variable Modelling and Applications to Causality (Los Angeles, CA, 1994)*, vol. 120 of Lecture Notes in Statistics. New York: Springer, 69–117.

Robins, J. (1999a) Marginal structural models versus structural nested models as tools for causal inference, in *Statistical Models in Epidemiology: the environment and clinical trials* (eds M. Halloran and D. Berry) New York: Springer-Verlag, 95–134.

Robins, J. (1999b) Testing and estimation of direct effects by reparameterizing directed acyclic graphs with structural nested models, in *Computation, Causation, and Discovery* (eds C. Glymour and G. Cooper), Menlo Park, CA/Cambridge, MA: AAAI Press/The MIT Press, 349–405.

Robins, J.M. (2003) Semantics of causal dag models and the identification of direct and indirect effects, in *Highly Structured Stochastic Systems* (eds, P. Green, N. Hjort and S. Richardson). New York: Oxford University Press, 70–81.

Robins, J. and Greenland, S. (1994) Adjusting for differential rates of prophylaxis therapy for PCP in high-dose versus low-dose AZT treatment arms in an aids randomized trial. *Journal of the American Statistical Association*, **89**, 737–749.

Robins, J.M. and Greenland, S. (1992) Identifiability and exchangeability for direct and indirect effects. *Epidemiology*, **3**, 143–155.

Robins, J. and Tsiatis, A. (1991) Correcting for non-compliance in randomized trials using rank-preserving structural failure time models. *Communications in Statistics*, **20**, 2609–2631.

Robins, J., Hernán, M. and Brumback, B. (2000) Marginal structural models and causal inference in epidemiology. *Epidemiology*, **11**, 550–560.

Robins, J., Rotnitzky, A. and Vansteelandt, S. (2007) Principal stratification designs to estimate input data missing due to death – discussion. *Biometrics*, **63**, 650–653.

Rosenblum, M., Jewell, N.P., van der Laan, M., Shiboski, S., van der Straten, A. and Padian, N. (2009) Analysing direct effects in randomized trials with secondary interventions: an application to human immunodeficiency virus prevention trials. *Journal of the Royal Statistical Society, Series A*, **172**, 443–465.

Rubin, D. (2004) Direct and indirect causal effects via potential outcomes. *Scandinavian Journal of Statistics*, **31**, 161–170.

Sjölander, A. (2009) Bounds on natural direct effects in the presence of confounded intermediate variables. *Statistics in Medicine*, **28**, 558–571.

Sobel, M.E. (2008) Identification of causal parameters in randomized studies with mediating variables. *Journal of Educational and Behavioral Statistics*, **33**, 230–251.

Ten Have, T. and Joffe, M. (2011) A review of causal estimation of effects in mediation analyses. *Statistical Methods in Medical Research*, **21**, 77–107.

Ten Have, T., Joffe, M., Lynch, K., Brown, G., Maisto, S. and Beck, A. (2007) Causal mediation analyses with rank preserving models. *Biometrics*, **63**, 926–934.

van der Laan, M.J. and Petersen, M.L. (2008) Direct effect models. *The International Journal of Biostatistics*, **4** (1), Article 23.

VanderWeele, T.J. (2008) Simple relations between principal stratification and direct and indirect effects. *Statistics and Probability Letters*, **78**, 2957–2962.

VanderWeele, T. (2009a) Mediation and mechanism. *European Journal of Epidemiology*, **24**, 217–224.

VanderWeele, T.J. (2009b) Marginal structural models for the estimation of direct and indirect effects. *Epidemiology*, **20**, 18–26.

VanderWeele, T.J. (2010) Bias formulas for sensitivity analysis for direct and indirect effects. *Epidemiology*, **21**, 540–551.

VanderWeele, T.J. and Vansteelandt, S. (2009) Conceptual issues concerning mediation, interventions and composition. *Statistics and Its Interface*, **2**, 457–468.

VanderWeele, T.J. and Vansteelandt, S. (2010) Odds ratios for mediation analysis for a dichotomous outcome. *American Journal of Epidemiology*, **172**, 1339–1348.

Vansteelandt, S. (2009) Estimating direct effects in cohort and case-control studies. *Epidemiology*, **20**, 851–860.

Vansteelandt, S. (2010) Estimation of controlled direct effects on a dichotomous outcome using logistic structural direct effect models. *Biometrika*, **97**, 921–934.

Vansteelandt, S. and VanderWeele, T. (2011) Natural direct and indirect effects on the exposed: effect decomposition under weaker assumptions. Technical Report, Ghent University.

Vansteelandt, S., Goetgeluk, S., Lutz, S., Waldman, I., Lyon, H., Schadt, E., Weiss, S. and Lange, C. (2009) On the adjustment for covariates in genetic association analysis: a novel, simple principle to infer causality. *Genetic Epidemiology*, **33**, 394–405.

12

The mediation formula: A guide to the assessment of causal pathways in nonlinear models

Judea Pearl
Computer Science Department, University of California, Los Angeles, California, USA

12.1 Mediation: Direct and indirect effects

12.1.1 Direct versus total effects

The target of many empirical studies in the social, behavioral, and health sciences is the *causal effect*, here denoted $P(y|do(x))$, which measures the *total* effect of a manipulated variable (or a set of variables) X on a response variable Y. In many cases, this quantity does not adequately represent the target of investigation and attention is focused instead on the *direct* effect of X on Y. The term 'direct effect' is meant to quantify an effect that is not mediated by other variables in the model or, more accurately, the sensitivity of Y to changes in X while all other factors in the analysis are held fixed. Naturally, holding those factors fixed would sever all causal paths from X to Y with the exception of the direct link $X \rightarrow Y$, which is not intercepted by any intermediaries.

A classical example of the ubiquity of direct effects involves legal disputes over race or sex discrimination in hiring. Here, neither the effect of sex or race on applicants' qualifications nor the effect of qualification on hiring are targets of litigation. Rather, defendants must prove that sex and race do not *directly* influence hiring decisions, whatever indirect effects they might have on hiring by way of an applicant qualification.

Causality: Statistical Perspectives and Applications, First Edition. Edited by Carlo Berzuini, Philip Dawid and Luisa Bernardinelli.

From a policy making viewpoint, an investigator may be interested in decomposing effects to quantify the extent to which weakening or strengthening specific causal pathways would impact the overall effect of X on Y. For example, the extent to which minimizing racial disparity in education would reduce racial disparity in earnings. Taking a health-related example, the extent to which efforts to eliminate side-effects of a given treatment are likely to weaken or enhance the efficacy of that treatment. More often, however, the decomposition of effects into their direct and indirect components carries theoretical scientific importance, for it tells us 'how nature works' and, therefore, enables us to predict behavior under a rich variety of conditions and interventions.

Structural equation models provide a natural language for analyzing path-specific effects and, indeed, considerable literature on direct, indirect, and total effects has been authored by SEM researchers (Alwin and Hauser, 1975; Graff and Schmidt, 1982; Sobel, 1987; Bollen, 1989), for both recursive and nonrecursive models. This analysis usually involves sums of powers of coefficient matrices, where each matrix represents the path coefficients associated with the structural equations.

Yet despite its ubiquity, the analysis of mediation has long been a thorny issue in the empirical sciences (Judd and Kenny, 1981; Baron and Kenny, 1986; Muller *et al.*, 2005; Shrout and Bolger, 2002; MacKinnon *et al.*, 2007a) primarily because structural equation modeling in those sciences were deeply entrenched in linear analysis, where the distinction between causal parameters and their regressional interpretations can easily be conflated (as in Holland, 1995; Sobel, 2008). The difficulties were further amplified in nonlinear models, where sums and products are no longer applicable. As demands grew to tackle problems involving binary and categorical variables, researchers could no longer define direct and indirect effects in terms of structural or regressional coefficients, and all attempts to extend the linear paradigms of effect decomposition to nonlinear systems produced distorted results (MacKinnon *et al.*, 2007b). These difficulties have accentuated the need to redefine and derive causal effects from first principles, uncommitted to distributional assumptions or a particular parametric form of the equations. The structural methodology presented in this paper adheres to this philosophy and it has produced indeed a principled solution to the mediation problem, based on the counterfactual reading of structural equations (Balke and Pearl, 1994a, 1994b; Pearl, 2009b, Chapter 7). The following subsections summarize the method and its solution, while Section 12.2 introduces the mediation formula, exemplifies its behavior, and demonstrates its usage in simple examples, including linear, quasi-linear, logistic, probit and nonparametric models. Finally, Section 12.4 compares the mediation formula to other methods proposed for effect decomposition and explains the difficulties that those methods have encountered in defining and assessing mediated effects.

12.1.2 Controlled direct effects

A major impediment to progress in mediation analysis has been the lack of a notational facility for expressing the key notion of 'holding the mediating variables fixed' in the definition of direct effect. Clearly, this notion must be interpreted as (hypothetically) setting the intermediate variables to constants by physical intervention, not by analytical means such as selection, regression conditioning, stratification matching, or adjustment. For example, consider the simple mediation models of Figure 12.1(a), where the error terms (not shown explicitly) are assumed to be mutually independent. To measure the direct effect of X on Y it is sufficient to measure their association conditioned on the mediator Z. In Figure 12.1(b), however, where

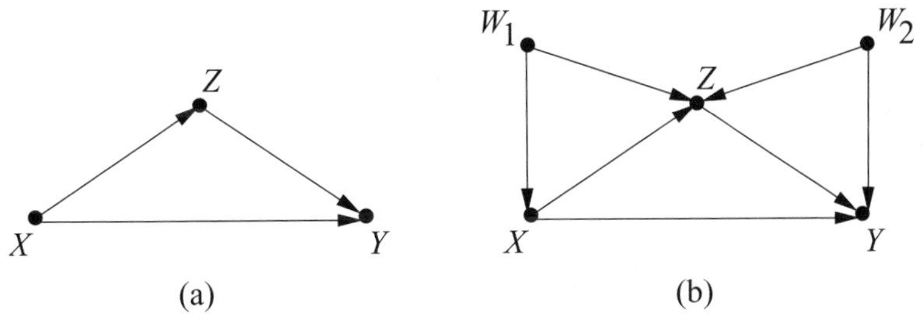

Figure 12.1 *(a) A generic model depicting mediation through Z with (a) no confounders and (b) two confounders, W_1 and W_2.*

the error terms are dependent, it will not be sufficient to measure the association between X and Y for a given level of Z because, by conditioning on the mediator Z, we create spurious associations between X and Y through W_2, even when there is no direct effect of X on Y (Pearl, 1998; Cole and Hernán, 2002).

Using the $do(x)$ notation enables us to express correctly the notion of 'holding Z fixed' and obtain a simple definition of the *controlled direct effect* of the transition from $X = x$ to $X = x'$ (Pearl, 2009b, p. 128):

$$CDE \triangleq E(Y|do(x'), do(z)) - E(Y|do(x), do(z)) \tag{12.1}$$

or, equivalently, using counterfactual notation:

$$CDE \triangleq E(Y_{x'z}) - E(Y_{xz})$$

where Z is the set of all mediating variables.[1] Readers can easily verify that, in linear systems, the controlled direct effect reduces to the path coefficient of the link $X \rightarrow Y$ regardless of whether confounders are present (as in Figure 12.1(b)) and regardless of whether the error terms are correlated or not.

This separates the task of definition from that of identification, and thus circumvents many pitfalls in this area of research (Pearl, 2009b). The identification of *CDE* would depend, of course, on whether confounders are present and whether they can be neutralized by adjustment, but these do not alter its definition. Nor should trepidation about infeasibility of the action $do(gender = male)$ enter the definitional phase of the study. Definitions apply to symbolic models, not to human biology.[2]

---

[1] Readers not familiar with this notation can consult Pearl (2009a, 2009b, 2010). Conceptually, $P(y|do(x))$ stands for the probability of $Y = y$ in a randomized experiment where the treatment level is set to $X = x$, while $Y_x(u)$ stands for the value that Y would attain in unit u, had X been x. Formally, $P(y|do(x))$ and $Y_x(u)$ are defined, respectively, as the probability and value of variable Y in a modified structural model, in which the equation for X is replaced by a constant $X = x$). This model encodes a system of natural laws that accommodates counterfactuals with nonmanipulable antecedants (e.g. race and sex), and is immune to the conceptual difficulties elaborated in Chapter 9 in this volume by Vanderweele and Hernán.

[2] In reality, it is the employer's perception of an applicant's gender and his/her assessment of gender–job compatibility that render gender a 'cause' of hiring – manipulation of gender is not needed.

Graphical identification conditions for expressions of the type $E(Y|do(x), do(z_1), do(z_2), \ldots, do(z_k))$ in the presence of unmeasured confounders were derived by Pearl and Robins (1995) and invoke sequential application of the back-door condition (Pearl, 2009a, pp. 252–254), which offers significant improvements over G-computation (Robins, 1986). Tian and Shpitser (2010) have further derived a necessary and sufficient condition for this task, and thus resolved the identification problem for controlled direct effects (see Equation (12.1)).

12.1.3 Natural direct effects

In linear systems, the direct effect is fully specified by the path coefficient attached to the link from X to Y; therefore, the direct effect is independent of the values at which we hold Z. In nonlinear systems, those values would, in general, modify the effect of X on Y and thus should be chosen carefully to represent the target policy under analysis. For example, it is not uncommon to find employers who prefer males for the high-paying jobs (i.e., high z) and females for low-paying jobs (low z). Focusing on one of these values of Z or averaging over all values would not capture the underlying pattern of discrimination.

When the direct effect is sensitive to the levels at which we hold Z, it is often more meaningful to define the direct effect relative to some 'natural' base-line level that may vary from individual to individual, and represents the level of Z just before the change in X. Conceptually, we can define the natural direct effect $DE_{x,x'}(Y)$ as the expected change in Y induced by changing X from x to x' while keeping all mediating factors constant at whatever value they *would have obtained* under $do(x)$. This hypothetical change, which Robins and Greenland (1992) conceived and called 'pure' and Pearl (2001) formalized and analyzed under the rubric 'natural', mirrors what lawmakers instruct us to consider in race or sex discrimination cases: 'The central question in any employment-discrimination case is whether the employer would have taken the same action had the employee been of a different race (age, sex, religion, national origin, etc.) and everything else had been the same.' (In *Carson vs. Bethlehem Steel Corp.*, 70 FEP Cases 921, 7th Cir. 1996.)

Thus, whereas the controlled direct effect measures the effect of X on Y while holding Z fixed, at a uniform level (z) for all units,[3] the natural direct effect allows z to vary from individual to individual and be fixed at the level that each individual held naturally, just before the change in X.

Pearl (2001) gave the following definition for the 'natural direct effect':

$$DE_{x,x'}(Y) = E(Y_{x',Z_x}) - E(Y_x) \tag{12.2}$$

Here, Y_{x',Z_x} represents the value that Y would attain under the operation of setting X to x' and, simultaneously, setting Z to whatever value it would have obtained under the setting $X = x$. For example, if one were to estimate that the natural direct effect of gender on hiring equals 20 % of the total effect, one can infer that 20 % of the current gender-related disparity in hiring can be eliminated by making hiring decision gender-blind, while keeping applicants' qualifications at their current values (which may be gender dependent).

We see from (12.2) that $DE_{x,x'}(Y)$, the natural direct effect of the transition from x to x', involves probabilities of *nested counterfactuals* and cannot be written in terms of the $do(x)$

[3] In the hiring discrimination example, this would amount, for example, to testing gender bias while marking all application forms with the same level of schooling and other skill-defining attributes.

operator. Therefore, the natural direct effect cannot in general be identified or estimated, even with the help of ideal, controlled experiments – a point emphasized in Robins and Greenland (1992).[4] However, aided by the formal definition of Equation (12.2) and the notational power of nested counterfactuals, Pearl (2001) was nevertheless able to derive conditions under which the natural direct effect can be expressed in terms of the $do(x)$ operator, implying identifiability from controlled experiments. For example, if a set W exists that deconfounds Y and Z, the natural direct effect can be reduced to[5]

$$DE_{x,x'}(Y) = \sum_{z,w}[E(Y|do(x',z),w) - E(Y|do(x,z),w)]P(z|do(x),w)P(w) \tag{12.3}$$

The intuition is simple; the W-specific natural direct effect is the weighted average of the controlled direct effect, using the causal effect $P(z|do(x),w)$ as a weighing function.[6]

In particular, it can be shown (Pearl, 2001) that the natural direct effect is identifiable in Markovian models (i.e., recursive equations with no unobserved confounders) where each *do*-expression can be reduced to a '*do*-free' expression by covariate adjustments (Pearl, 2009a) and then estimated by regression. For example, for the model in Figure 12.1(b), $DE_{x,x'}(Y)$ reduces to

$$DE_{x,x'}(Y) = \sum_{z}\sum_{w_2} P(w_2)[E(Y|x',z,w_2)) - E(Y|x,z,w_2))]\sum_{w_1} P(z|x,w_1,w_2)P(w_1) \tag{12.4}$$

while for the confounding-free model of Figure 12.1(a) we have

$$DE_{x,x'}(Y) = \sum_{z}[E(Y|x',z) - E(Y|x,z)]P(z|x) \tag{12.5}$$

Both (12.4) and (12.5) can be estimated by a regression.

When Z consists of multiple interconnected mediators, affected by an intricate network of observed and unobserved confounders, the adjustment illustrated in Equation (12.4) must be handled with some care. Theorems 1 and 2 of Pearl (2001) can then be used to reduce $DE_{x,x'}(Y)$ to a *do*-expression similar to (12.3) (see footnote 5). Once reduced, the machinery of *do*-calculus (Pearl, 1995) can be invoked, and the methods of Pearl and Robins (1995), Tian and Shpitser (2010), and Shpitser and VanderWeele (2011) can select the proper set of covariates and reduce the natural direct effect (12.3) to an expression estimable by regression, whenever such reduction is feasible. For example, if in Figure 12.1(b) W_1 is

[4] The reason being that we cannot rerun history and test individuals' response both before and after the intervention. Robins (2003) elaborates on the differences between the assumptions made in Pearl (2001) and the weaker assumptions made in Robins and Greenland (1992), which prevented the latter from identifying natural effects even in the simple case of no-confounding. (Figure 12.1(a)).

[5] The key condition for this reduction is the existence of a set W of covariates satisfying $Y_{xz}Z_{x'} \perp\!\!\!\perp W$, which simply states that W blocks all back-door paths from Z to Y (see Pearl, 2009a, p. 101). More refined counterfactual conditions for identification are derived in Petersen *et al.* (2006), Imai *et al.* (2010c), and Robins and Richardson (2011). However, none matches the clarity of the back-door condition above, and all are equivalent in the graphical language of nonparametric structural equations (Shpitser and VanderWeele, 2011).

[6] Throughout this paper we will use summation signs with the understanding that integrals should be used whenever the summed variables are continuous.

unobserved and another observed covariate, W_3, mediates the path $X \rightarrow Z$, the last term of Equation (12.5), $P(z|do(x), w_2)$, would then be identifiable through the front-door formula (Pearl, 1995, 2009a), thus rendering $DE_{x,x'}(Y)$ estimable by regression. This demonstrates that neither 'ignorability' (Rosenbaum and Rubin, 1983) nor 'sequential ignorability' (Imai *et al.*, 2010a, 2010c) is necessary for securing the identification of direct effects; transparent graph-based criteria are sufficient for determining when and how confounding can be controlled (Pearl, 2012). See Pearl (2009a, pp. 341–344) for a graphical interpretation of 'ignorability' assumptions.

12.1.4 Indirect effects

Remarkably, the definition of the natural direct effect (12.2) can be turned around and provide an operational definition for the *indirect effect* – a concept shrouded in mystery and controversy, because it is impossible, by controlling any of the variables in the model, to disable the direct link from X to Y so as to let X influence Y solely via indirect paths.

The *natural indirect effect*, *IE*, of the transition from x to x' is defined as the expected change in Y affected by holding X constant, at $X = x$, and changing Z to whatever value it would have attained had X been set to $X = x'$. Formally, this reads (Pearl, 2001)

$$IE_{x,x'}(Y) \triangleq E[(Y_{x,Z_{x'}}) - E(Y_x)] \tag{12.6}$$

which is almost identical to the direct effect (Equation (12.2)) save for exchanging x and x' in the first term.

Invoking the same conditions that led to the experimental identification of the direct effect, Equation (12.3), we obtain a parallel formula for the indirect effect:

$$IE_{x,x'}(Y) = \sum_{z,w} E(Y|do(x,z),w)[P(z|do(x'),w) - P(z|do(x),w)] \tag{12.7}$$

The intuition here is somewhat different and represents a nonlinear version of the 'product-of-coefficients' strategy in linear models (MacKinnon, 2008); the $E(Y|do(x,z),w)$ term encodes the effect of Z on Y for fixed $X = x$ and $W = w$, while $[P(z|do(x'),w) - P(z|do(x),w)]$ encodes the effect of X on Z. We see that what was a simple product-of-coefficients in linear models turns into a convolution type operation, involving all values of Z.

In nonexperimental studies, the *do*-operator needs to be reduced to a regression type expression using covariate adjustment or instrumental variable methods. For example, for the model in Figure 12.1(b), Equation (12.7) reads

$$IE_{x,x'}(Y) = \sum_{z}\sum_{w_2} P(w_2)E(Y|x,z,w_2)\sum_{w_1}[P(z|x',w_1,w_2) - P(z|x,w_1,w_2)]P(w_1) \tag{12.8}$$

while for the confounding-free model of Figure 12.1(a) we have

$$IE_{x,x'}(Y) = \sum_{z} E(Y|x,z)[P(z|x') - P(z|x)] \tag{12.9}$$

which, like Equation (12.5) can be estimated by a two-step regression.

12.1.5 Effect decomposition

Not surprisingly, owing to the nonlinear nature of the model, the relationship between the total, direct, and indirect effects is nonadditive. Indeed, it can be shown that, in general, the total effect *TE* of a transition is equal to the *difference* between the direct effect of that transition and the indirect effect of the *reverse* transition. Formally,

$$TE_{x,x'}(Y) \triangleq E(Y_{x'} - Y_x) = DE_{x,x'}(Y) - IE_{x',x}(Y) \tag{12.10}$$

In linear systems, where reversal of transitions amounts to negating the signs of their effects, we have the standard additive formula

$$TE_{x,x'}(Y) = DE_{x,x'}(Y) + IE_{x,x'}(Y) \tag{12.11}$$

Since each term above is based on an independent operational definition, this equality constitutes a formal justification for the additive formula used routinely in linear systems.[7]

Note that, although it cannot in general be expressed in *do*-notation, the indirect effect has clear policy-making implications. For example, in the hiring discrimination context, a policy maker may be interested in predicting the gender mix in the work force if gender bias is eliminated and all applicants are treated equally – say, in the same way that males are currently treated. This quantity will be given by the indirect effect of gender on hiring, mediated by factors such as education and aptitude, which may be gender-dependent.

More generally, a policy maker may be interested in the effect of issuing a directive to a select set of subordinate employees or in carefully selecting the routing of messages in a network of interacting agents. Such applications motivate the analysis of *path-specific effects*, that is, the effect of X on Y through a selected set of paths (Avin *et al.*, 2005), with all other paths *deactivated*. The operation of disabling a path can be expressed in nested counterfactual notation, as in Equations (12.2) and (12.6).

In all these cases, the policy intervention invokes the selection of signals to be sensed, rather than variables to be fixed. Pearl (2001, 2009a, p. 361) has suggested therefore that *signal sensing* is more fundamental to the notion of causation than *manipulation*, the latter being but a crude way of stimulating the former in an experimental setup. The mantra 'no causation without manipulation' must be rejected. (See Pearl, 2009a, Section 11.4.5.)

12.2 The mediation formula: A simple solution to a thorny problem

12.2.1 Mediation in nonparametric models

This subsection demonstrates how the solution provided in Equations (12.5) and (12.9) can be applied in assessing mediation effects in nonparametric, possibly nonlinear models. We

[7] Some authors (e.g., VanderWeele, 2009, and Vansteelandt, Chapter 11 in this volume) take Equation (12.11) as the definition of indirect effect (see footnote 8), which ensures additivity by definition, but presents a problem of interpretation; the resulting indirect effect, aside from being redundant, does not represent the same transition, from x to x', as do the total and direct effects. This prevents us from comparing the effect attributable to mediating paths with that attributable to unmediated paths, under the same conditions.

will use the simple mediation model of Figure 12.1(a), where all error terms (not shown explicitly) are assumed to be mutually independent, with the understanding that adjustment for appropriate sets of covariates W may be necessary to achieve this independence (as in (12.4) and (12.8)) and that integrals should replace summations when dealing with continuous variables (Imai *et al.*, 2010c).

Combining (12.5), (12.9), and (12.10), the expressions for the direct (DE), indirect (IE), and total (TE) effects, IE becomes:

$$DE_{x,x'}(Y) = \sum_z [E(Y|x', z) - E(Y|x, z)]P(z|x) \tag{12.12}$$

$$IE_{x,x'}(Y) = \sum_z E(Y|x, z)[P(z|x') - P(z|x)] \tag{12.13}$$

$$TE_{x,x'}(Y) = E(Y|x') - E(Y|x) \tag{12.14}$$

These three equations provide general formulas for mediation effects, applicable to any nonlinear system, any distribution, and any type of variables. Moreover, the formulas are readily estimable by regression. Owing to their generality and ubiquity, I have referred to these expressions as the 'mediation formula' (Pearl, 2009b).

The mediation formula (12.13) represents the average increase in the outcome Y that the transition from $X = x$ to $X = x'$ is expected to produce without any direct effect of X on Y. Though based on solid causal principles, it embodies no causal assumption other than the generic mediation structure of Figure 12.1(a). When the outcome Y is binary (e.g., recovery or hiring) the ratio $(1 - IE/TE)$ represents the fraction of responding individuals who owe their response to direct paths, while $(1 - DE/TE)$ represents the fraction who owe their response to Z-mediated paths.[8]

The mediation formula tells us that IE depends only on the expectation of the counterfactual Y_{xz}, not on its functional form $f_Y(x, z, u_Y)$ or its distribution $P(Y_{xz} = y)$. It therefore calls for a two-step regression which, in principle, can be performed nonparametrically. In the first step we regress Y on X and Z, and obtain the estimate

$$g(x, z) \triangleq E(Y|x, z)$$

for every (x, z) cell. In the second step we fix x and estimate the conditional expectation of $g(x,z)$ with respect to z, conditional on $X = x'$ and $X = x$, respectively, and take the difference:

$$IE_{x,x'}(Y) = E_{Z|x'}[g(x, z)] - E_{Z|x}[g(x, z)]$$

Nonparametric estimation is not always practical. When Z consists of a vector of several mediators, the dimensionality of the problem might prohibit the estimation of $E(Y|x, z)$ for every (x, z) cell, and the need arises to use parametric approximation. We can then choose

[8] For simplicity and clarity, we remove the subscripts from TE, DE, and IE, whenever no ubiquity arises. Robins (2003) and Hafeman and Schwartz (2009) refer to $TE - IE$ and $TE - DE$ as 'total direct' and 'total indirect' effects, respectively.

any convenient parametric form for $E(Y|x, z)$ (e.g., linear, quasi-linear logit, probit), estimate the parameters separately (e.g., by regression or maximum likelihood methods), insert the parametric approximation into (12.13), and estimate its two conditional expectations (over z) to get the mediated effect.

The power of the mediation formula was recognized by Petersen *et al.* (2006), Glynn (2009), Hafeman and Schwartz (2009), Mortensen *et al.* (2009), VanderWeele (2009), Kaufman (2010), and Imai *et al.* (2010c). Imai *et al.* (2010a, 2010c) have further shown that nonparametric identification of mediation effects under the no-confounding assumption (Figure 12.1(a)) allows for a flexible estimation strategy and illustrate this with various nonlinear models, quantile regressions, and generalized additive models. Imai *et al.* (2010b) describe an implementation of these extensions using a convenient *R* package. Sjölander (2009) provides a bound on *DE* in cases where the confounders between Z and Y cannot be controlled.

In the next section this power will be demonstrated on linear and nonlinear models, with the aim of explaining the distortions produced by conventional methods of parametric mediation analysis, and how they are rectified through the mediation formula.

12.2.2 Mediation effects in linear, logistic, and probit models

12.2.2.1 The linear case: Difference versus product estimation

Let us examine what the mediation formula yields when applied to the linear version of model 12.1(a), which reads:

$$
\begin{aligned}
x &= u_X \\
z &= \gamma_{xz} x + u_Z \\
y &= \gamma_{xy} x + \gamma_{zy} z + u_Y
\end{aligned}
\tag{12.15}
$$

Computing the conditional expectation in (12.13) gives

$$g(x, z) = E(Y|x, z) = E(\gamma_{xy} x + \gamma_{zy} z + u_Y) = a_0 + \gamma_{xy} x + \gamma_{zy} z$$

and yields

$$
\begin{aligned}
DE_{x,x'} &= \sum_z [(a_0 + \gamma_{xy} x' + \gamma_{zy} z) - (a_0 + \gamma_{xy} x + \gamma_{zy} z)] P(z|x) \\
&= \gamma_{xy}(x' - x) && (12.16) \\
IE_{x,x'}(Y) &= \sum_z (a_0 + \gamma_{xy} x + \gamma_{zy} z)[P(z|x') - P(z|x)] \\
&= \gamma_{zy}[E(Z|x') - E(Z|x)] \\
&= (\gamma_{zy}\gamma_{xz})(x' - x) && (12.17) \\
&= (\beta_{xy} - \gamma_{xy})(x' - x) && (12.18)
\end{aligned}
$$

where β_{xy} is the regression coefficient $\beta_{xy} = \frac{\partial}{\partial x} E(Y|x) = \gamma_{xy} + \gamma_{xz}\gamma_{zy}$;

$$\begin{aligned} TE_{x,x'}(Y) &= (E(Y|x') - E(Y|x)) \\ &= \sum_z E(Y|x', z)P(z|x') + \sum_z E(Y|x, z)P(z|x) \\ &= \sum_z (a_0 + \gamma_{zy}x' + \gamma_{zy}z)P(z|x') - \sum_z (a_0 + \gamma_{xy}x + \gamma_{zy}z)P(z|x) \\ &= \gamma_{xy}(x' - x) + \gamma_{zy}E(Z|x') - \gamma_{zy}E(Z|x) \\ &= (\gamma_{xy} + \gamma_{zy}\gamma_{xz})(x' - x) \end{aligned} \tag{12.19}$$

We thus obtained the standard expressions for effects in linear systems. In particular, we see that the indirect effect can be estimated either as a difference in two regression coefficients (Equation (12.18)) or a product of two regression coefficients (Equation (12.17)), with Y regressed on both X and Z.[9] When generalized to nonlinear systems, however, these two strategies yield conflicting results (MacKinnon and Dwyer, 1993; MacKinnon *et al.*, 2007b) and much controversy has developed as to which strategy should be used in assessing the size of mediation effects (MacKinnon and Dwyer, 1993; Freedman *et al.*, 1992; Molenberghs *et al.*, 2002; MacKinnon *et al.*, 2007b; Glynn, 2009; Green and Bullock, 2010).

We now show that neither of these strategies generalizes to nonlinear systems; direct application of (12.13) is necessary. Moreover, we will see that, though yielding identical results in linear systems, the two strategies represent legitimate intuitions in pursuit of two distinct causal quantities. The difference-in-coefficients method seeks to estimate $TE - DE$, while the product-of-coefficients method seeks to estimate IE. The former represents the reduction in TE if indirect paths were deactivated, while the latter represents the portion of TE that would remain if the direct path were deactivated. The choice between $TE - DE$ and IE depends of course on the specific decision making objectives that the study aims to inform. If the policy evaluated aims to prevent the outcome Y by ways of manipulating the mediating pathways, the target of analysis should be the difference $TE - DE$, which measures the highest prevention effect of any such manipulation. If, on the other hand, the policy aims to prevent the outcome by manipulating the direct pathway, the target of analysis should shift IE, for $TE - IE$ measures the highest preventive impact of this type of manipulation.

In the hiring discrimination example, $TE - DE$ gives the maximum reduction in racial earning disparity that can be expected from programs aiming to achieve educational parity. $TE - IE$, on the other hand, measures the maximum reduction in earning disparity that can be expected from eliminating hiring discrimination by employers. The difference-in-coefficients strategy is motivated by the former types of problem while the product-of-coefficients by the latter.

The next subsection illustrates how nonlinearities bring about the disparity between IE and $TE - DE$.

[9] Note that the equality $\beta_{xy} - \gamma_{xy} = \gamma_{xz}\gamma_{zy}$ established in (12.18) is a universal identity among regressional coefficients of any three variables, and has nothing to do with causation or mediation. It will therefore continue to hold regardless of whether confounders are present, whether the structural parameters are identifiable, whether the underlying model is linear or nonlinear and regardless of whether the arrows in the model of Figure 12.1(a) point in the right direction. Moreover, the equality will hold among the *OLS* estimates of these parameters, regardless of sample size. Therefore, the failure of parameters in nonlinear regression to obey similar equalities should not be construed as an indication of faulty standardization, as suggested by MacKinnon *et al.* (2007a, 2007b).

12.2.2.2 The logistic case

To see how the mediation formula facilitates nonlinear analysis, let us consider the logistic and probit models treated in MacKinnon *et al.* (2007b).[10] To this end, let us retain the linear model of (12.15) with one modification: the outcome of interest will be a threshold-based indicator of the linear outcome Y in (12.15). In other words, we regard

$$Y^* = \gamma_{xy}x + \gamma_{zy}z + u_Y \tag{12.20}$$

as a latent variable and define the outcome Y as

$$Y = \begin{cases} 1 & \text{if } \gamma_0 + \gamma_{xy}x + \gamma_{zy}z + u_Y > 0 \\ 0 & \text{otherwise} \end{cases} \tag{12.21}$$

where γ_0 is some unknown threshold level. We will assume that the error U_Y is governed by the logistic distribution

$$P(U_Y < u) = L(u) \triangleq \frac{1}{1+\mathrm{e}^{-u}} \tag{12.22}$$

and, consequently, $E(Y|x, z)$ attains the form

$$E(Y|x,z) = \frac{1}{1+\mathrm{e}^{-(\gamma_0+\gamma_{xy}x+\gamma_{zy}z)}} \tag{12.23}$$

$$= L(\gamma_0 + \gamma_{xy}x + \gamma_{zy}z) \tag{12.24}$$

We will further assume that U_Z is normal with zero mean and infinitesimal variance $\sigma_z^2 \ll 1$.

Given this logistic model and its parameter set $(\gamma_0, \gamma_{xy}, \gamma_{zy}, \gamma_{xz}, \sigma_z^2)$, we will now compute the direct (*DE*), indirect (*IE*) and total (*TE*) effects associated with the transition from $X = 0$ to $X = 1$. From the mediation formula (Equations (12.12) to (12.14)), we obtain

$$DE = \int_{z=-\infty}^{\infty} [L(\gamma_0 + \gamma_{xy} + \gamma_{zy}) - L(\gamma_0 + \gamma_{zy}z)] f_{Z|X}(z|X=0)\mathrm{d}z$$

$$= L(\gamma_0 + \gamma_{xy}) - L(\gamma_0) + 0(\sigma_z^2) \tag{12.25}$$

$$IE = \int_{z=-\infty}^{\infty} [L(\gamma_0 + \gamma_{zy}z)[f_{Z|X}(z|X=1) - f_{Z|X}(z|X=0)]\mathrm{d}z$$

$$= L(\gamma_0 + \gamma_{zy}\gamma_{xz}) - L(\gamma_0) + 0(\sigma_z^2) \tag{12.26}$$

$$TE = E(Y|X=1) - E(Y|X=0) = \int_{z=\infty}^{\infty} E(Y|X=1,z) f_{Z|X}(z|X=1)\mathrm{d}z$$

$$- \int_{z=\infty}^{\infty} E(Y|X=0,z) f_{Z|x}(z|X=0)\mathrm{d}z$$

$$= L(\gamma_0 + \gamma_{xy}x + \gamma_{zy}z) - L(\gamma_0) + 0(\sigma_z^2) \tag{12.27}$$

where $0(\sigma_z^2) \to 0$ as $\sigma_z \to 0$.

[10] Pearl (2010) analyzes Boolean models with Bernoulli noise.

It is clear that, due to the nonlinear nature of $L(u)$, none of these effects coincides with its corresponding effect in the linear model of Equation (12.15). In other words, it would be wrong to assert the equalities

$$\begin{aligned} DE_{0,1} &= \gamma_{xy} \\ IE_{0,1} &= \gamma_{zy}\gamma_{xz} \\ TE_{0,1} &= \gamma_{xy} + \gamma_{zy}\gamma_{xz} = \beta_{xy} \end{aligned}$$

as is normally assumed in the mediation literature (Prentice, 1989; Freedman *et al.*, 1992; MacKinnon and Dwyer, 1993; Fleming and DeMets, 1996; Molenberghs *et al.*, 2002; MacKinnon *et al.*, 2007b). In particular, the mediated fractions $1 - DE/TE$ and IE/TE may differ substantially from the fractions $\gamma_{xz}\gamma_{zy}/(\gamma_{xy} + \gamma_{xz}\gamma_{zy})$, $1 - \gamma_{xy}/\beta_{xy}$, or $\gamma_{xz}\gamma_{zy}/\beta_{xy}$ that have been proposed to evaluate mediation effects by traditional methods. The latter are heuristic ratios informed by the linear portion of the model, while the formers are derived formally from the counterfactual specifications of the target quantities, as in (12.2) and (12.6).

Figure 12.2 depicts DE, IE, and TE as a function of γ_0, the threshold coefficient that dichotomizes the outcome (as in Equation (12.21)). These were obtained analytically from Equations (12.25)–(12.27), using the values $\gamma_{xz} = \gamma_{xy} = \gamma_{zy} = 0.5$ for illustrative purposes. We see that all three measures vary with γ_0 and deviate substantially from the assumptions that equate DE with $\gamma_{xy} = 0.50$, IE with $\gamma_{xz}\gamma_{zy} = 0.25$, and TE with $\gamma_{xy} + \gamma_{xz}\gamma_{zy} = 0.75$ (MacKinnon and Dwyer, 1993; MacKinnon *et al.*, 2007b).

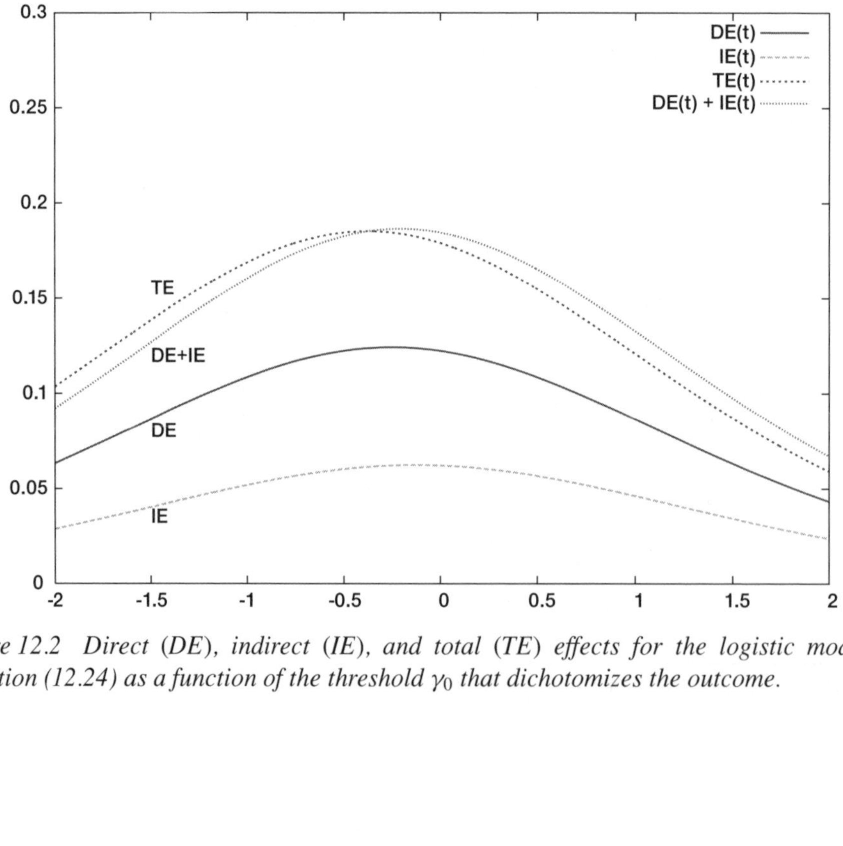

Figure 12.2 Direct (DE), indirect (IE), and total (TE) effects for the logistic model of Equation (12.24) as a function of the threshold γ_0 that dichotomizes the outcome.

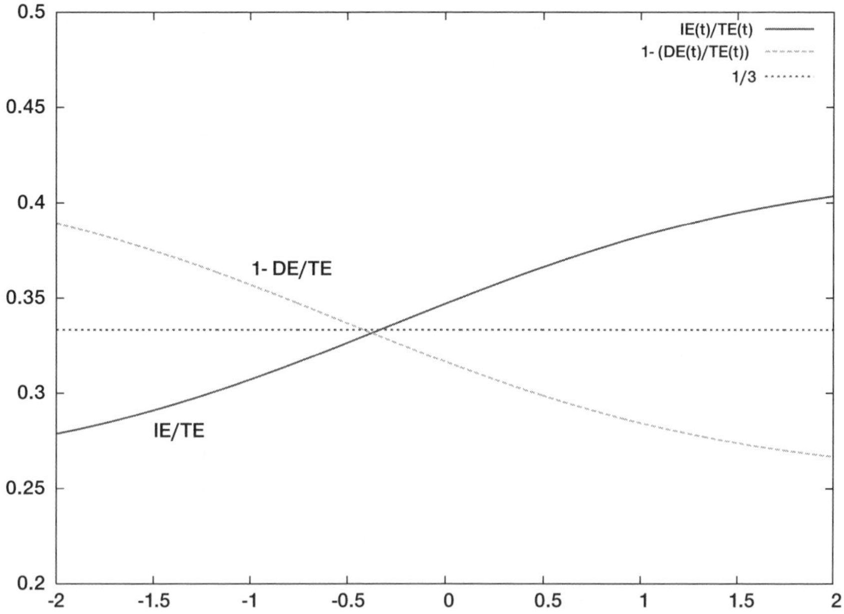

Figure 12.3 Necessary $(1 - DE/TE)$ *and sufficient* (IE/TE) *mediation proportions for the logistic model of Equation (12.24).*

The bias produced by such assumptions is further accentuated in Figure 12.3, which compares several fractions (or proportions) proposed to measure the relative contribution of mediation to the observed response. Recall that $1 - DE/TE$ measures the extent to which mediation was *necessary* for the observed response, while IE/TE measures the extent to which it was *sufficient*. Figure 12.3 shows that the necessary fraction $(1 - DE/TE)$ exceeds the sufficient fraction (IE/TE) as γ_0 becomes more negative. Indeed, in this region, both direct and indirect paths need be activated for Y^* to exceed the threshold of Equation (12.22). Therefore, the fraction of responses for which mediation was necessary is high and the fraction for which mediation was sufficient is low. The disparity between the two will be revealed by varying the intercept γ_0, a parameter that has hardly been given notice in traditional analyses and which will be shown to be important for understanding the interplay between DE and IE, and the role they play in shaping mediated effects. The opposite occurs for positive γ_0 (negative threshold), where each path alone is sufficient for activating Y, and it is unlikely therefore that the mediator becomes a necessary enabler of $Y = 1$.

None of this dynamics is represented in the fixed fraction $\gamma_{xz}\gamma_{zy}/(\gamma_{xy} + \gamma_{xz}\gamma_{zy}) = 0.25/0.75 = 1/3$, which standard logistic regression would report as the fraction of cases 'explained by mediation'. Some of this dynamics is reflected in the fraction $\gamma_{xz}\gamma_{zy}/[E(Y|X = 1) - E(Y|X = 0)] = 0.25/TE$ (not shown in Figure 12.3) which some researchers have recommended as a measure of mediation (MacKinnon and Dwyer, 1993; MacKinnon *et al.*, 2007b). However, this measure is totally incompatible with the correct fractions shown in Figure 12.3. The differences are accentuated again for negative γ_0 (positive threshold), where both direct and indirect processes must be activated for Y^* to exceed the threshold, and the

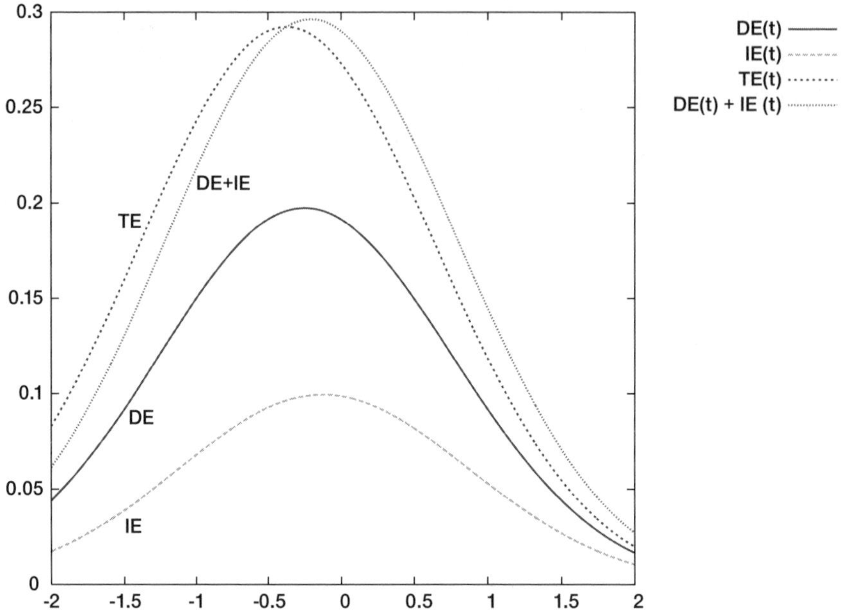

Figure 12.4 Direct (DE), indirect (IE), and total (TE) effects for a probit model.

fraction of responses for which mediation is necessary ($1 - DE/TE$) is high and the fraction for which mediation is sufficient (IE/TE) is low.

12.2.2.3 The probit case

Figure 12.4 displays the behavior of a probit model. It was computed analytically by assuming a probit distribution in Equation (12.22), which leads to the same expressions in (12.21) to (12.24), with Φ replacing L. Noticeably, Figures 12.4 and 12.5 reflect more pronounced variations of all effects with γ_0, as well as a more pronounced deviation of these curves from the constant $\gamma_{xz}\gamma_{zy}/(\gamma_{xy} + \gamma_{xz}\gamma_{zy}) = 1/3$ that regression analysis defines as the 'proportion mediated' measure (Sjölander, 2009). We speculate that the difference in behavior between the logistic and probit models is due to the latter sharper approach toward the asymptotic limits.

12.2.3 Special cases of mediation models

In this section we will discuss three special cases of mediation processes that lend themselves to simplified analysis.

12.2.3.1 Incremental causal effects

Consider again the logistic threshold model of Equation (12.21) and assume we are interested in assessing the response Y to an incremental change in the treatment variable X, say from

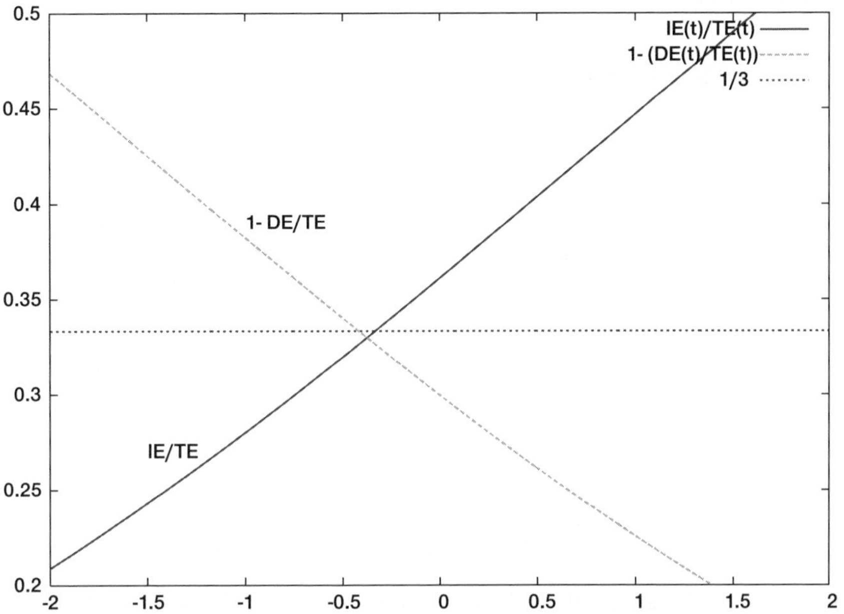

Figure 12.5 Necessary $(1 - DE/TE)$ *and sufficient* (IE/TE) *mediation proportions for a probit model.*

$X = x$ to $X = x + \delta$. In other words, our target quantities are the limits as $\delta \to 0$ of

$$DE_{inc}(x) = \frac{1}{\delta} DE_{x,x+\delta}$$

$$IE_{inc}(x) = \frac{1}{\delta} IE_{x,x+\delta}$$

$$TE_{inc}(x) = \frac{1}{\delta} TE_{x,x+\delta}$$

If we maintain the infinitesimal variance assumption $\sigma_z^2 \ll 1$, we obtain

$$\begin{aligned} DE_{inc}(x) &= \lim_{\delta \to 0} \frac{1}{\delta} DE_{x,x+\delta} \\ &= \lim_{\delta \to 0} \frac{1}{\delta} \int [E(Y|x+\delta, z) - E(Y|x, z)] f_{Z|X}(z|x) \mathrm{d}z \\ &= \frac{\partial}{\partial x} E(Y|x, z)|_{z=h(x)} + 0(\sigma_z^2) \end{aligned}$$

where $h(x) = E(Z|x)$.

Similarly, we have

$$\begin{aligned}
IE_{inc}(x) &= \lim_{\delta\to 0}\frac{1}{\delta}IE_{x,x+\delta} \\
&= \lim_{\delta\to 0}\frac{1}{\delta}\int_z E(Y|x,z)[f(z|x+\delta)-f(z|x)]\mathrm{d}z \\
&= \lim_{\delta\to 0}\frac{1}{\delta}E(Y|x,z)|_{z=h(x+\delta)} - E(Y|x,z)|_{z=h(x)} \\
&= \lim_{\delta\to 0}\frac{1}{\delta}[E(Y|x,h(x+\delta)) - E(Y|x,h(x))] + 0(\sigma_z^2) \\
&= \frac{\partial}{\partial z}E(Y|x,z)\frac{\mathrm{d}}{\mathrm{d}x}h(x)|_{z=h(x)}
\end{aligned}$$

and

$$TE_{inc}(x) = \lim_{\delta\to 0}\frac{1}{\delta}[E(Y|x+\delta) - E(Y|x)] = \frac{\mathrm{d}}{\mathrm{d}x}E(Y|x)$$

Using the rule of partial differentiation, we have $TE_{inc} = DE_{inc} + IE_{inc}$, a result obtained in Winship and Mare (1983), though starting from a different perspective.

12.2.3.2 Linear outcome with binary mediator

It is interesting to inquire how effects are decomposed when we retain the linear form of the outcome process, but let the intermediate variable Z be a binary variable that is related to X through an arbitrary nonlinear process $P(Z=1|x)$.

Considering a transition from $X = x_0$ to $X = x_1$ and writing

$$E(Y|x,z) = \alpha x + \beta z + \gamma$$

we readily obtain

$$\begin{aligned}
DE &= \sum_z [(\alpha x_1 + \beta z + \gamma)] - [(\alpha x_0 + \beta z + \gamma)]P(z|x_0) \\
&= \alpha(x_1 - x_0) && (12.28) \\
IE &= \sum_z (\alpha x_1 + \beta z + \gamma) - [P(z|x_1) - P(z|x_0)] \\
&= \beta[E(Z_0|x_1) - E(Z|x_0)] && (12.29) \\
TE &= \sum_z E(Y|x_1,z)P(z|x_1) - E(Y|x_0,z)P(z|x_0) \\
&= \sum_z (\alpha x_1 + \beta z + \gamma)P(z|x_1) - \sum_z (\alpha x_0 + \beta z + \gamma)P(z|x_0) \\
&= \alpha(x_1 - x_0) + \beta[E(Z|x_1) - E(Z|x_0)] && (12.30)
\end{aligned}$$

Again, we have $TE = DE + IE$.

We see that as long as the outcome processes is linear, nonlinearities in the mediation process do not introduce any surprises; effects are decomposed into their direct and indirect components in a textbook-like fashion. Moreover, the distribution $P(Z|x)$ plays no role in the analysis; it is only the expectation $E(Z|x)$ that needs to be estimated.

This result was obtained by Li *et al.* (2007) who estimated *IE* using a difference-in-coefficients strategy. It follows, in fact, from a more general property of the mediation formula (12.13), first noted by VanderWeele (2009), which will be discussed next.

12.2.3.3 Semi-linear outcome process

Suppose $E(Y|x, z)$ is linear in Z, but not necessarily in X. We can then write

$$E(Y|x, z) \triangleq g(x, z) = f(x) + t(x)z$$
$$E(Z|x) \triangleq h(x)$$

and the mediation formulas give (for the transition from $X = x_0$ to $X = x_1$)

$$\begin{aligned} DE &= \sum_z \{(f(x_1) + t(x_1)z) - (f(x_0) - t(x_0)z)\} P(z|x_0) \\ &= f(x_1) - f(x_0) + (t(x_1) - t(x_0))E(Z|x_0) \\ &= g(x_1, h(x_0)) - g(x_0, h(x_0)) \qquad (12.31) \\ IE &= \sum_z (f(x_0) + t(x_0)z)(P(z|x_1) - P(z|x_0)) \\ &= t(x_0)(E(Z|x_1) - E(Z|x_0)) \\ &= t(x_0)[h(x_1) - h(x_0)] \qquad (12.32) \\ TE &= \sum_z (f(x_1) + t(x_1)z)P(z|x_1) - \sum_z (f(x_0) + t(x_0)z)P(z|x_0) \\ &= f(x_1) - f(x_0) + t(x_1)E(Z|x_1) - t(x_0)E(Z|x_0) \\ &= g(x_1, h(x_1)) - g(x_0, h(x_0)) \qquad (12.33) \end{aligned}$$

We see again that only the conditional mean $E(Z|x)$ need enter the estimation of causal effects in this model, not the entire distribution $P(z|x)$. However, the equality $TE = DE + IE$ no longer holds in this case; the nonlinearities embedded in the interaction term $t(x)z$ may render Z an enabler or inhibitor of the direct path, thus violating the additive relationship between the three effect measures.

This becomes more transparent when we examine the standard linear model to which a multiplicative term xz is added, as is done, for example, in the analyses of Kraemer *et al.* (2008), Jo (2008), and Preacher *et al.* (2007). In this model we have

$$E(Y|x, z) \triangleq g(x, z)) = \beta_0 + \beta_1 x + \beta_2 z + \beta_3 xz$$
$$E(Z|x) \triangleq h(x) = \gamma_0 + \gamma_1 x$$

Substituting

$$f(x) = \beta_0 + \beta_1 x$$
$$t(x) = \beta_2 + \beta_3 x$$
$$h(x) = \gamma_0 + \gamma_1 x$$

into (12.31) to (12.33) and letting $x_1 - x_0 = 1$ gives

$$DE = \beta_1 + \beta_3(\gamma_0 + \gamma_1 x_0) \tag{12.34}$$
$$IE = \gamma_1(\beta_2 + \beta_3 x_0) \tag{12.35}$$
$$TE = \beta_1 + \beta_3\gamma_0 + \beta_2\gamma_1 + \beta_3\gamma_1(x_0 + x_1) \tag{12.36}$$

In particular, the relationships between *DE*, *IE*, and *TE* becomes

$$TE = DE + IE + \gamma_1\beta_3$$

which clearly identifies the product $\gamma_1\beta_3$ as the culprit for the nonadditivity $TE \neq TE + IE$. Indeed, when $\gamma_1\beta_3 \neq 0$, Z acts both as a moderator and a mediator, and both *DE* and *IE* are affected by the interaction term $\beta_3 xz$. Note further that the direct and indirect effects can both be zero and the total effect nonzero, a familiar nonlinear phenomenon that occurs when Z is a necessary enabler for the effect of X on Y. This dynamics has escaped standard analyses of mediation, which focused exclusively on estimating structural parameters, rather than effect measures, as in (12.34) to (12.36).

It is interesting to note that, due to interaction, a direct effect can exist even when β_1 vanishes, though β_1 is the path coefficient associated with the direct link $X \to Y$. This illustrates that estimating parameters in isolation tells us little about the problem until we understand the way they combine to form effect measures. More generally, mediation and moderation are inextricably intertwined and cannot be assessed separately, a position affirmed by Kraemer *et al.* (2008) and Preacher *et al.* (2007).

12.2.3.4 The binary case

To complete our discussion of models in which the mediation problem lends itself to a simple solution, we now address the case where all variables are binary, still allowing though for arbitrary interactions and arbitrary distributions of all processes. The low dimensionality of the binary case permits both a nonparametric solution and an explicit demonstration of how mediation can be estimated directly from the data. Generalizations to multivalued outcomes are straightforward.

Assume that the model of Figure 12.1(a) is valid and that the observed data is given by Table 12.1. The factors $E(Y|x, z)$ and $P(Z|x)$ in Equations (12.12) to (12.14) can be readily estimated, as shown in the two right-most columns of Table 12.1 and, when substituted into (12.12) to (12.14), yield

$$DE = (g_{10} - g_{00})(1 - h_0) + (g_{11} - g_{01})h_0 \tag{12.37}$$
$$IE = (h_1 - h_0)(g_{01} - g_{00}) \tag{12.38}$$
$$TE = g_{11}h_1 + g_{10}(1 - h_1) - [g_{01}h_0 + g_{00}(1 - h_0)] \tag{12.39}$$

Table 12.1 Computing the Mediation Formula for the model in Figure 12.1(a), with X, Y, Z binary.

<table>
<tr><th></th><th>X</th><th>Z</th><th>Y</th><th>$E(Y|x,z) = \boldsymbol{g}_{xz}$</th><th>$E(Z|x) = \boldsymbol{h}_x$</th></tr>
<tr><td>n_1</td><td>0</td><td>0</td><td>0</td><td rowspan="2">$\frac{n_2}{n_1+n_2} = g_{00}$</td><td rowspan="4">$\frac{n_3+n_4}{n_1+n_2+n_3+n_4} = h_0$</td></tr>
<tr><td>n_2</td><td>0</td><td>0</td><td>1</td></tr>
<tr><td>n_3</td><td>0</td><td>1</td><td>0</td><td rowspan="2">$\frac{n_4}{n_3+n_4} = g_{01}$</td></tr>
<tr><td>n_4</td><td>0</td><td>1</td><td>1</td></tr>
<tr><td>n_5</td><td>1</td><td>0</td><td>0</td><td rowspan="2">$\frac{n_6}{n_5+n_6} = g_{10}$</td><td rowspan="4">$\frac{n_7+n_8}{n_5+n_6+n_7+n_8} = h_1$</td></tr>
<tr><td>n_6</td><td>1</td><td>0</td><td>1</td></tr>
<tr><td>n_7</td><td>1</td><td>1</td><td>0</td><td rowspan="2">$\frac{n_8}{n_7+n_8} = g_{11}$</td></tr>
<tr><td>n_8</td><td>1</td><td>1</td><td>1</td></tr>
</table>

We see that logistic or probit regression is not necessary as simple arithmetic operations suffice to provide a general solution for any conceivable dataset.

12.2.4 Numerical example

To anchor these formulas in a concrete example, let us assume that $X = 1$ stands for a drug treatment, $Y = 1$ for recovery, and $Z = 1$ for the presence of a certain enzyme in a patient's blood, which appears to be stimulated by the treatment. Assume further that the data described in Tables 12.2 and 12.3 were obtained in a randomized clinical trial and that our research question is whether Z mediates the action of X on Y, and to what extent.

Substituting this data into Equations (12.37) to (12.39) yields

$$DE = (0.40 - 0.20)(1 - 0.40) + (0.80 - 0.30)0.40 = 0.32$$

$$IE = (0.75 - 0.40)(0.30 - 0.20) = 0.035$$

$$TE = 0.80 \times 0.75 + 0.40 \times 0.25 - (0.30 \times 0.40 + 0.20 \times 0.60) = 0.46$$

$$IE/TE = 0.07, \qquad DE/TE = 0.696, \qquad 1 - DE/TE = 0.304$$

Table 12.2 Typical frequency data showing the dependence of cure rate on both treatment (X) on the mediator (Z).

Treatment X	Enzyme present Z	Percentage cured $g_{xz} = E(Y|x,z)$
Yes	Yes	$g_{11} = 80\,\%$
Yes	No	$g_{10} = 40\,\%$
No	Yes	$g_{01} = 30\,\%$
No	No	$g_{00} = 20\,\%$

Table 12.3 Typical frequency data showing the effect of treatment (X) on the mediator (Z).

Treatment X	Percentage with Z present
No	$h_0 = 40\,\%$
Yes	$h_1 = 75\,\%$

We conclude that 30.4% of all recoveries is owed to the capacity of the treatment to enhance the secretion of the enzyme, while only 7% of recoveries would be sustained by enzyme enhancement alone. The policy implication of such a study would be that efforts to develop a cheaper drug, identical to the one studied, but lacking the potential to stimulate enzyme secretion would face a reduction of 30.4% in recovery cases. More decisively, proposals to substitute the drug with one that merely mimics its stimulant action on Z but has no direct effect on Y are bound for failure; the drug evidently has a beneficial effect on recovery that is independent of, though enhanced by enzyme stimulation.

For completeness, we note that the controlled direct effects are (using (12.1))

$$CDE_{z=0} = g_{10} - g_{00} = 0.40 - 0.20 = 0.20$$

and

$$CDE_{z=1} = g_{11} - g_{01} = 0.80 - 0.30 = 0.50$$

which are quite far apart. Their weighted average, governed by $P(Z = 1|X = 0) = h_0 = 0.40$, gives us $DE = 0.32$. These do not enter, however, into the calculation of IE, since the indirect effect cannot be based on controlling variables; it requires instead a path-deactivating operator, as mirrored in the definition of Equation (12.6).

12.3 Relation to other methods

12.3.1 Methods based on differences and products

Attempts to compare these results to those produced by conventional mediation analyses encounter two obstacles. First, conventional methods do not define direct and indirect effects in causal vocabulary, without committing to specific functional or distributional forms. MacKinnon (2008, Chapter 11), for example, analyzes categorical data using logistic and probit regressions and constructs effect measures using products and differences of the parameters in those regressional forms. Section 12.2 demonstrates that this strategy is not compatible with the causal interpretation of effect measures, even when the parameters are known precisely; IE and DE may be extremely complicated functions of those regression coefficients (see Equations (12.25) to (12.26)). Fortunately, those coefficients need not be estimated at all; effect measures can be estimated directly from the data, circumventing the parametric representation altogether.

Second, attempts to extend the difference and product heuristics to nonparametric analysis have encountered ambiguities that conventional analysis fails to resolve. The product-of-coefficients heuristic advises us to multiply the slope of Z on X,

$$C_\beta = E(Z|X = 1) - E(Z|X = 0) = h_1 - h_0$$

by the slope of Y on Z fixing X,

$$C_\gamma = E(Y|X = x, Z = 1) - E(Y|X = x, Z = 0) = g_{x1} - g_{x0}$$

but does not specify at what value we should fix X. Equation (12.38) resolves this ambiguity by determining that X should be fixed to $X = 0$; only then would the product $C_\beta C_\gamma$ yield the correct mediation measure, *IE*.

The difference-in-coefficients heuristics instructs us to estimate the direct effect coefficient

$$C_\alpha = E(Y|X = 1, Z = z) - E(Y|X = 0, Z = z) = g_{1z} - g_{0z}$$

and subtract it from the total effect, but does not specify on what value we should condition Z. Equation (12.37) determines that the correct way of estimating C_α would be to condition on both $Z = 0$ and $Z = 1$ and take their weighted average, with $h_0 = P(Z = 1|X = 0)$ as the weighting function.

To summarize, in calculating *IE* we should condition on both $Z = 1$ and $Z = 0$ and average while in calculating *DE* we should condition on only one value, $X = 0$, and no average need be taken.

Reiterating the discussion of Section 12.2, the difference and product heuristics are both legitimate, with each seeking a different effect measure. The difference-in-coefficients heuristics, leading to *TE* − *DE*, seeks to measure the percentage of units for which mediation was *necessary*. The product-of-coefficients heuristics, on the other hand, leading to *IE*, seeks to estimate the percentage of units for which mediation was *sufficient*. The former informs policies aiming to modify the direct pathway while the latter informs those aiming to modify mediating pathways.

The ability of the mediation formula to move across the linear–nonlinear barrier may suggest that all mediation-related questions can now be answered nonparametrically and, more specifically, that, similar to traditional path analysis in linear systems (Alwin and Hauser, 1975; Bollen, 1989), we can now assess the extent to which an effect is mediated through any chosen path or a bundle of paths in a causal diagram. This turned out not to be the case. Avin *et al.* (2005) showed that there are bundles of paths (i.e., subgraphs) in a graph whose mediation effects cannot be assessed from either observational or experimental studies, even in the absence of unobserved confounders. They proved that the effect mediated by a subgraph SG is estimable if and only if SG contains no 'broken fork', that is, a path p_1 from X to some W, and two paths, p_2 and p_3, from W to Y, such that p_1 and p_2 are in SG and p_3 is in G but not in SG. Such subgraphs may exist if any of the mediator–outcome confounders (W_2 in Figure 12.1) is a descendant of X. See Pearl (2012) for in-depth discussion of identification.

12.3.2 Relation to the principal-strata direct effect

The derivation of the mediation formula (Pearl, 2001) was made possible by the counterfactual interpretation of structural equations (see footnote 1) and the symbiosis between graphical and counterfactual analysis that this interpretation engenders.[11] In contrast, the structureless approach of Rubin (1974) has spawned other definitions of direct effects, normally referred to as 'principal-strata direct effect (PSDE)' (Frangakis and Rubin, 2002; Mealli and Rubin, 2003; Rubin, 2004, 2005; Egleston *et al.*, 2010). Whereas the natural direct effect measures the average effect that would be transmitted in the population with all mediating paths (hypothetically) *deactivated*, the PSDE is defined as the effect transmitted in those units only for whom mediating paths *happened to be deactivated* in the study. This definition leads to unintended results that stand contrary to common usage of direct effects (Robins *et al.*, 2007, 2009; VanderWeele, 2008), excluding from the analysis all individuals who are both directly and indirectly affected by the causal variable X (Pearl, 2009b). In linear models, as a striking example, a direct effect will be flatly undefined, unless β, the $X \rightarrow Z$ coefficient, is zero. In some other cases, the direct effect of the treatment will be deemed to be nil, if a small subpopulation exists for which treatment has no effect on both Y and Z.

To witness what the *PSDE* estimates in the example of Table 12.1, we should note that the *PSDE*, like the natural direct effect, also computes a weighted average of the two controlled direct effects,[12]

$$PSDE = \alpha CDE_{z=0} + (1 - \alpha)CDE_{z=1} \tag{12.40}$$

However, the weight, α, is not identifiable, and may range from zero to one depending on unobserved factors. If in addition to nonconfoundedness we also assume monotonicity (i.e., $Z_1 \geq Z_0$), α is identified and is given by the relative sizes of two extreme ends of the population:

$$\alpha = \frac{P(Z = 1|X = 0)}{P(Z = 1|X = 0) + P(Z = 0|X = 1)}$$

For example, if $P(Z = 1|X = 0) = 0.10$ and $P(Z = 0|X = 1) = 0.01$, 90 % of the weight will be placed on $CDE_{z=0}$ and the *PSDE* will be close to $CDE_{z=0}$. If, on the other hand, $P(Z = 1|X = 0) = 0.01$ and $P(Z = 0|X = 1) = 0.10$, the *PSDE* will be close to $CDE_{z=1}$. Such sensitivity to the exceptional cases in the population is not what we usually expect from direct effects.

---

[11] Such symbiosis is now standard in epidemiology research (Robins, 2001; Petersen *et al.*, 2006; VanderWeele and Robins, 2007; Hafeman and Schwartz, 2009; VanderWeele, 2009; Albert and Nelson, 2011) and is making its way slowly toward the social and behavioral sciences (e.g., Morgan and Winship, 2007; Imai *et al.*, 2010a, 2010c; Elwert and Winship, 2010; Chalak and White, 2011), despite islands of resistance (Wilkinson *et al.*, 1999, p. 600; Sobel, 2008; Rubin, 2010; Imbens, 2010).

[12] We take here the definition used in Gallop *et al.* (2009), $PSDE = E[Y_1 - Y_0|Z_0 = Z_1]$, and, invoking nonconfoundedness, $Y_{x,z} \perp\!\!\!\perp Z_x$, together with composition, $Y_x = Y_{x,Z_x}$, we obtain

$$\begin{aligned} PSDE &= \sum_z E(Y_{1,Z_1} - Y_{0,Z_0}|Z_1 = Z_0 = z)P(Z_1 = z|Z_0 = Z_1) \\ &= \sum_z E(Y_{1,z} - Y_{0,z}|Z_1 = Z_0 = z)P(Z_1 = z|Z_0 = Z_1) \\ &= \sum_z E(Y_{1,z} - Y_{0,z})P(Z_1 = z|Z_0 = Z_1) \end{aligned}$$

which reduces to Equation (12.40), with $\alpha = P(Z_1 = 1|Z_0 = Z_1)$.

In view of these definitional inadequacies we do not include 'principal-strata direct effect' in our discussion of mediation, though they may well be suited for other applications,[13] for example, when a stratum-specific property is genuinely at the focus of one's research.

Indeed, taking a 'principal strata' perspective, Rubin found the concept of mediation 'ill-defined.' In his words: 'The general theme here is that the concepts of direct and indirect causal effects are generally ill-defined and often more deceptive than helpful to clear statistical thinking in real, as opposed to artificial problems' (Rubin, 2004). Conversely, attempts to define and understand mediation using the notion of 'principal-strata direct effect' have encountered basic conceptual difficulties (Lauritzen, 2004; Robins *et al.*, 2007, 2009; Pearl, 2009b), concluding that 'it is not always clear that knowing about the presence of principal stratification effects will be of particular use' (VanderWeele, 2008). As a result, it is becoming widely recognized that the controlled, natural, and indirect effects discussed in this paper are of greater interest, both for the purposes of making treatment decisions and for the purposes of explanation and identifying causal mechanisms (Joffe *et al.*, 2007; Albert and Nelson, 2011; Mortensen *et al.*, 2009; Imai *et al.*, 2010a, 2010c; Geneletti, 2007; Robins *et al.*, 2007, 2009; Petersen *et al.*, 2006; Hafeman and Schwartz, 2009; Kaufman, 2010; Cai *et al.*, 2008).

The limitation of PSDE stems not from the notion of 'principal-strata' per se, which is merely a classification of units into homogeneously reacting classes, and has been used advantageously by many researchers (Balke and Pearl, 1994a, 1994b; Pearl, 1993; Balke and Pearl, 1997; Heckerman and Shachter, 1995; Pearl, 2000, p. 264; Lauritzen, 2004; Sjölander, 2009). Rather, the limitation results from strict adherence to an artificial restriction that prohibits one from regarding a mediator as a cause unless it is manipulable. This prohibition prevents one from defining the direct effect as it is commonly used in decision making and scientific discourse – an effect transmitted once all mediating paths are 'deactivated' (Pearl, 2001; Avin *et al.*, 2005; Albert and Nelson, 2011), and forces one to use statistical conditionalization instead. Path deactivation requires counterfactual constructs in which the mediator acts as an antecedent, as in Equations (12.1), (12.2) and (12.6), regardless of whether it is physically manipulable. After all, if our aim is to uncover causal mechanisms, it is hard to accept the PSDE restriction that nature's pathways should depend on whether we have the technology to manipulate one variable or another. (For a comprehensive public discussion of these issues, including opinions from enthusiastic and disappointed practitioners, see Pearl (2011), Sjölander (2011), Baker *et al.* (2011), and Egleston (2011).

12.4 Conclusions

Traditional methods of mediation analysis produce distorted estimates of 'mediation effects' when applied to nonlinear models or models with categorical variables. By focusing on parameters of logistic and probit estimators, instead of the target effect measures themselves, traditional methods produce consistent estimates of the former and biased estimates of the latter. This paper demonstrates that the bias can be substantial even in simple systems with

[13] Joffe and Green (2009) and Pearl and Bareinboim (2011) examine the adequacy of the 'principal-strata' definition of surrogate outcomes, a notion related, though not identical, to mediation. There, too, the restrictions imposed by the 'principal-strata' framework lead to surrogacy criteria that are incompatible with the practical aims of surrogacy (see Pearl, 2011).

all processes correctly parameterized and only the outcome dichotomized. The chapter offers a causally sound alternative that ensures bias-free estimates while making no assumption on the distributional form of the underlying process.

We distinguished between proportion of response cases for which mediation was *necessary* and those for which mediation would have been *sufficient*. Both measures play a role in mediation analysis and are given here a formal representation and effective estimation methods through the mediation formula.

In addition to providing causally sound estimates for mediation effects, the mediation formula also enables researchers to evaluate analytically the effectiveness of various parametric specifications relative to any assumed model. For example, it would be straightforward to investigate the distortion created by assuming a logistic model (as in (12.23)) when data are generated in fact by a probit distribution, or vice versa. This exercise would amount to finding the maximum-likelihood (ML) estimates of γ_0, γ_{xy}, and γ_{zy} in (12.24) for data generated by a probit distribution and comparing the estimated effect measures computed through (12.25) to (12.27) with the true values of those measures, as dictated by the probit model.[14] This type of analytical 'sensitivity analysis' has been used extensively in statistics for parameter estimation, but could not be adequately applied to mediation analysis, owing to the absence of an objective target quantity that captures the notion of indirect effect in nonlinear systems. MacKinnon *et al.* (2007b), for example, evaluated sensitivity to misspecifications by comparing the estimated parameters against their true values, though disparities in parameters may not represent disparity in effect measures (i.e., *ED* or *IE*). By providing such objective measures of effects, the mediation formula of Equation (12.13) enables us to measure directly the disparities in the target quantities.[15]

While the validity of the mediation formulas rests on the same assumptions (i.e., no unmeasured confounders) that are a standard requirement in linear mediation analysis, their appeal to general nonlinear systems, continuous and categorical variables, and arbitrary complex interactions render them a powerful tool for the assessment of causal pathways in many of the social, behavioral, and health-related sciences.

Acknowledgments

Portions of this paper are adapted from Pearl (2000, 2009a, 2009b, 2010). I am indebted to Elias Bareinboim, Adam Glynn, Donald Green, Booil Jo, Marshall Joffe, Helena Kraemer, David MacKinnon, Stephen Morgan, Patrick Shrout, Arvid Sjölander, Dustin Tingley, Tyler VanderWeele, Scott Weaver, Christopher Winship, Teppei Yamamoto, and readers of the UCLA Causality Blog (http://www.mii.ucla.edu/causality/) for discussions leading to these explorations. Send comments to <judea@cs.ucla.edu>. This research was supported in parts by NIH #1R01 LM009961-01, NSF #IIS-0914211 and #IIS-1018922, and ONR #N000-14-09-1-0665 and #N00014-10-1-0933.

[14] An alternative would be to find the ML estimates of *DE*, *IE*, and *TE* directly, through (12.25), (12.26), and (12.27), rather than going through (12.23) (van der Laan and Rubin, 2006).

[15] Sensitivity analysis using both analytical and simulation techniques are described in Imai *et al.* (2010a, 2010c). For questions of external validity (Shadish *et al.*, 2002; Heckman, 2005) see the formal criteria provided by graphical models (Pearl and Bareinboim, 2011; Petersen, 2011).

References

Albert, J.M. and Nelson, S. (2011) Generalized causal mediation analysis. *Biometrics*. DOI: 10.1111/j.1541–0420.2010.01547.x.

Alwin, D. and Hauser, R. (1975) The decomposition of effects in path analysis. *American Sociological Review*, **40**, 37–47.

Avin, C., Shpitser, I. and Pearl, J. (2005) Identifiability of path-specific effects, in *Proceedings of the Nineteenth International Joint Conference on Artificial Intelligence IJCAI-05*. Edinburgh, UK: Morgan-Kaufmann Publishers.

Baker, S.G., Lindeman, K.S. and Kramer, B.S. (2011) Clarifying the role of principal stratification in the paired availability design. *The International Journal of Biostatistics*, **7**, Article 25. DOI: 10.2202/1557-4679.1338. Available at: http://www.bepress.com/ijb/vol7/iss1/25.

Balke, A. and Pearl, J. (1994a) Counterfactual probabilities: computational methods, bounds, and applications, in *Uncertainty in Artificial Intelligence 10* (eds R. L. de Mantaras and D. Poole). San Mateo, CA: Morgan Kaufmann, pp. 46–54.

Balke, A. and Pearl, J. (1994b) Probabilistic evaluation of counterfactual queries, in *Proceedings of the Twelfth National Conference on Artificial Intelligence*, vol. I. Menlo Park, CA: MIT Press, pp. 230–237.

Balke, A. and Pearl, J. (1997) Bounds on treatment effects from studies with imperfect compliance. *Journal of the American Statistical Association*, **92**, 1172–1176.

Baron, R. and Kenny, D. (1986) The moderator–mediator variable distinction in social psychological research: conceptual, strategic, and statistical considerations. *Journal of Personality and Social Psychology*, **51**, 1173–1182.

Bollen, K. (1989) *Structural Equations with Latent Variables*. New York: John Wiley & Sons, Inc.

Cai, Z., Kuroki, M., Pearl, J. and Tian, J. (2008) Bounds on direct effect in the presence of confounded intermediate variables. *Biometrics*, **64**, 695–701.

Chalak, K. and White, H. (2011) Direct and extended class of instrumental variables for the estimation of causal effects. *Canadian Journal of Economics*, **44**, 1–51.

Cole, S. and Hernán, M. (2002) Fallibility in estimating direct effects. *International Journal of Epidemiology*, **31**, 163–165.

Egleston, B.L. (2011) Response to Pearl's comments on principal stratification. *The International Journal of Biostatistics*, **7**, Article 24. DOI: 10.2202/1557-4679.1330. Available at: http://www.bepress.com/ijb/vol7/iss1/24.

Egleston, B.L., Cropsey, K.L., Lazev, A.B. and Heckman, C.J. (2010) A tutorial on principal stratification-based sensitivity analysis: application to smoking cessation studies. *Clinical Trials*, **7**, 286–298.

Elwert, F. and Winship, C. (2010) Effect heterogeneity and bias in main-effects-only regression models, in *Heuristics, Probability and Causality: a tribute to Judea Pearl* (eds R. Dechter, H. Geffner and J. Halpern). UK: College Publications, pp. 327–336.

Fleming, T. and DeMets, D. (1996) Surrogate end points in clinical trials: are we being misled? *Annals of Internal Medicine*, **125**, 605–613.

Frangakis, C. and Rubin, D. (2002) Principal stratification in causal inference. *Biometrics*, **1**, 21–29.

Freedman, L., Graubard, B. and Schatzkin, A. (1992) Statistical validation of intermediate endpoints for chronic diseases. *Statistics in Medicine*, **8**, 167–178.

Gallop, R., Small, D.S., Lin, J.Y., Elliott, M.R., Joffe, M. and Ten Have, T.R. (2009) Mediation analysis with principal stratification. *Statistics in Medicine*, **28**, 1108–1130.

Geneletti, S. (2007) Identifying direct and indirect effects in a non-counterfactual framework. *Journal of the Royal Statistical Society, Series B (Methodological)*, **69**, 199–215.

Glynn, A. (2012) The product and difference fallacies for indirect effects. *American Journal of Political Science*, **56**, 257–269.

Graff, J. and Schmidt, P. (1982) A general model for decomposition of effects, in *Systems Under Indirect Observation: causality, structure, prediction* (eds K. Jöreskog and H. Wold). Amsterdam: North-Holland, pp. 131–148.

Green, D., Ha, S. and Bullock, J. (2010) Enough already about black box experiments: studying mediation is more difficult than most scholars suppose. *Annals of the American Academy of Political and Social Science*, **628**, 200–208.

Hafeman, D. and Schwartz, S. (2009) Opening the black box: a motivation for the assessment of mediation. *International Journal of Epidemiology*, **3**, 838–845.

Heckerman, D. and Shachter, R. (1995) Decision-theoretic foundations for causal reasoning. *Journal of Artificial Intelligence Research*, **3**, 405–430.

Heckman, J. (2005) The scientific model of causality. *Sociological Methodology*, **35**, 1–97.

Holland, P. (1995) Some reflections on Freedman's critiques. *Foundations of Science*, **1**, 50–57.

Imai, K., Keele, L. and Tingley, D. (2010a) A general approach to causal mediation analysis. *Psychological Methods*, **15**, 309–334.

Imai, K., Keele, L., Tingley, D. and Yamamoto, T. (2010b) Causal mediation analysis using R, in *Advances in Social Science Research Using R* (ed H. Vinod), Springer Lecture Notes in Statistics. New York: Springer, pp. 129–154; <http://imai.princeton.edu/research/mediationR.html>.

Imai, K., Keele, L. and Yamamoto, T. (2010c). Identification, inference, and sensitivity analysis for causal mediation effects. *Statistical Science*, **25**, 51–71.

Imbens, G. (2010) An economists perspective on of Shadish (2010) and West and Thoemmes (2010). *Psychological Methods*, **15**, 47–55.

Jo, B. (2008) Causal inference in randomized experiments with mediational processes. *Psychological Methods*, **13**, 314–336.

Joffe, M. and Green, T. (2009) Related causal frameworks for surrogate outcomes. *Biometrics*, 530–538.

Joffe, M., Small, D. and Hsu, C.-Y. (2007) Defining and estimating intervention effects for groups that will develop an auxiliary outcome. *Statistical Science*, **22**, 74–97.

Judd, C. and Kenny, D. (1981) Process analysis: estimating mediation in treatment evaluations. *Evaluation Review*, **5**, 602–619.

Kaufman, J. (2010) Invited commentary: decomposing with a lot of supposing. *American Journal of Epidemiology*, **172**, 1349–1351.

Kraemer, H., Kiernan, M., Essex, M. and Kupfer, D. (2008) How and why criteria defining moderators and mediators differ between the Baron and Kenny and MacArthur approaches. *Health Psychology*, **27**, S101–S108.

Lauritzen, S. (2004) Discussion on causality. *Scandinavian Journal of Statistics*, **31**, 189–192.

Li, Y., Schneider, J. and Bennett, D. (2007) Estimation of the mediation effect with a binary mediator. *Statistics in Medicine*, **26**, 3398–3414.

MacKinnon, D. (2008) *Introduction to Statistical Mediation Analysis*. New York: Lawrence Erlbaum Associates.

MacKinnon, D. and Dwyer, J. (1993) Estimating mediated effects in prevention studies. *Evaluation Review*, **4**, 144–158.

MacKinnon, D., Fairchild, A. and Fritz, M. (2007a) Mediation analysis. *Annual Review of Psychology*, **58**, 593–614.

MacKinnon, D., Lockwood, C., Brown, C., Wang, W. and Hoffman, J. (2007b) The intermediate endpoint effect in logistic and probit regression. *Clinical Trials*, **4**, 499–513.

Mealli, F. and Rubin, D. (2003) Assumptions allowing the estimation of direct causal effects. *Journal of Econometrics*, **112**, 79–87.

Molenberghs, G., Buyse, M., Geys, H., Renard, D., Burzykowski, T. and Alonso, A. (2002) Statistical challenges in the evaluation of surrogate endpoints in randomized trials. *Controlled Clinical Trials*, **23**, 607–625.

Morgan, S. and Winship, C. (2007) *Counterfactuals and Causal Inference: methods and principles for social research (analytical methods for social research).* New York: Cambridge University Press.

Mortensen, L., Diderichsen, F., Smith, G. and Andersen, A. (2009) The social gradient in birthweight at term: quantification of the mediating role of maternal smoking and body mass index. *Human Reproduction*, **24**, 2629–2635.

Muller, D., Judd, C. and Yzerbyt, V. (2005) When moderation is mediated and mediation is moderated. *Journal of Personality and Social Psychology*, **89**, 852–863.

Pearl, J. (1993) Aspects of graphical models connected with causality, in *Proceedings of the 49th Session of the International Statistical Institute*. Tome LV, Book 1, Florence, Italy.

Pearl, J. (1995) Causal diagrams for empirical research. *Biometrika*, **82**, 669–710.

Pearl, J. (1998) Graphs, causality, and structural equation models. *Sociological Methods and Research*, **27**, 226–284.

Pearl, J. (2000) *Causality: models, reasoning, and inference*. New York: Cambridge University Press. 2nd edn, 2009.

Pearl, J. (2001) Direct and indirect effects, in *Proceedings of the Seventeenth Conference on Uncertainty in Artificial Intelligence*. San Francisco, CA: Morgan Kaufmann, pp. 411–420.

Pearl, J. (2009a) *Causality: models, reasoning, and inference*, 2nd edn. New York: Cambridge University Press.

Pearl, J. (2009b) Causal inference in statistics: an overview. *Statistics Surveys*, **3**, 96–146; <http://ftp.cs.ucla.edu/pub/stat_ser/r350.pdf>.

Pearl, J. (2010) An introduction to causal inference. *The International Journal of Biostatistics*, **6**, Article 7. DOI: 10.2202/1557–4679.1203. Available at: http://www.bepress.com/ijb/vol6/iss2/7/.

Pearl, J. (2011) Principal stratification – a goal or a tool? *The International Journal of Biostatistics*, **7**, Article 20. DOI: 10.2202/1557-4679.1322. Available at: http://www.bepress.com/ijb/vol7/iss1/20.

Pearl, J. (2012) Interpretable conditions for identifying direct and indirect effects. Technical Report. R-389, <http://ftp.cs.ucla.edu/pub/stat_ser/r389.pdf>, Department of Computer Science, University of California, Los Angeles, CA.

Pearl, J. and Bareinboim, E. (2011) Transportability across studies: a formal approach. Technical Report. R-372, <http://ftp.cs.ucla.edu/pub/stat_ser/r372.pdf>, Department of Computer Science, University of California, Los Angeles, CA.

Pearl, J. and Robins, J. (1995) Probabilistic evaluation of sequential plans from causal models with hidden variables, in *Uncertainty in Artificial Intelligence 11* (eds P. Besnard and S. Hanks). San Francisco, CA: Morgan Kaufmann, pp. 444–453.

Petersen, M. (2011) Compound treatments, transportability, and the structural causal model: the power and simplicity of causal graphs. *Epidemiology*, **22**, 378–381.

Petersen, M., Sinisi, S. and van der Laan, M. (2006) Estimation of direct causal effects. *Epidemiology*, **17**, 276–284.

Preacher, K., Rucker, D. and Hayes, A. (2007) Addressing moderated mediation hypotheses: Theory, methods, and prescriptions. *Multivariate Behavioral Research*, **28**, 185–227.

Prentice, R. (1989) Surrogate endpoints in clinical trials: definition and operational criteria. *Statistics in Medicine*, **8**, 431–440.

Robins, J. (1986) A new approach to causal inference in mortality studies with a sustained exposure period – applications to control of the healthy workers survivor effect. *Mathematical Modeling*, **7**, 1393–1512.

Robins, J. (2001) Data, design, and background knowledge in etiologic inference. *Epidemiology*, **12**, 313–320.

Robins, J. (2003) Semantics of causal DAG models and the identification of direct and indirect effects, in *Highly Structured Stochastic Systems* (eds P. Green, N. Hjort and S. Richardson). Oxford: Oxford University Press, pp. 70–81.

Robins, J. and Greenland, S. (1992) Identifiability and exchangeability for direct and indirect effects. *Epidemiology*, **3**, 143–155.

Robins, J. and Richardson, T. (2011) Alternative graphical causal models and the identification of direct effects, in *Causality and Psychopathology, Finding the Determinants of Disorder and their Cures* (eds P. E. Shrout, K. M. Keyes and K. Ornstein). New York: Oxford University Press, pp. 103–158.

Robins, J., Rotnitzky, A. and Vansteelandt, S. (2007) Discussion of principal stratification designs to estimate input data missing due to death. *Biometrics*, **63**, 650–654.

Robins, J., Richardson, T. and Spirtes, P. (2009) On identification and inference for direct effects. Technical Report, Harvard University, MA.

Rosenbaum, P. and Rubin, D. (1983) The central role of propensity score in observational studies for causal effects. *Biometrika*, **70**, 41–55.

Rubin, D. (1974) Estimating causal effects of treatments in randomized and nonrandomized studies. *Journal of Educational Psychology*, **66**, 688–701.

Rubin, D. (2004) Direct and indirect causal effects via potential outcomes. *Scandinavian Journal of Statistics*, **31**, 161–170.

Rubin, D. (2005) Causal inference using potential outcomes: design, modeling, decisions. *Journal of the American Statistical Association*, **100**, 322–331.

Rubin, D. (2010) Reflections stimulated by the comments of Shadish (2010) and West and Thoemmes (2010). *Psychological Methods*, **15**, 38–46.

Shadish, W., Cook, T. and Campbell, D. (2002) *Experimental and Quasi-Experimental Design for Generalized Causal Inference*. Boston, MA: Houghton-Mifflin.

Shpitser, I. and VanderWeele, T. (2011) A complete graphical criterion for the adjustment formula in mediation analysis. *The International Journal of Biostatistics*, **7**, Article 16.

Shrout, P. and Bolger, N. (2002) Mediation in experimental and nonexperimental studies: new procedures and recommendations. *Psychological Methods*, **7**, 422–445.

Sjölander, A. (2009) Bounds on natural direct effects in the presence of confounded intermediate variables. *Statistics in Medicine*, **28**, 558–571.

Sjölander, A. (2011) Reaction to Pearl's critique of principal stratification. *The International Journal of Biostatistics*, **7**, Article 22. DOI: 110.2202/1557-4679.1324. Available at: http://www.bepress.com/ijb/vol7/iss1/22.

Sobel, M. (1987) Direct and indirect effects in linear structural equation models. *Sociological Methods and Research*, **16**, 1155–1176.

Sobel, M. (2008) Identification of causal parameters in randomized studies with mediating variables. *Journal of Educational and Behavioral Statistics*, **33**, 230–231.

Tian, J. and Shpitser, I. (2010) On identifying causal effects, in *Heuristics, Probability and Causality: a tribute to Judea Pearl* (eds R. Dechter, H. Geffner and J. Halpern). UK: College Publications, pp. 415–444.

van der Laan, M.J. and Rubin, D. (2006) Targeted maximum likelihood learning. *The International Journal of Biostatistics*, **2**, Article 11. Available at: http://www.bepress.com/ijb/vol2/iss1/11.

VanderWeele, T. (2008) Simple relations between principal stratification and direct and indirect effects. *Statistics and Probability Letters*, **78**, 2957–2962.

VanderWeele, T. (2009) Marginal structural models for the estimation of direct and indirect effects. *Epidemiology*, **20**, 18–26.

VanderWeele, T. and Robins, J. (2007) Four types of effect modification: a classification based on directed acyclic graphs. *Epidemiology*, **18**, 561–568.

Wilkinson, L., The Task Force on Statistical Inference and APA Board of Scientific Affairs (1999) Statistical methods in psychology journals: guidelines and explanations. *American Psychologist*, **54**, 594–604.

Winship, C. and Mare, R. (1983) Structural equations and path analysis for discrete data. *The American Journal of Sociology*, **89**, 54–110.

13

The sufficient cause framework in statistics, philosophy and the biomedical and social sciences

Tyler J. VanderWeele
Departments of Epidemiology and Biostatistics, Harvard School of Public Health, Boston, Massachusetts, USA

13.1 Introduction

The sufficient cause framework (Cayley, 1853; Mackie, 1965; Rothman, 1976) conceptualizes causation as a collection of different sufficient conditions or causes for the occurrence of an event. Each sufficient condition or cause is usually conceived of as consisting of various (necessary) components such that if all components are present the sufficient cause is complete and the event or outcome occurs. Whereas the principal focus of the counterfactual approach to conceiving of causation, dominant in statistics (Neyman, 1923; Lewis, 1973; Rubin, 1974), is the cause or intervention itself, the sufficient cause framework instead considers primarily the effect. Said another way, the counterfactual framework asks questions of the effects of causes, the sufficient cause framework of the causes of effects.

In this chapter we review the development and use of the sufficient cause framework in statistics, philosophy and the biomedical and social sciences. We will briefly mention some more recent developments and we will consider some alternative notions of sufficiency in causal inference (Pearl, 2009).

Causality: Statistical Perspectives and Applications, First Edition. Edited by Carlo Berzuini, Philip Dawid and Luisa Bernardinelli.

13.2 The sufficient cause framework in philosophy

Within the philosophical literature, the sufficient cause framework is most closely associated with the work of John Mackie (1965). Mackie himself, however, notes that two precedents to his own proposal are to be found in the writings of Marc-Wogau (1962) and Scriven (1964). Much of Mackie's (1965) paper offers refinements of this prior work and attempts to address problematic examples that are raised by these two prior authors. Mackie proposed that in general when we refer to something as a cause it is known to be a 'an *insufficient* but *necessary* part of a condition which is itself *unnecessary* but *sufficient* for the result'. Mackie used the term 'INUS condition' as a shorthand for the expression in quotations in the previous sentence. The term 'INUS' is derived from the first letter of each of the italicized words. Mackie notes that sometimes a cause A is itself sufficient for an event P and thus later refines his definition of an INUS condition so that A is an INUS condition if 'there is a condition which . . . is necessary and sufficient for P, and which is of one of these forms: (AX or Y), (A or Y), AX, A'. Mackie sometimes refers to this latter requirement as being that 'A is *at least* an INUS condition'.

A substantial portion of the philosophical literature on causation has been occupied with the problem of actual causation, the task of trying to give a characterization of the rules by which we make judgements of the form 'A caused P'. Instances in which we say 'A caused P' are sometimes referred to as instances of singular causation and indicate A caused P on this particular occasion. Mackie's interest was not simply in under what conditions we refer to something as a 'cause' but also in what we mean by expressions such as 'A caused P'. Mackie states that claims of singular causation, i.e. statements of the form 'A caused P', implicitly make the following claims: (i) A is at least an INUS condition, (ii) A was present, (iii) the factors represented by X were present, (iv) every disjunct in Y that does not contain A as a conjunct was absent. Mackie further notes that often with statements such as 'A caused P', there are a set of background characteristics that are assumed held fixed when we make these statements. Mackie refers to these as a field and claims our statements about causation are generally relative to a field. Mackie does not claim that (i) to (iv) contain everything we mean by 'A caused P', but claims that we generally mean at least (i) to (iv).

Although Mackie's attempt to characterize singular causation in terms of the INUS condition came under criticism for not adequately distinguishing between causes and effects and temporal ordering (Scriven, 1966; Pearl, 2009), the notion of an INUS condition exerted considerable influence on subsequent literature. Some of the subsequent attempts to provide an account for actual causation (Mackie, 1980; Hiddleston, 2005) have drawn on notions of INUS conditions.

13.3 The sufficient cause framework in epidemiology and biomedicine

Early development of the sufficient cause framework in epidemiology appears in the writings of MacMahon and Pugh (1967). The sufficient cause framework, however, came to popularity in epidemiology principally through the writing of Rothman (1976). Rothman provided a graphical schematic for the sufficient causes that have informally come to be known as 'causal pies'. If, for example, there were two sufficient causes for an outcome D, say ABC and EFG, we might present these sufficient causes as two 'causal pies', as in Figure 13.1.

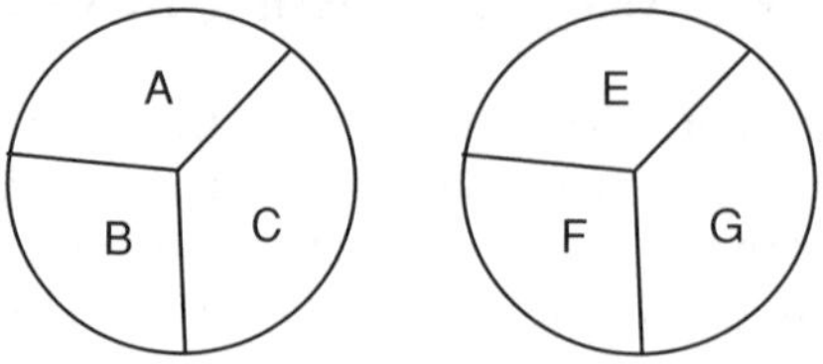

Figure 13.1 Visual depiction of sufficient causes; often referred to as Rothman's 'causal pies'.

Rothman conceived of each of the sufficient causes or causal pies as a mechanism for bringing about the outcome. Whenever a sufficient cause was complete, the mechanism was set in motion and the outcome would inevitably occur. Rothman referred to each component of a sufficient cause as a 'component cause' or simply as a 'cause'. Rothman's conceptualization thus essentially matches that of Mackie.

Rothman's conceptualization gained popularity not so much as an analytic tool but as a framework in which to understand more easily several important principles within epidemiology. For example, the representation of causal pies make evident that the strength of a causal association between a particular exposure and outcome will vary substantially with the prevalence of the other components of the sufficient cause(s) in which the exposure of interest is present. Causal pies also make clear that, for multiple exposures, the sum of attributable fractions (the proportion of the disease attributable to an exposure; see Miettinen, 1974) need not equal 1 because of the joint presence of multiple causes in the same sufficient cause. The sufficient cause framework also allowed epidemiologists to conceive of 'interaction' or 'synergism' in a manner that was not dependent on a particular statistical model. Rothman conceived of synergism between two factors as their joint presence in a particular sufficient cause. This understanding of synergism received considerable attention from epidemiologists in the decades that followed, a point we return to below. Because of the transparency of these observations when using 'causal pies', the sufficient cause framework came to be taught quite routinely in introductory courses within epidemiology.

Beyond its use in pedagogy, the sufficient cause framework has been the focus of attention within the epidemiologic literature in two other respects. First, starting within the work of Miettinen (1982) and Greenland and Poole (1988) there has been effort to relate sufficient causes to the potential outcomes or counterfactual framework. Following Miettinen (1982), Greenland and Poole (1988) considered a setting with two binary causes, X_1 and X_2, and a binary outcome D and considered the potential responses ('counterfactual outcomes') for individuals under different combinations of the exposures so that $D_{x_1x_2}$ denotes the potential outcome that would have occurred had X_1 been set to x_1 and X_2 to x_2. With binary X_1 and X_2, because there are four possible exposure combinations, each individual has four potential outcomes $(D_{11}, D_{10}, D_{01}, D_{00})$. The four potential outcomes defined what was called an individual's 'response type'; because there were four different potential outcomes, there were $2^4 = 16$ different possible response types. Within this setting of two binary causes of interest, Greenland and Poole (1988) also noted that one could conceive of nine different possible sufficient causes, U_0, U_1X_1, $U_2\overline{X_1}$, U_3X_2, $U_4\overline{X_2}$, $U_5X_1X_2$, $U_6\overline{X_1}X_2$, $U_7X_1\overline{X_2}$ and $U_8\overline{X_1}\,\overline{X_2}$, each involving the presence of X_i or the absence of X_i (denoted by $\overline{X_i}$) or being unrelated to X_i, and each also possibly involving an additional background cause U_j. Greenland and Poole noted the presence or absence of different sufficient causes (indicated by $U_j = 1$ or $U_j = 0$) would give rise to different response types (cf. Greenland and Brumback, 2002).

More formal treatments were given by Flanders (2006) for one-cause sufficient models, VanderWeele and Robins (2008) for two-cause models and VanderWeele and Richardson (2012) for sufficient cause models with an arbitrary number of causes. For example, let $\omega \in \Omega$ index individuals and let $\vee$ denote the disjunctive operator defined for binary A and B as $A \vee B = 1$ if and only if at least one of A or B is 1. In the case of two binary causes, it can be shown (VanderWeele and Robins, 2008) that for any collection of potential outcomes $((D_{00}(\omega), D_{01}(\omega), D_{10}(\omega), D_{11}(\omega) : \omega \in \Omega)$ there exist binary variables $U_0(\omega), \ldots, U_8(\omega)$ that are functions of the potential outcomes $\{D_{00}(\omega), D_{01}(\omega), D_{10}(\omega), D_{11}(\omega)\}$ such that

$$\begin{aligned} D_{x_1x_2} &= U_0 \vee U_1x_1 \vee U_2(1-x_1) \vee U_3x_2 \vee U_4(1-x_2) \vee U_5x_1x_2 \\ &\vee U_6(1-x_1)x_2 \vee U_7x_1(1-x_2) \vee U_8(1-x_1)(1-x_2) \end{aligned} \tag{13.1}$$

Any set of potential outcomes can be replicated by a sufficient cause model. Any set of variables $(U_0, \ldots, U_8)$ satisfying (13.1) VanderWeele and Robins (2008) called a sufficient cause representation. Such a formalization was useful in deriving results for synergism within the sufficient cause framework. A sufficient cause interaction was said to be present between X_1 and X_2 if for every sufficient cause representation there exists an ω such that $U_5(\omega) \neq 0$, i.e., if there is a sufficient cause with both X_1 and X_2 present. It was moreover shown that a sufficient cause interaction was present if and only if there existed an individual ω such that $D_{11}(\omega) = 1$ but $D_{10}(\omega) = D_{01}(\omega) = 0$. We assume no measurement error or selection bias. Under a further assumption of exogeneity (also called 'no confounding' or 'ignorability') that $D_{x_1x_2} \perp\!\!\!\perp (X_1, X_2)$ then the empirical condition

$$p_{11} - p_{10} - p_{01} > 0 \tag{13.2}$$

implies a sufficient cause interaction (VanderWeele and Robins, 2007, 2008) where $p_{x_1x_2} = P(D = 1|X_1 = x_1, X_2 = x_2)$. Under a monotonicity assumption that $D_{x_1x_2}(\omega)$ is nondecreasing in x_1 and x_2 for all ω,

$$p_{11} - p_{10} - p_{01} + p_{00} > 0$$

implies a sufficient cause interaction (Rothman and Greenland, 1998; VanderWeele and Robins, 2007). The condition under monotonicity corresponds to standard tests for additive interaction but the condition without monotonicity in (13.2) gives rise to tests that differ from tests for nonzero interactions in standard statistical models (VanderWeele, 2009a).

One point of ambiguity within Rothman's description concerned the latent period and this point of ambiguity has led to somewhat different interpretations of the model. The latent period may be conceived of as the time between all the components of a sufficient cause being present (i.e., the sufficient cause is complete) and when the outcome actually occurs. Rothman (1976, p. 591) wrote 'At the point in time at which the sufficient cause is completed, the disease process is set in motion, though usually not yet manifest. The latent period is the interval during which a sufficient cause accumulates plus the time it takes for the disease to become manifest'. It is unclear from this passage whether the outcome is thought to be the 'disease process being set in motion' or the 'disease process becoming manifest'. If the former, then the outcome occurs as soon as the sufficient cause is completed and essentially at most one sufficient cause can be complete for nonrecurrent events. If the latter, then the disease process could be set in motion but before the disease is manifest and a second sufficient cause might become complete, which perhaps has a shorter latent period, and is then the actual cause of the event. This latter scenario is referred to as one of preemption in the philosophical literature. Conceived of another way, Rothman's original paper left open the question of whether the completion of the latent period should itself be viewed as a component of the sufficient cause.

The distinction between the two leads to subtle differences within the sufficient cause framework. In the former conceptualization, if the outcome occurs as soon as the first sufficient cause is complete then there ought in general only be one sufficient cause that is ever complete. In the latter conceptualization a sufficient cause may be complete but not the actual cause of the outcome. This would occur if the sufficient cause would result in the outcome (had all other sufficient causes been blocked) but was in fact preempted by the action of another sufficient cause. Suppose, for example, an individual dies if either of two members of a firing squad pulls the trigger of his gun; in actual fact, one individual pulls the trigger before the other but both do in the end fire. In the former conceptualization, only one sufficient cause is complete (that involving the gunman pulling his trigger first) and the other sufficient cause (the second gunman killing the individual) would include as a component the first gunman's not firing. In the second conceptualization, both sufficient causes are complete because either would be sufficient for the outcome in the absence of the other (though the first sufficient cause is the actual cause of the individual's death). A disadvantage of the first conceptualization is that one must have information about the actual cause of an event to say whether a sufficient cause was completed and the notion of actual causation has proved difficult to characterize, as discussed below. In the second conceptualization, this issue is essentially sidestepped since a sufficient cause simply needs to be such that its completion would imply the outcome even if all other sufficient causes were blocked. The formulation given by VanderWeele and Robins (2008) corresponds to the second conceptualization of the framework. See Poole (2010) for a formulation of the first. Note also that the formulation given by VanderWeele and Robins (2008) includes within it the first conceptualization as one possible representation of the potential outcomes. One of the advantages of the tests for sufficient cause interactions such as that in (13.2) above is that the conclusions apply to all possible sufficient cause representations for the potential outcomes; one need not decide between the two conceptualizations nor is it necessary to have any knowledge about actual causes.

More recent work on the sufficient cause framework in epidemiology has focused on methods for estimating an attributable fraction for specific sufficient causes (Hoffmann *et al.*, 2006; Heidemann *et al.*, 2007; VanderWeele, 2010a; Liao and Lee, 2010), where the attributable fraction for a cause X is generally defined as $[P(Y = 1) - P(Y_{x=0} = 1)]/P(Y = 1)$, i.e., the proportion of the outcome that could be eliminated by eliminating the exposure X (Miettinen, 1974; Robins and Greenland, 1988). Other work in epidemiology has related the sufficient cause framework to notions of antagonism (VanderWeele and Knol, 2011) and to questions of mediation (Hafeman, 2008; VanderWeele, 2009b; Suzuki *et al.*, 2011), i.e., assessing the extent to which an exposure affects the outcome through some intermediate and the extent to which it is through other pathways.

Another notable development concerning sufficient causation in epidemiology and biomedicine was a book length treatment of the subject given by Aickin (2002). Some of Aickin's treatment parallels that of the epidemiologic literature that preceded it; however, Aickin goes a good deal further in considering multiple causes and the effects of specific sufficient causes. Unfortunately, Aickin does not in any way tie his work to counterfactuals and as a result many of the assumptions made are not transparent and often implausible and the temporal aspects of causation are generally unclear (VanderWeele and Robins, 2009).

13.4 The sufficient cause framework in statistics

The origins of the sufficient cause framework in statistics can be traced fairly far back to the work of Cayley (1853). Cayley presented a simple sufficient cause model with two causes of the outcome such that (i) the two causes occur independently of one another, (ii) the causes occur independently of what was called above the background causes (what Cayley calls 'the cause acting efficaciously') and (iii) the background causes are independent of one another. Cayley then goes on to relate the probabilities of the causes and the background causes occurring to the observed probability that the effect occurs when one or the other of the causes of interest is present and to the overall probability of the effect.

In spite of its early beginnings in Cayley's work, there has been relatively little development within statistics on the sufficient cause framework until recently and the recent work and developments in the statistics literature have been largely derivative on that in the epidemiologic literature. The dominance and effectiveness of the potential outcomes framework and its relatively early appearance in the contemporary statistics literature (Rubin, 1974) arguably made it the focus of causal reasoning in the field of statistics. The sufficient cause framework was essentially neglected.

More recent work on the sufficient causes framework in statistics has focused on multiply robust inference for tests for sufficient cause interactions (Vansteelandt *et al.*, 2008), relating the sufficient cause framework to structural models and causal diagrams (VanderWeele and Robins, 2009) and extensions of the framework to allow for n-variable sufficient cause models (VanderWeele and Richardson, 2012), for categorical and ordinal rather than binary causes (VanderWeele, 2010b), for ordinal or continuous exposures that have been dichotomized (VanderWeele et al., 2011; Berzuini and Dawid, 2012) and for stochastic sufficient causes and stochastic counterfactuals (VanderWeele and Robins, 2012).

13.5 The sufficient cause framework in the social sciences

The sufficient cause framework has also exerted some influence on causal thinking in the social sciences, principally within psychology and law. Within the psychology literature a specific class of sufficient cause models was considered by Cheng (1997) and Novick and Cheng (2004) which they refer to as a 'causal powers' approach. In their model, the background causes, U_j above, are assumed statistically independent of one another and the causes of interest, X_1 and X_2, are assumed independent of the background causes. Similar assumptions were in fact made in some of the early literature in epidemiology on the sufficient cause framework and synergism (Rothman, 1974; Weinberg, 1986). The model of Cheng (1997) and Novick and Cheng (2004) is also closely related to that considered by Cayley (1853).

Cheng (1997) and Novick and Cheng (2004) also make a monotonicity assumption that the causes of interest never prevent the outcome (or alternatively never cause, only prevent, the outcome). Thus, in the case of one cause of interest, the model considered by Novick and Cheng (2004) is essentially $D_x = U_0 \vee U_1 x$ with (i) U_0 statistically independent of U_1 and (ii) U_1 statistically independent of X. In the case of two causes, the model is $D_{x_1 x_2} = U_0 \vee U_1 x_1 \vee U_2 x_2 \vee U_3 x_1 x_2$ with (i) (U_0, U_1, U_2, U_3) all mutually independent and (ii) (U_1, U_2, U_3) independent of (X_1, X_2).

In the case of one cause of interest with the model, $D_x = U_0 \vee U_1 x$, the generative 'causal power' of X is then defined as $P(U_1 = 1)$. Cheng (1997) shows that under the model considered,

$$P(U_1=1)=\frac{P(D=1|X=1)-P(D=1|X=0)-\{P(U_0=1|X=1)-P(U_0=1|X=0)\}}{1-P(U_0=1|X=1)}$$

When U_0 and X are also statistically independent (a condition referred to as 'no-confounding' by Cheng, 1997) they obtain $P(U_0 = 1) = P(D = 1|X = 0)$ and

$$P(U_1 = 1) = \frac{P(D = 1|X = 1) - P(D = 1|X = 0)}{1 - P(D = 1|X = 0)}$$

For the case of two causes of interest the generative causal powers for X_1 and X_2 are $P(U_1 = 1)$ and $P(U_2 = 1)$ respectively and the generative power of the conjunction X_1X_2 can be defined to be $P(U_3 = 1)$. If, in addition to the model assumptions that (i) (U_0, U_1, U_2, U_3) are all mutually independent and (ii) (U_1, U_2, U_3) is independent of (X_1, X_2), we also have a 'no-confounding' assumption that (iii) U_0 is independent of (X_1, X_2); then Novick and Cheng (2004) show that

$$P(U_0 = 1) = P(D = 1|X_1 = 0, X_2 = 0)$$

$$P(U_1 = 1) = \frac{P(D = 1|X_1 = 1, X_2 = 0) - P(D = 1|X_1 = 0, X_2 = 0)}{P(D = 0|X_1 = 0, X_2 = 0)}$$

$$P(U_2 = 1) = \frac{P(D = 1|X_1 = 0, X_2 = 1) - P(D = 1|X_1 = 0, X_2 = 0)}{P(D = 0|X_1 = 0, X_2 = 0)}$$

$$P(U_3 = 1) = 1 - \frac{P(D = 0|X_1 = 1, X_2 = 1)P(D = 0|X_1 = 0, X_2 = 0)}{P(D = 0|X_1 = 1, X_2 = 0)P(D = 0|X_1 = 0, X_2 = 1)}$$

Novick and Cheng (2004) also consider cases in which one or both of X_1 and X_2 are preventive rather than causative and consider relations among causal powers when the no-confounding assumption does not hold.

An interesting feature of the model proposed by Novick and Cheng (2004) in the context of the psychology of learning is that it is principally descriptive rather than normative. The proposal concerns principally how we ordinarily go about causal reasoning, not necessarily how we ought to. Thus, although the assumption that the background causes are independent of one another will often be violated, an important aspect of the proposal of Novick and Cheng (2004) is simply that the description is how we generally reason about causality in everyday life nonetheless. They propose that reasoners approximate the causal power calculation when evaluating whether an event might be a cause and in assessing whether different factors interact. Their proposal is subject to empirical study. Novick and Cheng (2004) note that their theory is normative for when covariation implies causation, i.e., which occurs when U_0 is independent of the causes of interest ('no-confounding'). They note also, however, that their account also requires assumptions concerning the independence of the background causes with one another (assumption (i) above) and of the background causes other than U_0 with the causes of interest (assumption (ii) above).

Within academic law, the sufficient cause framework has been used as a tool in determining whether a given factor was involved in the bringing about of the event in question. Building on the linguistic analysis of Hart and Honoré (1959), Wright (1988) proposed that the equivalent of an 'INUS condition' in Mackie's (1965) language or a 'component cause' in Rothman's (1976) language be used as a standard for causation in legal reasoning. Wright referred to such a condition as an 'NESS factor', where the term 'NESS' is derived from the first letter of the italicized words in the expression, '*necessary element* for the sufficiency of a *sufficient set*'. Wright proposed that a condition contributed to some consequence if and only if it was a NESS factor. In some of Wright's writings, NESS is taken not only as a practical test for establishing causation but as the meaning of causation itself. This latter position has come under critique (Fumerton and Kress, 2001) in the legal literature. More recently, Stapleton (2009) has argued that although the 'NESS test' does not capture the meaning of causation it is still useful as a practical algorithm to determine whether a particular factor was involved in the occurrence of some phenomenon and that it can accommodate conditions, omissions and communications of information that are all taken as causes in legal reasoning.

13.6 Other notions of sufficiency and necessity in causal inference

Rather than pursuing the sufficient cause framework itself, Pearl (2009) introduces alternative notions for the necessity and sufficiency of a cause. For outcome D and exposure X, Pearl (2009) defines the probability of necessary causation (PN), the probability of sufficient causation (PS) and the probability of necessary and sufficient causation (PNS), respectively, by

$$\begin{aligned} PN &= P(D_{x=0} = 0 | X = 1, D = 1) \\ PS &= P(D_{x=1} = 1 | X = 0, D = 0) \\ PNS &= P(D_{x=1} = 1, D_{x=0} = 0). \end{aligned}$$

The probability of necessary causation captures the probability, for an individual with the exposure and outcome, that the outcome would have been absent if the exposure had been removed. The probability of sufficient causation captures the probability, for an individual without the exposure and without the outcome, that the outcome would have occurred if the exposure had been present. The probability of necessary and sufficient causation captures the probability that the outcome occurs with the exposure but not without.

Pearl goes on to define the probability of disablement as $PD = P(D_{x=0} = 0 | D = 1)$ and the probability of enablement as $PE = P(D_{x=1} = 1 | D = 0)$. Pearl shows that under monotonicity, $D_{x=0} \leq D_{x=1}$, and exogeneity, $D_x \perp\!\!\!\perp X$ (i.e. 'no-confounding' or 'ignorability') these various probabilities of causation are identified. Under these assumptions we have

$$\begin{aligned} PN &= \frac{P(D = 1|X = 1) - P(D = 1|X = 0)}{P(D = 1|X = 1)} \\ PS &= \frac{P(D = 1|X = 1) - P(D = 1|X = 0)}{P(D = 0|X = 0)} \\ PNS &= P(D = 1|X = 1) - P(D = 1|X = 0) \end{aligned}$$

Also under exogeneity alone $PD = P(X = 1)PNS/P(D = 1)$ and $PE = P(X = 0)PNS/P(D = 0)$, thus also implying that PD and PE are identified under monotonicity and exogeneity. Pearl (2009) also gives bounds for these probabilities under the assumption of exogeneity without requiring monotonicity (cf. Tian and Pearl, 2000).

The 'probabilities of causation' defined by Pearl are useful in attributing responsibility. The probability of necessity is arguably of particular importance in legal reasoning since, for a particular instance of an event with a particular exposure present, it captures the probability that the event would not have occurred if the exposure had been absent. It effectively provides a lower bound on the probability that the exposure was the actual cause of the outcome. Although these probabilities of causation are useful in reasoning about responsibility, they are less useful in evaluating prospective policy interventions. The measure used for policy-relevant attribution most commonly employed in the medical and epidemiologic literature is the attributable fraction defined as $AF = [P(D = 1) - P(D_{x=0} = 1)]/P(D = 1)$, i.e., the proportion of the outcome that could be eliminated by eliminating the exposure X (Miettinen, 1974; Greenland and Robins, 1988). Somewhat surprisingly, although Pearl (2009) briefly mentions attributable fractions in a footnote, he does not consider them explicitly in his account, in spite of their importance. This is perhaps in part because the attributable fraction, AF, defined above, cannot be expressed in terms of Pearl's five probabilities of causation, PN, PS, PNS, PE and PD without additional assumptions. Under the assumption of exogeneity that $D_x \perp\!\!\!\perp X$, one can express the attributable fraction in terms of PNS and PS. For example, by using exogeneity and Theorem 9.2.11 of Pearl (2009) one can show that

$$AF = 1 - \frac{1 - PNS/PS}{P(D = 1)}$$

The relation does not have a particularly intuitive interpretation. Moreover, under exogeneity one has the simple result that $AF = [P(D = 1) - P(D = 1|X = 0)]/P(D = 1)$. The difficulty relating the attributable fraction to the probabilities of causation arises because all of the probabilities of causation condition on an event having occurred (or having not occurred) whereas when policy or public health interventions are considered prospectively, one is interested in the effects on the population as a whole.

In considering the issue of actual causation, Pearl (2009) goes on to criticize the sufficient cause framework (Mackie, 1965; Rothman, 1976) on the grounds that language about logical necessity and sufficiency is inadequate for describing causal notions; a logical account is not able to distinguish between stable mechanisms and circumstantial conditions (Pearl, 2009; pp. 313–316). Pearl's critique, however, does not take into account recent work which explicitly ties the sufficient cause framework to counterfactual and structural semantics (VanderWeele and Robins, 2008, 2009; VanderWeele and Richardson, 2011).

More recent work on the probabilities of causation has concerned variance estimators and inference for these probabilities (Cai and Kuroki, 2005; Kuroki and Cai, 2011) and has extended the results on bounds for PNS to settings that involve multiple causes (VanderWeele and Richardson, 2012).

13.7 Conclusion

Notions of necessity and sufficiency have historically played a relatively minor role in formal reasoning in the causal inference literature in statistics. This is probably in part due to what was, for a long time, a lack of empirical measures to capture these notions. However, recent developments in attributable fractions for sufficient causes (e.g. Hoffmann *et al*., 2006) in synergism between sufficient causes (VanderWeele and Robins, 2008) and in the probabilities of causation (Pearl, 2009) have provided formal quantitative measures that can be used in inference. That such measures can be used in empirical applications may increase the prominence that notions of necessity and sufficiency have in causal reasoning within statistics.

An issue that has been briefly touched upon a couple of times in this overview is that of 'actual' or 'singular' or 'token' causation, i.e. statements of the form 'X caused D' - not simply 'X causes D' is general but 'X caused D' in this particular instance. The problem of characterizing actual causation has long eluded philosophers (Hall and Paul, 2003; Collins *et al*., 2004). Every time a proposal is made as to the rules by which we intuitively make judgements of the form 'X caused D', a counterexample is provided shortly thereafter demonstrating that the proposal disagrees with common intuitions. We saw that one of Mackie's central motivations was to provide an account of actual causation, but the account came under criticism for not adequately distinguishing cause and effect and temporal ordering (Scriven, 1966; Pearl, 2009). A recent attempt employing the counterfactuals often used in the causal inference literature in statistics was made by Halpern and Pearl (2005), which was more subtle but ultimately also unsuccessful (Glymour *et al*., 2009; VanderWeele, 2009c). An open question is whether the sufficient cause framework/INUS conditions, when explicitly tied to potential outcomes (Rubin, 1974) or the structural models of Pearl (2009), perhaps along with other recent developments (Glymour *et al*., 2009; Halpern and Hitchcock, 2010; Voortman *et al*., 2010), might prove useful in addressing the question of actual causation.

Acknowledgements

The author thanks Charles Poole and the book's editors for helpful comments.

References

Aickin, M. (2002) *Causal Analysis in Biomedicine and Epidemiology Based on Minimal Sufficient Causation*. New York: Marcel Dekker.

Berzuini, C. and Dawid, A.P. (2012) Inference about biological mechanism on the basis of epidemiological data, in *Causality: Statistical Perspectives and Applications* (eds. C. Berzuini, A.P. Dawid and L. Bernardinelli). Chapter 14. J. Wiley.

Cai, Z. and Kuroki, M. (2005) Variance estimators for three 'probabilities of causation'. *Risk Analysis*, **25**, 1611–1620.

Cayley, A. (1853) Note on a question in the theory of probabilities. *London, Edinburgh and Dublin Philosophical Magazine*, **VI**, 259.

Cheng, P. W. (1997) From covariation to causation: a causal power theory. *Psychological Review*, **104**, 367–405.

Collins, J., Hall, N. and Paul, L.A. (2004) Counterfactual and causation: history, problems and prospects, in, *Causation and Counterfactuals* (eds. J. Collins, N. Hall and L. A. Paul). Cambridge, MA: MIT Press, pp. 1–58.

Flanders, D. (2006) Sufficient-component cause and potential outcome models. *Europeon Journal of Epidemiology*, **21**, 847–853.

Fumerton, R. and Kress, K. (2001) Causation and the law: preemption, lawful sufficiency, and causal sufficiency. *Law and Contemporary Problems*, **64**, 83–105.

Glymour, C., Danks, D., Glymour, B., Eberhardt, F., Ramsey, J., Scheines, R., Spirtes, P., Teng, C.M. and Zhang, J. (2010) Actual causation: a stone soup essay. *Synthese*, **75**, 169–192.

Greenland, S. and Brumback, B. (2002) An overview of relations among causal modelling methods. *International Journal of Epidemiology*, **31**, 1030–1037.

Greenland, S. and Poole, C. (1988) Invariants and noninvariants in the conceptof interdependent effects. *Scandinavian Journal of Work Environmental Health*, **14**, 125–129.

Greenland, S. and Robins, J.M. (1988) Conceptual problems in the definition and interpretation of attributable fractions. *American Journal of Epidemiology*, **128**, 1185–1197.

Hafeman, D. (2008) A sufficient cause based approach to the assessment of mediation. *European Journal of Epidemiology*, **23**, 711–721.

Hall, N. and Paul, L.A. (2003) Causation and preemption, in *Philosophy of Science Today* (eds. P. Clark and K. Hawley). Oxford: Oxford University Press, pp. 100–129.

Halpern, J.Y. and Hitchcock, C. (2010) Actual causation and the art of modeling, in *Heuristics, Probability and Causality: a* tribute to Judea Pearl (eds. R. Detchter, H. Geffner and J.Y. Halpern). London: College Publications, pp. 383–406.

Halpern, J. Y. and Pearl, J. (2005) Causes and explanations: a structural-model approach. Part I: causes. *British Journal of the Philosophy of Science*, **56**, 843–887.

Hart, H.L.A. and Honoré, A.M. (1959) *Causation in the Law*. Oxford: Oxford University Press.

Heidemann, C., Hoffmann, K., Klipstein-Grobusch, K., Weikert, C., Pischon, T., Hense, H.W. and Boeing, H. (2007) Potentially modifiable classic risk factors and their impact on incident myocardial infarction: results from the EPIC-Potsdam study. *European Journal of Cardiovascular Prevention and Rehabilitation*, **14**, 65–71.

Hiddleston, E. (2005) Causal powers. *British Journal of the Philosophy of Science*, **56**, 27–59.

Hoffmann, K., Heidemann, C., Weikert, C., Schulze, M.B. and Boeing, H. (2006) Estimating the proportion of disease due to classes of sufficient causes. *American Journal of Epidemiology*, **163**, 76–83.

Koopman, J.S. (1981) Interaction between discrete causes. *American Journal of Epidemiology*, **113**, 716–724.

Kuroki, M. and Cai, Z. (2011) Statistical analysis of 'probabilities of causation' using covariate information. *Scandinavian Journal of Statistics*, **38**, 564–577.

Liao, S.F. and Lee, W.C. (2010) Weighing the causal pies in case-control studies. *Annals of Epidemiology*, **20**, 568–573.

Mackie, J.L. (1965) Causes and conditions. *American Philosophical Quarterly*, **2**, 245–255.

Mackie, J.L. (1980) *The Cement of the Universe: a study of causation*. Oxford: Clarendon Press.

Marc-Wagau, K. (1962) On historical explanation. *Theoria*, **28**, 213–233.

Miettinen, O.S. (1974) Proportion of disease caused or prevented by a given exposure, trait or intervention. *American Journal of Epidemiology*, **99**, 325–332.

Miettinen, O.S. (1982) Causal and preventive interdependence: elementary principles. *Scandinavian Journal of Work Environmental Health*, **8**, 159–168.

Neyman, J. (1923) Sur les applications de la thar des probabilities aux experiences agaricales: essay des principle, in *Statistical Science 5* (eds. D. Dabrowska and T. Speed). Excerpts in English, pp. 463–472.

Novick, L.R. and Cheng, P.W. (2004) Assessing interactive causal influence. *Psychological Review*, **111**, 455–485.

Pearl, J. (2009) *Causality: Models, Reasoning, and Inference*, 2nd edn. Cambridge: Cambridge University Press.

Poole, C. (2010) How many are affected? A real limit of epidemiology. *Epidemiologic Perspectives and Innovations*, **7**, Article 6.

Rothman, K.J. (1974) Synergy and antagonism in cause–effect relationships. *American Journal of Epidemiology*, **99**, 385–388.

Rothman, K.J. (1976) Causes. *American Journal of Epidemiology*, **104**, 587–592.

Rothman, K.J. and Greenland, S. (1998) *Modern Epidemiology*. Philadelphia, PA: Lippincott-Raven.

Rubin, D.B. (1974) Estimating causal effects of treatments in randomized and nonrandomized studies. *Journal of Educational Psychololy*, **66**, 688–701.

Scriven, M. (1964) Review of Nagel's 'Structure of Science.' *Review of Metaphysics*, **17**, 403–424.

Scriven, M. (1966) Defects of the necessary condition analysis of causation, in *Philosophical Analysis and History* (ed. W. Dray). Harper Collins.

Stapleton, J. (2009) Causation in the law, in *The Oxford Handbook of Causation* (eds. H. Beebee, C. Hitchcock and P. Menzies). Oxford: Oxford University Press.

Suzuki, E., Yamamoto, E. and Tsuda, T. (2011) Identification of operating mediation and mechanism in the sufficient-component cause framework. *European Journal of Epidemiology*, **26**, 347–357.

Tian, J. and Pearl, J. (2000) Probabilities of causation: bounds and identification. *Annals of Mathematics and Artificial Intelligence*, **28**, 287–313.

VanderWeele, T.J. (2009a) Sufficient cause interactions and statistical interactions. *Epidemiology*, **20**, 6–13.

VanderWeele, T.J. (2009b) Mediation and mechanism. *European Journal of Epidemiology*, **24**, 217–224.

VanderWeele, T.J. (2009c) Criteria for the characterization of token causation. *Logic and Philosophy of Science*, 115–127.

VanderWeele, T.J. (2010a) Attributable fractions for sufficient cause interactions. *International Journal of Biostatistics (Special Issue for the Proceedings of the BIRS Workshop on Causal Inference in Statistics and the Quantitative Sciences)*, **10** (2), Article 5:1–26.

VanderWeele, T.J. (2010b) Sufficient cause interactions for categorical and ordinal exposures with three levels. *Biometrika*, **97**, 647–659.

VanderWeele, T.J. and Richardson, T.S. (2012) General theory for interactions in sufficient cause models with dichotomous exposures. *Annals of Statistics*, conditionally accepted.

VanderWeele, T.J. and Robins, J.M. (2007) The identification of synergism in the sufficient-component cause framework. *Epidemiology*, **18**, 329–339.

VanderWeele, T.J. and Robins, J.M. (2008). Empirical and counterfactual conditions for sufficient cause interactions. *Biometrika*, **95**, 49–61.

VanderWeele, T.J. and Robins, J.M. (2009) Minimal sufficient causation and directed acyclic graphs. *Annals of Statistics*, **37**, 1437–1465.

VanderWeele, T.J., Chen, Y. and Ahsan, H. (2011) Inference for causal interactions for continuous exposures under dichotomization. *Biometrics*, **67**, 1414–1421.

VanderWeele, T.J. and Robins, J.M. (2012) Stochastic counterfactuals and sufficient causes. Revised for *Statistica Sinica*, **22**, 379–392.

Vansteelandt, S., VanderWeele, T.J., Tchetgen, E.J. and Robins, J.M. (2008) Multiply robust inference for statistical interactions. *Journal of the American Statistical Association*, **103**, 1693–1704.

Voortman, M., Dash, D. and Druzdzel, M. (2010) Learning why things change: the difference-based causality learner, in *Proceedings of the 26th Conference on Uncertainty in Artificial Intelligence*.

Weinberg, C.R. (1986) Applicability of the simple independent action model to epidemiologic studies involving two factors and a dichotomous outcome. *American Journal of Epidemiology*, **123**, 162–173.

Wright, R.W. (1988) Causation, responsibility, risk, probability, naked statistics, and proof: pruning the bramble bush by clarifying the concepts. *Iowa Law Review*, **73**, 1001–1077.

14

Analysis of interaction for identifying causal mechanisms

Carlo Berzuini[1], Philip Dawid,[1] Hu Zhang[2] and Miles Parkes[2]

[1]*Statistical Laboratory, Centre for Mathematical Sciences, University of Cambridge, Cambridge, UK*

[2]*Department of Gastroenterology, Addenbrookes Hospital, Cambridge, UK*

14.1 Introduction

Great attention has been paid, in this book, to questions about the effect of actions (EOA), for example, 'What is the effect of causal action X?' With the exception of the preceding three chapters, little has been done to address questions about the 'mechanisms of effects' (MOE), that is, questions concerning the 'how' of effects, how and why they occur, how can we explain their occurrence and what are the mechanisms through which these effects operate. The following examples should help see the distinction:

(EOA) 'What is the effect of physical inactivity on risk of infarction, in an individual who carries the A variant of gene X?'

(MOE 1) 'Is the effect of the A variant of gene X on infarction entirely mediated by an increase in blood pressure?'

(MOE 2) 'Has frequent physical activity the power of nullifying the effect of the A variant of gene X on infarction?'

Questions of type MOE1 are tackled in Chapters 11 and 12 in this volume. Questions of type MOE2 represent the main theme of the present chapter. What they ask is whether, in some individuals or circumstances, two variables of interest interfere with each other's effect on

Causality: Statistical Perspectives and Applications, First Edition. Edited by Carlo Berzuini, Philip Dawid and Luisa Bernardinelli.

a specific outcome. Such interference we call *mechanistic interaction*. Looking for variables that interact mechanistically may be useful because it helps us to understand which causal factors (e.g. genes) are part of a common (e.g. physical, biological) mechanism affecting the outcome of interest. In this chapter, we discuss ways of assessing mechanistic interaction on the basis of observational data.

Our illustrations are largely epidemiological; the relevance of the ideas is much wider. This chapter owes much to the work of Tyler VanderWeele, who has pioneered the theoretical foundations of mechanistic interaction and the application of the concepts in observational epidemiology.

14.2 What is a mechanism?

In this chapter we regard a 'mechanism' as consisting of component *parts* and *operations*, orchestrated to achieve a specific *task*. An example of a biological mechanism in the above sense is the ion channel, whose task is to modulate nerve signal current. Ion channels are studied from the point of view of their pathogenetic import in multiple sclerosis in Chapter 15, by Luisa Bernardinelli and colleagues. Constituent parts of an ion channel are various types of protein, some of which form a tiny 'door' across the membrane of the neuron (for the ion current to pass through). Proteins of many different types participate as constituent parts of an ion channel and interact with each other. Chapter 15 by Bernardinelli and colleagues investigates the hypothesis that multiple sclerosis (or a specific stage of this disease) may be promoted/caused by a dysfunctional interaction between two constituents of the ion channel mechanism, resulting in a defective control of the 'door' opening.

There are many ways of advancing the understanding of biological mechanisms through an analysis of epidemiological data. One of these – this chapter's theme – is to use observational data to detect patterns of interaction between parts of a pathogenetic mechanism that may impact on disease susceptibility. Typically, but not necessarily, we do this after we have an idea of what kind of mechanism we are talking about. What epidemiologists know as a 'test for statistical interaction' is in general *not* an appropriate way of doing the thing. We shall later discuss why this is so, and we shall describe suitable methods for testing for mechanistic interaction on the basis of observational data. In the next section, we start by comparing the concepts of 'statistical' and 'biological' interaction.

14.3 Statistical versus mechanistic interaction

While mechanistic interaction is a property of biological mechanisms, *statistical* interaction is nothing more than a departure of the data from some statistical model of additive effects of causal factors – a departure from 'additivity'. This can be better understood in the context of regression analysis. Suppose we regress the binary outcome variable Y on binary factors A and B. One possibility is to use the *linear risk* model

$$\begin{cases} Y \mid A, B \sim \text{Bernoulli}(\pi) \\ \pi = \eta_0 + \eta_1 A + \eta_2 B + \eta_3 AB \end{cases} \tag{14.1}$$

where the symbol $\sim$ stands for 'distributed as' and $\eta_{0\ldots3}$ are unknown parameters to be estimated from the data. Large absolute values of η_3 signal departure of the data from additive effects on a linear risk scale; that is they signal presence of statistical interaction between these effects on a linear risk scale.

Another possibility is to use the following *linear odds* regression model:

$$\begin{cases} Y \mid A, B \sim \text{Bernoulli}(\pi) \\ \frac{\pi}{1-\pi} \qquad = \gamma_0 + \gamma_1 A + \gamma_2 B + \gamma_3 AB \end{cases} \tag{14.2}$$

where $\gamma_{0...3}$ are unknown parameters to be estimated from the data. A departure of γ_3 from zero is described as a 'statistical interaction' between the effects of A and B on Y, on an odds scale. If we choose instead the *logistic* regression model

$$\begin{cases} Y \mid A, B \sim \text{Bernoulli}(\pi) \\ \log(\frac{\pi}{1-\pi}) = \phi_0 + \phi_1 A + \phi_2 B + \phi_3 AB \end{cases} \tag{14.3}$$

where $\phi_{0...3}$ are unknown parameters to be estimated from the data, then a non-null ϕ_3 coefficient is described as the presence of $A \times B$ statistical interaction on a logit scale.

For a discussion of model choice in relation to a causal interpretation of the obtained estimates, see Greenland (1993). In many applicative contexts, it will be reasonable to assume that the disease under study is rare over the entire study period, and in all strata of the population, conditional on whatever level of (A, B) and remaining explanatory variables. If, in addition, the collapsibility conditions for the odds ratio of Section 14.9 are satisfied, then the linear odds and the logistic regression models will both provide causally meaningful effect estimates under either a prospective and a retrospective sampling regime.

It may happen that deviation from the additivity of effects is evident in one or two of the models above, but not in all three. More in general, evidence of statistical interaction will depend on aspects of the model (e.g. the linear predictor scale) that we have chosen for mathematical convenience, rather than in consideration of the nature of the phenomenon under study. For these reasons, evidence of statistical interaction between two variables does not per se imply that these interact in a deeper, biological, sense (Clayton, 2009). We may, in fact, go as far as saying that statistical interaction is devoid of empirical meaning. This is illustrated by the following illustrative study.

14.4 Illustrative example

We now illustrate the above argument with the aid of a study of the role of autophagy in Crohn's disease (CD), a major form of inflammatory disease of the bowel. What is autophagy? It is a cellular mechanism devoted to the necessary degradation of intracellular pathogens. Recognition of its possible role in inflammatory disease came in 2007, with the discovery of a CD association signal at a single nucleotide polymorphism (SNP; see the definition in the book index) called `rs2241880`, and located in the coding region of the human autophagy gene ATG16L1 (Hampe *et al.*, 2007; Rioux *et al.*, 2007). Signals of association with CD were also found at SNPs `rs26538` and `rs573775`, located in putative regulatory regions of autophagy genes ATG12 and ATG5, respectively. The protein products of the latter two genes, we denote as atg12 and atg5, following the convention of naming a protein from its corresponding coding gene, after turning capital into small fonts. Proteins atg12 and atg5 are known (Mizushima

Table 14.1 Sample genotypic frequencies in the dataset analysed in the autophagy study.

	Autophagy dataset	
Number of copies of minor allele	rs26538	rs573775
0 (wild type homozygous)	818	1184
1 (heterozygous)	789	956
2 (rare homozygous)	219	189

et al., 1999) to co-participate in the formation of a complex, the autophagosome, which represents an intermediate stage of the autophagy conjugation pathway.

Does the hypothesis of a pathogenetic role of autophagy and its component proteins make biological sense? An impaired autophagy may lead to an altered mucosal immune response to gut bacteria, perhaps triggering inappropriate activation of adaptive immune mechanisms. The latter might, in turn, be unable fully to clear the bacterial material, and therefore lead to a persistent state of gut inflammation and consequently to onset of CD. Therefore it makes sense to explore the hypothesis that susceptibility to CD depends on inherited dysfunctions of the component proteins that prevent these proteins from interacting with each other in the right way. Below we investigate this via a standard analysis of interaction of SNP effects on susceptibility to CD.

We analyse data including the `rs26538` and `rs573775` genotypes from 1826 controls and 2329 CD cases from East Anglia, United Kingdom (see sample genotypic frequencies in Table 14.1). No marked departure from Hardy–Weinberg equilibrium was detected in the controls, at either locus. Analysis was based on regression models of the binary disease indicator on binary variables indicating rare homozygosity at `rs26538` and `rs573775` respectively. (The term 'rare homozygosity' here indicates the presence of the less frequent allele at each of the two homologous copies of an SNP. In a more thorough analysis, the effect of different SNP codings ought to be explored and the consequent inflation of type-1 error taken care of.)

The fitting of the linear odds regression model to the data gave the estimates shown in Table 14.2. The subsequent fitting of a logistic regression model gave the estimates shown in Table 14.3. What do these estimates tell us? They indicate that the interaction coefficient is significantly different from zero (at a 5 % level) on a linear odds scale, but not on a log-odds scale. This illustrates our earlier point, that statistical interaction may come and go as we switch from one model to another. Even in the case of significant interaction under both models, we could not have concluded in favour of a 'biological' interaction between the two SNPs. This may not necessarily be a reason of concern if our aim is merely one of *predicting*

Table 14.2 Summary of results from a linear odds regression analysis of the autophagy dataset, when allowing for interaction between rare homozygosity indicators.

	Estimate	Std error	z-value	p-value
(Intercept)	0.5	0.01	34.5	$< 2e^{-16}$
rs573775 rare homozygous	0.0015	0.056	0.027	0.97
rs26538 rare homozygous	0.06	0.052	1.29	0.19
(rs573775 rare h.) × (rs26538 rare h.)	1.06	0.44	2.4	0.01

Table 14.3 Summary of results from a logistic regression analysis of the autophagy dataset, when allowing for interaction between rare homozygosity indicators.

	Estimate	Std error	z-value	p-value
(Intercept)	−0.53	0.05	−9.86	$< 2e^{-16}$
rs573775 rare homozygous	0.0026	0.18	0.015	0.98
rs26538 rare homozygous	0.11	0.15	0.7	0.48
(rs573775 rare h.) × (rs26538 rare h.)	0.96	0.528	1.8	0.06

risk of disease. However, it will prevent us from translating the results of the analysis into clues about the underlying biology. For the latter aim to be fulfilled, we need:

1. A mathematical definition of what we mean by *mechanistic interaction.*
2. Statistical tests and assumptions that allow mechanistic interaction to be detected in observational data.

This we are going to discuss in the remaining part of the chapter.

14.5 Mechanistic interaction defined

We continue to use symbols A and B to denote putative causal factors for a binary outcome variable Y, where

A is ordered categorical.
B is ordered categorical or continuous.

In general, there will be further variables that collude with A and B in causing Y. Let these consist of a (possibly empty) set W of observed variables, and of a set U of unobserved variables. We make a *deep determinism* assumption, that Y depends on (A, B, W, U) *deterministically*:

$$Y = f(A, B, W, U) \tag{14.4}$$

where f is a deterministic, typically unknown, function. Then B is said to *interfere with A in producing the event $Y = 1$* if:

> There exists a set C of observed variables, with $C \supseteq W$, and there exist values (c, u, b_1, b_2) such that, when $C = c$ and $U = u$, the action of setting $B = b_1$ causes Y to be 0, no matter how we manipulate A; whereas, had we set $B = b_2$, the value of A would have made a difference to Y.

The effects of A and B on Y are said to *interact mechanistically* if A interferes with B, and/or B interferes with A, in causing $Y = 1$. The above definition has clear relationships with definitions of interdependence given by VanderWeele in some of his works and with the concept of epistasis discussed in the next section.

14.6 Epistasis

The purpose of this section, which is not essential to the understanding of the method, is to discuss relationships between the concept of mechanistic interaction of the preceding section and the notion of *epistasis* in the genetic epidemiology literature, originally introduced by Bateson (1909), although different authors use the term somewhat differently. The two concepts are closely related. The term 'epistasis' is typically used by geneticists to describe the phenomenon where a specific variant of one gene nullifies the effect of a specific variant of another gene. The term is used to describe a property of a biological system, rather than a mere statistical phenomenon.

Evidence of epistasis may reflect a variety of possible mechanisms. For example, two genes may be epistatic to each other, in causing a specific disease, because they interact at a protein level within a mechanism of pathogenetic import. In this case, it is not infrequent to find evidence of mechanistic interaction between loci in the coding regions of the two genes (i.e. located in those portions of the gene sequence that are translated into messenger RNA and hence reflected in the structure and folding of the protein products).

As another example, gene-to-gene epistasis may reflect precedence relationships between the two genes in a pathway of pathogenetic import. This may mean that the protein product of gene A, say, operates downstream of the protein product of another gene, B, say, within the pathway. Thus, a functional (or disregulatory) alteration of gene B will result in the functional deactivation of the downstream portion of the pathway, and consequently in the nullification of the impact of gene A.

Another example is where the deleterious allele of an SNP affects disease risk by switching off the expression of a key gene. At least equally important is the study of epistasis between genes and environmental variables or genes and treatments. The latter are important for (i) the understanding of the mechanisms behind a treatment's effect, (ii) the development of more specific and harmless drugs and (iii) the development of personalized therapy protocols.

14.7 Excess risk and superadditivity

We are now going to discuss statistical tests for inferring mechanistic (as opposed to mere statistical) interaction from observational data. The method requires that the causal factors of main interest, A and B, be dichotomized, unless they arise as binary variables at the outset. Thus, for generality, we shall hereafter assume that A has been dichotomized into α, and B into β. We shall moreover assume that A has been dichotomized in such a way that α is equal to 1 only when A takes on its highest value, $a^{\max}$, and otherwise is equal to 0. Let the symbol C denote a generic set of observed variables. Then let the *prospective* risk of disease in an individual with $\alpha = i$, $\beta = j$, conditional on C taking a specific value c, be denoted

$$R_{ijc} := P(Y = 1 \mid \alpha = i, \beta = j, C = c) \qquad (14.5)$$

under the general convention that subscript c is dropped whenever C is empty and under the assumption that, for all i, j,

$$P(\alpha = i, \beta = j, C = c) > 0 \qquad (14.6)$$

Let

$$O_{ijc} := \frac{R_{ijc}}{1 - R_{ijc}} \tag{14.7}$$

denote the corresponding *odds* of disease. Rothman (1976) introduced an important statistic called *relative excess risk due to interaction* (RERI), with

$$\text{RERI} := \frac{R_{11c} - R_{10c} - R_{01c} + R_{00c}}{R_{00c}} \tag{14.8}$$

and discussed the relevance of RERI in the assessment of mechanistic interaction. As noted by Skrondal (2003) and VanderWeele (2009), RERI is a function of relative risks and therefore, under appropriate assumptions (such as 'rare disease' and collapsibility of the odds ratio, see Section 14.9), it can be estimated under case-control sampling. Many mechanistic interaction tests discussed in the literature are based on the idea of finding a stratum $C = c$ of the population in which RERI satisfies one of the following inequalities. When the inequality RERI > 1 is satisfied, we say there is *excess risk*, which is equivalent to saying that

$$R_{11c} - R_{10c} - R_{01c} > 0 \tag{14.9}$$

When the inequality RERI > 0 is satisfied, we say there is *superadditivity*, which is equivalent to

$$R_{11c} - R_{10c} - R_{01c} + R_{00c} > 0 \tag{14.10}$$

Note that excess risk implies superadditivity. In the next sections we show that, under specific conditions, excess risk or superadditivity implies mechanistic interference in the sense of the previous section, and therefore they provide a basis for statistical tests of mechanistic interaction from observational data. Such tests are not novel. They have been discussed by several authors, including Rothman and Greenland (1988), VanderWeele (2009), VanderWeele and Robins (2008, 2009) and Skrondal (2003). In particular, VanderWeele (2010) tackles categorical and multilevel ordinal exposures. These authors use a definition of mechanistic interaction that is very close to the one given in the previous section, albeit on the basis of a different theoretical justification of the method. Some authors, notably VanderWeele (2009) and VanderWeele and Robins (2008), develop a justification of RERI-based tests of mechanistic interaction by using Rothman's sufficient causation framework (1976, 1998), discussed in the preceding chapter. VanderWeele and Robins have also proposed a justification of the tests based on the potential response framework of Rubin (2005), introduced in Chapter 2 by Sjölander.

From a practical point of view, conditions (14.9) and (14.10) can be checked via regression analysis, possibly by applying the linear odds regression model (14.2) or the logistic regression model (14.3) in the relevant $C = c$ stratum, in either case with A and B replaced by α and β, respectively. Consider a case-control study (with $Y = 1$ indicating case) where we are able to assume that the disease is rare. Then, under the former model, the excess risk condition (14.9) is approximated by

$$\gamma_3 - \gamma_0 > 0 \tag{14.11}$$

and under the latter by

$$\exp(\phi_1 + \phi_2 + \phi_3) - \exp(\phi_1) - \exp(\phi_2) > 0 \tag{14.12}$$

Similarly, under the linear odds model the superadditivity condition (14.10) is approximated by

$$\gamma_3 > 0 \tag{14.13}$$

and under the logistic regression model by

$$\exp(\phi_1 + \phi_2 + \phi_3) - \exp(\phi_1) - \exp(\phi_2) + 1 > 0 \tag{14.14}$$

If the $C = c$ stratum within which we conduct the analysis is numerically small, the data information provided by the $C = c$ stratum of the population will have little power to detect significant deviations in the direction of the above inequalities. In such a case, one may consider fitting a regression model to the whole sample, this time by including a term for the effect of C in the model linear predictor. If no interaction is assumed between C and either α or β on the particular response scale of the model, the linear odds model becomes

$$\begin{cases} Y \mid \alpha, \beta, C = c \sim \text{Bernoulli}(\pi) \\ \frac{\pi}{1-\pi} \qquad\qquad = \gamma_0 + \gamma_1\alpha + \gamma_2\beta + \gamma_3\alpha\beta + \gamma_c \end{cases} \tag{14.15}$$

where the effect of C is represented by γ_c, whereas the logistic model becomes

$$\begin{cases} Y \mid \alpha, \beta, C = c \sim \text{Bernoulli}(\pi) \\ \log(\frac{\pi}{1-\pi}) \qquad = \phi_0 + \phi_1\alpha + \phi_2\beta + \phi_3\alpha\beta + \phi_c \end{cases} \tag{14.16}$$

where the effect of C is represented by ϕ_c. Condition (14.11) will correspondingly change into

$$\gamma_3 - \gamma_0 - \gamma_c > 0 \tag{14.17}$$

whereas conditions (14.13) to (14.14) do not change. A final, last, option is to perform a separate analysis of each different stratum (corresponding to different values of C) in which case one will usually consider correction for multiple testing. In this case, evidence of mechanistic interaction may not – and need not necessarily – be homogeneous across strata.

As originally pointed out by Greenland (1993), models (14.15) and (14.16) represent different assumptions about the data. The former assumes that lack of interaction with C occurs on an odds scale, whereas the latter assumes it occurs on a logistic scale. This leads to a problem of model selection. Just one possible – pragmatic – approach to this is to test for mechanistic interaction using different models, and then examine concordance of results. Approaches that help deal with issues of model selection are discussed in VanderWeele *et al.* (2010) and Vansteelandt *et al.* (2008).

Finally, note that conditions (14.11), (14.13), (14.12) and (14.14)) involve the comparison of a (possibly nonlinear) function of the model parameters with zero. A confidence interval for the estimated value of this function can, in the most general case, be calculated by using the delta method or via bootstrapping, and we shall not give details of this.

14.8 Conditions under which excess risk and superadditivity indicate the presence of mechanistic interaction

In general, confounding or selection effects will prevent excess risk and superadditivity from being valid indicators of mechanistic interaction, except under special assumptions and conditions; even then, their validity will – in general – depend on an appropriate choice of the stratifying variables C. In any specific application, an important task is therefore to select an appropriate set of stratification variables C – if it exists – such that, conditional on C, excess risk and/or superadditivity are valid indicators of mechanistic interaction. Depending on the particular perspective on causality, one generally obtains different approaches, which usually do not differ in the mathematical form of the tests, as much as in the conditions and assumptions under which those tests are *believed to be* valid.

Take, for example, the approach by VanderWeele and colleagues (VanderWeele, 2009). Their approach is based on the sufficient causation philosophical framework expounded in the preceding chapter in this volume. The approach proposed by these authors revolves around a 'no confounding' requirement, expressed in terms of conditional independence between counterfactuals. In other words, according to this approach, the main concern is to find a set C of variables such that, conditionally on C, we have no confounding. We propose a different approach to – and justification of – the method, which is based on the decision-theoretic perspective of Chapter 4 in this volume. This leads to a different (and more detailed) set of validity assumptions, expressed in terms of conditional independencies between problem variables rather than of counterfactuals.

We start by acknowledging that the joint distribution of the problem variables (A, B, Y, W, U) will in general vary across 'regimes'. Among these regimes we distinguish the observational one, in which we passively observe the variables as they are generated by Nature, and interventional regimes, in which we (perhaps only hypothetically) intervene to set the values of factors A and B. We let these regimes be indexed by the *regime indicator*, denoted F. In other words, the value of F is taken to indicate the way in which the values of A and B are generated, for example, both by intervention, or both by passive observation, or one by intervention and the other observationally. As explained by Philip Dawid in Chapter 4 in this volume, conditional independence relationships involving F make sense. In the context of interest here, for example, the statement $Y \perp\!\!\!\perp F \mid (A, B)$ tells us that the conditional distribution of the outcome variable Y, given the values of the causal factors A, B and of F, depends only on the values of A and B and not further on the regime under which these values have been generated. You may take this as a formal way of saying that the relationship between (A, B) and Y is not confounded.

We are now going to present a set of conditions under which excess risk and superadditivity are valid indicators of mechanistic interaction. In the following we continue to use symbols A, B, U, W, C, Y in the same way as above. Here is the main theorem.

Assume that the deep determinism condition (14.4) is satisfied. Further assume that, for some set C of observed variables, with $C \supseteq W$, and for some value c of C, that the following *core conditions* are satisfied:

1. $Y \perp\!\!\!\perp F \mid (A, B, C = c, U)$.
2. $U \perp\!\!\!\perp (A, B, F) \mid C = c$.

3. (Unnecessary if $\beta = 1$ only when B takes on a single value b^*). When A and B arise naturally, without intervention, then $A \perp\!\!\!\perp B \mid (C = c)$.

4. Y is either a non-decreasing function of A for all fixed values of the other variables or a non-increasing function of A for all such values.

Then, *if the excess risk condition* (14.9) *is satisfied for* $C = c$, *we can conclude that variable* B *interferes with* A *in producing the event* $Y = 1$, *in the sense of Section 14.5.*

A proof of the above theorem is given in Berzuini and Dawid (2010). The condition $C \supseteq W$ simply tells us that W should not vary within the group of people characterized by $C = c$. We do not need to make any assumptions about nonmembers of the $C = c$ group, corresponding to other values for C. The first two core conditions imply, but are not equivalent to, what some would call 'no confounding between (A, B) and Y'.

In certain applications, one may replace the *monotonicity* assumption by the stronger assumption of *positive monotonicity* about the effects of A and B: that an increase in A or B does not induce a decrease in Y, whatever the value configuration of the other variables. Under positive monotonicity, the excess condition (14.9) can be relaxed into the weaker superadditivity condition (14.10).

Note that, in our approach, the conditions for the validity of the method are expressed in terms of conditional independence relationships between problem variables (rather than, say, in terms of counterfactual independencies). One advantage of this approach, illustrated in Section 14.10, is the possibility of checking these conditions on a causal diagram of the problem. In a related paper, Berzuini and Dawid (2010) show that this approach sets the ground for the analysis of mechanistic interaction between direct effects.

14.9 Collapsibility

The discussion so far has left aside problem aspects that are sensitive to *sampling*. Most studies in epidemiology, especially in genetic epidemiology, use retrospective sampling. Most of them are, in fact, case-control studies, where the probability that a generic member of the underlying study cohort enters the study depends on his/her disease outcome. In such a situation, one will typically use the mentioned linear odds or logistic regression models. This is a good idea when the odds ratio of disease is unaffected by case-control sampling – a property that holds under specific conditions. What are these conditions? Following Didelez and colleagues (2010), we tackle this by introducing a random variable called the *selection indicator*, S, defined to take on value 1 in all individuals who are selected into the study (and 0 otherwise), as in Geneletti *et al.* (2009). Due to the tautological fact that analysis is performed on (all and only the) individuals who have been selected into the study, analysis takes place conditional on $S = 1$. This shifts attention from the odds of Equation (14.7) to the following conditional odds:

$$O^*_{ijc} = \frac{R^*_{ijc}}{1 - R^*_{ijc}} \tag{14.18}$$

where $R^*_{ijc} := P(Y = 1 \mid \alpha = i, \beta = j, C = c, S = 1)$. What we require is that the O_{1jc}/O_{0jc} odds ratio measure of the effect of A be *collapsible* over S, conditionally on B and C, in the sense that it is not affected by the sampling, formally:

$$\frac{O^*_{1jc}}{O^*_{0jc}} = \frac{O_{1jc}}{O_{0jc}} \tag{14.19}$$

Didelez and colleagues rephrase Whittemore's theorem (1978) into the following sufficient condition for the above collapsibility property:

$$S \perp\!\!\!\perp A \mid (C, Y, B) \tag{14.20}$$

In order that the method discussed in this paper be applicable in a case-control study, we also require that the odds ratio for the effect of B on Y be collapsible over S, conditionally on A and C, and a sufficient condition for this is obtained from (14.20) by swapping B and A. Violation of collapsibility is not unfrequent. Consider, for example, a matched study where each control is selected to match the age of the corresponding case at the time of the failure event. The selection event represented by S will, in this case, depend on time to failure, and hence on (A, B), conditionally on the remaining observed variables, thereby violating (14.20). For an advanced and more general discussion of collapsibility, see Greenland and Pearl (2011).

14.10 Back to the illustrative study

Let us go back to our illustrative study, where Y denotes the occurrence of Crohn's disease, within a specified age threshold, and A and B represent the `rs26538` and `rs573775` variants of genes ATG12 and ATG5, respectively. In this example, we take W and C to be empty. In order to apply the method, we dichotomize A and B into binary indicators of rare homozygosity at `rs26538` and `rs573775`, respectively. Call these indicators α and β. Our causal assumptions about the problem are represented by the diagram in Figure 14.1. The F node in this diagram is not new to the reader of this book. It represents the *regime indicator* introduced by Philip Dawid in Chapter 4. The diagram of Figure 14.1 satisfies $Y \perp\!\!\!\perp F \mid (A, B, U)$, representing the assumption that dependence of Y on (A, B, U) is the same, regardless of the way the values of (A, B) have been generated, whether observationally or experimentally. Condition 1 of our theorem is thus satisfied. Condition 2 is also satisfied in the diagram.

Assumption (14.4), that Y depends on its parents in the graph in a deterministic way, is represented in the diagram of Figure 14.1 through the use of full (as opposed to indented) arrowheads. Another assumption in Figure 14.1 is that `rs26538` and `rs573775` are independent in the population, a reasonable[1] one if we consider that these two loci lie in different chromosomes. Condition 3 is therefore also satisfied. The graph also assumes that selection of sample individuals was blind to genotype and satisfies the conditions for collapsibility of the odds ratio.

[1] One should, however, be aware that a marginal correlation between the two SNPs might – in principle – be induced by a dependence of pre-natal survival on the values of both.

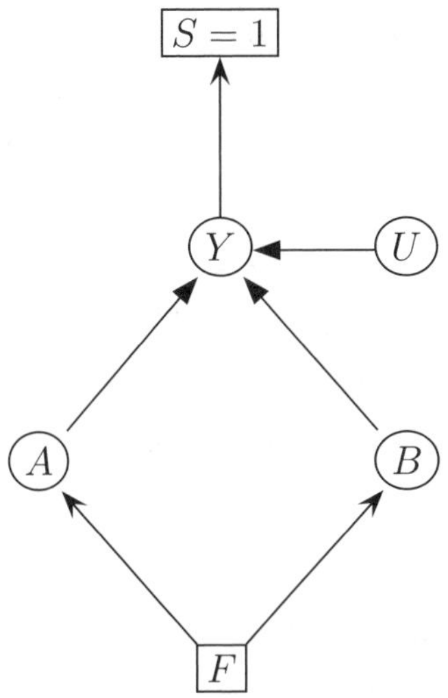

Figure 14.1 Causal diagram for our Crohn's Disease illustrative study. Node U represents a collection of unknown or unobserved variables that collude with (A, B) in causing Y. This diagram assumes that U is independent of (A, B) in the population. In principle, the diagram is a representation of the underlying study base cohort, from which study individuals are drawn. The binary variable S indicates whether the generic individual of this cohort (either a case or a control) is selected into the study. The top square node is therefore interpreted to indicate that, in the analysis, variable S is fixed at the value corresponding to the individual 'being selected'. Node F represents the regime indicator introduced in Section 14.8. This diagram assumes that dependence of Y on (A, B) is not affected by the regime, F, under which the values of these two variables are generated. Note that two types of arrowheads are used in this figure. Full arrowheads indicate deterministic dependence, according to the 'deep determinism' assumption introduced in the main text of this chapter. Indented arrowheads represent stochastic dependence.

Method validity further requires the validity of Condition 4 of our theorem, which in this case corresponds to the assumption that if an additional copy of the deleterious allele, at either locus, changes the risk of the event $Y = 1$, it does so in the same direction, whatever the values of the remaining variables. This assumption was, in our case, nonevidential (in the sense discussed by Sander Greenland in Chapter 5), although it could have been supported by evidence from previous studies. All the theorem conditions are satisfied at this point, so that, according to the theorem in that section, excess risk is a valid basis for inferring mechanistic interaction between *A* and *B* in causing *Y*.

According to Table 14.2, obtained from the fitting of a linear odds model to the data, the parameter δ_3 has an estimated value of 1.06, with a standard error of 0.44, and hence is significantly different from 0 at a 0.01 level of significance, in the positive direction, which represents fair evidence of superadditivity. At the same time, because the estimated difference $(\delta_3 - \delta_0)$ is, 0.47, with a 95 % confidence interval of $(-0.39, 1.34)$, we have modest evidence of excess risk. we conclude that only under an assumption of positive monotonicity of the effects of rs573775 and rs26538 can we interpret our results to represent strong evidence that the genetic variants related to these two SNPs interfere with each other in causing CD.

Strong evidence in favour of a mechanistic interaction between the effects of SNPs rs573775 and rs26538 on Crohn's disease would offer a number of substantive hypotheses for future investigation. First, an alteration of the atg12.atg5 balance/binding could underlie susceptibility to Crohn's disease in some patients, perhaps because immune fitness with respect to Crohn's disease requires coordinated changes in the regulation of the corresponding genes. Second, the atg12.atg5 mechanism should be best experimentally interrogated via *multifactorial* perturbation, for example, by simultaneously knocking down ATG5 and ATG12 – a point that echoes that of Jansen (2003). Third, other proteins that interact with atg12 and/or atg5 might as well be implicated in susceptibility. Fourth, individuals with rare homozygosity at both rs26538 and rs573775 may be more susceptible to an altered response to a given pharmacologic treatment.

14.11 Alternative approaches

The method presented so far may be described as using a *no-interaction* regression model as a reference model, and testing for departures in certain directions from such a reference. The no-interaction model has structure

$$g(\pi) = \gamma_0 + \gamma_1 A + \gamma_2 B \tag{14.21}$$

where as usual the symbol π denotes probability of disease and $g(.)$ is an appropriate link function, so as to embrace the three models (linear, linear odds and logistic) considered above.

Alternative approaches are available, one of them discussed by Berrington de Gonzalez and Cox (2007). For A and B coded as binary, (0, 1), variables, the idea is to use:

$$g(\pi) = \nu + \lambda AB \tag{14.22}$$

where ν and λ are unknown parameters to be estimated from the data. This is a two-parameter model, as contrasted with the three-parameter no-interaction model above. Yet model (14.22) is not a special case of (14.21). Assuming the data reject $\lambda = 0$, a test for epistasis may be based on a comparison between the goodness-of-fit of (14.22) and (14.21).

14.12 Discussion

The method discussed in this chapter is valid only under the deep determinism assumption that Y depends deterministically on its parents in the causal diagram, pa_Y. This implies that, for each generic jth configuration of pa_Y, a single value of Y, call it Y_j, gets all the probability. In other words, what deep determinism assumes is that the joint distribution of the $\{Y_j\}$ degenerates to a collection of point masses, in the face of the fact that, because it involves jointly unobservable variables, such distribution is experimentally unidentifiable, as discussed in (Dawid, 2000). Even in our simple Crohn's disease example, where A and B are three-level genotype variables, the joint distribution of the $\{Y_j\}$ lives in a nine-dimensional space, which could become infinite dimensional in the presence of continuous factors. These considerations suggest that deep determinism is, indeed, a strong assumption. Deep determinism should be regarded as a scientific hypothesis, rather than 'part of the method'. From this point of view, deep determinism is defendable only in special applicative situations. Many such situations arguably arise in the study of molecular mechanisms.

The above considerations should be contrasted with those areas of causal inference, notably structural equation modelling, where deep determinism is taken for granted, and rarely questioned, at *each* response node of the model (see Chapter 3 by Ilja Shpitser).

Our method encourages the use of causal diagrams to check the required assumptions and conditions. One *caveat*, here, is that the simplicity of a causal diagram may obscure the multitude of assumptions it embodies. The diagram in Figure 14.1, for example, assumes that, conditional on some unobserved set of biologically meaningful variables U, that are independent of (A, B), the outcome Y is a deterministic function of (A, B). Just a possible story is that (i) the U variables describe geographical and diet conditions, and molecular factors affecting the condition of the intestinal barrier, (ii) the studied SNPs have nothing to do with these variables, (iii) particular configurations of these variables may induce excess bacterial infiltration from the intestinal lumen into the blood and (iv) the excess infiltration sets the ground for a deterministic impact of alterations of autophagy on risk of Crohn's disease. These considerations suggest that, even in a simple example like that of Figure 14.1, the validity of the method rests on strong – nonevidential – assumptions about the causal structure of the problem. Many students feel uneasy about this. They perceive the circularity of the argument. The circularity exists, but rather than being a shortcoming of the method, it is in the nature of scientific discovery: you cannot improve our mechanistic understanding of a process without some pre-existing mechanistic understanding. Our analysis of Crohn's disease data has generated the hypothesis of a possible role of a role of autophagy, and more specifically of the atg5–atg12 interaction, in the disease. Sure, this inference is strictly valid only under the strong assumptions of Figure 14.1. However, the result provides real motivation for future experiments to confirm the truth of the hypothesis.

To conclude, we would emphasize that the method discussed in this chapter is relevant to the study of mechanistic interaction between variables of a diverse nature. Consider, for example, studies of genes interfering with the effects of epidemiological or environmental exposures, or of medical treatments, when these genes affect the cellular pathways through which those exposures, or treatments, operate. Many examples of this kind arise in personalized medicine, where one tries to match therapy with the genetic set-up of the patient.

Ethics statement

The Crohn's disease motivating study was approved by the Research Ethics Committee of the NHS National Research Ethics Service, Cambridgeshire 1.

Financial disclosure

This research has been partially supported by the UK Medical Research Council Grant No. G0802320 (www.mrc.ac.uk) and by the Cambridge Statistics Initiative.

References

Bateson, W. (1909) *Mendel's Principles of Heredity*. Cambridge, UK: Cambridge University Press.

Berrington de Gonzalez, A. and Cox, D.R. (2007) Interpretation of interaction: a review. *The Annals of Applied Statistics*, **1** (2), 371385.

Berzuini, C. and Dawid, A.P. (2010) Deep determinism and the assessment of mechanistic interaction. http://arxiv.org/abs/1012.2340v1 (submitted).

Clayton, D.G. (2009) Prediction and interaction in complex disease genetics: experience in type 1 diabetes. *PLoS Genet*, **5** (7), e1000540, 07.

Dawid, A.P. (2000) Causal inference without counterfactuals. *Journal of the American Statistical Association*, **95** (450), 407–424.

Didelez, V., Kreiner, S. and Keiding, N. (2010) Graphical models for inference under outcome dependent sampling. *Statistical Science*, **25**, 368–387.

Geneletti, S., Richardson, S. and Best, N. (2009) Adjusting for selection bias in retrospective, case-control studies. *Biostatistics*, **10** (1), 17–31.

Greenland, S. (1993) Additive risk versus additive relative risk models. *Epidemiology*, **4** (1), 32–36, January.

Greenland S. and Pearl. J. (2011) Adjustments and their consequences – collapsibility analysis using graphical models. *International Statistical Review,* **79** (3), 401–426, December.

Hampe, J., Franke, A., Rosenstiel, P., Till, A., Teuber, M., Huse, K., Albrecht, M., Mayr, G., De La Vega, F.M., Briggs, J., Genther, S., Prescott, N.J., Onnie, C.M., Hsler, R., Sipos, B., Flsch, U.R., Lengauer, T., Platzer, M., Mathew, C.G., Krawczak, M. and Schreiber. S. (2007) A genome-wide association scan of nonsynonymous snps identifies a susceptibility variant for Crohn disease in atg16l1. *Nature Genetics*, **39** (2), 207–211.

Jansen, R. (2003) Studying complex biological systems using multifactorial perturbation. *Nature Reviews Genetics*, **4**, 145–151.

Mizushima, N., Noda, T. and Ohsumi. Y. (1999) Apg16p is required for the function of the apg12papg5p conjugate in the yeast autophagy pathway. *The EMBO Journal*, **18**, 3888–3896.

Rioux, J., Rioux, J.D., Xavier, R.J., Taylor, K.D., Silverberg, M.S., Goyette, P., Huett, A., Green, T., Kuballa, P., Barmada, M.M., Datta, L.W., Shugart, Y.Y., Griffiths, A.M., Targan, S.R., Ippoliti, A.F., Bernard, E.J., Mei, L., Nicolae, D.L., Regueiro, M., Schumm, L.P., Steinhart, A.H., Rotter, J.I., Duerr, R.H., Cho, J.H., Brant S.R. and Daly M.J. (2007) Genome-wide association study identifies new susceptibility loci for crohn disease and implicates autophagy in disease pathogenesis. *Nature Genetics*, **39** (5), 596604.

Rothman, K.J. (1976) Causes. *American Journal of Epidemiology*, **104**, 587–592.

Rothman, K.J. and Greenland, S. (1998) *Modern epidemiology*. 2nd edn. Lippincott-Raven, pp. 737.

Rubin, D. (2005) Causal inference using potential outcomes: design, modeling, decisions. *Journal of the American Statistical Association*, **100**, 322–331.

Skrondal, A. (2003) Interaction as departure from additivity in case-control studies: a cautionary note. *American Journal of Epidemiology*, **158**, (3).

VanderWeele, T.J. (2009) Sufficient cause interactions and statistical interactions. *Epidemiology*, **20**, 6–13.

Vanderweele, T.J. (2010) Sufficient cause interactions for categorical and ordinal exposures with three levels. *Biometrika*, **97** (3), 647–659.

Vanderweele, T.J. and Robins, J.M. (2008) Empirical and counterfactual conditions for sufficient cause interactions. *Biometrika*, **95** (1), 4961.

Vanderweele, T.J. and Robins, J.M. (2009) Minimal sufficient causation and directed acyclic graphs. *Ann. Stat.*, **37** (3), 1437–1465.

VanderWeele, T.J., Vansteelandt, S. and Robins, J.M. (2010) Marginal structural models for sufficient cause interactions. *American Journal of Epidemiology*, **171**, 506–514.

Vansteelandt, S., VanderWeele, T.J., Tchetgen, E.J. and Robins, J.M. (2008) Multiply robust inference for statistical interactions. *Journal of the American Statistical Association*, **103**, 1693–1704.

Whittemore, A.S. (1978) Collapsibility of multidimensional contingency tables. *Journal of the Royal Statistical Society, Series B*, **40**, 328–340.

15

Ion channels as a possible mechanism of neurodegeneration in multiple sclerosis

Luisa Bernardinelli[1], Carlo Berzuini[1], Luisa Foco[2], and Roberta Pastorino[2]

[1]*Statistical Laboratory, Centre for Mathematical Sciences, University of Cambridge, Cambridge, UK*

[2]*Department of Applied Health Sciences, University of Pavia, Pavia, Italy*

15.1 Introduction

Genomewide studies of genetic association may prove less useful for individual risk estimation than for discovering and characterizing pathogenetic molecular pathways (Hirschhorn, 2009). In the latter task, they complement the traditional tools of bioinformatics and experimental biology. They contribute the empirical muscle that only the epidemiological observation of the disease process in humans may provide. This chapter illustrates this with the aid of a study of the types of ion channel dysfunction that may cause multiple sclerosis (MS).

From a methodological point of view, this chapter complements Chapter 14 in this volume by tackling a form of interaction where the value of a causal factor determines a *change of direction* in the effect of another factor (Azzalini and Cox, 1984). This form of interaction, known as *qualitative*, is interpretable in terms of mechanism. We propose a method for testing for qualitative interaction between genetic effects, in the context of the analysis of family structured genetic data. We illustrate the method with the aid of the above mentioned study.

Causality: Statistical Perspectives and Applications, First Edition. Edited by Carlo Berzuini, Philip Dawid and Luisa Bernardinelli.

The next section provides minimal background knowledge on the genetics of MS. Knowledge of elementary concepts of genetics is assumed.

15.2 Background

Multiple sclerosis (MS), a neuroinflammatory disease, leads to the demyelination and degeneration of the neurons of the central nervous system. MS is almost certainly causally dependent on dysfunctions of the immune response. Much credit is given to the hypothesis that, in certain stages of the disease and individuals, a dysfunction in *ion channels* may further contribute to development of the disease. Does this make sense?

Ion channels generate the ionic currents required for the transmission of the nervous signal (which is what goes wrong in MS), in response to changes in the extracellular space. Biologists have long hypothesized that an excessive flow of ionic current may cause neurodegeneration. Hence the suspected MS involvement of ion channels. This explains our enthusiasm when, years ago, we discovered a novel MS association within the gene that codes for the ion channel protein called ASIC2 (Bernardinelli *et al.*, 2007). ASIC2 belongs to the family of proteins known as ASICs (Acid Sensing Ion Channels) that are constituents of ion channels (Wemmie *et al.*, 2006). This association had its strongest signal at SNP rs28936, located in the 3′ untranslated region (3′UTR) of the ASIC2 (sometimes called ACCN1) messenger ribonucleic acid (mRNA) of the gene. The fact that the signal was found in this particular region matches recent evidence of a dependence of susceptibility to autoimmune disorders (a larger family of diseases that includes MS) on variation in 3′UTR (Ueda *et al.*, 2003; Morahan *et al.*, 2001). This is also in accord with the known involvement of 3′UTRs in the modification of the localization, stability or splicing of the mRNAs.

Further experimental findings support our hypothesis of an ASICs involvement in MS, from mouse model experiments (Friese *et al.*, 2007), gene expression localization studies and mice knock-out experiments (Lewin *et al.*, 2000). It is also noteworthy the fact that mutations of the MDEG gene, structurally related to the human ACCN1, cause neurodegeneration in Caenorhabditis elegans (Waldmann et al., 1996). Our hypothesis also matches growing awareness that the primary insult in MS might be of a neurodegenerative (rather than inflammatory) kind. A possible story is the following: an original neurodegenerative process brings in its wake a harmful autoimmunity, itself causing further collateral damage to neurons, in the context of a positive feedback link likely to set the pace of a progressive neuronal injury at certain stages of MS, and/or in some MS patients (Zipp and Aktas, 2006). Indeed, an involvement of a deregulated ion channel in MS has been reported in relation to cerebellar ataxia (Waxman, 2001). See Bernardinelli *et al.* (2007) for a more detailed discussion of the relevant biologic scenario.

15.3 The scientific hypothesis

Which molecular actors cooperate with ASIC2 in determining susceptibility to MS? A partial answer to this question are the experimental results of Baron *et al.* (2002) and Deval *et al.* (2004) (see also the beautiful thesis by Kerstin Imrell, 2009). These results suggests that the alpha isoform of protein kinase C, denoted PRKCA, regulates the activity of ASIC2 via shared interaction with the adaptor protein PICK1. A diagrammatic representation of the concept is

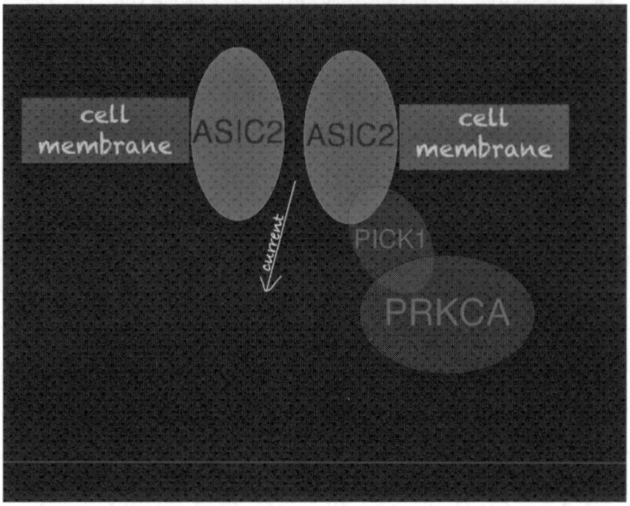

Figure 15.1 Diagrammatic representation of the mechanism under study. Two copies of the ASIC2 protein bind to each other to form a door through the cell membrane. This door allows the nerve signal electrical current to enter the cell. A protein called PRKCA activates the opening/closing of the door through the mediating action of the protein PICK1.

given in Figure 15.1. In this figure, you see two copies of the ASIC2 protein bind to each other to form a sort of door through the cell membrane, which conveys the ion current (the crucial ingredient of nerve signal transmission), into the cell. The protein PRKCA activates the opening/closing of the door, through the mediating action of PICK1, in such a way as to modulate the current. The hypothesis we are going to study is that an as yet uncharacterized dysfunction of the interaction between PRKCA and ASIC2 may be responsible for the development of MS.

Our approach to the above hypothesis involves a test for mechanistic interaction between the effects of variants of PRKCA and ASIC2 on susceptibility to MS. Our test will be based on genetic data collected from individual MS cases and from their respective parents.

One *caveat* in the analysis is the possible presence of a *parent-of-origin* effect. This is a well-documented biological phenomenon (Ebers *et al.*, 2004), whereby the effect of a particular allelic configuration on disease susceptibility depends on whether it was transmitted by the mother or by the father, and is independent of the gender of the proband, conditionally on the transmitting parent. Further discussion of this issue is given at the end of the discussion section of this chapter (Section 15.7).

15.4 Data

We have 78 cases of MS. They were mainly recruited when their disease was still in the relapsing-remitting phase, which is characterized by relapses followed by the regression of symptoms to the baseline. These cases were selected from the MS Register of Nuoro, capital of a genetically isolated and homogeneous province of western Sardinia, Italy. We have drawn these cases mainly from families with multiple affecteds. An advantage of this choice is that the cases are thus more likely to carry genetic variants that are strongly related to MS. Whenever possible, each case (also called a *proband*) was admitted into the sample together with his/her parents or close relatives, so as to form a *family nucleus* or, occasionally, together with a sex/age matched population control. For simplicity, in the description of the method, we shall hereafter pretend that the data consist entirely of *trios* formed by a proband and his/her biological parents.

All sample individuals were genotyped at:

locus A: the `rs28936` locus in the ASIC2 region of chromosome

locus B: the `rs3890137` locus in the PRKCA region of chromosome

Both loci are examples of a *single nucleotide polimorphism* (SNP). A SNP carries, by definition, two possible values called *alleles*. Of the two alleles of locus `rs28936`, G is the rarer in the studied population, and hence is called the *minor* allele. Allele G is also the minor allele of locus `rs3890137`. Recall from elementary genetics that each individual carries two homologous copies of each chromosome – hence, in particular, two copies of the chromosome containing `rs28936` and two copies of the chromosome containing `rs3890137`.[1] To protect from the possible presence of a parent-of-origin effect of Section 15.3, our analysis will restrict attention to maternal (hence ignoring paternal) information.

Genotype information about loci in the vicinity of `rs28936` and about loci in the vicinity of `rs3890137` was used (from each mother, father and child, in each trio) to estimate individual haplotypes in the chromosomal regions surrounding `rs28936` and `rs3890137`, as a vehicle to (i) imputing the missing alleles at the two SNP loci and (ii) identifying (albeit with uncertainty)

- the maternal allele of locus A that was transmitted to the child;
- the remaining, untransmitted, maternal allele at locus A;
- the maternal allele at locus B that was transmitted to the child;
- the remaining, untransmitted, maternal allele at locus B.

Hence each mother is characterized by a *set of transmitted alleles*, of size two, and by a *set of untransmitted alleles*, of size two. Computations were performed by using Markov chain Monte Carlo (MCMC) and importance sampling methods of simulation. For a general overview of allele imputation methods see Marchini and Howie (2010). For a description of the imputation method used in our study see Bernardinelli and colleagues (2007).

15.5 A simple preliminary analysis

Consider the analogy between one-to-one matched case-control studies and our analysis situation, with the transmitted set of maternal alleles acting as 'case' and the untransmitted one as a matched 'control'. In our study, treating the transmitted and the untransmitted sets of each mother as a matched pair in the analysis has the purpose of protecting against the potential

[1] Hence each individual, at both loci, is characterized by the number, (0,1,2), of copies of the G allele.

confounding effect of between-family variation of disease-causing factors. We shall in fact, analyse our data via conditional logistic regression analysis, in much the same way as this model is used in matched case-control studies. In our analysis, the conditioning is on the trio; each 'case' consists of the transmitted maternal *A* allele and transmitted maternal *B* allele. Each 'control' consists of the *un*transmitted maternal *A* allele and *un*transmitted maternal *B* allele. Because the studied loci are diallelic, this results in each 'case' and each 'control' being characterized by the following pair of binary 'covariates':

A: presence of a copy of the *G* allele at `rs28936`

B: presence of a copy of the *G* allele at `rs3890137`

In the preliminary analysis described in this section, we pretend that the above variables are observed with certainty. In reality, as mentioned earlier, this is not true: these variables are affected by the uncertainty introduced by the phasing stage of the analysis. What we really did here is to carry into the analysis only the a posteriori most likely configuration of the above variables, from among all the configurations generated by the Markov chain of the algorithm. We then performed a conditional logistic regression analysis of the dependence of the case-control (transmitted-untransmitted) indicator on the (*A*, *B*) allelic configuration. The results of the fitting are shown in Table 15.1.

The results of Table 15.1 suggest the presence of $A \times B$ *qualitative* interaction, the interpretation being that, in absence of the *G* allele at `rs3890137`, the *G* allele at `rs28936` has a beneficial, risk reduction, effect, whereas in the presence of the *G* allele at `rs3890137`, the *G* allele at `rs28936` has a deleterious effect. In other words, the maternal `rs28936` variant of ASIC2 appears to determine the sign of the effect of the maternal `rs3890137` variant of PRKCA. We have not checked whether the same pattern of interaction arises in paternal transmissions (see a discussion of this issue at the end of the discussion in Section 15.7). This form of interaction, where one factor has the power to reverse the sign of the effect of the remaining factor, is sometimes called *qualitative*. This is a form of interaction that cannot be removed by transformation (see Scheffe, 1959, for a discussion of this). There is therefore substantial consensus that qualitative interaction can be interpreted to indicate that the effect of the causal factors, *A* and *B*, on disease is mediated by a *common mechanism*, most likely the ion channel in this case. Note that the tests discussed by Berzuini and Dawid in Chapter 14 are not sensitive to qualitative interaction, despite the evident mechanistic interpretability of this type of evidence.

A discussion of the substantive ramifications of the results of Table 15.1 is given in Section 15.7. In the next section we present further analysis of the data.

Table 15.1 Results from a conditional logistic regression analysis of the data.

Effect	Coefficient	exp(coefficient)	Error	*Z*	*p*-value
A: presence of a copy of the *G* allele at `rs28936`	−0.69	0.5	0.3	−2.19	0.02
B: presence of a copy of the *G* allele at `rs3890137`	−0.42	0.6	0.34	−1.23	0.21
$A \times B$ interaction	1.64	5.16	0.6	2.72	0.0063

15.6 Testing for qualitative interaction

The above described analysis does not rigorously test for the presence of a qualitative, sign-switching, form of interaction and, moreover, it does not take into account the uncertainty involved in the MCMC phasing stage of the analysis. We remedy this in the following.

For simplicity, assume the absence of recombination. Let t_{kij}, with values between zero and one, denote the relative frequency, over the iterations of the MCMC phasing algorithm, with which the mother in trio k carries and transmits to her child an $(A = i, B = j)$ allelic configuration. For example, $t_{k01} = 0.5$ indicates that one MCMC iteration out of two depicts the mother in trio k as carrying and transmitting to her affected child a copy of the frequent `rs28936` allele plus a copy of the minor `rs3890137` allele. Let the binary variable u_{kij} be defined in the same way as t_{kij}, when 'carries and transmits' is replaced by 'carries but does not transmit'. Downstream of the phasing algorithm, the generic kth trio is characterized by a table of the form:

$$\begin{bmatrix} t_{k00} & t_{k01} & u_{k00} & u_{k01} \\ t_{k10} & t_{k11} & u_{k10} & u_{k11} \end{bmatrix}$$

By summing over the nuclei/trios, we obtain

$$d_{ij} = \sum_k (t_{kij} - u_{kij}) \tag{15.1}$$

Let our null hypothesis, H_0, state that, conditionally on A, locus B has no effect whatsoever on disease risk, and hence on chromosomal transmission (note the asymmetric role of A and B in this definition). Evidence of a departure from this null, in the direction of a qualitative interaction would then be provided if, for a sufficiently large t, we had

$$\begin{aligned} d_{01} - d_{00} &> t \\ d_{11} - d_{10} &< t \end{aligned} \tag{15.2}$$

or

$$\begin{aligned} d_{01} - d_{00} &< t \\ d_{11} - d_{10} &> t \end{aligned} \tag{15.3}$$

Therefore, define the statistic S to be the largest value of t such that either (15.2) or (15.3) is true. A large value of S can be taken as evidence of a departure from H_0 in the direction of a qualitative interaction. It would be straightforward to modify the test in such a way to make it sensitive to only (15.2) or only (15.3), not both, or even sensitive to a more general null hypothesis in which the two loci play a symmetric role. However, we shall not pursue this here.

Under H_0 we can freely interchange t_{ki1} with t_{ki0} and u_{ki1} with u_{ki0}, with probability 0.5, independently in each kth trio, for $i = 0, 1$. A test of H_0 can be based on the long term proportion of datasets generated by the mentioned random interchanges, which yield a larger S statistic than that obtained from the data. Such a proportion, in fact, provides an empirical

p-value for departures from H_0 in the direction of a causal influence of A upon the sign of the effect of B on the disease.

Application of the test of the above data yielded strong evidence of qualitative interaction between the studied loci, with a p-value of 0.009, which substantially confirms the conclusion from the previous analysis.

15.7 Discussion

This will be a discussion of the substantive aspects of the obtained results.

From a substantive point of view, this chapter originates from the hypothesis that specific alterations in the pattern of interaction between proteins ASIC2 and PRKCA may affect susceptibility to multiple sclerosis. The results presented above support this hypothesis, but tell us nothing about the functional aspects of the relevant alterations. These functional aspects are still open to investigation. A possible – in a sense indirect – first step in this direction is to analyse the genomic context of `rs28936` and `rs3890137`. Such an analysis may provide further support to the hypothesis that these two SNPs have, indeed, a functional role, and point to the nature of those alterations of the ASIC2–PRKCA mechanism that may induce susceptibility to MS.

Maybe the two studied SNPs have no functional role at all, in which case the data analysis results could be ascribable to linkage disequilibrium, and physical closeness, between the studied SNPs and unobserved causative loci. That is why we are not exempted from performing a bioinformatic analysis – possibly in conjunction with biological assays – of the chromosomal regions surrounding the studied SNPs, in search for as yet undescribed genomic features of pathogenetic import.

However, there is some evidence that the role of `rs28936` could – indeed – be functional. This SNP lies in the 3′ untranslated region of the ASIC2 gene, a typical location of loci of regulatory import. In addition, `rs28936` has been predicted to lie within the putative binding sites for some micro-RNAs. We conclude that this SNP, given its position, cannot cause structural – and functional – alterations of the ASIC2 protein, but is likely to control the expression of the ASIC2 gene via micro-RNA-mediated silencing (a hypothesis confirmed by unpublished results of luciferase-reported assays), thus contributing to determination the amounts of proteins ASIC2 in the cell. Concerning `rs3890137`, all we know is that this SNP lies in a non-better-characterized intronic region of the PRKCA gene, so that, again, all we can do is to hypothesize that this SNP controls the expression of PRKCA. Various types of biological experiment, such as reporter gene assays, could be performed towards a definitive elucidation of the functional roles.

We should then look for direct functional evidence about the involved proteins. The earlier mentioned paper by Baron *et al.*, (2002) shows that the PRKCA protein activates the 'door opening of ASIC2'. More precisely, this protein induces the phosphorylation of ASIC2 in a PICK1-dependent manner, an event that increases the channel open rate.[2]

The above evidence items, combined with our detected mechanistic interaction between ASIC2 and PRKCA, open the door to various explanations of the studied phenomenon.

[2] The phosphorylation is a chemical modification consisting of the addition of a phosphate group to some amino acids as tyrosine, serine and threonine: the phosphorylation is a crucial signalling event observed in cells, leading to activation or inactivation of proteins.

Considering that this is an educational chapter and not a biological paper, we shall now formulate an explanation that is realistic, consistent with the known facts, albeit not yet supported by conclusive evidental support.

Statistical evidence from our analysis (see Table 15.1) indicates that the presence of a *G* allele at both `rs28936` and `rs3890137` may be associated with an increased susceptibility to MS. Does this make sense? It is not unreasonable (albeit justifiably simplistic) to hypothesize that MS-related neurodegeneration is a possible consequence of an imbalance between the following two tendencies:

(1) an increased expression of ASIC2, which means a higher concentration of ion channels, leading to a high ionic current;

(2) excess concentration of PRKCA, leading to excess of activity ('door opening') of ion channels, which may cause excess ionic current.

Let us make the provisional assumption that condition (1) is a possible consequence of the presence of a *G* allele at `rs28936` and condition (2) a possible consequence of the presence of a *G* allele at `rs3890137`. Under this assumption, the results of Table 15.1 may be interpreted as saying that a high concentration of overactive ion channels, a condition induced by the simultaneous presence of a *G* allele at `rs28936` and `rs3890137`, may cause MS-related neurodegeneration.[3] In fact, according to Table 15.1, the simultaneous presence of a *G* allele at `rs28936` and `rs3890137` is associated with an estimated fivefold increase in the risk of MS, with respect to any other allelic combination at the two loci. It would be interesting to compare these estimates with the results of related and relevant *in vitro* or *in vivo* experiments. The results of Table 15.1 also suggest that an excess concentration of ion channels alone, not accompanied by excess expression of PRKCA, has, marginally, a beneficial effect on susceptibility to MS.

The above interpretation should not be intended to represent a solid scientific conclusion, as yet. The only aim here is to give an idea of the way statistical analysis of mechanistic interaction cooperates with the traditional tools of experimental biology in the effort to elucidate disease mechanisms. In particular, in our study example, analysis of mechanistic interaction has allowed rescuing an important actor in the studied phenomenon – PRKCA – from the ocean of genomewide nonsignificance. In this way, analysis of mechanistic interaction helps focus future biological experiments on the 'right' components of the system under study.

Pairwise mechanistic interactions may not provide adequate guidance in those situations where the crucial interactions are of order three or higher. Methods for detecting mechanistic interactions between three causal factors exist. We have not experimented with them. We envisage possible difficulties in the interpretation of three-way interactions, and direct the reader to Berrington de Gonzalez and Cox (2007) for relevant ideas.

Much in the same way as classical genetic experiments on epistasis have been able to predict the existence of entire pathways, before their actual biological description could become available, genomewide data could be mined to identify a priori unexpected interactions between genes, hopefully suggestive of novel relevant pathways. Suitable methods to tame the type 1 error inflation problem arising in this context need to be developed.

[3] Interestingly, excess current may also cause pain, and this is in accord with a detected association between ASIC2 and neuropathic pain [],and with painbeing a symptom of MS.

The ideas in this chapter are relevant to the study of gene–environment interaction and of the tightly linked aspects that fall under the 'epigenetics' heading. Epigenetic modifications such as DNA methylation and histone modifications do not directly alter the DNA sequence, but have a profound impact on DNA chromatin organization and on the regulation of gene expression, and they are possibly promoted by environmental factors, mostly unknown. In the context of our analysis, the minor `rs28936` allele appears to reverse the sign of the effect of the minor `rs3890137` allele. We have found this in maternal transmissions. Had we performed the same analysis on paternal transmissions, we might have detected a difference, interpretable as an example of what is sometimes called the parent-of-origin effect. Evidence of such an effect would suggest the possible influence of some epigenetic factor. Bioinformatic analysis of the regions around `rs28936` and `rs3890137` in search for marks of epigenetic variation would in this case be an appropriate step in the elucidation of these complex aspects of the genetics of disease.

We conclude by mentioning that *in vitro* experiments or (carefully designed) epidemiological studies may identify environmental exposures that interact with ASIC2 mechanistically – another important aspect of the understanding of the origin of disease – one suggestion being exposure to heavy metals.

Acknowledgments

The blood sample collection was performed at the Division of Neurology, S. Francesco Hospital, Nuoro (I); the buffy coat preparation and the DNA extraction, according to classical salting out protocol, were performed at the Centro di Tipizzazione Tissutale of the Azienda Sanitaria Locale N3, Nuoro, where the biological bank resides. Further DNA extraction from buffy coat, microsatellite typing and gene sequencing were performed at the Centre National de Genotypage, Evry, France.

References

Azzalini, A. and Cox, D.R. (1984) Two new tests associated with analysis of variance. *Journal of the Royal Statistical Society, Series B (Methodological)*, **46** (2), 335–343.

Baron, A., Deval, E., Salinas, M., Lingueglia, E., Voilley, N. and Lazdunski, M. (2002) Protein kinase C stimulates the acid-sensing ion channel ASIC2a via the PDZ domain-containing protein PICK1. *Journal of Biological Chemistry*, **277** (52), 50463–50468.

Bernardinelli, L., Murgia, S.B., Bitti, P.P., Foco, L., Ferrai, R., Musu, L., Prokopenko, I., Pastorino, R., Saddi, V., Ticca, A., Piras, M.L., Cox, D.R. and Berzuini, C. (2007) Association between the ACCN1 gene and multiple sclerosis in central east Sardinia. PLoS ONE **2**(5): e480.

Berrington de Gonzalez, A. and Cox, D.R. (2007) Interpretation of interaction: a review. *The Annals of Applied Statistics*, **1** (2), 371–385.

Deval, E., Salinas, M., Baron, A., Lingueglia, E. and Lazdunski, M. (2004) ASIC2b-dependent regulation of ASIC3, an essential acid-sensing ion channel subunit in sensory neurons via the partner protein PICK-1. *Journal of Biology and Chemistry*, **279** (19), 19531–19539.

Ebers, G.C., Sadovnick, A.D., Dyment, D.A., Yee, I.M.L., Willer, C.J. and Risch, N. (2004) Parent-of-origin effect in multiple sclerosis: observations in half-siblings. *The Lancet*, **363**, 1773–1774, May.

Friese, M.J., Craner, M.A., Etzensperger, R., Vergo, S., Wemmie, J.A., Welsh, M.J., Vincent, A. and Fugger, L. (2007) Acid-sensing ion channel-1 contributes to axonal degeneration in autoimmune inflammation of the central nervous system. *Nature Medicine*, **13** (12), 1483–1489.

Hirschhorn, J.N. (2009) Genomewide association studies illuminating biologic pathways. *New England Journal of Medicine*, **360** (17),1699–1701.

Imrell, K. (2009) *Conquering Complexity: successful strategies for finding disease genes in multiple sclerosis*. Department of Clinical Neuroscience, Karolinska Institute, Stockholm, Sweden.

Lewin, G.R., McIlwrath, S.L., Cheng, C., Xie, J., Price, M.P. *et al.* (2000) The mammalian sodium channel BNC1 is required for normal touch sensation. *Nature*, **407**, 1007–1011.

Marchini, J. and Howie, B. (2010) Genotype imputation for genome-wide association studies. *National Review of Genetics*, **11**, 499–511.

Morahan, G., Huang, D., Ymer, S.I., Cancilla, M.R., Stephen, K. *et al.* (2001) Linkage disequilibrium of a type 1 diabetes susceptibility locus with a regulatory il12b allele. *Nature Genetics*, **27**, 218–221.

Scheffe, H. (1959) *Analysis of Variance*. New York: John Wiley & Sons, Inc.

Ueda, H., Howson, J.M., Esposito, L., Heward, J., Snook, H. *et al.* (2003) Association of the T-cell regulatory gene CTLA 4 with susceptibility to autoimmune disease. *Nature*, **423**, 506–511.

Waxman, S.G. (2001) Transcriptional channelopathies: an emerging class of disorders. *National Review of Neuroscience*, **2**, 652–659.

Wemmie, J.A., Price, M.P. and Welsh, M.J. (2006) Acid-sensing ion channels: advances, questions and therapeutic opportunities. *Trends in Neuroscience*, **29**, 578–586.

Zipp, F. and Aktas, O. (2006) The brain as a target of inflammation: common pathways link inflammatory and neurodegenerative diseases. *Trends in Neuroscience*, **29**, 518–527.

16

Supplementary variables for causal estimation

Roland R. Ramsahai
Statistical Laboratory, Centre for Mathematical Sciences, University of Cambridge, Cambridge, UK

16.1 Introduction

Causal inference from observational studies quite often involves roughly two stages, identification and estimation. Identification involves obtaining an expression that relates intervention and observational distribution parameters. The function of the observational parameters in the identification expression is then estimated, resulting in an estimate of the intervention parameter of interest. The conditions required for the existence of any such expression or identifiability of the parameter are discussed at a very general level in Chapter 6 in this volume.

Supplementary variables are considered here as variables that are not of interest in themselves but aid causal inference and their roles vary. An instrumental variable (IV) is an example of a supplementary variable. Intervention parameters are identifiable in the presence of unobserved confounders by observing additional unconfounded variables, the instruments (Durbin, 1954; Bowden and Turkington, 1984). Their use can be traced back to at least Wald (1940), Reiersøl (1941, 1945) and Geary (1942, 1943).

Supplementary variables can also be used to improve the efficiency of estimators. Analysis of covariance, which involves observing a covariate to improve the precision and/or remove the bias of an estimator, originates from Fisher (1932) and has been well developed throughout the literature (Cochran, 1957; Cox and McCullagh, 1982). The similar concept of observing intermediate variables, possibly of no practical importance, to increase the precision of estimators was introduced by Cox (1960).

Causality: Statistical Perspectives and Applications, First Edition. Edited by Carlo Berzuini, Philip Dawid and Luisa Bernardinelli.

Figure 16.1 Augmented DAG for the model discussed in Cox (1960), which involves a treatment (T)*, intermediate* (L) *and response* (R) *variable.*

Here and throughout, we use graphical representations of causality. We favour the augmented directed acyclic graphs, or 'augmented DAGs', introduced in Chapters 4 and 8 in this volume. Recall that such diagrams are obtained by supplementing a standard causal diagram with a regime indicator node, which we denote by F_T, and draw as square. See the mentioned chapters for a discussion of the interpretation of these diagrams. Cox (1960) considers a model that involves a treatment variable, T, an intermediate variable, L, and a response variable, R. The total regression of R on T is of interest and the model is represented by the augmented DAG in Figure 16.1. Consider linear regressions of the form

$$R \mid L = \gamma_{rl} L + \epsilon_{r|l}, \qquad L \mid T = \gamma_{lt} T + \epsilon_{l|t}$$

where $\epsilon_{a|b} \sim N(0, \sigma_{aa|b})$ and $\sigma_{ab|c} = \text{cov}(A, B \mid C)$, equivalently stated as

$$R \mid L \sim N(\gamma_{rl} L, \sigma_{rr|l}), \qquad L \mid T \sim N(\gamma_{lt} T, \sigma_{ll|t}) \tag{16.1}$$

where γ_{ab} is the coefficient of B in the regression of A on B. From Equation (16.1)

$$\text{ACE}(T \to R) = \partial \mathbb{E}(C \parallel B)/\partial B = \gamma_{rt} = \gamma_{rl}\gamma_{lt} \tag{16.2}$$

Cox (1960) compared the estimators based on (R, T) and (R, L, T) data and showed that the maximum likelihood (ML) estimators always satisfy the relationship $\mathbb{V}_n^{\infty}(\hat{\gamma}_{rt}) \geq \mathbb{V}_n^{\infty}(\hat{\gamma}_{rl} . \hat{\gamma}_{lt})$, where $\mathbb{V}_n^{\infty}(\cdot)$ is the asymptotic variance sequence and $\hat{\gamma}$ is the ML estimator of γ, and that appreciable additional precision can be potentially obtained from observing the intermediate variable L, in addition to (R, T). Consider the generalisation of Figure 16.1 in Figure 16.2, where $(X_1, \ldots, X_k)$ follow a multivariate Gaussian distribution.

When all of the X's are recorded in Figure 16.2,

$$1/\mathbb{I}_n(\gamma_{k,1}) = \sum_{i=1}^{k-1} 1/\mathbb{I}_n(\gamma_{i+1,i}) \tag{16.3}$$

where

$$\mathbb{I}_n(\gamma) = \gamma^2 I_n(\gamma)/n = \left\{n\mathbb{V}_n^{\infty}(\hat{\gamma})/\gamma^2\right\}^{-1} = \rho^2/1 - \rho^2 \tag{16.4}$$

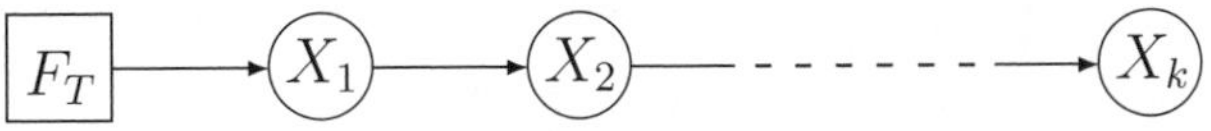

Figure 16.2 Augmented DAG for a more general version of the model in Cox (1960).

$\mathbb{I}_n(\cdot)$ is the information matrix, ρ is the correlation coefficient and $\gamma_{j,i}$ is the regression coefficient of X_i in the regression of X_j on X_i (see the proof in Appendix 16.A). $\mathbb{I}_n(\cdot)$ is the inverse of the square of the asymptotic standard error of the ML estimate per unit value of the parameter (standardised asymptotic variance) or the standardised Fisher information. It represents the ratio of the proportion of the variation in the dependent variable explained to unexplained by the independent variable. $\mathbb{I}_n(\cdot)$ is equivalent to *Cohen's* f (Cohen, 1988), which is a dimensionless measure of effect size commonly used in the behavioural sciences.

Equation (16.3) demonstrates additivity of the standardised variation in the estimators and will be used later to provide an intuitive interpretation of expressions for relative efficiency. The discussion of Cox (1960) was aimed at regression coefficient estimators but the parameters are equivalent to causal effects and will be interpreted as the latter. In models where multiple expressions for causal effects exist, the expression chosen can affect the efficiency of the estimator. Analogously to Cox (1960), it is demonstrated here that increased precision can be obtained by observing supplementary variables, whether intermediate or covariate. For ethical, funding or other reasons, there may be restrictions on the variables that can be recorded and it might only be possible to measure covariates or intermediate variables but not both. Criteria are given for choosing which supplementary variables to record in such circumstances.

16.2 Multiple expressions for causal effect

Consider the model represented by the augmented DAG in Figure 16.3, where C is a confounder (or covariate) and the joint distribution of (R, L, T, C) is multivariate normal. It is assumed throughout that there exists a known intermediate variable, L, and a known covariate, or concomitant variable, C, which satisfy the assumptions of the model in Figure 16.3.

Without loss of generality, assume $\mathbb{E}(C) = 0$ and let

$$\begin{aligned} R \mid L, C &\sim N(\gamma_{rl|c} L + \gamma_{rc|l} C, \sigma_{rr|l,c}) \\ L \mid T &\sim N(\gamma_{lt} T, \sigma_{ll|t}) \\ T \mid C &\sim N(\gamma_{tc} C, \sigma_{tt|c}) \end{aligned} \tag{16.5}$$

denote the relationships between the nodes and their parents in Figure 16.3. Using the 'back-door' formula (Pearl, 1993), 'front-door' formula (Pearl, 1995) and 'extended back-door' formula (Lauritzen, 2001), multiple expressions for ACE($T \to R$) can be obtained in terms of observational parameters

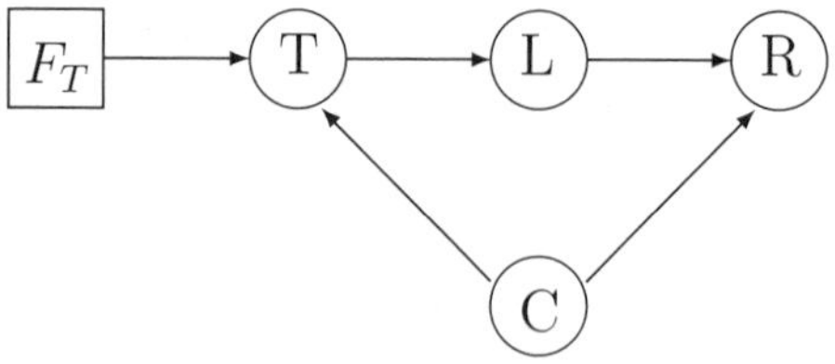

Figure 16.3 Augmented DAG for a model with an intermediate (L) *and covariate* (C) *variable.*

- Back-door (R, C, T distribution):

$$\begin{aligned}\mathbb{E}(R \parallel T^*) &= \mathbb{E}_c\{\mathbb{E}(R \mid C, T^*)\} \\ &= \gamma_{rc|t}\mu_c + \gamma_{rt|c}T^* \\ \Rightarrow \text{ACE}(T \to R) &= \gamma_{rt|c}\end{aligned}$$

- Front-door (R, L, T distribution):

$$\begin{aligned}\mathbb{E}(R \parallel T^*) &= \mathbb{E}_l[\mathbb{E}_t\{\mathbb{E}(R \mid L, T)\} \mid T^*] \\ &= \mathbb{E}_l[\gamma_{rl|t}L + \gamma_{rt|l}\mu_t \mid T^*] \\ &= \gamma_{rl|t}\mathbb{E}(L \mid T^*) + \gamma_{rt|l}\mu_t \\ &= \gamma_{rl|t}\gamma_{lt}T^* + \gamma_{rt|l}\mu_t \\ \Rightarrow \text{ACE}(T \to R) &= \gamma_{rl|t}\gamma_{lt}\end{aligned}$$

- Extended back-door (R, L, C, T distribution):

$$\begin{aligned}\mathbb{E}(R \parallel T^*) &= \mathbb{E}_l[\mathbb{E}_c\{\mathbb{E}(R \mid L, C)\} \mid T^*] \\ &= \mathbb{E}_l[\gamma_{rl|c}L + \gamma_{rc|l}\mu_c \mid T^*] \\ &= \gamma_{rl|c}\mathbb{E}(L \mid T^*) + \gamma_{rc|l}\mu_c \\ &= \gamma_{rl|c}\gamma_{lt}T^* + \gamma_{rc|l}\mu_c \\ \Rightarrow \text{ACE}(T \to R) &= \gamma_{rl|c}\gamma_{lt}\end{aligned}$$

where $\mu_{a|c} = \mathbb{E}(A \mid C)$ and $\gamma_{ab|c}$ is the partial regression coefficient of B in the regression of A on B and C. Hence, in particular, when $\gamma_{lt} \neq 0$, $\gamma_{rl|c} = \gamma_{rl|t}$. It can be easily deduced that, unlike the model in Figure 16.1, Equation (16.2) does not hold because of the confounding by C. There is no expression for ACE($T \to R$) in terms of the joint observational distribution of R and T only, so the concern here is with the efficiency of the estimator when L, C or both are recorded in addition to the (R, T) data.

Similarly to Cox (1960), Lauritzen (2001) suggests a comparison of the efficiency of the estimators based on the multiple expressions

$$\gamma_{rt|c} = \gamma_{rl|t}\cdot\gamma_{lt} = \gamma_{rl|c}\cdot\gamma_{lt} \tag{16.6}$$

which are all path coefficients (Wright, 1921) for the $T \to R$ path.

Graphical criteria for choosing between the 'back-door' formula and 'conditional IV' method (Brito and Pearl, 2002) to minimise the asymptotic variance of the corresponding estimators are developed in Kuroki and Cai (2004). They note that, in general, the difference in asymptotic variance between these two methods and the 'front-door' formula cannot be determined from the graph structure.

16.3 Asymptotic variance of causal estimators

The asymptotic variance sequence of the estimators are:

- *R*, *C* and *T* recorded:

$$\mathbb{V}_n^{\infty}(\hat{\gamma}_{rt|c}) = \frac{1}{n}\left(\frac{\sigma_{rr|l,c}}{\sigma_{tt|c}} + \gamma_{rl|c}^2\frac{\sigma_{ll|t}}{\sigma_{tt|c}}\right) \tag{16.7}$$

- *R*, *L* and *T* recorded:

$$\mathbb{V}_n^{\infty}(\hat{\gamma}_{rl|t}\cdot\hat{\gamma}_{lt}) = \frac{1}{n}\left(\gamma_{lt}^2\frac{\sigma_{rr|l,t}}{\sigma_{ll|t}} + \gamma_{rl|t}^2\frac{\sigma_{ll|t}}{\sigma_{tt}}\right) \tag{16.8}$$

- *R*, *L*, *C* and *T* recorded:

$$\mathbb{V}_n^{\infty}(\hat{\gamma}_{rl|c}\cdot\hat{\gamma}_{lt}) = \frac{1}{n}\left(\gamma_{lt}^2\frac{\sigma_{rr|l,c}}{\sigma_{ll|c}} + \gamma_{rl|c}^2\frac{\sigma_{ll|t}}{\sigma_{tt}}\right) \tag{16.9}$$

with proofs in Appendix 16.B. To facilitate an intuitive comparison of these expressions, they are expressed in terms of correlation coefficients:

- *R*, *C* and *T* recorded:

$$n\mathbb{V}_n^{\infty}(\hat{\gamma}_{rt|c}) = \gamma_{rl|c}^2\cdot\gamma_{lt}^2\left[\frac{1-\rho_{rl|c}^2}{\rho_{rl|c}^2}\left\{1+\frac{1-\rho_{lt}^2}{\rho_{lt}^2(1-\rho_{tc}^2)}\right\}+\frac{1-\rho_{lt}^2}{\rho_{lt}^2(1-\rho_{tc}^2)}\right] \tag{16.10}$$

- *R*, *L* and *T* recorded:

$$n\mathbb{V}_n^{\infty}(\hat{\gamma}_{rl|t}\cdot\hat{\gamma}_{lt}) = \gamma_{rl|t}^2\cdot\gamma_{lt}^2\left(\frac{1-\rho_{rl|t}^2}{\rho_{rl|t}^2}+\frac{1-\rho_{lt}^2}{\rho_{lt}^2}\right) \tag{16.11}$$

- *R*, *L*, *C* and *T* recorded:

$$n\mathbb{V}_n^{\infty}(\hat{\gamma}_{rl|c}\cdot\hat{\gamma}_{lt}) = \gamma_{rl|c}^2\cdot\gamma_{lt}^2\left(\frac{1-\rho_{rl|c}^2}{\rho_{rl|c}^2}+\frac{1-\rho_{lt}^2}{\rho_{lt}^2}\right) \tag{16.12}$$

where $\rho_{ab|c}$ is the correlation between *A* and *B* conditioning on a fixed value of *C*. The derivations are given in Appendix 16.C and a comparison in Section 16.4.

16.4 Comparison of causal estimators

In this section the asymptotic variance of the various estimators in Equations (16.10), (16.11) and (16.12) are compared to determine which variables should ideally be recorded in addition to R and T. For two estimators $\hat{\gamma}_1$ and $\hat{\gamma}_2$, define

$$\nabla\mathbb{I}(\hat{\gamma}_1;\hat{\gamma}_2) = \frac{\mathbb{I}_n(\gamma_1)}{\mathbb{I}_n(\gamma_2)}, \qquad \Delta\mathbb{I}(\hat{\gamma}_1;\hat{\gamma}_2) = \frac{1}{\mathbb{I}_n(\gamma_2)} - \frac{1}{\mathbb{I}_n(\gamma_1)}$$

where $\nabla\mathbb{I}(\cdot)$ is the relative information and $\Delta\mathbb{I}(\cdot)$ is the difference in the inverse of the Fisher information. Both are dimensionless measures of contrast in effect size and will be used for formal comparisons of the efficiency of two estimators. The particular measure used will be chosen according to which has a more intuitive interpretation. If $\gamma_1 = \gamma_2$ then $\nabla\mathbb{I}(\hat{\gamma}_1;\hat{\gamma}_2)$ is the relative asymptotic efficiency of $\hat{\gamma}_1$ to $\hat{\gamma}_2$ and

$$\nabla\mathbb{I}(\hat{\gamma}_1;\hat{\gamma}_2) \geq 1 \quad \Leftrightarrow \quad \Delta\mathbb{I}(\hat{\gamma}_1;\hat{\gamma}_2) \geq 0$$

implies that $\hat{\gamma}_1$ is more asymptotically efficient. The larger the value of both $\nabla\mathbb{I}(\hat{\gamma}_1;\hat{\gamma}_2)$ and $\Delta\mathbb{I}(\hat{\gamma}_1;\hat{\gamma}_2)$ the more efficient $\hat{\gamma}_1$ is compared to $\hat{\gamma}_2$.

16.4.1 Supplement C with L or not

In a study in which R, C and T are recorded, the purpose of additionally measuring L is to remove variation entering the system after T has exerted its effect. If an investigator suspects that L contains little information about ACE($T \to R$) that is not already available in R, C and T then they would opt to omit it. Measures of the importance of L are the relative asymptotic efficiency of the estimators

$$\nabla\mathbb{I}(\hat{\gamma}_{rl|c}.\hat{\gamma}_{lt};\hat{\gamma}_{rt|c}) = \frac{\dfrac{1}{\mathbb{I}_n(\gamma_{rl|t})} + \dfrac{1}{\mathbb{I}_n(\gamma_{lt})}}{\dfrac{1}{\mathbb{I}_n(\gamma_{rl|c})} + \dfrac{1}{\mathbb{I}_n(\gamma_{lt})}} \tag{16.13}$$

and

$$\Delta\mathbb{I}(\hat{\gamma}_{rl|c}.\hat{\gamma}_{lt};\hat{\gamma}_{rt|c}) = \frac{1}{\mathbb{I}_n(\gamma_{rl|c})}\left\{\frac{1+\mathbb{I}_n(\gamma_{tc})}{\mathbb{I}_n(\gamma_{lt})}\right\} + \frac{\mathbb{I}_n(\gamma_{tc})}{\mathbb{I}_n(\gamma_{lt})} \tag{16.14}$$

from Equations (16.10) and (16.12), where $\mathbb{I}_n(\cdot)$ is defined in Equation (16.4):

$$\mathbb{I}_n^*(\gamma_{rl|c}) = \frac{\mathbb{I}_n(\gamma_{rl|c})}{1 + 1/\mathbb{I}_n^*(\gamma_{lt})}, \qquad \mathbb{I}_n^*(\gamma_{lt}) = \frac{\mathbb{I}_n(\gamma_{lt})}{1 + \mathbb{I}_n(\gamma_{tc})}$$

From Equations (16.13) and (16.14), since $\mathbb{I}_n(\cdot) \geq 0$,

$$\nabla\mathbb{I}(\hat{\gamma}_{rl|c}.\hat{\gamma}_{lt};\hat{\gamma}_{rt|c}) \geq 1, \qquad \Delta\mathbb{I}(\hat{\gamma}_{rl|c}.\hat{\gamma}_{lt};\hat{\gamma}_{rt|c}) \geq 0$$

which agrees with intuition since it means that the omission of an intermediate variable does not increase the efficiency of the estimator.

Intuitively, the denominator in Equation (16.13) follows the additive formula of Equation (16.3) but the quantities in the numerator are inflated by the omission of L, i.e. less information about the $T \rightarrow L$ and $L \rightarrow R$ edges are available. As expected, the inclusion of L cannot worsen the efficiency of the estimator because L removes variation entering the system after treatment allocation.

The larger the value of $\Delta\mathbb{I}(\hat{\gamma}_{rl|c}.\hat{\gamma}_{lt};\hat{\gamma}_{rt|c})$ the more beneficial it is to observe L. Therefore, from Equation (16.14), as $\mathbb{I}_n(\gamma_{rl|c})$ or $\mathbb{I}_n(\gamma_{lt})$ increases the benefit of L decreases. This is because L contains less additional information than what is already available in T and R.

When $\rho_{tc} = 0$ and $\rho_{rl|c} = \rho_{rl} \Leftrightarrow \rho_{rc|l} = 0$, i.e. there is no confounding, Equation (16.13) is equivalent to Equation (16.18) of Cox (1960).

16.4.2 Supplement L with C or not

In a study in which R, L and T are recorded, additionally measuring C serves the purpose of removing variation entering the system before T has exerted its effect, just as in analysis of covariance. In addition to increasing the precision of treatment contrasts, analysis of covariance can also serve to remove bias (Cochran, 1957; Cox and McCullagh, 1982) but observing C is not necessary to remove bias here because L alone is sufficient for that. If L was not recorded then C would also be needed for the bias adjustment. It may be decided not to record C if it does not improve the precision of the estimator of ACE($T \rightarrow R$) much. The importance of C is quantified by comparing Equations (16.8) and (16.9) via

$$\nabla\mathbb{I}(\hat{\gamma}_{rl|c}.\hat{\gamma}_{lt};\hat{\gamma}_{rl|t}.\hat{\gamma}_{lt}) = \frac{\dfrac{1}{\mathbb{I}_n(\gamma_{rl|t})} + \dfrac{1}{\mathbb{I}_n(\gamma_{lt})}}{\dfrac{1}{\mathbb{I}_n(\gamma_{rl|c})} + \dfrac{1}{\mathbb{I}_n(\gamma_{lt})}} \tag{16.15}$$

from Equations (16.11) and (16.12), and

$$\Delta\mathbb{I}(\hat{\gamma}_{rl|c}.\hat{\gamma}_{lt};\hat{\gamma}_{rl|t}.\hat{\gamma}_{lt}) = \frac{1}{\mathbb{I}_n(\gamma_{rl|c})}\left\{\frac{\mathbb{I}_n(\gamma_{lt})}{1+\mathbb{I}_n(\gamma_{tc})}\right\} + \frac{\mathbb{I}_n(\gamma_{rc|l})}{\mathbb{I}_n(\gamma_{rl|c})}\left\{\frac{1+\mathbb{I}_n(\gamma_{lt})}{1+\mathbb{I}_n(\gamma_{tc})}\right\} \tag{16.16}$$

with proof in Appendix 16.D. From Equations (16.15) and (16.16), since $\rho^2_{rl|c} \geq \rho^2_{rl|t}$ (see the proof in Appendix 16.E), which implies that $\mathbb{I}_n(\gamma_{rl|c}) \geq \mathbb{I}_n(\gamma_{rl|t})$,

$$\nabla\mathbb{I}(\hat{\gamma}_{rl|c}.\hat{\gamma}_{lt};\hat{\gamma}_{rl|t}.\hat{\gamma}_{lt}) \geq 1, \qquad \Delta\mathbb{I}(\hat{\gamma}_{rl|c}.\hat{\gamma}_{lt};\hat{\gamma}_{rl|t}.\hat{\gamma}_{lt}) \geq 0$$

This agrees with intuition since it means that omission of a covariate cannot increase the efficiency of the estimator. Recording C removes variation entering the system before treatment allocation.

From Equation (16.16), the benefit of recording C increases as $\mathbb{I}_n(\gamma_{rc|l})$ increases and decreases as $\mathbb{I}_n(\gamma_{tc})$ increases. The importance of C reflects the two competing effects and agrees with classical results (cf. Robinson and Jewell, 1991; Section 9.5 of Jewell, 2003). The effect of the strength of the $C \rightarrow T$ relationship exists because the increasing collinearity increases the redundant information or reduces the information that C contains that is not already available in T. Thus, similarly to L, the less redundant the information contained in the supplementary variable the greater the benefit of recording it. The effect of the strength of the $C \rightarrow R$ relationship occurs because the stronger the relationship the greater the proportion of residual variation, in the regression of R on L, that is removed by conditioning on C.

16.4.3 Replace C with L or not

For the scenarios in Sections 16.4.1 and 16.4.2 the more supplementary variables recorded the greater the efficiency of the estimator. A study with all variables measured is ideal but such studies may be expensive or impossible to conduct. Thus it is often necessary to choose whether to record L or C only, in addition to R and T.

The relative efficiency of the corresponding estimators is very important and is useful to assess the trade-off between cost and efficiency of the estimator. Such an analysis provides criteria for judging whether it is more beneficial for covariates or intermediate variables to be recorded in a particular study. Although $\nabla\mathbb{I}(\hat{\gamma}_{rl|t}.\hat{\gamma}_{lt};\hat{\gamma}_{rt|c})$ can be derived from Equations (16.10) and (16.11) and used to determine formal conditions for choosing whether to record L or C, the comparison is simpler by considering the quantity

$$\Delta\mathbb{I}(\hat{\gamma}_{rl|t}.\hat{\gamma}_{lt};\hat{\gamma}_{rt|c}) = \frac{1}{\mathbb{I}_n(\gamma_{rl|c})}\left\{\frac{1+\mathbb{I}_n(\gamma_{tc})}{\mathbb{I}_n(\gamma_{lt})} - \frac{\mathbb{I}_n(\gamma_{lt})}{1+\mathbb{I}_n(\gamma_{tc})} + \mathbb{I}_n(\gamma_{rl|c})\frac{\mathbb{I}_n(\gamma_{tc})}{\mathbb{I}_n(\gamma_{lt})} - \mathbb{I}_n(\gamma_{rc|l})\frac{1+\mathbb{I}_n(\gamma_{lt})}{1+\mathbb{I}_n(\gamma_{tc})}\right\} \tag{16.17}$$

with proof in Appendix 16.D. When $\nabla\mathbb{I}(\hat{\gamma}_{rl|t}.\hat{\gamma}_{lt};\hat{\gamma}_{rt|c}) > 1$, which is equivalent to $\Delta\mathbb{I}(\hat{\gamma}_{rl|t}.\hat{\gamma}_{lt};\hat{\gamma}_{rt|c}) > 0$, and $\nabla\mathbb{I}(\hat{\gamma}_{rl|t}.\hat{\gamma}_{lt};\hat{\gamma}_{rt|c})$ can take values anywhere on the positive real line, it is more beneficial to record L. Based on Equation (16.17), sufficient conditions to choose to record L instead of C are

$$\mathbb{I}_n(\gamma_{lt}) < \mathbb{I}_n(\gamma_{tc}), \qquad \mathbb{I}_n(\gamma_{rc|l}) < \mathbb{I}_n(\gamma_{rl|c})$$

Thus it is more beneficial to record L if the strength of each of the $T \to L$ and $C \to R$ relationships is smaller than the $C \to T$ and $L \to R$ relationships respectively. In other words, it is better to record L when L contains less redundant information than C from T and L contains more information than C about R. These conditions are represented in Figure 16.4. Interestingly, the reverse conditions are not sufficient for C to be more useful. Sufficient conditions for recording C are actually

$$\mathbb{I}_n(\gamma_{lt}) > \mathbb{I}_n(\gamma_{tc}) + 1, \qquad \mathbb{I}_n(\gamma_{rc|l}) > \mathbb{I}_n(\gamma_{rl|c})$$

They are similar, but not exactly opposite, to those required for recording L instead and are represented in Figure 16.4. Therefore, it is possible for the observation of C to be less beneficial

Figure 16.4 DAGs representing conditions for choosing to record L *instead of* C *(left) and vice versa (right). The thicker arrow represents a stronger relationship, as measured by* $\mathbb{I}_n(\cdot)$.

if it contains only slightly less redundant information than L from T:

$$\mathbb{I}_n(\gamma_{tc}) < \mathbb{I}_n(\gamma_{lt}) < \mathbb{I}_n(\gamma_{tc}) + 1$$

even though $\mathbb{I}_n(\gamma_{rc|l}) > \mathbb{I}_n(\gamma_{rl|c})$. In such borderline cases the behaviour of $\Delta\mathbb{I}(\hat{\gamma}_{rl|t}.\hat{\gamma}_{lt}; \hat{\gamma}_{rt|c})$ is determined by $\mathbb{I}_n(\gamma_{rl|c})$ and $\mathbb{I}_n(\gamma_{rc|l})$.

At first glance the asymmetry may seem a bit strange since the conditional independence relations represented by the DAG in Figure 16.4, $L \perp\!\!\!\perp C \mid T$ and $R \perp\!\!\!\perp T \mid (L, C)$, are symmetric in L and C. However, nontrivial causal assumptions are being made by augmenting the DAG with the regime indicator node to form Figure 16.3. The augmented DAG in Figure 16.3 represents additional causal assumptions (Dawid, 2002), such as $L \perp\!\!\!\perp F_T \mid T$ and $C \perp\!\!\!\perp F_T$, which are not symmetric in L and C.

16.5 Discussion

Sufficient criteria have been derived for choosing which supplementary variables to record to improve the efficiency of estimators. Although the precision of the resulting estimator only was taken into account, there is much room for extension by modelling the cost of recording variables as well in a decision framework. Other worthwhile directions of research include the analysis of non-Gaussian models and other supplementary variables. Robinson and Jewell (1991) investigate the impact on efficiency of the strength of the relationships with covariates in a logistic model. Future work will involve a comparison of efficiency when recording confounders versus instrumental variables.

Acknowledgements

This research was conducted under the supervision of Steffen Lauritzen and the author has benefited from many detailed discussions with Sir David Cox.

A Appendices

16.A Estimator given all *X*'s recorded

Theorem A.1 *The relation*

$$\frac{1}{\mathbb{I}_n(\gamma_{k,1})} = \sum_{i=1}^{k-1} \frac{1}{\mathbb{I}_n(\gamma_{i+1,i})}$$

is true for the DAG in Figure 16.2 when all of the variables are recorded.

Proof of Theorem A.1. When $k = 2$ the result is trivially true. Assume it is true for $k - 1$. By the delta method

$$n\mathbb{V}_n^{\infty}(\hat{\gamma}_{k,1}) \approx \gamma_{k,k-1}^2 n\mathbb{V}_n^{\infty}(\hat{\gamma}_{k-1,1}) + \gamma_{k-1,1}^2 n\mathbb{V}_n^{\infty}(\hat{\gamma}_{k,k-1})$$

since $\hat{\gamma}_{i+1,1} = \hat{\gamma}_{i+1,i} \cdots \hat{\gamma}_{2,1}$ for $i = 1, \ldots, n-1$ and

$$\mathrm{cov}_n^{\infty}(\hat{\gamma}_{i+1,i}, \hat{\gamma}_{j+1,j}) = 0 \text{ for } i \neq j$$

from the factorisation of the likelihood function. From the inductive hypothesis

$$\begin{aligned}
n\mathbb{V}_n^{\infty}(\hat{\gamma}_{k,1}) &= \gamma_{k,k-1}^2 n\mathbb{V}_n^{\infty}(\hat{\gamma}_{k-1,1}) + \gamma_{k-1,1}^2 n\mathbb{V}_n^{\infty}(\hat{\gamma}_{k,k-1}) \\
\gamma_{k,1}^{-2} n\mathbb{V}_n^{\infty}(\hat{\gamma}_{k,1}) &= \gamma_{k-1,1}^{-2} n\mathbb{V}_n^{\infty}(\hat{\gamma}_{k-1,1}) + \gamma_{k,k-1}^{-2} n\mathbb{V}_n^{\infty}(\hat{\gamma}_{k,k-1}) \\
\frac{1}{\mathbb{I}_n(\gamma_{k,1})} &= \sum_{i=1}^{k-2} \frac{1}{\mathbb{I}_n(\gamma_{i+1,i})} + \frac{1}{\mathbb{I}_n(\gamma_{k,k-1})}
\end{aligned}$$

since $\gamma_{i+1,1} = \gamma_{i+1,i} \cdots \gamma_{2,1}$ for $i = 1, \ldots, n-1$.

16.B Derivations of asymptotic variances

The derivations of Equations (16.7), (16.8) and (16.9) in Section 16.3 are given here.

16.B.1 Estimator given *R*, *C* and *T* recorded

Let $f(\cdot)$ be the probability density function; then the likelihood function is

$$\mathcal{L}(\gamma_{rt|c}) = f(R \mid T, C) f(T \mid C) f(C) \propto f(R \mid T, C)$$

Therefore the log-likelihood function is

$$l(\gamma_{rt|c}) \propto \ln f(R \mid T, C) \propto -\frac{n}{2} \ln(2\pi\sigma_{rr|t,c}) - \sum_{i=1}^{n} \frac{(r_i - \gamma_{rt|c} t_i - \gamma_{rc|t} c_i)^2}{2\sigma_{rr|t,c}}$$

where $\sigma_{ab|t,c} = \text{cov}(A, B \mid T, C)$ and the information matrix of $(\gamma_{rt|c}, \gamma_{rc|t})$ is

$$\mathbb{I}_n(\gamma_{rt|c}, \gamma_{rc|t}) = \begin{pmatrix} \dfrac{\partial^2 l}{\partial \gamma_{rt|c}^2} & \dfrac{\partial^2 l}{\partial \gamma_{rt|c} \partial \gamma_{rc|t}} \\ \dfrac{\partial^2 l}{\partial \gamma_{rt|c} \partial \gamma_{rc|t}} & \dfrac{\partial^2 l}{\partial \gamma_{rc|t}^2} \end{pmatrix} = \frac{1}{\sigma_{rr|t,c}} \begin{pmatrix} -\sum t_i^2 & -\sum t_i c_i \\ -\sum t_i c_i & -\sum c_i^2 \end{pmatrix}$$

The information matrix of the vector $(\gamma_{rt|c}, \gamma_{rc|t})$ is considered because of the factorisation of the likelihood function. Since $T \mid C \sim N(\gamma_{tc} C, \sigma_{tt|c})$ and $\sigma_{tt|c}$ is functionally independent of C,

$$\sigma_{tt} = \mathbb{E}_c(\sigma_{tt|c}) + \mathbb{V}_c(\gamma_{tc} C) = \sigma_{tt|c} + \gamma_{tc}^2 \sigma_{cc}$$

Therefore, since $\gamma_{tc} = \sigma_{tc}/\sigma_{cc}$,

$$\sigma_{tt|c} = (\sigma_{tt}\sigma_{cc} - \sigma_{tc}^2)/\sigma_{cc} = \sigma_{tt}(1 - \rho_{tc}^2) \tag{16.18}$$

Equation (16.18) can also be obtained from the properties of the concentration matrix of a multivariate Gaussian distribution. It follows that

$$\mathbb{I}_n(\gamma_{rt|c}, \gamma_{rc|t}) = \frac{n}{\sigma_{rr|t,c}} \begin{pmatrix} \sigma_{tt} & \sigma_{tc} \\ \sigma_{tc} & \sigma_{cc} \end{pmatrix} \Rightarrow n\mathbb{V}_n^\infty(\hat{\gamma}_{rt|c}) = \frac{n}{\mathbb{I}_n(\gamma_{rt|c}, \gamma_{rc|t})_{tt}} = \frac{\sigma_{rr|t,c}}{\sigma_{tt|c}}$$

from Equation (16.18). It is true that

$$\begin{aligned} \sigma_{rr|t,c} &= \mathbb{E}_{l|t,c}(\sigma_{rr|l,t,c} \mid T, C) + l|t, c(\mu_{r|l,t,c} \mid T, C) \\ &= \sigma_{rr|l,t,c} + l|t, c(\gamma_{rl|c} L + \gamma_{rc|l} C \mid T, C) \\ &= \sigma_{rr|l,c} + \gamma_{rl|c}^2 \sigma_{ll|t} \end{aligned}$$

since $R \perp\!\!\!\perp T \mid (L, C)$, $L \perp\!\!\!\perp C \mid T$ and $\sigma_{rr|l,t,c}$ is functionally independent of L. Therefore

$$n\mathbb{V}_n^\infty(\hat{\gamma}_{rt|c}) = \frac{\sigma_{rr|l,c}}{\sigma_{tt|c}} + \gamma_{rl|c}^2 \frac{\sigma_{ll|t}}{\sigma_{tt|c}}$$

16.B.2 Estimator given *R*, *L* and *T* recorded

The likelihood function for $(\gamma_{rl|t}, \gamma_{lt})$ is

$$\mathcal{L}(\gamma_{rl|t}, \gamma_{lt}) = f(R \mid L, T) f(L \mid T) f(T) \propto f(R \mid L, T) f(L \mid T)$$

and the log-likelihood function is

$$l(\gamma_{rl|t}, \gamma_{lt}) \propto -\frac{n}{2} \ln(4\pi^2 \sigma_{rr|l,t}.\sigma_{ll|t}) - \sum_{i=1}^{n} \frac{(r_i - \gamma_{rl|t} l_i - \gamma_{rt|l} t_i)^2}{2\sigma_{rr|l,t}} - \sum_{i=1}^{n} \frac{(l_i - \gamma_{lt} t_i)^2}{2\sigma_{ll|t}}$$

The information matrix of $(\gamma_{rl|t}, \gamma_{rt|l}, \gamma_{lt})$ is

$$\mathbb{I}_n(\gamma_{rl|t}, \gamma_{rt|l}, \gamma_{lt}) = n \begin{pmatrix} \sigma_{ll}/\sigma_{rr|l,t} & \sigma_{lt}/\sigma_{rr|l,t} & 0 \\ \sigma_{lt}/\sigma_{rr|l,t} & \sigma_{tt}/\sigma_{rr|l,t} & 0 \\ 0 & 0 & \sigma_{tt}/\sigma_{ll|t} \end{pmatrix}$$

from which it follows by the delta method that

$$\begin{aligned} \mathbb{V}(\hat{\gamma}_{rl|t}.\hat{\gamma}_{lt}) &\approx \mathbb{E}(\hat{\gamma}_{lt})^2\mathbb{V}(\hat{\gamma}_{rl|t}) + \mathbb{E}(\hat{\gamma}_{rl|t})^2\mathbb{V}(\hat{\gamma}_{lt}) + 2\mathbb{E}(\hat{\gamma}_{rl|t})\mathbb{E}(\hat{\gamma}_{lt})\mathrm{cov}(\hat{\gamma}_{rl|t}, \hat{\gamma}_{lt}) \\ n\mathbb{V}_n^{\infty}(\hat{\gamma}_{rl|t}.\hat{\gamma}_{lt}) &\approx \gamma_{lt}^2 \left\{ \frac{\sigma_{rr|l,t}.\sigma_{tt}}{\sigma_{ll}\sigma_{tt} - \sigma_{lt}^2} \right\} + \gamma_{rl|t}^2 \left(\frac{\sigma_{ll|t}}{\sigma_{tt}} \right) \\ &\approx \gamma_{lt}^2 \left(\frac{\sigma_{rr|l,t}}{\sigma_{ll|t}} \right) + \gamma_{rl|t}^2 \left(\frac{\sigma_{ll|t}}{\sigma_{tt}} \right) \end{aligned}$$

The derivation of Equation (16.9) is very similar and is omitted (Ramsahai, 2008).

16.C Expressions with correlation coefficients

The derivations of Equations (16.10), (16.11) and (16.12) are given here.

16.C.1 Estimator given *R*, *C* and *T* recorded

Since $L \perp\!\!\!\perp C \mid T$ from the causal DAG in Figure 16.3,

$$\begin{aligned} \sigma_{ll|c} &= \mathbb{E}_{t|c}\{\mathbb{V}(L \mid T, C)\} + \mathbb{V}_{t|c}\{\mathbb{E}(L \mid T, C)\} \\ &= \sigma_{ll|t} + \gamma_{lt}^2\sigma_{tt|c} \\ \frac{\sigma_{ll|c} - \sigma_{ll|t}}{\sigma_{ll|t}} &= \frac{\sigma_{lt}^2\sigma_{tt}(1-\rho_{tc}^2)}{\sigma_{tt}^2\sigma_{ll}(1-\rho_{lt}^2)} \end{aligned}$$

from Equation (16.18), which implies that

$$\frac{\gamma_{lt}^2\sigma_{tt|c}}{\sigma_{ll|t}} = \frac{\sigma_{ll|c} - \sigma_{ll|t}}{\sigma_{ll|t}} = \frac{\rho_{lt}^2}{1-\rho_{lt}^2}(1-\rho_{tc}^2) = \frac{\mathbb{I}_n(\gamma_{lt})}{1+\mathbb{I}_n(\gamma_{tc})} \tag{16.19}$$

The following equation also holds for the model:

$$\sigma_{rr|c} = \mathbb{E}_{l|c}\{\mathbb{V}(R \mid L, C)\} + \mathbb{V}_{l|c}\{\mathbb{E}(R \mid L, C)\} = \sigma_{rr|l,c} + \gamma_{rl|c}^2\sigma_{ll|c}$$

which implies that

$$\sigma_{rr|l,c}/(\gamma_{rl|c}^2\sigma_{ll|c}) = (1-\rho_{rl|c}^2)/\rho_{rl|c}^2 \tag{16.20}$$

From Equations (16.7), (16.19) and (16.20),

$$
\begin{aligned}
n\mathbb{V}_n^{\infty}(\hat{\gamma}_{rt|c}) &= \frac{\sigma_{rr|l,c}}{\sigma_{tt|c}} + \gamma_{rl|c}^2 \frac{\sigma_{ll|t}}{\sigma_{tt|c}} \\
&= \gamma_{rl|c}^2 \cdot \gamma_{lt}^2 \left\{ \frac{\sigma_{rr|l,c}}{\gamma_{rl|c}^2 \sigma_{ll|c}} \left(\frac{\sigma_{ll|c}}{\gamma_{lt}^2 \sigma_{tt|c}} \right) + \frac{\sigma_{ll|t}}{\gamma_{lt}^2 \sigma_{tt|c}} \right\} \\
&= \gamma_{rl|c}^2 \cdot \gamma_{lt}^2 \left[\frac{1-\rho_{rl|c}^2}{\rho_{rl|c}^2} \left\{ 1 + \frac{1}{1-\rho_{tc}^2} \left(\frac{1-\rho_{lt}^2}{\rho_{lt}^2} \right) \right\} + \frac{1}{1-\rho_{tc}^2} \left(\frac{1-\rho_{lt}^2}{\rho_{lt}^2} \right) \right]
\end{aligned}
$$

16.C.2 Estimator given *R*, *L* and *T* recorded

Similarly to the derivation of Equation (16.20),

$$
\sigma_{rr|t} = \mathbb{E}_{l|t}\{\mathbb{V}(R \mid L, T)\} + \mathbb{V}_{l|t}\{\mathbb{E}(R \mid L, T)\} = \sigma_{rr|l,t} + \gamma_{rl|t}^2 \sigma_{ll|t}
$$

which implies that

$$
\frac{\sigma_{rr|l,t}}{\gamma_{rl|t}^2 \sigma_{ll|t}} = \frac{\sigma_{rr|l,t}}{\sigma_{rr|t} - \sigma_{rr|l,t}} = \frac{1-\rho_{rl|t}^2}{\rho_{rl|t}^2} \tag{16.21}
$$

From Equations (16.8) and (16.21),

$$
\begin{aligned}
n\mathbb{V}_n^{\infty}(\hat{\gamma}_{rl|t} \cdot \hat{\gamma}_{lt}) &= \gamma_{lt}^2 \frac{\sigma_{rr|l,t}}{\sigma_{ll|t}} + \gamma_{rl|t}^2 \frac{\sigma_{ll|t}}{\sigma_{tt}} \\
&= \gamma_{rl|t}^2 \cdot \gamma_{lt}^2 \left(\frac{\sigma_{rr|l,t}}{\gamma_{rl|t}^2 \sigma_{ll|t}} + \frac{\sigma_{ll|t}}{\gamma_{lt}^2 \sigma_{tt}} \right) \\
&= \gamma_{rl|t}^2 \cdot \gamma_{lt}^2 \left(\frac{1-\rho_{rl|t}^2}{\rho_{rl|t}^2} + \frac{1-\rho_{lt}^2}{\rho_{lt}^2} \right)
\end{aligned}
$$

16.C.3 Estimator given *R*, *L*, *C* and *T* recorded

From Equations (16.9), (16.18), (16.19) and (16.20),

$$
\begin{aligned}
n\mathbb{V}_n^{\infty}(\hat{\gamma}_{rl|c} \cdot \hat{\gamma}_{lt}) &= \gamma_{lt}^2 \frac{\sigma_{rr|l,c}}{\sigma_{ll|c}} + \gamma_{rl|c}^2 \frac{\sigma_{ll|t}}{\sigma_{tt}} \\
&= \gamma_{rl|c}^2 \cdot \gamma_{lt}^2 \left(\frac{\sigma_{rr|l,c}}{\gamma_{rl|c}^2 \sigma_{ll|c}} + \frac{\sigma_{ll|t}}{\gamma_{lt}^2 \sigma_{tt}} \right) \\
&= \gamma_{rl|c}^2 \cdot \gamma_{lt}^2 \left(\frac{1-\rho_{rl|c}^2}{\rho_{rl|c}^2} + \frac{1-\rho_{lt}^2}{\rho_{lt}^2} \right)
\end{aligned}
$$

16.D Derivation of $\Delta\mathbb{I}$'s

The expressions in Equations (16.16) and (16.17) are derived here. Since $R \perp\!\!\!\perp T \mid (L, C)$,

$$
\sigma_{rr|l,t} = \mathbb{E}_{c|l,t}(\sigma_{rr|l,t,c}) + \mathbb{V}_{c|l,t}(\mu_{r|l,t,c}) = \sigma_{rr|l,c} + \gamma_{rc|l}^2 \sigma_{cc|t} \tag{16.22}
$$

because $\sigma_{rr|l,c}$ is functionally independent of C. From Equations (16.8), (16.9) and (16.22)

$$\begin{aligned}
n\mathbb{V}_n^{\infty}(\hat{\gamma}_{rl|t}.\hat{\gamma}_{lt}) - n\mathbb{V}_n^{\infty}(\hat{\gamma}_{rl|c}.\hat{\gamma}_{lt}) &= \gamma_{lt}^2\frac{\sigma_{rr|l,t}}{\sigma_{ll|t}} + \gamma_{rl|t}^2\frac{\sigma_{ll|t}}{\sigma_{tt}} - \gamma_{lt}^2\frac{\sigma_{rr|l,c}}{\sigma_{ll|c}} - \gamma_{rl|c}^2\frac{\sigma_{ll|t}}{\sigma_{tt}} \\
&= \gamma_{lt}^2\left(\frac{\sigma_{rr|l,t}}{\sigma_{ll|t}} - \frac{\sigma_{rr|l,c}}{\sigma_{ll|c}}\right) \\
&= \gamma_{lt}^2\left(\frac{\sigma_{rr|l,c}}{\sigma_{ll|t}} - \frac{\sigma_{rr|l,c}}{\sigma_{ll|c}}\right) + \gamma_{rl|c}^2.\gamma_{lt}^2\left\{\frac{\mathbb{I}_n(\gamma_{rc|l})}{\mathbb{I}_n(\gamma_{rl|c})}\right\}\frac{1+\mathbb{I}_n(\gamma_{lt})}{1+\mathbb{I}_n(\gamma_{tc})} \\
&= \gamma_{rl|c}^2.\gamma_{lt}^2\left[\frac{\sigma_{rr|l,c}}{\gamma_{rl|c}^2\sigma_{ll|c}}\left(\frac{\gamma_{lt}^2\sigma_{tt|c}}{\sigma_{ll|t}}\right) + \frac{\mathbb{I}_n(\gamma_{rc|l})}{\mathbb{I}_n(\gamma_{rl|c})}\left\{\frac{1+\mathbb{I}_n(\gamma_{lt})}{1+\mathbb{I}_n(\gamma_{tc})}\right\}\right]
\end{aligned}$$

since

$$\begin{aligned}
\gamma_{lt}^2\frac{\gamma_{rc|l}^2\sigma_{cc|t}}{\sigma_{ll|t}} &= \gamma_{lt}^2\frac{\gamma_{rc|l}^2\sigma_{cc|l}}{\sigma_{rr|l,c}}\frac{\sigma_{rr|l,c}\sigma_{cc|t}}{\sigma_{ll|t}\sigma_{cc|l}} \\
&= \gamma_{rl|c}^2.\gamma_{lt}^2\frac{\gamma_{rc|l}^2\sigma_{cc|l}}{\sigma_{rr|l,c}}\frac{\sigma_{rr|l,c}}{\gamma_{rl|c}^2\sigma_{ll|c}}\frac{\sigma_{cc|t}\sigma_{ll|c}}{\sigma_{ll|t}\sigma_{cc|l}} \\
&= \gamma_{rl|c}^2.\gamma_{lt}^2\frac{\gamma_{rc|l}^2\sigma_{cc|l}}{\sigma_{rr|l,c}}\frac{\sigma_{rr|l,c}}{\gamma_{rl|c}^2\sigma_{ll|c}}\frac{1-\rho_{tc}^2}{1-\rho_{lt}^2} \\
&= \gamma_{rl|c}^2.\gamma_{lt}^2\left\{\frac{\mathbb{I}_n(\gamma_{rc|l})}{\mathbb{I}_n(\gamma_{rl|c})}\right\}\frac{1+\mathbb{I}_n(\gamma_{lt})}{1+\mathbb{I}_n(\gamma_{tc})}
\end{aligned}$$

It follows that

$$\Delta\mathbb{I}(\hat{\gamma}_{rl|c}.\hat{\gamma}_{lt};\hat{\gamma}_{rl|t}.\hat{\gamma}_{lt}) = \frac{1}{\mathbb{I}_n(\gamma_{rl|c})}\left\{\frac{\mathbb{I}_n(\gamma_{lt})}{1+\mathbb{I}_n(\gamma_{tc})}\right\} + \frac{\mathbb{I}_n(\gamma_{rc|l})}{\mathbb{I}_n(\gamma_{rl|c})}\left\{\frac{1+\mathbb{I}_n(\gamma_{lt})}{1+\mathbb{I}_n(\gamma_{tc})}\right\}$$

and

$$\begin{aligned}
\Delta\mathbb{I}(\hat{\gamma}_{rl|t}.\hat{\gamma}_{lt};\hat{\gamma}_{rt|c}) &= \Delta\mathbb{I}(\hat{\gamma}_{rl|c}.\hat{\gamma}_{lt};\hat{\gamma}_{rt|c}) - \Delta\mathbb{I}(\hat{\gamma}_{rl|c}.\hat{\gamma}_{lt};\hat{\gamma}_{rl|t}.\hat{\gamma}_{lt}) \\
&= \frac{1}{\mathbb{I}_n(\gamma_{rl|c})}\left\{\frac{1+\mathbb{I}_n(\gamma_{tc})}{\mathbb{I}_n(\gamma_{lt})} - \frac{\mathbb{I}_n(\gamma_{lt})}{1+\mathbb{I}_n(\gamma_{tc})}\right. \\
&\quad \left. + \mathbb{I}_n(\gamma_{rl|c})\frac{\mathbb{I}_n(\gamma_{tc})}{\mathbb{I}_n(\gamma_{lt})} - \mathbb{I}_n(\gamma_{rc|l})\frac{1+\mathbb{I}_n(\gamma_{lt})}{1+\mathbb{I}_n(\gamma_{tc})}\right\}
\end{aligned}$$

16.E Relation between $\rho_{rl|t}^2$ and $\rho_{rl|c}^2$

Theorem E.1

$$\rho_{rl|c}^2 \geq \rho_{rl|t}^2$$

Proof of Theorem E.1. From Equations (16.6), (16.20), (16.21) and (16.22)

$$\frac{\sigma_{rr|l,t}}{\gamma^2_{rl|t}\gamma^2_{lt}\sigma_{ll|t}} = \frac{\sigma_{rr|l,c}}{\gamma^2_{rl|c}\gamma^2_{lt}\sigma_{ll|t}} + \frac{\gamma^2_{rc|l}\sigma_{cc|t}}{\gamma^2_{rl|c}\gamma^2_{lt}\sigma_{ll|t}}$$

$$\frac{1-\rho^2_{rl|t}}{\rho^2_{rl|t}} = \frac{\sigma_{rr|l,c}}{\gamma^2_{rl|c}\sigma_{ll|c}}\left(\frac{\sigma_{ll|c}}{\sigma_{ll|t}}\right)\left\{1+\frac{\gamma^2_{rc|l}\sigma_{cc|l}}{\sigma_{rr|l,c}}\left(\frac{\sigma_{cc|t}}{\sigma_{cc|l}}\right)\right\}$$

$$= \frac{1-\rho^2_{rl|c}}{\rho^2_{rl|c}}\left(\frac{\sigma_{ll|c}}{\sigma_{ll|t}}\right)\left\{1+\frac{\rho^2_{rc|l}}{1-\rho^2_{rc|l}}\left(\frac{\sigma_{cc|t}}{\sigma_{cc|l}}\right)\right\}$$

It follows from Equation (16.19) that

$$(1-\rho^2_{rl|t})/\rho^2_{rl|t} \geq (1-\rho^2_{rl|c})/\rho^2_{rl|c}$$

References

Bowden, R.J. and Turkington, D.A. (1984) *Instrumental Variables*. Cambridge, MA: Cambridge University Press.

Brito, C. and Pearl, J. (2002) Generalized instrumental variables. *Uncertainty in Artificial Intelligence*, **18**, 85–93.

Cochran, W.G. (1957) Analysis of covariance: its nature and uses. *Biometrics*, **13**, 261–281.

Cohen, J. (1988) *Statistical Power Analysis for the Behavioral Sciences*, 2nd edn. Hillsdale, NJ: Lawrence Erlbaum Associates.

Cox, D.R. (1960) Regression analysis when there is prior information about supplementary variables. *Journal of the Royal Statistical Society, Series B*, **22**, 172–176.

Cox, D.R. and McCullagh, P. (1982) Some aspects of analysis of covariance. *Biometrics*, **38**, 541–561.

Dawid, A.P. (2002) Influence diagrams for causal modelling and inference. *International Statistical Review*, **70**, 161–189. Corrigenda, p. 437.

Durbin, J. (1954) Errors in variables. *Review of the International Statistical Institute.*, **22**, 23–32.

Fisher, R.A. (1932) *Statistical Methods for Research Workers*. Edinburgh: Oliver and Boyd.

Geary, R.C. (1942) Inherent relations between random variables. *Proceedings of the Royal Irish Academy*, **47**, 63–76.

Geary, R.C. (1943) Relations between statistics: the general and the sampling problem. *Proceedings of the Royal Irish Academy*, **49**, 177–196.

Jewell, N.P. (2003) *Statistics for Epidemiology*. London: Chapman & Hall/CRC Press.

Kuroki, M. and Cai, Z. (2004) Selection of identifiability criteria for total effects by using path diagrams, in *Proceedings of the* 20*th Annual Conference on Uncertainty in Artifical Intelligence*, pp. 333–340.

Lauritzen, S.L. (2001) Causal inference from graphical models, in *Complex Stochastic Systems* (eds O.E. Barndorff-Nielsen, D.R. Cox and C. Klüppelberg). London: CRC Press.

Pearl, J. (1993) Comment: graphical models, causality and interventions. *Statistical Science*, **8**, 266–269.

Pearl, J. (1995) Causal diagrams for empirical research. *Biometrika*, **82**, 669–710.

Ramsahai, R.R. (2008) Causal inference with instruments and other supplementary variables. DPhil Thesis, University of Oxford.

Reiersøl, O. (1941) Confluence analysis by means of lag moments and other methods of confluence analysis. *Econometrica*, **9**, 1–24.

Reiersøl, O. (1945) Confluence analysis by means of instrumental sets of variables. *Arkiv för Matematik, Astronomi och Fysik*, **32A**, 1–119.

Robinson, L.D. and Jewell, N.P. (1991) Some surprising results about covariate adjustment in logistic regression models. *International Statistical Review*, **59**, 227–240.

Wald, A. (1940) The fitting of straight lines if both variables are subject to error. *Annals of Mathematical Statistics*, **11**, 284–300.

Wright, S.S. (1921) Correlation and causation. *Journal of Agricultural Research*, **20**, 557–585.

17

Time-varying confounding: Some practical considerations in a likelihood framework

Rhian Daniel, Bianca De Stavola and Simon Cousens
Centre for Statistical Methodology, London School of Hygiene and Tropical Medicine, London, UK

17.1 Introduction

Two previous chapters in this volume, Chapter 7 by Arjas and Chapter 8 by Berzuini, Dawid and Didelez, discuss the problem of estimating the causal effect of a time-varying exposure on an outcome using observational data, with a particular focus on dynamic treatment regimes. The authors set out the assumptions under which this is possible and use posterior predictive distributions to make inferences about the effects of interest within a Bayesian framework.

In this chapter, we complement these accounts with some more practical considerations facing applied researchers when working with such data. While the two earlier chapters focus more on aspects of identifiability of causal effects than on statistical inference and computation, this chapter mainly addresses the latter two issues. We diverge from Arjas and Berzuini *et al.* in our framework and notation; while they formulate their work using Bayesian decision theory (see Chapter 4 in this volume), we take a likelihood approach using potential outcomes (see Chapter 2 in this volume) and highlight the similarities between these two approaches.

We discuss how to deal with two very common features of longitudinal studies, namely time-to-event outcomes and losses to follow-up. A common feature of both the Bayesian and likelihood approaches is the need, in almost all realistic settings, to specify parametric

Causality: Statistical Perspectives and Applications, First Edition. Edited by Carlo Berzuini, Philip Dawid and Luisa Bernardinelli.

models for many aspects of the distribution of the observational data. We review the choice of such models and discuss how causal inferences are achieved in practice using Monte Carlo simulation. Finally, we briefly discuss the challenges posed by competing events and unbalanced measurement times. These issues are illustrated using a simulated dataset based on the HIV example introduced by Berzuini *et al.* in Chapter 8.

17.2 General setting

17.2.1 Notation

In keeping with Berzuini *et al.*, but in contrast with Arjas, we consider a fixed sequence of time points $\{T_k\}$ for $k = 0, 1, \ldots, N + 1$, with $0 \equiv T_0 < T_1 < \cdots < T_{N+1}$. As in Berzuini *et al.* we assume that these times are measured on a suitable scale (such as follow-up time) such that they are the same for each subject in the study. In Section 17.7.3 we discuss how this may be relaxed. We also adopt the same notation as in the earlier chapters for the covariates $\{X_k\}$ and the treatments[1] $\{A_k\}$, so that, at each time T_k ($k = 0, 1, \ldots, N + 1$), a vector of covariates X_k is observed, and, for $k = 0, 1, \ldots, N$, the value of the treatment A_k immediately follows X_k. Y is the outcome of interest and, in the simplest case, is included in X_{N+1}. See Section 17.2.5 for the extension to a time-to-event outcome when the event can occur at any time $T > T_0$.

We use $\bar{A}_k = (A_0, \ldots, A_k)$ to denote the *treatment history* and $\bar{X}_k = (X_0, \ldots, X_k)$ the *covariate history* up to and including time T_k. We also use $\bar{X}^k = (X_k, \ldots, X_{N+1})$ to denote the covariate *future* from time T_k onwards.

17.2.2 Observed data structure

The assumed causal structure for the observed data is summarised by the directed acyclic graph (DAG) shown in Figure 17.1. The key features are that the treatment and covariates mutually affect one another and that the covariates are additionally correlated over time by being affected by an unmeasured process $\{U_k : 0 \leq k \leq N + 1\}$. The circle around each U_k denotes that it is unmeasured (i.e. latent). Arrows from each U_k to $\{X_l : k + 2 \leq l \leq N + 1\}$, from each X_k to $\{A_l : k + 1 \leq l \leq N\}$, from each A_k to $\{X_l : k + 2 \leq l \leq N\}$, and from each A_k to $\{U_l : k < l \leq N + 1\}$ are omitted to increase the readability of the diagram. The absence of arrows from each U_k to $\{A_l : k \leq l \leq N\}$ is integral, however, and represents the *conditional exchangeability* assumption (see Section 17.3). In the observational data, therefore, the value of each A_k can depend on $(\bar{A}_{k-1}, \bar{X}_k)$, as well as on a stochastic perturbation at time T_k that is independent of all variables shown in Figure 17.1. Thus $\{A_k : 0 \leq k \leq N\}$ is a discrete time stochastic process; likewise $\{X_k : 0 \leq k \leq N + 1\}$.

Note that, since X_k is, in general, a vector, the DAG shown in Figure 17.1 implies a chain graph (Cox and Wermuth, 1996) over the set of all variables. More specifically, each X_k is a component consisting of a number of potentially correlated variables, and the DAG over these components shown in Figure 17.1 is agnostic as to the causal structure giving rise to

[1] More generally, A_k could be any exposure. For the remainder of this chapter, with our motivating example in mind, we refer to A_k as a treatment.

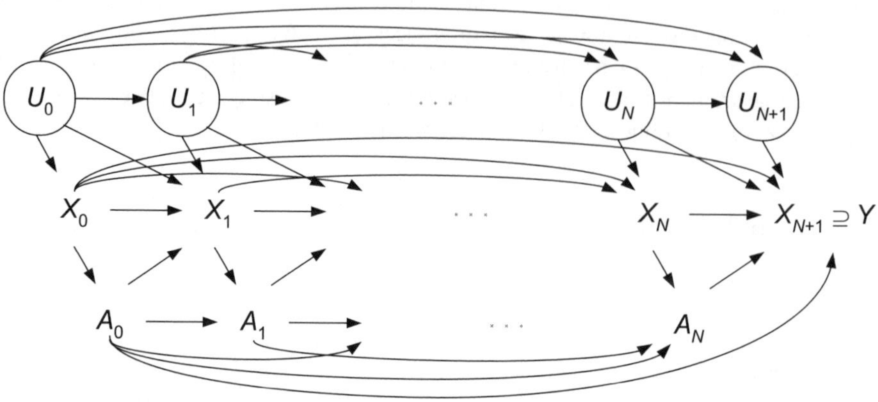

Figure 17.1 A directed acyclic graph representing the causal structure of the observational data.

these within-component correlations. As such, each component X_k could be re-drawn as a complete undirected graph, with the edges representing potential correlations between every pair of variables. We return to the consequences of this in Section 17.4.2.

The joint density $f_V(v)$ of the observed data $V = (\bar{A}_N, \bar{X}_{N+1})$ can be factorised as

$$f_V(v) = f_{X_0}(x_0) f_{A_0|X_0}(a_0|x_0) \prod_{k=1}^{N+1} f_{X_k|\bar{A}_{k-1},\bar{X}_{k-1}}(x_k|\bar{a}_{k-1}, \bar{x}_{k-1}) \prod_{k=1}^{N} f_{A_k|\bar{A}_{k-1},\bar{X}_k}(a_k|\bar{a}_{k-1}, \bar{x}_k) \tag{17.1}$$

17.2.3 Intervention strategies

An *intervention strategy* (or *intervention regime*, henceforth used interchangeably) is a rule for determining the values of $\{A_k : 0 \leq k \leq N\}$, possibly based on the values of $(\bar{A}_{k-1}, \bar{X}_k)$, in a hypothetical experiment. As was made clear in the two previous chapters by Arjas and Berzuini *et al.*, causal (as opposed to descriptive) inference is based on knowledge of how Y behaves under various intervention strategies.

A particular intervention strategy is denoted as $\bar{s}_N = (s_0, \ldots, s_N)$ (or simply s, for short) where s_0 is a function that assigns the value of a_0 based on x_0:

$$a_0 = s_0(x_0)$$

and, for $k > 0$, each s_k is a function that assigns the value of a_k based on $(\bar{a}_{k-1}, \bar{x}_k)$:

$$a_k = s_k(\bar{a}_{k-1}, \bar{x}_k)$$

A *static* strategy is one in which $s_0(x_0)$ does not depend on x_0 and, for $k > 0$, $s_k(\bar{a}_{k-1}, \bar{x}_k)$ does not depend on $(\bar{a}_{k-1}, \bar{x}_k)$; all other strategies are *dynamic*. Examples of static strategies (for binary treatments) are 'never treat' ($s_0(x_0) = s_k(\bar{a}_{k-1}, \bar{x}_k) = 0 \ \forall k > 0, \bar{a}_{k-1}, \bar{x}_k$), 'always treat' ($s_0(x_0) = s_k(\bar{a}_{k-1}, \bar{x}_k) = 1 \; \forall k > 0, \bar{a}_{k-1}, \bar{x}_k$) and 'treat from visit 3 onwards':

$$s_0(x_0) = 0 \; \forall x_0$$
$$s_k(\bar{a}_{k-1}, \bar{x}_k) = \begin{cases} 0 & \text{if} \quad 0 < k < 3 \\ 1 & \text{if} \quad k \geq 3 \end{cases} \qquad \forall \bar{a}_{k-1}, \bar{x}_k$$

Examples of dynamic strategies are 'treat if and only if $x_k > 5$':

$$s_0(x_0) = \begin{cases} 0 & \text{if } x_0 \leq 5 \\ 1 & \text{if } x_0 > 5 \end{cases}$$
$$s_k(\bar{a}_{k-1}, \bar{x}_k) = \begin{cases} 0 & \text{if } x_k \leq 5 \\ 1 & \text{if } x_k > 5 \end{cases} \quad \forall k > 0, \bar{a}_{k-1}, \bar{x}_{k-1}$$

and 'start treatment as soon as $x_k > 5$, and then continue on treatment':

$$s_0(x_0) = \begin{cases} 0 & \text{if } x_0 \leq 5 \\ 1 & \text{otherwise} \end{cases}$$
$$s_k(\bar{a}_{k-1}, \bar{x}_k) = \begin{cases} 0 & \text{if } x_k \leq 5 \text{ and } a_{k-1} = 0 \\ 1 & \text{otherwise} \end{cases} \quad \forall k > 0, \bar{a}_{k-2}, \bar{x}_{k-1}$$

17.2.4 Potential outcomes

For each possible realisation $\bar{a}_N$ of the treatment history $\bar{A}_N$, let $Y(\bar{a}_N)$ be the *potential outcome* associated with $\bar{a}_N$, i.e. $Y(\bar{a}_N)$ is the value of Y that would have been observed had $\bar{A}_N$ been set to $\bar{a}_N$ by a (static) intervention. Likewise, for each k, $X_k(\bar{a}_N)$ is the potential value of X_k that would have been observed had $\bar{A}_N$ been set to $\bar{a}_N$.

More generally, let $Y(\bar{s}_N)$ be the potential outcome associated with a strategy s; i.e. $Y(\bar{s}_N)$ is the value of Y that would have been observed had $\bar{A}_N$ been determined according to the (possibly dynamic) strategy s. Similarly, let $X_k(\bar{s}_N)$ be the potential value of X_k associated with strategy s.

17.2.5 Time-to-event outcomes

In longitudinal studies, rather than being measured at the end of the study, the outcome of interest is often the time to a particular event, which can occur at most once for each subject and at any time during the follow-up period (such as death or some other form of failure). For simplicity, we assume that the exact time of the event is not known, but that, at each visit k, it is ascertained whether or not the event has occurred in the time interval $(T_{t-1}, T_k]$. As such, the outcome data comprise a sequence of binary variables $Y_1, \ldots, Y_{N+1}$ measured at visits $1, \ldots, N+1$, where $Y_k = 1$ if and only if the event of interest has occurred during the interval $(T_0, T_k]$. Note that all subjects are assumed to be event-free at time T_0 and that if $Y_k = 1$ for some k, then $Y_l = 1$ for all $l \geq k$. In this case, the time-to-event outcome Y is defined as

$$Y = \min(k : Y_k = 1)$$

Each Y_k is included in X_k and is assumed to occur before $X_k \backslash Y_k$; i.e. first it is ascertained whether or not the event of interest has occurred in the interval $(T_{k-1}, T_k]$, and immediately after this $X_k \backslash Y_k$ is observed, followed by A_k.

17.2.6 Causal estimands

In the setting discussed in this chapter, causal inference involves making a comparison between some feature(s) of the distributions of the potential outcomes $Y(\bar{s}_N)$ and $Y(\bar{s}'_N)$ for (at least) two different strategies s and s'. For example, if Y were a univariate outcome measured at the end of the study, we might define the causal estimand of interest to be the difference between the expected value of Y under the two strategies:

$$E\{Y(\bar{s}_N)\} - E\{Y(\bar{s}'_N)\}$$

More generally, we may seek to find the strategy s^* among a set of candidate strategies $\mathcal{S}$ that optimises some feature(s) of the distribution of $Y(\bar{s}_N)$. For example, if our aim were to maximise the expectation of Y, then the inferential problem becomes one of finding s^* such that

$$E\{Y(\bar{s}^*_N)\} = \max_{s \in \mathcal{S}}[E\{Y(\bar{s}_N)\}]$$

Either way, if we define the causal parameter θ_s associated with strategy s to be

$$\theta_s = E\{Y(\bar{s}_N)\}$$

then, if inference about θ_s can be made from the observed data, our causal problem can be tackled.[2]

In the case where Y is the time to an event, then, using the dicrete-time formulation introduced in Section 17.2.5, the above definitions generalise naturally to represent the discrete-time hazard function:

$$\theta_s = \left\{\theta_s^k = E\left[Y_k(\bar{s}_{k-1})\,|Y_{k-1}(\bar{s}_{k-2}) = 0\,\right] : k = 1, \ldots, N+1\right\}$$

17.3 Identifying assumptions

The inferential approach proposed by Robins (1986), summarised in Sections 17.4 and 17.5, differs from that proposed in the two previous chapters by Arjas and Berzuini *et al.* only in the sense that Robins seeks to find a maximum likelihood estimator for θ_s, whereas Arjas and Berzuini *et al.* compute its Bayesian posterior distribution.[3] Unsurprisingly, the identifying assumptions are the same in both approaches. We re-visit these below.

1. *Consistency*. This assumption states that for each strategy s to be considered, if $A_k = s_k\left(\bar{A}_{k-1}, \bar{X}_k\right)\ \forall k = 0, \ldots, N$, then $\bar{X}_{N+1}(\bar{s}_N) = X_{N+1}$ (and thus $Y(\bar{s}_N) = Y$). In words, if we consider two scenarios for a particular subject, one in which the realised values of $\bar{A}_N$ happen to be $\bar{a}_N$ and another in which the values of $\bar{A}_N$ are set to $\bar{a}_N$ as a result of an intervention strategy s, the value of X_{N+1} (and hence of Y), if consistency holds, should be the same in both settings. This is a necessary condition, although – and

[2] Note that θ_s is not a causal parameter in the usual sense, since it does not represent a *comparison* between two or more hypothetical interventions. However, by summarising the distribution of Y under one possible intervention, causal comparisons are made by comparing θ_s for different strategies s.

[3] More strictly, the two approaches differ also in the way in which θ_s is defined (with or without potential outcomes), but we take the pragmatic view that this is a linguistic rather than a conceptual difference and that the two definitions can be considered – at least from an applied perspective – each as shorthand for the other.

here we disagree with Arjas – it is often not natural.[4] For example, the mechanism by which body mass index (BMI) changes in an observational setting is almost impossible to mimic in an experimental setting. If a gastric band is fitted to achieve a BMI of 25, this will likely have very different implications for a number of subsequent variables to be considered (e.g. coronary heart disease) than if a similar reduction in BMI were achieved by changing diet and exercise habits. In observational epidemiology, therefore, the plausibility of the consistency assumption must always be carefully considered (VanderWeele, 2009; Hernán and VanderWeele, 2011). In the HIV example introduced by Berzuini *et al.* in Chapter 8, however, it is more plausible that the effect of taking a particular antiretroviral therapy will be the same irrespective of whether the therapy was assigned by intervention or chosen in the observational setting.

2. *Conditional exchangeability.* Corresponding to 'unconfoundedness' (and 'local independence') in Arjas in Chapter 7, to 'stability' in Berzuini *et al.* in Chapter 8, and to 'no unmeasured confounders' and 'sequential ignorability' elsewhere in the literature, the conditional exchangeability assumption states that

$$A_k \perp\!\!\!\perp \bar{X}^{k+1}(\bar{s}_N) | (\bar{A}_{k-1} = \bar{s}_{k-1}(\bar{x}_{k-1}), \bar{X}_k = \bar{x}_k) \; \forall \bar{s}_N \in \mathcal{S}, \; \forall \bar{x}_k \in \mathcal{X}_k, \; \forall k = 0, \ldots, N$$

where $\bar{s}_k(\bar{x}_k)$ is defined recursively as

$$\bar{s}_k(\bar{x}_k) = \{\bar{s}_{k-1}(\bar{x}_{k-1}), s_k(\bar{s}_{k-1}(\bar{x}_{k-1}), \bar{x}_k)\}$$

That is, conditional exchangeability states that the treatment chosen at time T_k (in the observational setting) is conditionally independent of all future potential outcomes under any of the regimes to be considered ($\mathcal{S}$), given the covariate and treatment history up to time T_k. More colloquially, this assumption states that the covariate and treatment history up to time T_k are sufficiently rich such that any further variation in the choice of A_k can be considered to be random. As such, this condition holds by design in a sequentially randomised trial, when randomisation probabilities at time T_k depend only on $(\bar{A}_{k-1}, \bar{X}_k)$. Subtle variations on this assumption have been discussed in the literature (Dawid and Didelez, 2010) but are not discussed further here.

3. *Positivity.* Expressed in the current notation, this assumption states that, given any particular history $(\bar{a}_{k-1}, \bar{x}_k)$ such that $f_{\bar{A}_{k-1}, \bar{X}_k}(\bar{a}_{k-1}, \bar{x}_k) > 0$, then the conditional density that A_k attains a particular value a_k in the observational setting ($f_{A_k|\bar{A}_{k-1}, \bar{X}_k}(a_k|\bar{a}_{k-1}, \bar{x}_k)$) must be nonzero for all a_k in the range of s_k in strategy s, and for all such strategies s to be considered. More colloquially, this assumption states that the treatments that could arise from any strategy to be considered must also have a nonzero probability of occurring in the observational setting.

17.4 G-computation formula

17.4.1 The formula

Under the assumptions of the previous section, Robins (1986) showed how the distribution of the potential outcome $Y(\bar{s}_N)$ can be estimated using the g-computation formula. This

[4] Arjas writes the following on the consistency assumption: 'It seems both natural and necessary to assume that otherwise the probabilistic structure and description of the see-study is lifted to the, perhaps only hypothetical, do-experiment without change.'

corresponds to Theorem 3 in Arjas (Chapter 7) and to Equation (8.9) in Berzuini *et al.* (Chapter 8).

First, the distribution of V under strategy s is given by

$$f_V^s(v) = \begin{cases} f_{X_0}(x_0) \prod_{k=1}^{N+1} f_{X_k|\bar{A}_{k-1},\bar{X}_{k-1}}(x_k|\bar{a}_{k-1},\bar{x}_{k-1}) & \text{if } s_0(x_0) = a_0 \text{ and} \\ & s_k(\bar{a}_{k-1},\bar{x}_k) = a_k \ \forall k > 0 \\ 0 & \text{otherwise} \end{cases}$$

This is the same as (17.1), except that the conditional distribution of each A_k given $(\bar{A}_{k-1}, \bar{X}_k)$ has been replaced by a degenerate distribution, taking value 1 for the value of A_k assigned by the regime and 0 otherwise.

Thus, the expectation of Y under strategy s is given by

$$E\{Y(\bar{s}_N)\} = \int_{x_0 \in \mathcal{X}_0} \int_{x_1 \in \mathcal{X}_1} \cdots \int_{x_N \in \mathcal{X}_N} E\{Y|\bar{A}_N = \bar{s}_N(\bar{x}_N), \bar{X}_N = \bar{x}_N\}$$
$$\times f_{X_0}(x_0) \prod_{k=1}^{N} f_{X_k|\bar{A}_{k-1},\bar{X}_{k-1}}(x_k|\bar{s}_{k-1}(\bar{x}_{k-1}), \bar{x}_{k-1}) dx_N \, dx_{N-1} \cdots dx_0 \qquad (17.2)$$

where the kth integral ($k = 0, \ldots, N$) is over the domain $\mathcal{X}_k$ of X_k.

Equation (17.2) is known as the *g-computation formula*.[5]

When Y is a time-to-event outcome, equation (17.3) naturally extends to a sequence of $N + 1$ equations of the form

$$E\{Y_k(\bar{s}_{k-1})\} = \int_{x_0 \in \mathcal{X}_0} \int_{x_1 \in \mathcal{X}_1} \cdots \int_{x_{k-1} \in \mathcal{X}_{k-1}} E\{Y_k \,|\bar{A}_{k-1} = \bar{s}_{k-1}(\bar{x}_{k-1}), \bar{X}_{k-1} = \bar{x}_{k-1}, Y_{k-2} = 0\}$$
$$\times f_{X_0}(x_0) \prod_{l=1}^{k-1} f_{X_l|\bar{A}_{l-1},\bar{X}_{l-1}}(x_l\,|\bar{s}_{l-1}(\bar{x}_{l-1}), \bar{x}_{l-1}) \, dx_{k-1} \, dx_{k-2} \cdots dx_0 \qquad (17.3)$$

for each $k = 1, \ldots, N + 1$.

Note that if some components of X_k are discrete, the corresponding Lebesgue measures in (17.2) and (17.3) would be replaced with counting measures.

17.4.2 Plug-in regression estimation

Each term in (17.2) and (17.3) can be estimated from appropriately fitted associational models.

If X_k is univariate, then $f_{X_k|\bar{A}_{k-1},\bar{X}_{k-1}}(x_k|\bar{s}_{k-1}(\bar{x}_{k-1}), \bar{x}_{k-1})$ can be estimated from a parametric regression model fitted to X_k given $\bar{A}_{k-1}$ and $\bar{X}_{k-1}$. Write $g_{X_k|\bar{A}_{k-1},\bar{X}_{k-1}}(x_k|\bar{a}_{k-1}, \bar{x}_{k-1}; \alpha_k)$ for the density implied by the posited regression model, parameterised in terms of a vector of parameters α_k.[6] By fitting this model to the observed data on X_k, $\bar{X}_{k-1}$ and $\bar{A}_{k-1}$ we obtain (maximum likelihood) estimates $\hat{\alpha}_k$ of α_k. An estimate of $f_{X_k|\bar{A}_{k-1},\bar{X}_{k-1}}(x_k|\bar{s}_{k-1}(\bar{x}_{k-1}), \bar{x}_{k-1})$ is then obtained by substituting $\bar{s}_{k-1}(\bar{x}_{k-1})$ for $\bar{a}_{k-1}$ in $g_{X_k|\bar{A}_{k-1},\bar{X}_{k-1}}(x_k|\bar{a}_{k-1}, \bar{x}_{k-1}; \hat{\alpha}_k)$.

---

[5] That the expectation of Y is not written conditional on $X_{N+1} \backslash Y$ is a consequence of the assumed ordering discussed in Section 17.2.5: Y occurs fractionally prior to $X_{N+1} \backslash Y$.

[6] We use $g(\cdot)$ rather than $f(\cdot)$ to stress that the density implied by the posited regression model does not necessarily coincide with the true density that gave rise to the data.

More generally, X_k consists of $p_k > 1$ (potentially correlated) variables. In this case, we choose an arbitrary[7,8] order $X_k = (X_{k,1}, X_{k,2}, \ldots, X_{k,p_k})$ for the p_k variables constituting X_k and then fit a series of univariate regression models:

(1) to $X_{k,1}$ given $\bar{A}_{k-1}$ and $\bar{X}_{k-1}$,

(2) to $X_{k,2}$ given $X_{k,1}$, $\bar{A}_{k-1}$ and $\bar{X}_{k-1}$,

...

(p_k) to X_{k,p_k} given $X_{k,1}, X_{k,2}, \ldots, X_{k,p_k-1}$, $\bar{A}_{k-1}$ and $\bar{X}_{k-1}$.

We write:

(1) $g_{X_{k,1}|\bar{A}_{k-1},\bar{X}_{k-1}}(x_{k,1}|\bar{a}_{k-1}, \bar{x}_{k-1}; \alpha_{k,1})$,

(2) $g_{X_{k,2}|\bar{A}_{k-1},\bar{X}_{k-1},X_{k,1}}(x_{k,2}|\bar{a}_{k-1}, \bar{x}_{k-1}, x_{k,1}; \alpha_{k,2})$,

...

(p_k) $g_{X_{k,p_k}|\bar{A}_{k-1},\bar{X}_{k-1},X_{k,1},\ldots,X_{k,p_k-1}}(x_{k,p_k}|\bar{a}_{k-1}, \bar{x}_{k-1}, x_{k,1}, \ldots, x_{k,p_k-1}; \alpha_{k,p_k})$

for the densities implied by these posited regression models. By fitting each of these models to the observed data we obtain estimates $\hat{\alpha}_k = (\hat{\alpha}_{k,1}, \hat{\alpha}_{k,2}, \ldots, \hat{\alpha}_{k,p_k})$ of $\alpha_k = (\alpha_{k,1}, \alpha_{k,2}, \ldots, \alpha_{k,p_k})$.

The product of these p_k estimated densities

$$\begin{aligned}
&g_{X_k|\bar{A}_{k-1},\bar{X}_{k-1}}(x_k|\bar{a}_{k-1}, \bar{x}_{k-1}; \hat{\alpha}_k)\\
&\quad = g_{X_{k,1}|\bar{A}_{k-1},\bar{X}_{k-1}}(x_{k,1}|\bar{a}_{k-1}, \bar{x}_{k-1}; \hat{\alpha}_{k,1})\\
&\quad \times g_{X_{k,2}|\bar{A}_{k-1},\bar{X}_{k-1},X_{k,1}}(x_{k,2}|\bar{a}_{k-1}, \bar{x}_{k-1}, x_{k,1}; \hat{\alpha}_{k,2})\\
&\quad \times \cdots \times g_{X_{k,p_k}|\bar{A}_{k-1},\bar{X}_{k-1},X_{k,1},\ldots,X_{k,p_k-1}}(x_{k,p_k}|\bar{a}_{k-1}, \bar{x}_{k-1}, x_{k,1}, \ldots, x_{k,p_k-1}; \hat{\alpha}_{k,p_k})
\end{aligned}$$

is our estimate of the joint density $f_{X_k|\bar{A}_{k-1},\bar{X}_{k-1}}(x_k|\bar{s}_{k-1}(\bar{x}_{k-1}), \bar{x}_{k-1})$.

In the case of a single outcome Y measured at the end of follow-up, $E(Y|\bar{A}_N = \bar{s}_N(\bar{x}_N), \bar{X}_N = \bar{x}_N)$ can similarly be estimated. A model for the conditional distribution of Y given $\bar{X}_N$ and $\bar{A}_N$ is specified, and the estimates $\hat{\beta}$ of the parameters β from $g_{Y|\bar{A}_N,\bar{X}_N}(y|\bar{a}_N, \bar{x}_N; \beta)$ are obtained from the observed data.

When the outcome is time-to-event, i.e. described by a series of binary variables $Y_1, Y_2, \ldots, Y_{N+1}$, then, for each $k \in [1, N+1]$, a model for the conditional probability of $Y_k = 1$ given $\bar{X}_{k-1}$, $\bar{A}_{k-1}$ and $Y_{k-1} = 0$ is specified, and the estimates $\hat{\beta}_k$ of the parameters β_k in $g_{Y_k|\bar{A}_{k-1},\bar{X}_{k-1}}(y_k|\bar{a}_{k-1}, \bar{x}_{k-1}; \beta_k)$ are obtained from the subset of the observed data with $Y_{k-1} = 0$ (i.e. those still at risk of experiencing the event).

The models can either be fitted separately at each time-point or the data can be pooled across all time-points and the (shared) parameters estimated.

[7] That this choice can be made arbitrarily is due to the same argument that allows us to marginalise over $\{U\}$. See Section 17.5.1.

[8] In the time-to-event setting, where Y_k is included in X_k, we set $X_{k,1} = Y_k$, i.e. Y_k is the first variable, and the others are ordered arbitrarily.

In theory, the integral in (17.2) or (17.3) can then be evaluated, with the regression estimates substituted for the unknown expectations and densities, and a corresponding estimate of θ_s obtained.

In practice, however, the integrals in (17.2) and (17.3) are typically intractable, and hence θ_s is estimated by Monte Carlo simulation, as outlined in the next section.

Naturally, the validity of the causal parameter estimates obtained is dependent on the appropriateness of the selected parametric models, as discussed in Section 17.7.1.

17.5 Implementation by Monte Carlo simulation

17.5.1 Simulating an end-of-study outcome

Rather than calculating the distribution of $Y(\bar{s}_N)$ analytically, we can instead simulate values from this distribution using a sequential simulation scheme. Let $X_0^*, X_1^*, \ldots, X_{N+1}^*$ be the simulated values of $X_0, X_1, \ldots, X_{N+1}$ under the intervention strategy being considered. X_0 precedes A_0 and is therefore unaffected by the strategy. Thus, $X_0^* = X_0$.

X_1^* is simulated from the distribution defined by $g_{X_1|A_0,X_0}(x_1|s_0(X_0^*), X_0^*; \hat{\alpha}_1)$. In other words, we take the conditional distribution of X_1 given X_0 and A_0 as estimated from the observed data; we then simulate X_1^* from this distribution, after replacing X_0 by X_0^* and A_0 by $s_0(X_0^*)$, i.e. the values of A_0 and X_0 under the intervention being considered. When X_1 is a vector, this simulation is done sequentially for each of its constituent variables.

Similarly, X_k^* is simulated from $g_{X_k|\bar{A}_{k-1},\bar{X}_{k-1}}(x_k|\bar{s}_{k-1}(\bar{X}_{k-1}^*), \bar{X}_{k-1}^*; \hat{\alpha}_k)$ for each $k \in [2, N+1]$. As part of X_{N+1}^*, Y^* is simulated from $g_{Y|\bar{A}_N,\bar{X}_N}(y|\bar{s}_N(\bar{X}_N^*), \bar{X}_N^*; \hat{\beta})$.

Since U is unmeasured, all model fitting and simulation is done marginally over the unobserved distribution of U, but since U is not a confounder of the A–Y relationships (under the conditional exchangeability assumption), this does not introduce bias, and the distribution of the simulated Y^* is an empirical estimate of the distribution of $Y(\bar{s}_N)$ provided that the modelling assumptions made during the model-fitting stage are valid. This follows from Lemma 1 in Chapter 7 by Arjas.

17.5.2 Simulating a time-to-event outcome

In the case of a time-to-event outcome, observable only at times $\{T_k : 1 \le k \le N+1\}$, Y_1^* is simulated from a Bernoulli distribution with probability $g_{Y_1|A_0,X_0}(1|s_0(X_0^*), X_0^*; \hat{\beta}_1)$. For those with $Y_1^* = 0$, Y_2^* is simulated from a Bernoulli distribution with probability $g_{Y_2|\bar{A}_1,\bar{X}_1}(1|\bar{s}_1(\bar{X}_1^*), \bar{X}_1^*; \hat{\beta}_2)$, etc. Finally, for those with $Y_N^* = 0$, Y_{N+1}^* is simulated from a Bernoulli distribution with probability $g_{Y_{N+1}|\bar{A}_N,\bar{X}_N}(1|\bar{s}_N(\bar{X}_N^*), \bar{X}_N^*; \hat{\beta}_{N+1})$. Together, $\{Y_1^*, \ldots, Y_{N+1}^*\}$ represent the potential time-to-event Y^* outcome simulated under strategy s.

17.5.3 Inference

The variation across simulated values of Y^* does not represent the full uncertainty about our knowledge of this distribution, since it fails to acknowledge that the parameter values $\hat{\alpha}$ and $\hat{\beta}$ used to generate these simulations are themselves estimated from the data. For this reason, bootstrapping is required for valid statistical inference about our causal parameter θ_s

of interest.[9] This is in contrast to the approach outlined in the earlier two chapters, which, by virtue of being Bayesian, incorporate both sources of uncertainty simultaneously. The approach advocated by Arjas and Berzuini *et al.* can be seen as an extension of that described above, in which the model parameters are also drawn from their posterior distributions after each associational model is fitted and before the values of the simulated variables are drawn.

17.5.4 Losses to follow-up

In longitudinal studies such as the ones considered here, it is always likely that some subjects drop out before the end of follow-up. Under the assumption that this drop-out occurs at random (Little and Rubin, 2002), i.e. that drop-out is conditionally independent of the unobserved data given the observed data (observed prior to drop-out), then such loss to follow-up can be very easily allowed for in the g-computation formula analysis. Dropping out can be seen as one of the potential treatment trajectories, and then the simulations are made for trajectories such as 'follow strategy s and do not drop out'. The missing at random assumption is then implicit in the conditional exchangeability assumption. The fact that drop-out need not be explicitly modelled is an example of *ignorability* as defined for likelihood analyses under MAR (Little and Rubin, 2002).

17.5.5 Software

This Monte Carlo simulation procedure can be carried out using the gformula command in Stata (Daniel *et al.*, 2011). A similar macro has been written in SAS and is described in Taubman *et al.* (2009).

17.6 Analyses of simulated data

17.6.1 The data

A dataset consisting of 10 000 patients is simulated to mimic the HIV example described by Berzuini *et al.* in Chapter 8. Interest lies in comparing rules for switching between two antiretroviral therapies, A and B. Initially, at T_0, all subjects take therapy A (coded as $A_0 = 1$). Subsequently, from T_1 onwards, some subjects switch to therapy B (coded 0). Switches back to A, and even back again to B, also occur.

In a slight departure from the description given by Berzuini *et al.* in Chapter 8, we simulate the data at seven visits: at the beginning of follow-up and then at six 3-monthly visits thereafter, i.e. $(T_0, T_1, T_2, T_3, T_4, T_5, T_6) = (0, 3, 6, 9, 12, 15, 18)$ months after start of follow-up.

The covariates $X = (R, G)$ measured at each visit are viral RNA (R) and triglyceride (G), both measured on a logarithmic (base 10) scale. The unlogged values are measured in copies per mL plasma and in mg/dL, respectively.

Therapy A is more efficacious (in terms of lowering viral RNA) on average than therapy B, but it also causes dyslipidaemia (typified by elevated triglyceride levels) in some patients, whereas therapy B does not. There is a trade-off therefore between efficacy on the one hand and safety on the other.

[9] In Section 17.6, we report results based on the nonparametric bootstrap. See Davison and Hinkley (1997) for a discussion of potentially better-suited bootstrapping techniques, particularly for the case of a time-to-event outcome.

Failure is deemed to have occurred in the interval $(T_{k-1}, T_k]$ for $k \geq 1$ if either $R_k > 6.2$ or $G_k > 2.5$ (with both $R_l \leq 6.2$ and $G_l \leq 2.5$ for all $l < k$, i.e. no failure prior to this interval) or if the patient dies in the interval $(T_{k-1}, T_k]$.

The data are generated according to the structure of Figure 17.1. A latent class variable U is generated that categorises the patients into one of four types: in type 1 patients (labelled 'lucky'), treatment A is more efficacious than treatment B, and does not cause dyslipidaemia; in type 2 patients (labelled 'responsive'), treatment A is more efficacious, but it also causes dyslipidaemia; treatment A and B are equally efficacious in type 3 patients ('unresponsive'), and treatment A does not cause dyslipidaemia in these patients either; finally, for type 4 ('unlucky') patients, the two treatments are equally efficacious, and treatment A causes dyslipidaemia.

The latent classes are used in the simulation of X_1–X_6; more precisely, they define how treatment A_0–A_5 affects covariates X_1–X_6. However, U does not (directly) affect the treatment variables A_0–A_5, i.e. conditional exchangeability holds.

The decisions to switch treatments are generated to be sensible in the sense that elevated triglycerides raises the probability of switching to (or staying on) B and elevated viral RNA tends to result in switching to (or staying on) therapy A. The decisions are made with sufficient noise, however, to ensure that positivity holds for all the regimes to be compared. For more details, see Section 17.6.2.

Loss to follow-up is generated according to a missing-at-random mechanism, in which the probability of dropping out at time k depends on the histories of A and X up to time $k - 1$, such that patients who are less well are more likely to drop out. A similar process is used to simulate death.

17.6.2 Regimes to be compared

We compare two static strategies, *regime 1* (always treat with A) and *regime 2* (from T_1 onwards, always treat with B) and three dynamic strategies.

The first of the dynamic regimes, *regime 3* (labelled 'sensible'), states that $A_0 = 1$, and then, for $k > 0$, $A_k = A_{k-1}$ unless:

$$A_{k-1} = 1,\ R_k < 6.0 \text{ and } G_k > 2.4;\ \text{in which case } A_k = 0$$

or

$$A_{k-1} = 0,\ R_k > 6.0 \text{ and } G_k < 2.4;\ \text{in which case } A_k = 1$$

This is sensible, since it states that, provided that RNA levels are acceptable, elevated triglycerides should lead to a switch to treatment B and that, provided that triglyceride levels are acceptable, elevated RNA should lead to a switch to treatment A.[10]

The second of the dynamic regimes, *regime 4* (labelled 'injudicious'), states that $A_0 = 1$, and then, for $k > 0$, $A_k = A_{k-1}$ unless:

$$A_{k-1} = 1 \text{ and either } R_k > 6.0 \text{ or } G_k < 2.4 \text{ or both; in which case } A_k = 0$$

[10] Treatment choices in our simulated dataset were generated according to a stochastic version of this regime. Rather than determining that $A_k = 0$, if $A_{k-1} = 1$, $R_k < 6.0$ and $G_k > 2.4$, then $A_k = 0$ with probability 0.7; and similarly for the second condition.

or

$$A_{k-1} = 0,\ R_k < 6.0 \text{ and } G_k > 2.4;\ \text{ in which case } A_k = 1$$

This is injudicious since it is effectively the opposite of regime 3.

Another regime under which we may want to simulate potential outcomes is the *observational regime*, *regime 5*. To do this we must first specify additional parametric models for each A_k conditional on $(\bar{A}_{k-1}, \bar{X}_k)$; i.e. we model the treatment process, whereas for the static and dynamic regimes discussed thus far, only the covariate process need be modelled. The values of $\{A_k : k > 0\}$ are then drawn from the distributions implied by these models after fitting to the observed data, and after substituting the already simulated values of $\bar{A}_{k-1}$ and $\bar{X}_{k-1}$, similarly to what was described in Section 17.5. Note that the observational regime is a *stochastic* dynamic strategy, since the values of A_k are determined, not only by $(\bar{A}_{k-1}, \bar{X}_k)$ but also by independent random noise. If these additional models are correctly specified, and if the assumptions of Section 17.3 hold, then the distribution of the Y-values simulated under this regime should be the same as that in the observed data. In the absence of significant loss to follow-up, a comparison of the observed data and the simulated data under the observational regime can give some indication as to the success of the g-computation procedure. If these two were very different, it would indicate that at least some of the parametric modelling assumptions, or the identifying assumptions discussed in Section 17.3, do not hold. Good agreement, however, does not guarantee that the assumptions hold.

In practice, of course, we would want to compare far more regimes than are listed above. In particular, let regime $r(\phi, \psi)$ be the same as regime 3 defined above, but with 6.0 replaced with ϕ and 2.4 replaced with ψ. A typical analysis would search through a range of values of ϕ and ψ until an optimal regime in this class is found. For simplicity, however, we report only the comparison of regimes 1 to 4 above, with regime 5 added for model-checking purposes.

17.6.3 Parametric modelling choices

Other than dealing with time-varying covariates that are affected by earlier treatment, and comparing dynamic treatment regimes, another attractive feature of the g-computation formula is that the parameters of each individual associational model to be fitted (used in the simulation stage) are not the direct focus of interpretation. We can therefore include interactions and nonlinear terms in these associational models, as long as the sample size is large enough to support their estimation without unacceptable imprecision. For smaller datasets, power can be gained by pooling the fitting of models across time-points, but with a sample size of 10 000 and no strong biological basis for assuming that the relationships between the variables are stable over time, we do not conduct any pooled associational analyses here.

Specifically, a separate linear regression model is fitted at each visit k, with R_k as the outcome, and all previous values of R and all previous treatment variables included as covariates; in addition, interaction terms are included between R_{k-1} and A_{k-1} and between R_{k-1} and A_{k-2}. The inclusion of interaction terms is motivated by the belief that there is an underlying (latent) U defining how A affects R (and G) in different patients. A regime based on U would be potentially very effective, but since U is unobserved, allowing for its manifestation in the associational models is likely to benefit the search for an optimal regime.

Similarly, linear regression models are fitted to each G_k given past values of A and G and the interactions between G_{k-1} and A_{k-1} and between G_{k-1} and A_{k-2}; in addition, to allow for a correlation between R_k and G_k, R_k is included as a predictor in the model for G_k.

A logistic regression model is specified for the failure variable Y_k at each k, with R_{k-1}, G_{k-1}, A_{k-1}, A_{k-2}, A_{k-3}, $R_{k-1} * A_{k-1}$, $R_{k-1} * A_{k-2}$, $G_{k-1} * A_{k-1}$ and $G_{k-1} * A_{k-2}$ as predictors.

Finally, for regime 5 only, a model for each A_k is specified, also as a logistic regression with R_k, G_k, A_{k-1}, $R_k * A_{k-1}$ and $G_k * A_{k-1}$ as predictors.

These models correspond to the densities $g(\cdot)$ in Section 17.4.2, and their appropriateness should be carefully checked using conventional model-checking procedures. This is discussed further in Section 17.7.1.

17.6.4 Results

17.6.4.1 Descriptive analyses

Of the 10 000 simulated patients, 4656 suffered an event (there were 63 deaths, 4038 failures due to lack of efficacy, 476 failures due to dyslipidaemia and 79 failures due to both lack of efficacy and dyslipidaemia), while 175 were lost to follow-up.

A selection of profiles of observed $\log_{10}$(RNA) and $\log_{10}$(triglyceride) are shown in Figure 17.2. The eight most frequently observed treatment trajectories are shown for a random selection of patients to aid readability. Each plot includes partial profiles for patients who either suffered the event or were lost to follow-up before 18 months, but whose observed partial treatment trajectory is consistent with that shown. For example, a patient with treatment trajectory AAAB who is subsequently not observed can contribute to AAABAA and/or AAABBB, if selected. In addition to the eight most popular trajectories, trajectory ABBBBB is also shown, due to its connection with static regime 2. The same trajectories are listed in Table 17.1, which gives the associated observed failure rates.

A naïve comparison of the above trajectories would suggest that therapy B is inferior to therapy A, but such a comparison is of course confounded by factors influencing both the choice of trajectory and failure time.

Another way to describe the data is to classify events according to the last (i.e. most recent, or 'lagged') treatment. According to this classification there are 2.1 times more total months of exposure to A than to B (97,800 versus 46 308), with the former ending more frequently in failure (as reflected by the rates of 3.84 and 2.44 $\times$ 100 months, respectively). Again, this comparison is confounded. Fitting a time-varying Cox proportional hazards model to the event times, adjusting for R and G as time-dependent covariates, does not satisfactorily control for this confounding, since part of the effect of earlier treatments on the survival time will be blocked by conditioning on subsequent values of the covariates that lie on the causal pathway from the earlier treatments to the outcome. In addition, standard methods do not allow us to estimate the effects of dynamic regimes. We turn next, therefore, to the analysis based on the g-computation formula.

17.6.4.2 G-computation formula analysis

The results obtained using the g-computation formula can be viewed in Figure 17.3. These Nelson–Aalen cumulative hazard estimates are obtained nonparametrically from the simulated

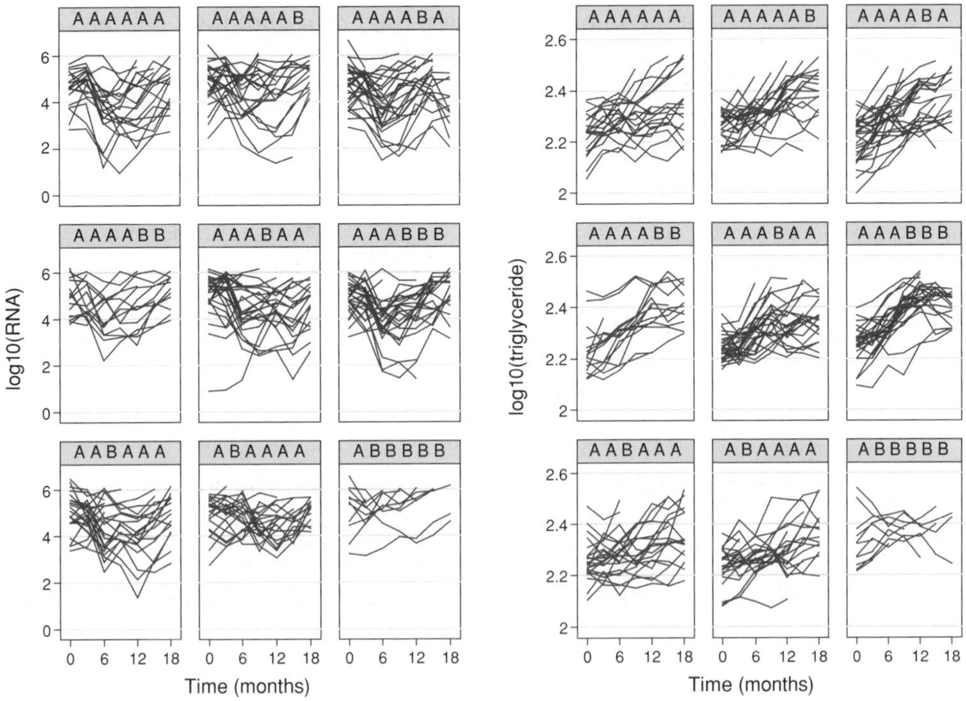

Figure 17.2 Observed $\log_{10}$*(RNA) and* $\log_{10}$*(triglyceride) profiles according to the 8 most common actual treatment trajectories, together with ABBBBB, including partial treatment trajectories consistent with these. A random selection of 200 patients are shown to aid readability.*

Table 17.1 Distribution of a selection of actual treatment trajectories and associated failure rates (per 100 months) in 10 000 simulated profiles.

Trajectory consistent with	Rate × 100	95 % confidence intervals
A A A A A A	4.57	4.26, 4.90
A A A A A B	5.49	5.11, 5.89
A A A A B A	5.53	5.14, 5.94
A A A A B B	5.39	5.02, 5.79
A A A B A A	5.32	4.91, 5.77
A A A B B B	5.49	5.08, 5.94
A A B A A A	5.33	4.89, 5.81
A B A A A A	5.56	5.10, 6.06
A B B B B B	8.75	7.96, 9.63
⋮		
All trajectories	3.20	3.11, 3.29

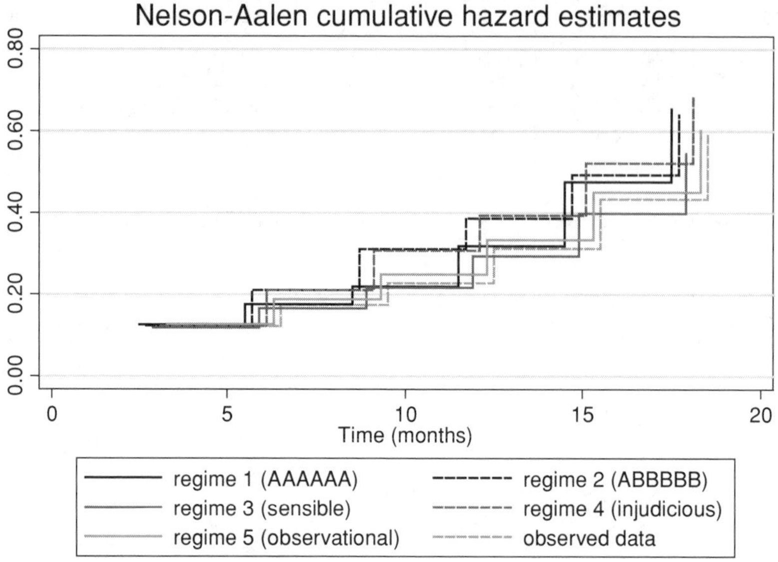

Figure 17.3 Estimated cumulative hazards for each simulated regime and for the observed data.

data under each of the regimes, as well as for the observed data. Although there is some crossing and no measure of precision, a tentative interpretation of this plot is that the sensible strategy 3 is best, followed by the observed data and the observational regime 5. This is as we would expect, since the strategy followed in the observed data is itself sensible (but with some added noise). It is difficult to distinguish between the two static regimes and the injudicious dynamic strategy on the basis of this plot.

For a more formal comparison we must make additional assumptions, for example that the hazards under different regimes are proportional. Under this assumption, we can fit a Cox proportional hazards model to the simulated data for all five regimes and estimate a hazard ratio for each regime compared to regime 1, i.e. the static regime (AAAAAA). Standard errors are estimated using the nonparametric bootstrap, with 1000 bootstrap samples, and normal-based confidence intervals for the log hazard ratios obtained (and then transformed to the HR scale). These results are given in Table 17.2.

Table 17.2 Estimated hazard ratios (HR) comparing each regime to regime 1 (static, AAAAAA) using Cox proportional hazards regression following the g-computation formula Monte Carlo simulations.

Regime	HR	Normal-based bootstrap 95% confidence intervals
2 (static, ABBBBB)	1.007	0.933, 1.087
3 (dynamic, sensible)	0.847	0.801, 0.895
4 (dynamic, injudicious)	1.064	1.000, 1.133
5 (dynamic, observational)	0.937	0.890, 0.986

17.7 Further considerations

17.7.1 Parametric model misspecification

As argued by Vansteelandt in Chapter 11 of this volume, for the related problem of estimating direct and indirect effects, the g-computation formula is 'greedy' in terms of the extent of parametric assumptions required, and as such it is more prone to bias caused by parametric model misspecification than rival semiparametric approaches. Readers are referred to Moodie *et al.* (2007), Orellana *et al.* (2010) and Cain *et al.* (2010) for details of these alternative approaches based on g-estimation of structural nested models and inverse probability weighting of marginal structural models.

In practice, therefore, when the g-computation formula is used, considerable effort should be invested in checking the plausibility of the parametric modelling assumptions using conventional model-checking techniques.

Take, as one example, the model for R_3 conditional on $(\bar{A}_2, \bar{X}_2)$ used as one step of many in the g-computation formula analysis. We have many choices for this model, and in the analysis reported in Section 17.6 we used a linear regression with $R_2, A_2, R_1, A_1, R_2 * A_2$ and $R_2 * A_1$ as predictors. Suppose, instead, we had used the raw RNA values (10^{R_2}, 10^{R_1}, $10^{R_2} * A_2$ and $10^{R_2} * A_1$). We could compare these two models using quantile plots of the residuals from each regression, to see which gives the better fit to the data. These plots are shown in Figure 17.4. Unsurprisingly (given our knowledge of how the data were generated), we find that the model based on the logged (to base 10) values are a much better fit to the observed data. Of course, in practice, a very large number of such plots would need to be checked for model misspecification.

As outlined in Section 17.6.2, another check of the parametric modelling assumptions made is to compare data simulated under the observational regime (regime 5) with the observed data themselves. Figure 17.3 is somewhat reassuring in this respect, since it shows both cumulative hazards to be evolving similarly, with a slight divergence at later time-points, which may be due to losses to follow-up. The observed data contain losses to follow-up, whereas the data simulated under the observational regime are simulated as if no losses to follow-up would occur.

17.7.2 Competing events

In our simulated example, we considered a composite outcome of 'failure or death'; this is the norm in HIV studies. In other settings, we may wish instead to view death (or another event that precludes the occurrence of the event of interest) as a competing event. The occurrence of the competing event can then be viewed as a source of loss to follow-up and can in theory be dealt with, under the missing-at-random assumption, as described in Section 17.5.4. It seems unnatural (and indeed potentially misleading), however, to simulate data under a hypothetical intervention for a subject after the time at which she would have died under that intervention. Therefore, survival can be seen as an additional outcome process to be simulated in the same way as described for event-free survival above.

Suppose a subject survives event-free up to and including time T_{k-1}. By time T_k, she may or may not be alive, i.e. she may or may not have died during the period $(T_{k-1}, T_k]$. If she is alive, then we can ascertain whether or not she experienced the event of interest in the interval $(T_{k-1}, T_k]$. (If she died at a time D during the interval $(T_{k-1}, T_k]$, then we assume it to be impossible to ascertain whether or not she experienced the event of interest in the interval

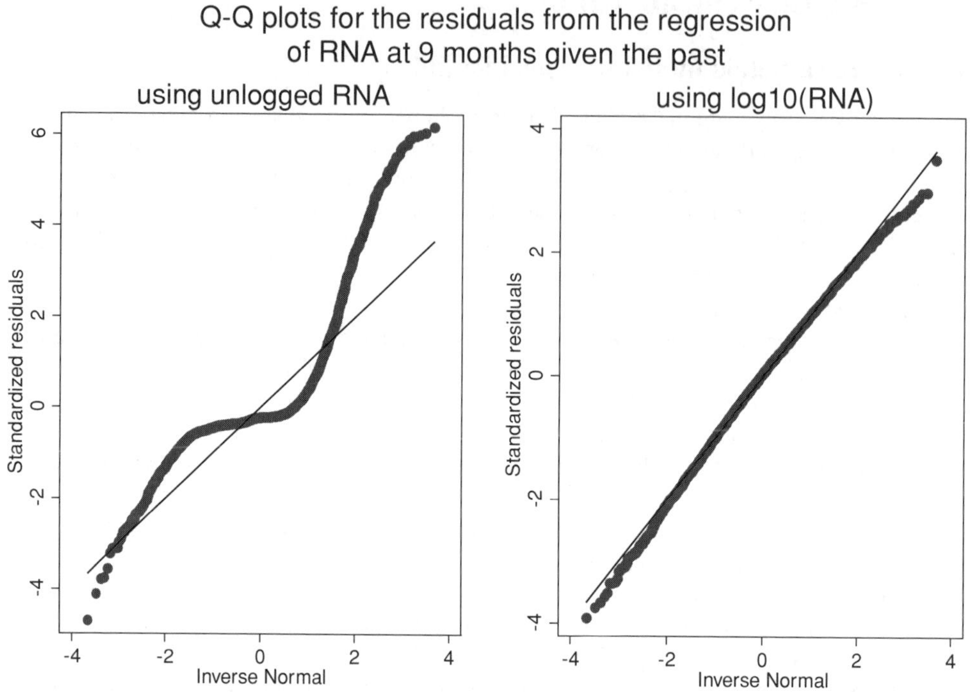

Figure 17.4 Two different quantile–quantile plots for the residuals from possible regression models fitted during the g-computation formula analysis. The left-hand plot shows the residuals obtained when the raw (unlogged) values of RNA are used, whereas the right-hand plot shows the residuals from the analysis that was actually used, i.e. that based on the logarithm to base 10 of RNA.

$(T_{k-1}, D]$.) To reflect this, the data could be simulated as follows. First, the following question is asked: 'Did this subject survive the interval $(T_{k-1}, T_k]$?' Then, conditional on its answer being simulated as 'yes', the second question: 'Did the subject experience the event of interest in the interval $(T_{k-1}, T_k]$?' is asked, and the answer simulated.

17.7.3 Unbalanced measurement times

An important assumption in the implementation of the g-computation formula described above is that of common fixed measurement times $\{T_k : 0 \leq k \leq N + 1\}$ for all subjects. In many realistic (e.g. clinical) settings, it is implausible that visit times will follow such a pattern even approximately, let alone exactly. Arjas in Chapter 7 hints at the fact that this assumption is not necessary in his formulation, and yet it is unclear how the necessary 'smoothing' over unbalanced times of measurement would be achieved in practice without imposing yet stronger parametric assumptions. For more on this issue, see Orellana and Rotnitzky (2007).

Suppose in our example that, rather than being fixed, the kth visit time T_k for each subject is distributed randomly in continuous time. One practical approach would be then to divide follow-up into $m > 7$ visits (recall that there were 7 visits in our example), with the data

(A, X) consisting entirely of missing values for at least $m - 7$ visits for any given subject. A stochastic imputation procedure could be employed to deal with these missing data before carrying out the Monte Carlo simulation technique under the various intervention strategies to be compared. Inspecting the stability of the conclusions drawn as m increases would constitute a possible sensitivity analysis. An option for stochastic imputation of such intermittent patterns of missing data is included in the gformula command in Stata (Daniel *et al.*, 2011).

17.8 Summary

In this chapter, we have re-visited the methods outlined in two earlier chapters for estimating the causal effects of time-varying exposures and for comparing dynamic treatment strategies, using potential outcomes in a likelihood framework.

In particular, we have focused on the implementation of the approach in practice, illustrated using a simplified but realistic simulated example, and outlined how some of the practical considerations, such as loss to follow-up, competing events and model-checking, can be approached.

While we have discussed possible violations of parametric modelling assumptions, another important concern is the validity of the structural (i.e. causal) assumptions on which the enterprise relies, namely those depicted by the DAG of Figure 17.1. The central rôle of the conditional exchangeability assumption should guide the design of longitudinal observational studies. In settings such as the HIV study discussed above, it is clearly important that factors likely to influence a clinician's choice of treatment for a given patient be measured, whenever these factors are thought to be related (causally or otherwise) to the outcome of interest other than through their effect on treatment. On the other hand, we hope that the doctors' decisions are made in part on a whim, so that positivity holds.

Sensitivity analyses to these structural assumptions could (and should) be performed along the lines suggested by Lash *et al.* (2009) and others.

References

Cain, L.E., Robins, J.M., Lanoy, E., Logan, R., Costagliola, D. and Hernán, M.A. (2010) When to start treatment? A systematic approach to the comparison of dynamic regimes using observational data. *The International Journal of Biostatistics*, **6** (2), Article 18.

Cox, D.R. and Wermuth, N. (1996) *Multivariate Dependencies: models, analysis and interpretation*. Chapman&Hall/CRC.

Daniel, R.M., De Stavola, B.L. and Cousens, S.N. (2011) gformula: estimating causal effects in the presence of time-varying confounding or mediation using the g-computation formula. *The Stata Journal*, **11** (4), 479–517.

Davison, A.C. and Hinkley, D.V. (1997) *Bootstrap Methods and Their Application*. Cambridge University Press.

Dawid, A.P. and Didelez, V. (2010) Identifying the consequences of dynamic treatment strategies: a decision theoretic overview. *Statistics Surveys*, **4**, 184–231.

Hernán, M.A. and VanderWeele, T.J. (2011) Compound treatments and transportability of causal inference. *Epidemiology*, **22**, 368–377.

Lash, T.L., Fox, M.P. and Fink, A.K. (2009) *Applying Quantitatvie Bias Analysis to Epidemiologic Data*. Springer.

Little, R.J.A. and Rubin, D.B. (2002) *Statistical Analysis with Missing Data*, 2nd edn. John Wiley & Sons, Ltd.

Moodie, E.E., Richardson, T.S. and Stephens, D.A. (2007) Demystifying optimal dynamic treatment regimes. *Biometrics*, **63**, 447–455.

Orellana, L. and Rotnitzky, A. (2007) Estimation of the effect of dynamic treatment regimes under flexible dynamic visit regimes. Liliana Orellana's PhD Thesis, Department of Biostatistics, Harvard University, Chapter 2.

Orellana, L., Rotnitzky, A. and Robins, J.M. (2010) Dynamic regime marginal structural mean models for estimation of optimal dynamic treatment regimes, Part I: main content. *The International Journal of Biostatistics*, **6** (2), Article 8.

Robins, J.M. (1986) A new approach to causal inference in mortality studies with a sustained exposure period – application to control of the healthy worker survivor effect. *Mathematical Modelling*, **7**, 1393–1512.

Taubman, S.L., Robins, J.M., Mittleman, M.A. and Hernán, M.A. (2009) Intervening on risk factors for coronary heart disease: an application of the parametric g-formula. *International Journal of Epidemiology*, **38**, 1599–1611.

VanderWeele, T.J. (2009) Concerning the consistency assumption in causal inference. *Epidemiology*, **20**, 880–883.

18

‘Natural experiments’ as a means of testing causal inferences

Michael Rutter
MRC SGDP Centre, Institute of Psychiatry, London, UK

18.1 Introduction

In this chapter many different forms of ‘natural experiments’ that can be helpful in testing causal inferences are considered. Although there are general principles for good research designs, they take rather different forms according to the nature of the sample and the purpose of the study. In discussing the science, it is crucial to specify the measures and the target population within which cause is to be inferred. In all cases, the analysis needs to complement and take advantage of the design. Accordingly, the presentation of specific examples is essential for considering what makes for better and for less justified causal inferences.

18.2 Noncausal interpretations of an association

There are several rather different reasons why statistically significant correlations or associations or differences cannot be assumed to imply a causative influence (Academy of Medical Sciences, 2007; Jaffee *et al.*, 2011 Rutter, 2007). Because different research strategies are needed to deal with each of the problems, it is necessary to begin with a brief summary of the nature of each hazard. Throughout this chapter, all of the examples come from the fields of psychology and psychopathology, but the principles apply across the whole of biomedicine (even though the relative magnitude of the hazard varies with different outcomes and different causal influences).

Causality: Statistical Perspectives and Applications, First Edition. Edited by Carlo Berzuini, Philip Dawid and Luisa Bernardinelli.

A key issue concerns the possibility of reverse causation. Bell (1968) first raised this matter in relation to socialisation effects. He queried whether associations between, say, marked parental punishment and children's mental disorder reflected the influence of children's disruptive behaviour on parental disciplinary practices, rather than the other way round. The reality of child effects has been shown using a range of different research strategies (Bell and Harper, 1977), including experimental methods (Anderson *et al.*, 1986) and adoptee studies (Ge *et al.*, 1996; O'Connor *et al.*, 1998). Longitudinal studies, too, have shown that the rate of environmental hazards in adult life is strongly associated with prior behaviour in childhood (Robins, 1966; Champion *et al.*, 1995). Second, there is the possibility of social selection – meaning that the outcome reflects the *origin* of the putative risk factor, rather than its effects. Thus, is the marked increase in psychopathologic risk associated with being born to a teenage mother (Moffitt and The E-Risk Study Team, 2002) due to the quality of rearing provided by an adolescent parent or to the sort of person likely to have a child during the teenage years (Maynard, 1997)? Similar questions arise with respect to the association between social disadvantage and some forms of psychopathology (see Miech *et al.*, 1999).

A third possibility is that, although there is indeed a true environmentally mediated causal effect, its nature has been misidentified. For example, the early literature on the supposed criminogenic or depressogenic effects of 'broken homes' focused on the putative causal effects of parent–child separation. Subsequent research, however, showed that the main risk effect came from family discord or conflict rather than separation (Fergusson *et al.*, 1992) or from the consequent effects of family break-up in leading to poor parental care (Brown *et al.*, 1986). Similarly, it has been widely accepted that maternal smoking exercises a causal influence on antisocial behaviours, but mothers who smoke differ from those who do not in being more likely to have shown antisocial behaviours, to have had less education and to have a lower income (Wakschlag *et al.*, 2003). Which of these features (if any) is causal?

Alternatively, the misidentification may concern the *timing* of the effects; this matters when the timing influences inferences on the causal mechanism. The most obvious situation of this kind arises with respect to whether the effect is prenatal or postnatal. Thus, women who smoke or drink heavily during pregnancy are likely to do the same after the baby is born. The question then is whether any causal effects stem from the numerous postnatal risks associated with parental alcoholism or nicotine dependence rather than from any form of prenatal biological programming.

Fourth, there is the possibility that the mediation is genetic rather than environmental. That is the case even if the putative cause is truly environmental. Thus, for example, using a twin design with individually measured specific features, Pike *et al.* (1996) showed that about a third of the variance attributable to the effects of family discord were genetically, rather than environmentally, mediated. This possibility arises with any environmental feature that can be influenced by the behaviour of the individuals in shaping or selecting environments involved. This applies to most environmental risk factors associated with the cause or course of mental disorders – such as family conflict, child abuse or neglect, parental divorce or acute or chronic stresses carrying long-term threat (Kendler and Baker, 2007; Jaffee and Price, 2007; Jaffee, 2011). It might be thought that when DNA methods can be used to identify individual susceptibility genes, this obviates the problem, but it does not. That is because, although such methods can rule in, or out, the effects of the identified genes, they cannot rule out the effects of other unmeasured genes.

Fifth, the statistically significant association might reflect an artefactual effect attributable to some confounder (meaning any feature whose effect might misleadingly mimic the supposed causal effect). Often the confounder will operate via one of the aforementioned four possibilities, but sometimes it may reflect biases in ascertainment or measurement, or it may simply constitute a 'third variable' (such as age or sex) that just happens to be correlated with the supposed causal influence (see Academy of Medical Sciences, 2007). The traditional way of dealing with confounders has been to introduce some form of statistical control or adjustment to determine whether the significant effect remains after taking into account the confounder, or confounders (Cook and Campbell, 1979).

18.3 Dealing with confounders

Unfortunately, these methods involve several serious problems (see Rutter *et al.*, 2001). To begin with, there is the logical problem of having to use an invariant set of associations to infer what would happen if a key association was different or was changed (McKim and Turner, 1997). That necessitates the assumption that the effects within the existing sample would apply to a different set of circumstances (i.e. without the confounder). That is problematic if the two groups being compared do not overlap in key cells (see Melhuish *et al.*, 2008, for an example of this kind). Propensity scores are particularly useful in identifying where there is a lack of overlap and the problem can be, at least partially, circumvented by dropping the nonoverlap cells before undertaking the statistical controls. They are also helpful in reducing the biasing effects down to a single balancing factor. Care is also needed in the conclusions that can and cannot be drawn from statistical adjustment for confounders. For example, it is commonly assumed that if a multivariate analysis shows an effect of 'A' that is maintained after taking account of the effects of 'B' or 'C', this means that there is a true effect of 'A' in the absence of 'B' or 'C', but it does not (see Rutter, 1983).

An even greater problem is presented by the uncertainties regarding unmeasured confounders. Any statistical control for confounders will necessarily be influenced by the quality of the measures obtained. A distinction needs to be made between studies in which major confounders were considered in advance and deliberately assessed with care, and those where that was not the case. In addition, there are ways of somewhat reducing the problems, of which five may be noted (see Rutter, 2009). First, when there has been good measurement of a wide range of features associated with the *exposure* to the putative causal influences, propensity scores can be used to calculate an inverse probability to treatment weighting (IPTW) to create two comparable groups (Robins *et al.* 2000), the technique involving a minimum of statistical assumptions (see Sampson *et al.*, 2006, for a particularly clear account of its use), but its value is critically dependent on the identification of all the most important features influencing risk exposure.

An alternative approach is to use growth curve trajectories to take account of an individual's overall liability to the disease/disorder outcome (Nagin *et al.*, 2008). The logic here is that if the putative causal influence alters a person's growth curve with respect to the outcome being considered, a causal inference is more plausible. The use of this approach requires at least four measurement points over a substantial period of time (in order to create a valid growth curve) and requires a time-specific exposure to the putative causal influence. This was the case in the Nagin *et al.* (2008) example of testing for the effects of gang membership on criminal behaviour. It should be appreciated, however, that, as with any other form of cluster analysis, the curves represent a model (rather than a 'real' phenomenon) and the number of curves selected for analysis is a matter of judgement on their likely meaning rather than the result of some single statistical solution (although, of course, statistical criteria should play a crucial part in the choices).

Structural equation models (SEMs) can be used to treat confounders as a latent variable, hence allowing a better assessment of their effects after taking into account measurement error. If multimeasurement longitudinal data are available, a two-stage approach can be employed in which the first step involves a regression of effects on exposure and the second step uses the predicted values of the exposure covariate in a regression of the putative cause on outcome (see Dawid and Pickles, 2007).

A fourth approach is to use statistical modelling based on direct acyclic graphs (DAGs) (Pearl, 1995; Gill and Robins, 2001). These use preexisting background knowledge to specify the ways in which a possible confounder might operate. The focus then is on concrete likely alternative pathways that, through appropriate statistical modelling, can go a long way down the path of testing the causal inference. Of course, the value of the approach is crucially dependent on the plausibility of the model.

A fifth method is provided by mediation analyses (Baron and Kenny, 1986; MacKinnon *et al.*, 2007; Fairchild and MacKinnon, 2009). In themselves, they do not provide a test of causation, but what they do is to isolate a putative key mechanism that could account for the observed statistically significant association. The logic is that when the postulated mediator for 'C' for the effects of 'A' on 'B' is introduced into the overall model, if mediation is occurring, the original correlation of A on B should drop to zero (if there is complete mediation) or be substantially reduced (if there is partial mediation; see Rutter and Sonuga-Barke, 2010, for a discussion of the issues). The value of the approach lies more in ruling out some causal possibilities rather than proving the causal effect of others. Care in mediation analyses is required in order to ensure that there is no confounding between the mediator and the outcome.

Finally, whatever the approaches used, there is value in using sensitivity analyses to quantify how strong a confounding effect would have to be to overturn a causal inference based on a case-control comparison. For example, Cornfield *et al.* (1959) used such an analysis to show that if the association between smoking and lung cancer was due to a confounder, such a confounder would have to be nine times as frequent in heavy smokers as in nonsmokers to undermine the causal inference. Clearly, that was extremely implausible.

18.4 'Natural experiments'

'Natural experiments' differ fundamentally from many of the conventional statistical approaches in that they do not *adjust* or *control* for confounders. Rather, they use naturally occurring circumstances that serve to separate variables that ordinarily go together and thereby provide *manipulations*, and not just analysis of crucial variables. Of course, each of the natural experiments involves essential statistical implications but this chapter focuses on the design issues. The brief introductory section on 'dealing with confounders' serves as a reminder that many of the procedures noted in that section apply to 'natural experiments' as much as to naturalistic observational studies. Cook and Campbell (1979) and Shadish *et al.* (2002) were pioneers in developing quasi-experiments, or 'natural experiments', and they discussed the vital interplay between designs and statistical analysis. It is necessary to add that the principles of 'natural experiments' may be applied to the analysis of data sets gathered for other purposes. Thus, researchers may work with cohort studies but construct estimators that exploit natural experiments that happen to be embedded within the sample (for example siblings and twins).

18.4.1 Genetically sensitive designs

Several types of 'natural experiments' focus on the need to differentiate between genetic and environmental mediation of risks. They do this through a variety of twin and adoptee designs or by making use of migration strategies. Twin designs use the contrast between monozygotic (MZ) and dizygotic (DZ) twin pairs to partition the population variance into additive genetic and shared or nonshared environmental effects. This contrast capitalises on the fact that MZ pairs share 100 % of their segregating genes whereas DZ pairs share on average just 50 %. By treating some postulated environmental causal factor as a trait or phenotype, it is possible to estimate the heritability of the risk factor trait. By means of cross-twin, cross-trait analyses (the cross being between the risk factor and the outcome), the extent to which mediation is genetically or environmentally mediated can be quantified. In order to check temporal order, longitudinal data will be required.

18.4.1.1 Multivariate twin designs

The longest established twin strategy is the use of multivariate twin analyses (see Kendler and Prescott, 2006). By this approach, doubt was raised with respect to the causal inference made with respect to the substantial association found between an early age of first drinking alcohol and the later development of alcoholism (Grant and Dawson, 1997). Multivariate twin analyses showed no environmentally mediated effect deriving from the age of first drinking. Rather, both the early age of drinking and the alcoholism reflected the same genetic liability. Similarly, Jaffee *et al.* (2004) used the same strategy to examine the observed associations between both corporal punishment and physical child abuse, and children's propensity to engage in antisocial behaviour. The findings confirmed environmental mediation in the case of abuse but not for corporal punishment, for which the mediation was largely genetic.

While there is no doubt that multivariate twin analyses have proved very informative, it is necessary to note several important limitations. Although there have been doubts expressed with respect to the assumptions of twin comparisons and to sampling biases, the best modern studies have dealt adequately with these issues (see Rutter, 2006; Kendler and Prescott, 2006). However, three other rather different hazards remain. First, the operation of parsimony in the modelling sometimes means that potentially important features are dropped from the model because they just fail to reach significance; this arises most frequently with shared environmental effects. On occasion, the consequence may be a misleading model (see, for example, Maes *et al.* 1999). In the Maes *et al.* (1999) study, the majority of young people who ever used alcohol also did so without permission. Nevertheless, modelling showed a zero genetic influence on the former but a preponderant influence on the latter. Second, twin studies assume that mating is random, whereas often it may not be. Third, there is rarely the power to deal with the confounding of genetic effects by the inclusion of GxE (when the E involves shared environmental effects).

18.4.1.2 Discordant twin pairs

Discordant same-sex twin pairs reared together represent a different twin strategy. The rationale is that by examining pairs that are discordant for both the hypothesised environmental risk factor and the outcome being studied in order to determine how the two coincide, there is a substantial degree of control for shared genetic and environmental liabilities (Kendler and Prescott, 2006). For example, Prescott and Kendler (1999) used this design to test the supposed causal effect of early drinking on later alcoholism. Statistical modelling showed that the association largely reflected a shared genetic liability (although there was some effect from shared environmental factors). Two limitations need to be noted. First, the discordance within MZ pairs raises the question of *why* they were treated differently within the family or had different experiences outside the family. Second, although the design provides a strong pointer to the likelihood that the within-pair difference in risk exposure exerted a causal influence it cannot completely rule out the possibility that some *other* experience was responsible (Vitaro *et al.*, 2009)

As the study by Caspi *et al.* (2004) illustrates well, a combination of other design features can do much to increase the power of the discordant twin pair design. The putative environmental cause being investigated was negative expressed emotion (EE) focused on one twin, with the outcome being disruptive behaviour. EE and disruptive behaviour were assessed at age 5 with the outcome measured at age 7. Thus, the prospective longitudinal design provided measurement of within-individual change and determination of the temporal sequence. EE was measured from maternal videotapes and disruptive behaviour from teacher reports. This removed the possibility of rater bias and it also examined effects that spanned a radical change in environment (namely starting school). The contrast within MZ pairs meant that it was highly unlikely that there was any family-wide social confound. The findings showed a statistically significant effect of EE that could not reflect genetic mediation. The causal inference is therefore well justified. Nevertheless, two limitations need to be noted. First, it remains uncertain whether the EE itself was causal, as distinct from it indexing some other associated environmental feature. Second, it leaves open the question of what set of features caused the mother to respond so differently to her two MZ twins (i.e. raising the possibility of reverse causation).

18.4.1.3 Discordant sib pairs

Discordant twin pairs cannot be used to examine postulated prenatal causal influences, but discordant sib pairs can be used for this purpose, as shown by the example of the effects of prenatal exposure to maternal smoking. D'Onofrio *et al.* (2008) compared disruptive behaviour in a sibling born after a pregnancy in which the mother smoked with that of a sibling born after a nonsmoking pregnancy. The findings showed no environmentally mediated effect of prenatal smoking exposure on disruptive behaviour but did show such an effect on birth weight. Importantly, structural equation modelling of the same data set designed to control for confounders had suggested a prenatal causal effect. This serves as yet another reminder of the potential inadequacy of statistical control for confounders. This study, however, had the weakness of reliance on maternal report for both the independent (postulate causal) and dependent (outcome variables).

A Scandinavian study by Obel and colleagues (2011) did not have this limitation. Smoking was assessed from maternal report at the time of the child's birth and attention deficit/hyperactivity disorder (ADHD) was determined from clinical record register data. In the sample as a whole there was a highly significant odds ratio of 2.2 but the discordant sibling comparison showed a much reduced, statistically nonsignificant, ratio of 1.2, casting serious doubt on the causal inference with respect to ADHD, but again confirming the causal effect on birth weight (supported by all research strategies). Because there is a random allocation of genes across sibs there is good control for genetic liability. The main circumstance that could undermine that control is when there is a shared genetic liability between the discordance in the maternal smoking and individual differences in the liability to disruptive behaviour. That is an implausible situation if only because the former concerns maternal genes and the latter child genes plus the requirement that the former should apply to a *difference* between pregnancies rather than an enduring feature. A rather greater limitation is the uncertainty over the factors that led mothers to smoke in one pregnancy but not in the other. Could those include a possible risk effect for the child?

18.4.2 Children of twins (CoT) design

The children of twins (CoT) design provides a useful extension of twin strategies (D'Onofrio *et al.*, 2003; Silberg and Eaves, 2004). The rationale is that the offspring of adult MZ twins will be social cousins but genetic half siblings (half because one parent is part of an MZ pair and the other parent is not). By contrast, the offspring of DZ twins are cousins both genetically and socially. Put simply, if the postulated causal parental feature involves a genetically mediated risk, the risk should apply as much to the offspring of both the MZ twins regardless of whether one is providing the social rearing and the other not. This will not apply as much to the offspring of DZ twins because, on average, they share just 50 % of their segregating genes. Using this strategy, it has been found that harsh forms of physical punishment have an environmentally mediated influence on both disruptive behaviour and substance use/abuse (Lynch *et al.*, 2006). Similarly, use of the CoT design showed no effect on ADHD of prenatal exposure to maternal smoking, at least not in alcoholics (D'Onofrio *et al.*, 2008), although there was an effect on birth weight. Also, Silberg *et al.* (2010) showed that parental depression has an environmentally mediated effect on depression in the offspring. There are three main limitations to the CoT strategy. First, because the genetic difference between cousins and half-siblings is quite small, a very large sample size is required. Second, it is quite difficult to take adequate account of the unknowns in both the genetic background and rearing environment provided by the parent who is not a twin. Third, unless the spouse background and rearing behaviours are measured and included in the analysis, the findings could be misleading (see Eaves *et al.*, 2005).

18.4.2.1 Offspring of women who conceived through assisted reproduction technologies (ARTs)

Yet another way to exclude genetic mediation of risks is provided by the use of assisted reproduction technologies (ARTs) (see Thapar *et al.*, 2007). Some of these retain the genetic link between mother and child (as with homologous *in vitro* fertilisation and sperm donations) and others do not (as with egg or embryo donations). If the maternally provided prenatal risk factor has an environmentally mediated effect on the child outcome, this should give rise to the same associations irrespective of whether or not the mother and child are genetically related. Conversely, if the risk is genetically mediated, it should be evident only where there is a genetic link between mother and child. The method has been used to investigate the postulated prenatal risk for maternal smoking associated with both antisocial behaviour (Rice *et al.*, 2009) and ADHD (Thapar *et al.*, 2009). The findings showed genetic mediation in relation to both outcomes but environmental mediation in relation to birth weight. Importantly, as with other studies, the findings showed that statistical control for confounders (including parental ADHD and maternal alcohol consumption) was inadequate.

Several limitations apply. First, it is no easy matter to acquire appropriate ART samples of adequate size. This is especially the case when secular trends have meant that the number of women smoking in pregnancy is low, with an inevitable reduction in the size of key cells. On the other hand, the sample size was adequate to show a true causal effect on birth weight. Second, the analyses were based on symptom counts rather than clinical diagnoses. Third, it is possible that families using ART are unusual in ways that could create bias. However, at least in the UK, the evidence suggested that ART families were reasonably representative of the general population.

18.4.2.2 Adoption studies

Because adoption separates the genetic heritage from the rearing environment, it provides a further way of excluding genetic mediation. Influences from the rearing environment imply a true environmental effect. Thus, Duyme *et al.* (1999) studied the effects of the socioeconomic (SES) level of the adoptive family on the IQ gains following adoption between the ages of 4 and 6 years of children who had been compulsorily removed from their families during infancy. The IQ gains were significantly greater for the children placed in high SES families than for those placed in low SES families. While adoption does indeed separate genetic and rearing influences, there are considerable problems in using the design because: (i) selective placement needs to be excluded; (ii) the range of risk environments is markedly restricted in most adoptee samples (Stoolmiller, 1999); (iii) at least in the UK, there has been a marked reduction in the number of early adopted healthy babies; and (iv) often the information on the qualities of the adoptive family environment is weak. Accordingly, despite its theoretical attractions, practical issues have meant that the design has been little used.

18.4.2.3 Migration strategies

The last strategy to mention, which seeks to exclude genetic mediation, is migration. Individuals carry their genes with them when they move to a different country but they may change their lifestyles. The natural experiment arises when high-risk (or low-risk) ethnic groups move to a different country and adopt entirely different lifestyles. The value of the strategy has been best demonstrated by medical examples (see, for example, Marmot and Syme, 1976; Valencia *et al.*, 1999), but it has also been applied with good effect in relation to the raised rate of schizophrenia spectrum disorders in people of Afro-Caribbean origin living in the UK or the Netherlands (Jones and Fung, 2005). The natural experiment is provided by a comparison of rates from those found in the countries of origin, as well as with those of a different ethnicity in the countries of destination. Careful methodological checks have shown that the marked differences in rates of schizophrenia spectrum disorders are valid, that it is highly unlikely that direct genetic mediation is involved and that some form of environmentally mediated effect is implicated (Coid *et al.*, 2008). Quite what this involves, however, remains uncertain. Possibilities include racial discrimination, isolation from within-group support and early parent–child separation. Migration as such is not likely to constitute the causal factor because the findings apply to both second and first generation migrants and because, in other circumstances, migration may even bring benefits (Stillman *et al.*, 2009).

18.4.3 Strategies to identify the key environmental risk feature

Twin and adoptee strategies tend to be thought about solely in terms of their power to exclude genetic mediation of risk effects. However, in certain circumstances, they provide the 'natural experiment' opportunity to compare two contrasting postulated environmental causes by holding genetic influences constant rather than by determining the possibility of genetic mediation. Two examples serve to illustrate the strategy.

18.4.3.1 Twin–singleton comparisons

The starting point for this strategy is the well-established finding that, as a group, twins lag behind singletons in their language development by about 3 months at 3 years of age (Rutter *et al.*, 2003; Thorpe *et al.*, 2003). The 'natural experiment' arises from two main features: (1) although genetic influences will operate on individual differences in language, there is no reason to suppose that they will differ in either strength or type between twins and singletons and (2) overall social disadvantage is unlikely to be responsible for the language impairment because it is not particularly associated with twinning. Accordingly, the comparison virtually rules out the two major confounding factors found in the general population. The two leading contenders for the postulated environmental cause were obstetric complications (known to be much more common in twins) and altered patterns of parent–child interaction arising from having to deal with two children of the same age at the same time. In order to focus on influences likely to be important in the general population, children born before 34 weeks of gestation were excluded (because other research had shown that this was associated with a substantial increase in brain damage). In order to focus on within-individual change, the analysis of language performance at 3 years took into account the children's language at age 20 months. The findings showed no effect of obstetric complications but a significant effect of parent–child interaction/communication. When the latter was introduced into the overall model, it eliminated the twin–singleton difference (i.e. it showed complete mediation). The several methodological steps taken made the causal inference plausible. The main limitations concern the uncertainties over the measurement of key features of the parent–child interaction and the uncertainties over the generalisation of the findings to populations of singletons. While there is no reason to doubt the exclusion of any major effect of obstetric complications, questions remain on the key causal elements of mother–child interaction, particularly as father–child interaction was not measured.

18.4.3.2 Adoption as a radical change in environment

The English and Romanian Adoptees (ERA) study set out to examine the long-term effects of profound institutional deprivation (Rutter and Sonuga-Barke, 2010). It constituted a 'natural experiment' because (i) the children were admitted into institutional care in early infancy (thereby eliminating the usual confound of children being admitted when older because of identified impairment); (ii) scarcely any children were adopted or returned to their biological families before the Ceauceşcu regime fell in 1989 (thereby avoiding the usual biases associated with which children remained in institutional care); (iii) it was possible to pinpoint accurately the timing of leaving institutional care; (iv) the move from profoundly depriving institutional care to generally well-functioning adoptive families involved an exceptionally marked environmental change; and (v) the design involved a follow-up to at least the age of 15 years (with multimodal measurements at 6, 11 and 15). Although the adoption created a quite exceptional natural experiment, it was necessary to go on to check whether there could be a bias resulting from variations in the parents' main reasons for adoption (altruism versus responding to infertility). Because none of these variations were associated with outcome, it appeared unlikely that these constituted a significant bias.

The postulated causal effect of institutional deprivation could be tested in two different ways. First, there was the test of developmental recovery following adoption. Findings showed major recovery; these indicated that the initial deficits at the time of leaving institutional care were likely to have been caused by the institutional deprivation. Second, because the recovery was incomplete in many instances, it was necessary to determine whether the deficits were a function of the preadoption institutional environment or variations in the qualities of the postadoption family environment. Several key findings pointed strongly to an institutional effect: (i) the associations with institutional deprivation lasting beyond the age of 6 months were as strong at 15 years as they had been at 6 and 11 years; (ii) the effects were largely of a kind (such as quasi-autism or disinhibited attachment) that are rare in children not experiencing institutional deprivation; (iii) the variations in outcome were not accounted for by variations in the postadoption environment; and (iv) the institutional deprivation, even in the absence of subnutrition, was associated with a large and significant reduction in brain size that showed partial mediation with respect to the psychological outcomes. Various other methodological checks were also undertaken to rule out nondeprivation causes. As a result of these multiple design features, a causal inference seems well justified. Nevertheless, limitations remain. To begin with, no valid evidence on the biological parents was available, making it difficult to rule out all possibilities of genetic mediation. Also, although the findings were compellingly consistent, there was noteworthy heterogeneity in outcome and the causes of this variation remain uncertain. Most crucially, although there is good reason to judge the internal validity as very high, questions remain on the extent to which the findings can be generalised to other forms of deprivation in other populations (Rutter and Azis-Clauson, 2011).

18.4.3.3 Effects of juvenile gangs on delinquent activities

Numerous studies have shown that members of delinquent gangs tend to commit serious and violent offences at a high frequency (Spergel, 1990). The key question is whether this is because individuals with a high antisocial liability are more likely to join gangs (i.e. a selection effect) or rather being a member of a gang fosters increased crime (i.e. a socialisation effect). The two possibilities can be contrasted by examining within-individual change over time in relation to delinquent acts before, during and after gang membership (Thornberry *et al.*, 1993). Strictly speaking, this does not constitute a 'natural experiment' because it does not pull apart variables that ordinarily go together. Nevertheless, it is included here because the hypothesis-testing approach has many features in common with 'natural experiments'. In brief, with transient gang members there was no evidence of a selection effect (i.e. they did not differ from non-gang members before going into the gang), but there was substantial evidence of social facilitation (i.e. delinquent activities were highest during the period of gang membership). By contrast, for stable gang members, there was evidence of both selection and social facilitation effects. Because the findings were derived from a longitudinal study, there were many variables available to test for selection, and it was possible to exploit within-individual change associated with a change of environment (i.e. joining a gang). Moreover, the division between transient and stable gang members was valuable in identifying the groups to which effects applied. The finding that social facilitation (i.e. more crime during gang membership) applied to *both* groups suggests that the results may be generalisable. The finding that social selection operated only with respect to stable gang membership raises questions on the reasons why that was so. The combination of trajectory analyses and propensity score matching provides a powerful alternative approach (Haviland *et al.*, 2008).

18.4.4 Designs for dealing with selection bias

In all ordinary circumstances, selection biases seriously interfere with any attempts to test environmental causes. This is because most risk experiences derive from situations influenced by people's behaviour in selecting or shaping their environments. The strategies to deal with selection biases rely on the study of experiences that are not open to personal choice because influences of personal choice have been excluded. These may involve either the introduction of risk experiences or their removal. Two examples of each serve to illustrate the strategy.

18.4.4.1 Universal introduction of risk

Both the risk-introduction examples involve famine. The first concerns the Dutch famine during World War II (Stein *et al.*, 1975). Its strengths as a 'natural experiment' derive from the fact that the famine was both severe and time-limited and that it was imposed on the total population and not just on a socially deprived subgroup. It was found that a twofold increase in the risk for schizophrenia (as assessed on a national registry) was associated with exposure to famine conditions in early gestation. The possibilities of selection bias and selective survival were examined in detail and no support for their operation was found. The causal inference was substantially strengthened by a similar effect on congenital anomalies of the central nervous system – thereby suggesting a possible biological pathway (see also McClellan *et al.*, 2006), by the gestational period specificity and by the diagnostic specificity. The main limitation concerned the reliance on group, rather than individual, exposure.

The second example is provided by the Chinese famine in 1959–1961 (St Clair *et al.*, 2005), which showed a very similar twofold increase in the risk for schizophrenia. The causal inference was again strengthened by the biological parallels – the reduction in birth rate and increase in mortality rate associated with the famine. Once more, however, the main limitation lay in the reliance on group differences, as well in this case, with uncertainties on the exact timing of the prenatal exposure.

18.4.4.2 Universal removal of risk

Two examples may be given of the effects of a universal removal of a postulated environmental cause. First, Honda *et al.* (2005) used the 'natural experiment' of Japan stopping the use of the measles, mumps and rubella (MMR) vaccine (which had been claimed to cause autism) at a time when MMR use was continuing in most other countries. In this case, the strength of the 'natural experiment' derived from four main features: (i) systematic standardised diagnostic data for a defined geographical area; (ii) the exact timing of withdrawal of MMR usage; (iii) follow-up to age 7 years; and (iv) the availability of time trends in rates of autism diagnosis. The findings showed that the withdrawal of MMR had no effect on the rising rate of diagnosed autism, a rise that followed the same pattern as that in other countries still using MMR. Because the withdrawal of MMR was total in Japan (and, therefore, not subject to personal choice), the findings effectively ruled out any general population causal effect. As with the famine examples, the plausibility of the findings was strengthened by the weight of other negative evidence based on different research strategies (Rutter, 2005).

The second example concerns the effects of the opening of a casino on an American 'Indian' reservation (Costello *et al.*, 2003). In this case, the 'natural experiment' arose from the Federal requirement that a particular portion of the profits of the casino had to be distributed to *all* those living on the reservation without any action from the individuals themselves (thereby eliminating the possibility of allocation bias). The other key elements making this a 'natural experiment' were: (i) the availability of data from a prospective longitudinal study of child mental health that spanned some 5 years before and 5 years after the casino opening; (ii) data at an individual level on whether or not the casino proceeds lifted individuals above the poverty line; and (iii) parallel time trends data on non-Indians who received no casino proceeds. The findings showed that the casino proceeds did diminish the rate of poverty (there being no such trend in non-Indians) and that those lifted out of poverty did show a significant reduction in some (but not all) forms of psychopathology. More detailed analyses indicated that the benefits were likely to have been mediated by changes in the family (this inference was inevitably less certain because it did not benefit from the universal intervention control).

The considerable strengths of all four examples lay in the effective avoidance of selection bias and the availability of systematic data on time trends as they applied to a large population. The lack of individual data did not seriously detract from the power of the study to test a causal hypothesis but it did limit the power to focus on specific individual biological mechanisms.

18.4.5 Instrumental variables to rule out reverse causation

In the previous examples, the grouping variable of the cohort coincided with a change in the level of exposure to an exogenous risk. It is an example of variables now commonly referred to as 'instruments'. An instrumental variable is a feature that does not affect the disease/disorder outcome being examined, but does influence the putative environmental cause under investigation (Foster and McLanahan, 1996). The 'cause' needs to be one that involves an intermediate phenotype of some kind. Two such examples serve to illustrate the strategy.

18.4.5.1 Mendelian randomisation (MR)

Mendelian randomisation (MR) was first described as a 'natural experiment' strategy by Katan (1986), but it has been more fully developed by Davey Smith and Ebrahim (2003, 2005). The strategy was first put forward as a means of ruling out reverse causation. The term MR refers to the use of a genotype to investigate an hypothesised environmental cause. The 'natural experiment' is operative because the genotype is unaffected by the disease/disorder being studied and because the allelic variations are likely to be randomly distributed with respect to the environmental cause being investigated. The psychopathological example involves drinking alcohol (the intermediate phenotype) and the causal hypothesis concerns the supposed 'gateway' effect of early drinking of alcohol in predisposing to 'hard' drug usage and antisocial behaviour. Irons *et al.* (2007) were astute in recognising that MR could be used in their study of children born in Korea but adopted by US parents (none of whom was of East Asian descent). The relevant genotype was responsible for a greatly reduced aldehyde dehydrogenase (ALDH) enzyme deficiency. This leads to an unpleasant flushing response with the ingestion of alcohol. Much research has shown that this results in a marked probabilistic reduction in the rate of alcoholism. If the gateway hypothesis was correct it should follow that there should be a parallel reduction in the rates of disruptive behaviours supposed to be fostered by early heavy drinking. The findings ran counter to this hypothesis in that the ALDH deficient and nondeficient groups did differ markedly in their rate of alcoholism but did *not* differ with respect to either drug abuse or antisocial behaviour.

As already noted, it is always desirable to include more than one design element in any single 'natural experiment'. Adoption provided that extra element in the Irons *et al.* (2007) study. The findings showed no effect of adoptive parent alcoholism (because there was no genetic link) but a significant effect of drinking by siblings (suggesting some form of environmental influence).

MR works best when there is no pleiotropic effect of the regulatory allelic variation, when the gene has a strong, highly focused, effect and when there is a well-defined intermediate phenotype. If the risk effect is weak, huge samples will be required – with all of the hazards involved in combining samples from different sources. Accordingly, in spite of its numerous attractions, the MR strategy has important practical limitations.

18.4.5.2 Early puberty

A different instrumental variable approach is provided by unusually early puberty. It constitutes a 'natural experiment' because an unusually early puberty in girls has been shown to be associated with an increased rate of drinking alcohol and of drunkenness during the teenage years (Grant and Dawson, 1997). In the general population, early use of alcohol has been strongly associated with an increased risk of alcoholism in adult life. The causal hypothesis, therefore, is that early alcohol use contributes to the causation of later alcoholism. Early puberty constitutes an instrumental variable because it is outside the control of the individual (to avoid allocation bias) and because it affects the outcome by some means that is independent of the usual liability for alcoholism (in order to avoid both genetic mediation and possible social confounding). Both alcoholism and the timing of puberty are genetically influenced (the latter especially so), but the genetic influences in the two cases are likely to be quite different.

Three separate large-scale epidemiological longitudinal studies in Sweden, New Zealand and Finland were consistent in their finding that early puberty was indeed accompanied by an increased heavy use of alcohol in adolescence, but there was no such increase in use of alcohol or alcohol abuse in adult life (see, for example, Stattin and Magnusson, 1990). The indication is that early use was not a cause; rather, other evidence suggested that early alcohol use and alcoholism were associated through a shared genetic liability that incorporated a wide range of problem behaviours (McGue and Iacano, 2005).

18.4.6 Regression discontinuity (RD) designs to deal with unmeasured confounders

The only strategy (other than randomisation) that can deal adequately with unmeasured confounders is the regression discontinuity (RD) design (see Thistlethwaite and Campbell, 1960). Because it was designed for planned interventions, it has rather limited utility as a 'natural experiment' to evaluate naturally occurring causes. Nevertheless, it has been used for this purpose. The rationale is that the treatment allocation uses a predetermined cut-off rather than randomisation. The effects are measured in terms of a discontinuity in regression lines, instead of a difference in means. More than most other designs, it involves quite strict sampling and statistical criteria. All participants must be from the same population prior to assignment;

the assignment must adhere to a strict predetermined cut-off not open to manipulation; there must be accurate specification of the form of the intervention effect (e.g. whether linear or curvilinear); and the analysis must include an interaction term when that is relevant.

A single example of RD usage illustrates its potential, but also underlines the key methodological steps that must be taken. Cahan and Cohen (1989) used the design to pit against each other two competing alternatives for causal influences on cognitive performance. First, the main effect could come from extent of school experiences and, second, it could come from increasing biological maturity (as indexed by age). The ‘natural experiment’ opportunity arose with a school system in which all admissions occurred on a single date in the year. The consequence was that, within each school year group, there was a 12 month age span from the oldest to the youngest (all of whom would have received the same amount of schooling). Equally, between school year groups there was a 12 month difference in duration of schooling (but a similar within-group variation in age). In order to exclude biases due to the rare instances of grade retention and grade skipping, all above – and below – age children were excluded and also all children born in the two months prior to the admission date. The findings showed that the school effect exceeded the age effect for 10 out of the 12 cognitive measures used. In other words, the RD design provided support for the postulated effect of duration of schooling on cognitive performance.

18.5 Overall conclusion on ‘natural experiments’

It would be a mistake to view ‘natural experiments’ simply as a list of specific designs that can provide a predetermined set of techniques to test causal inferences. Rather they constitute a diverse range of problem-solving procedures that need to be employed in a rigorous hypothesis-testing fashion, specifically tailored to be appropriate for both the causal question to be investigated and the particular methodological hazards that are implicit in the samples and circumstances of the individual research project. As discussed, none of the ‘natural experiment’ strategies are free of limitations but, fortunately, the advantages and the limitations are not the same for each design. Accordingly, as in the whole of science, greater weight can be placed on findings that are consistent across different strategies. This may be illustrated by considering several examples of both confirmation and disconfirmation of the postulated causal influences.

18.5.1 Supported causes

The effects on birth weight of prenatal exposure to maternal smoking has been confirmed through all the designs that have been employed – including discordant siblings, children of twins (CoT) and assisted reproduction techniques (ARTs). It is also relevant that animal models have both confirmed the reduction of birth weight and indicated the likely biological mechanisms. The strongly positive findings for birth weight contrast sharply with the consistently negative findings in the case of antisocial behaviour and ADHD (see below).

The effect of family experiences on language/cognition has been shown by both a twin–singleton design and an adoption design. The RD design also showed the impact of the duration of schooling. The effects of different forms of seriously abusive or depriving experiences has been shown in multivariate twin designs, discordant twin designs, a CoT design and adoption following a radical change of environment following institutional deprivation. Care needs to be taken in interpreting these findings because both the postulated causal influence and the target outcome differ across studies. Nevertheless, what has been shown is the viability of using 'natural experiments' to study socialisation effects. It should be added that the strength of the causal inference has been increased in several instances by the use of biological data (as in the adoptee study of children experiencing profound institutional deprivation) and animal models (again in relation to deprivation).

18.5.2 Disconfirmed causes

Prenatal exposure to maternal smoking as a cause of antisocial behaviour and ADHD constitutes the first example of disconfirmation. The negative findings have been consistent across three different designs – CoT, ART and discordant siblings (see Jaffee *et al.*, 2011 for a fuller description). What has clearly been shown is that the conventional regression techniques to take account of confounders were inadequate for the purpose. Each of the 'natural experiment' designs showed that the effects dropped to a statistically nonsignificant level once the key comparison was undertaken. It is highly unlikely that this effect arose by chance or was artefactual because the findings pointed to the positive effect of genetic mediation. What has to be less certain is whether there has been exclusion of a small effect from the prenatal exposure to maternal smoking. Much larger sample sizes would be required to eliminate the possibility of very small effects, but what is evident is that the confident claims of substantial effects based on naturalistic observational studies should be viewed with some scepticism. Deterring women from smoking in pregnancy would bring benefits with respect to birth weight but it should not be expected to make much difference to the rate of either antisocial behaviour or ADHD. It should be added that the animal studies demonstrating effects on birth weight did not show comparable effects on behaviour.

Much the same applies to the findings disconfirming the supposed gateway effects of early drinking of alcohol on later alcoholism, substance misuse and antisocial behaviour. The negative findings were consistent across 'natural experiment' designs based on multivariate twin analyses, discordant twins, Mendelian randomisation and unusually early puberty as an instrumental variable. The strength of the disconfirmation lies as much in positively showing the effects of a shared genetic liability as in disconfirming the gateway hypothesis. Note, however, that this negative finding refers to the lack of a gateway effect of early alcohol drinking. The findings for early drug use are rather different (Rutter, 2007; Odgers *et al.*, 2008).

Throughout the whole of the discussion of 'natural experiments' emphasis has been laid on the difficulties in identifying the key causal element, even when a causal effect has been identified. It is not that such accurate identification is impossible in 'natural experiments' but it requires careful planning. Also, it will always be necessary to use a combination of appropriate research strategies including animal models and human experiments when they are feasible. 'Natural experiments' constitute invaluable elements in testing causal hypotheses, particularly when several different strategies can be integrated. In this chapter, the focus has been on design, but all designs have to be combined with appropriate statistics in order to capitalise on the design opportunities. Such statistical approaches are discussed in other chapters of this volume.

Acknowledgement

The author thanks Andrew Pickles, Professor of Biostatistics and Psychological Methods at the Institute of Psychiatry, for his helpful suggestions and comments on an earlier draft.

References

Academy of Medical Sciences (2007) *Identifying the Environmental Causes of Disease: How should we decide what to believe and when to take action*? London: Academy of Medical Sciences.

Anderson, K.E., Lytton, H. and Romney, D.M. (1986) Mothers' interactions with normal and conduct-disordered boys: Who affects whom? *Developmental Psychology*, **22**, 604–609.

Baron, R.M. and Kenny, D.A. (1986) The moderator–mediator variable distinction in social psychological research: conceptual, strategic, and statistical considerations. *Journal of Personality and Social Psychology*, **51**, 1173–1182.

Bell, R.Q. (1968) A reinterpretation of the direction of effects in studies of socialization. *Psychological Review*, **75**, 81–95.

Bell, R.Q. and Harper, L.V. (1977) *Child Effects on Adults*. Hillsdale, NJ: Erlbaum.

Brown, G.W., Harris, T.O. and Bifulco, A. (1986) Long-term effects of early loss of parent, in *Depression in Childhood: developmental perspectives* (eds M. Rutter, C. Izard and P. Reed). New York: Guilford Press, pp. 251–296.

Cahan, S. and Cohen, N. (1989) Age versus schooling effects on intelligence development. *Child Development*, **60**, 1239–1249.

Caspi, A., Moffitt, T.E., Morgan, J., Rutter, M., Taylor, A., Arseneault, L., *et al.* (2004) Maternal expressed emotion predicts children's antisocial behavior problems: Using monozygotic-twin differences to identify environmental effects on behavioral development. *Developmental Psychology*, **40**, 149–161.

Champion, L. A., Goodall, G. and Rutter, M. (1995). Behavioural problems in childhood and stressors in early adult life. I: A 20-year follow-up of London school children. *Psychological Medicine*, **25**, 231–246.

Coid, J., Kirkbride, J.B., Barker, D., Cowden, F., Stamps, R., Yang, M. and Jones, P. B. (2008) Raised incidence rates of all psychoses among migrant groups: findings from the East London first episode psychosis study. *Archives of General Psychiatry*, **65** (11), 1250–1258.

Cook, T.D. and Campbell, D.T. (1979) *Quasi-experimentation: design and analysis issues for field settings*. Chicago: Rand-McNally.

Cornfield, J., Haenszel, W., Hammond, E., Lilienfeld, A., Shimkin, M. and Wynder, E. (1959) Smoking and lung cancer: recent evidence and a discussion of some questions. *Journal of the National Cancer Institute*, **22**, 173–203.

Costello, E. J., Compton, F. N., Keeler, G. and Angold, A. (2003) Relationships between poverty and psychopathology: A natural experiment. *Journal of American Medical Association*, **290**, 2023–2029.

Davey Smith, G. and Ebrahim, S. (2003) 'Mendelian randomization': Can genetic epidemiology contribute to understanding environmental determinants of disease? *International Journal of Epidemiology*, **32**, 1–22.

Davey Smith, G. and Ebrahim, S. (2005) What can Mendelian randomization tell us about modifiable behavioural and environmental exposures. *British Medical Journal*, **330**, 1076–1079.

Dawid, P. and Pickles, A. (2007) Appendix I: Statistics, in *Identifying the Environmental Causes of Disease: How should we decide what to believe and when to take action*? (ed. M. Rutter). London: Academy of Medical Sciences, pp. 93–102.

D'Onofrio, B.M., Turkheimer, E., Eaves, L.J., Corey, L.A., Berg, K., Solaas, M.H. and Emery, R.E. (2003) The role of the children of twins design in elucidating causal relations between parent characteristics and child outcomes. *Journal of Child Psychology and Psychiatry*, **44**, 1130–1144.

D'Onofrio, B.M., Van Hulle, C.A., Waldman, I.D., Rodgers, J.L., Harden, K.P., Rathouz, P.J. and Lahey, B.B. (2008) Smoking during pregnancy and offspring externalizing problems: an exploration of genetic and environmental confounds. *Development and Psychopathology*, **20**, 139–164.

Duyme, M., Dumaret, A.-C. and Tomkiewicz, S. (1999). How can we boost IQs of 'dull children'? A late adoption study. *Proceedings of the National Academy of Sciences, USA*, **96**, 8790–8794.

Eaves, L.J., Silberg, J.L. and Maes, H.H. (2005) Revisiting the children of twins: Can they be used to resolve the environmental effects of dyadic parental treatment on child behaviour? *Twin Research and Human Genetics*, **8**, 283–290.

Fairchild, A.J. and MacKinnon, D.P. (2009) A general model for testing mediation and moderation effects. Preventative Sci*ence*, **10**, 87–99.

Fergusson, D.M., Horwood, L.J. and Lynskey, M.T. (1992) Family change, parental discord and early offending. *Journal of Child Psychology and Psychiatry*, **33**, 1059–1075.

Foster, E.M. and McLanahan, S. (1996) An illustration of the use of instrumental variables: Do neighborhood conditions affect a young person's chance of finishing high school? *Psychological Methods*, **1**, 249–260.

Ge, X., Conger, R.D., Cadoret, R.J., Neiderhiser, J.M., Yates, W., Troughton, E. and Stewart, M.A. (1996) The developmental interface between nature and nurture: a mutual influence model of child antisocial behavior and parenting. *Developmental Psychology*, **32**, 574–589.

Gill, R.D. and Robins, J.M. (2001) Causal inference for complex longitudinal data: the continuous case. *Annals of Statistics*, **29**, 1785–1811.

Grant, B.F. and Dawson, D.A. (1997) Age at onset of alcohol use and its association with DSM-IV alcohol abuse and dependence: results from the National Longitudinal Alcohol Epidemiologic Survey. *Journal of Substance Abuse*, **9**, 103–110.

Haviland, A.M., Rosenbaum, P.R., Nagin, D.S. and Tremblay, R.E. (2008) Combining group-based trajectory modelling and propensity score matching for causal inferences in nonexperimental longitudinal data. *Developmental Psychology*, **44**, 422–436.

Honda, H., Shimizu, Y. and Rutter, M. (2005) No effect of MMR withdrawal on the incidence of autism: a total population study. *Journal of Child Psychology and Psychiatry*, **46**, 572–579.

Irons, D.E., McGue, M., Iacono, W.G. and Oetting, W.S. (2007) Mendelian randomization: a novel test of the gateway hypothesis and models of gene–environment interplay. *Development and Psychopathology*, **19**, 1181–1195.

Jaffee, S.R. (2011) Genotype–environment correlations: definitions, methods of measurement, and implications for research on adolescent psychopathology, in *The Dynamic Genome and Mental Health: the role of genes and environments in youth development* (eds K.K. Kendler, S.R. Jaffee and D. Romer), New York: Oxford University Press, pp. 79–102.

Jaffee, S. R. and Price, T. S. (2007) Gene–environment correlations: a review of the evidence and implications for prevention of mental illness. *Molecular Psychiatry*, **12**, 432–442.

Jaffee, S.R., Caspi, A., Moffitt, T.E., Polo-Tomas, M., Price, T.S. and Taylor, A. (2004) The limits of child effects: evidence for genetically mediated child effects on corporal punishment but not on physical maltreatment. *Developmental Psychology*, **40**, 1047–1058.

Jaffee, S.R., Strait, L.B. and Odgers, C.L. (2011) From correlates to causes: Can quasi-experimental studies and statistical innovations bring us closer to identifying the causes of antisocial behaviour? *Psychological Bulletin* [Oct 24 Epub ahead of print].

Jones, P.B. and Fung, W.L.A. (2005) Ethnicity and mental health – the example of schizophrenia in the African Caribbean population in Europe, in *Ethnicity and Causal Mechanisms* (eds M. Rutter and M. Tienda). Cambridge: Cambridge University Press, pp. 227–261

Katan, M.B. (1986) Apolipoprotein E isoforms, serum cholesterol, and cancer. *Lancet*, **1**, 507–508.

Kendler, K.S. and Baker, J.H. (2007) Genetic influences on measures of the environment: a systematic review. *Psychological Medicine*, **37**, 615–626.

Kendler, K.S. and Prescott, C.A. (2006) *Genes, Environment, and Psychopathology: understanding the causes of psychiatric and substance use disorders*. New York: Guilford Press.

Lynch, S.K., Turkheimer, E., D'Onofrio, B.M., Mendle, J. and Emery, R.E. (2006) A genetically informed study of the association between harsh punishment and offspring behavioural problems. *Journal of Family Psychology*, **20**, 190–198.

McClellan, J.M., Susser, E. and King, M.C. (2006) Maternal famine, de novo mutations, and schizophrenia. *Journal of the American Medical Association*, **296**, 582–584.

McGue, M. and Iacono, W.G. (2005) The association of early adolescent problem behavior with adult psychopathology. *American Journal of Psychiatry*, **162**, 1118–1124.

McKim, V.R. and Turner, S.P. (eds) (1997) *Causality in Crisis?: statistical methods and the search for causal knowledge in the social sciences*. Notre Dame, IN: University of Notre Dame Press.

MacKinnon, D.P., Fairchild, A.J. and Fritz, M.S. (2007) Mediation analysis. *Annual Review of Psychology*, **58**, 593–614.

Maes, H. H., Woodard, C.E., Murrelle, L., Meyer, J.M., Silberg, J.L., Hewitt, J.K., *et al.* (1999) Tobacco, alcohol and drug use in eight- to 16-year-old twins: the Virginia twin study of adolescent behavioral development. *Journal of Studies on Alcohol*, **60**, 293–305.

Marmot, M.G. and Syme, S.L. (1976) Acculturation and coronary heart disease in Japanese–Americans. *American Journal of Epidemiology*, **104**, 225–247.

Maynard, R.A. (ed.) (1997) *Kids Having Kids: economic costs and social consequences of teen pregnancy*. Washington, DC: Urban Institute Press.

Melhuish, E., Belsky, J., Leyland, A.H. and Barnes, J. (2008) Effects of fully-established Sure Start Local Programmes on 3-year-old children and their families living in England: a quasi-experimental observational study. National Evaluation of Sure Start Research Team. *Lancet*, **8**, 1641–1647.

Miech, R.A., Caspi, A., Moffitt, T.E., Entner Wright, B.R. and Silva, P.A. (1999) Low socio-economic status and mental disorders: a longitudinal study of selection and causation during young adulthood. *American Journal of Sociology*, **104**, 1096–1131.

Moffitt, T.E. and The E-Risk Study Team. (2002) Teenaged mothers in contemporary Britain. *Journal of Child Psychology and Psychiatry*, **43**, 727–742.

Nagin, D.S., Barker, T., Lacourse, E. and Tremblay, R.E. (2008) The interrelationship of temporally distinct risk markers and the transition from childhood physical aggression to adolescent violent delinquency, in Applied Data Techniques for Turning Points Research. New York: Routledge, Chapter 2, pp. 17–36.

Obel, C., Olsen, J., Brink Heriksen, T., Rodriguez, A., Jarvelin, M-R., Moilanen, R., Parner, E., Markussen Linnet, K., Taanila, A., Ebeling, H., Heiervang, E. and Gissler, M. (2011) Is maternal smoking during pregnancy a risk factor for hyperkinetic disorder? Findings from a sibling design. *International Journal of Epidemiology*, **40**, 338–345.

O'Connor, T.G., Deater-Deckard, K., Fulker, D., Rutter, M. and Plomin, R. (1998) Genotype–environment correlations in late childhood and early adolescence: antisocial behavioral problems and coercive parenting. *Developmental Psychology*, **34**, 970–981.

Odgers, C.L., Caspi, A., Nagin, D.S., Piquero, A.R., Slutske, W.S., Milne, B.J., Dickson, N., Poulton, R. and Moffitt, T.E. (2008) Is it important to prevent early exposure to drugs and alcohol among adolscents? *Psychological Science*, **19**, 1037–1044.

Pearl, J. (1995) Causal diagrams for empirical research, *Biometrika*, **82**, 669-688.

Pike, A., McGuire, S., Hetherington, E.M., Reiss, D. and Plomin, R. (1996) Family environment and adolescent depression and antisocial behaviour: a multivariate genetic analysis. *Developmental Psychology*, **32**, 590–603.

Prescott, C.A. and Kendler, K.S. (1999) Age at first drink and risk for alcoholism: a noncausal association. *Alcoholism: Clinical and Experimental Research*, **23**, 101–107.

Rice, F., Harold, G.T., Boivin, J., Hay, D.F., van den Bree, M. and Thapar, A. (2009) Disentangling prenatal and inherited influences in humans with an experimental design. *Proceedings of the National Academy of Science USA*, **106**, 2464–2467.

Robins, L. (1966) *Deviant Children Grown Up*. Baltimore: Williams & Wilkins.

Robins, J.M., Hernán, M.A. and Brumback, B. (2000) Marginal structural models and causal inference in epidemiology. *Epidemiology*, **11**, 550–560.

Rutter, M. (1983) Statistical and personal interactions: facets and perspectives, in: *D. Human Development: An Interactional Perspective* (eds D. Magnusson and V. Allen). New York: Academic Press, pp. 295–319.

Rutter, M. (2005) Incidence of autism spectrum disorders: changes over time and their meaning. *Acta Paediatrica*, **94**, 2–15.

Rutter, M. (2006) *Genes and Behavior: nature–nurture interplay explained*. Oxford, UK: Blackwell.

Rutter, M. (2007) Proceeding from observed correlation to causal inference: the use of natural experiments. *Perspectives on Psychological Science*, **2**, 377–395.

Rutter, M. (2009) Epidemiological methods to tackle causal questions. *International Journal of Epidemiology*, **38**, 3–6.

Rutter, M. and Azis-Clauson, C. (2011) Extreme deprivation, in *The SAGE Handbook of Developmental Disorders* (eds P. Howlin, T. Charman and M. Ghaziuddin), SAGE Publications Ltd., pp. 529–550.

Rutter, M. and Sonuga-Barke, E.J. (eds) (2010) Deprivation-specific psychological patterns: effects of institutional deprivation. *Monographs of the Society for Research in Child Development*, **75** (1).

Rutter, M., Pickles, A., Murray, R. and Eaves, L. (2001) Testing hypotheses on specific environmental causal effects on behaviour. *Psychological Bulletin*, **127**, 291–324.

Rutter, M., Thorpe, K., Greenwood, R., Northstone, K. and Golding, J. (2003). Twins as a natural experiment to study the causes of mild language delay: I. Design; twin–singleton differences in language, and obstetric risks. *Journal of Child Psychology and Psychiatry*, **44**, 326–334.

Sampson, R.J., Laub, J.H. and Wimer, C. (2006) Does marriage reduce crime? A counterfactual approach to within-individual causal effects. *Criminology*, **44**, 465–508.

Shadish, W.R., Cook, T.D. and Campbell, D.T. (2002). *Experimental and Quasi-experimental Designs for Generalized Causal Inference*. Boston, MA: Houghton Mifflin.

Silberg, J.L. and Eaves, L.J. (2004) Analysing the contributions of genes and parent–child interaction to childhood behavioural and emotional problems: a model for the children of twins. *Psychological Medicine*, **34**, 347–356.

Silberg, J.L., Maes, H. and Eaves, L.J. (2010) Genetic and environmental influences on the transmission of parental depression to children's depression and conduct disturbance: an extended children of twins study. *Journal of Child Psychology and Psychiatry*, **51**, 734–744.

Spergel, I.A. (1990) Youth gangs: continuity and change, in *Crime and Justice: a review of research* (eds M. Tonry and N. Morris), vol. 12. Chicago: University of Chicago Press, pp. 171–275.

St Clair, D., Xu, M., Wang, P., Yu, Y., Fang, Y., Zhang, F., *et al.* (2005) Rates of adult schizophrenia following prenatal exposure to the Chinese famine of 1959–1961. *Journal of the American Medical Association*, **294**, 557–562.

Stattin, J. and Magnusson, D. (1990) *Paths Through Life*, vol. 2, *Pubertal Maturation in Female Development*. Hillsdale, NJ: Erlbaum.

Stein, Z.A., Susser, M., Saenger, G. and Marolla, F. (1975) Famine and human development: the Dutch hunger winter of 1944–1945. New York: Oxford University Press.

Stillman, S., McKenzie, D. and Gibson, J. (2009) Migration and mental health: evidence from a natural experiment. *Journal of Health Economics*, **3**, 677–687.

Stoolmiller, M. (1999) Implications of the restricted range of family environments for estimates of heritability and nonshared environment in behaviour-genetic adoption studies. *Psychological Bulletin*, **125**, 392–409.

Thapar, A., Harold, G., Rice, F., Ge, X., Boivin, J., Hay, D., van den Bree, M. and Lewis, A. (2007) Do intrauterine or genetic influences explain the foetal origins of chronic disease? A novel experimental method for disentangling effects. *BMC Medical Research Methodology*, **7**, 25.

Thapar, A., Rice, F., Hay, D., Boivin, J., Langley, K., van den Bree, M., Rutter, M., & Harold, G. (2009) Prenatal smoking might not cause attention-deficit/hyperactivity disorder: evidence from a novel design. *Biological Psychiatry*, **66**, 722–727.

Thistlewaite, D.L. and Campbell, D.T. (1960) Regression-discontinuity analysis: an alternative to the ex-post facto experiment. *Journal of Educational Psychology*, **51**, 309–317.

Thornberry, T.P., Krohn, M.D., Lizotte, A.J. and Chard-Wiershem, D. (1993). The role of juvenile gangs in facilitating delinquent behavior. *Journal of Research in Crime and Delinquency*, **30**, 55–87.

Thorpe, K., Rutter, M. and Greenwood, R. (2003) Twins as a natural experiment to study the causes of mild language delay: II. Family interaction risk factors. *Journal of Child Psychology and Psychiatry*, **44**, 342–355.

Valencia, M.E., Bennett, P.H., Ravussin, E., Esparza, J., Fox, C. and Schulz, L.O. (1999) The Pima Indians in Sonora, Mexico. *Nutrition Reviews*, **57**, S55–S58.

Vitaro, F., Brendgen, M. and Arseneault, L. (2009) The discordant MZ-twin method: one step closer to the holy grail of causality. *International Journal of Behavioral Development*, **33**, 376–382.

Wakschlag, L.S., Pickett, K.E., Middlecamp, M.K., Walton, L.L., Tenzer, P. and Leventhal, B.L. (2003) Pregnant smokers who quit, pregnant smokers who don't: Does history of problem behaviour make a difference? *Social Science and Medicine*, **56**, 2449–2460.

19

Nonreactive and purely reactive doses in observational studies

Paul R. Rosenbaum
Wharton School, University of Pennsylvania, Philadelphia, Pennsylvania, USA

19.1 Introduction: Background, example

19.1.1 Does a dose–response relationship provide information that distinguishes treatment effects from biases due to unmeasured covariates?

In an observational study, an effort is made to draw inferences about the effects caused by a treatment when subjects are not randomly assigned to treatment or control as they would be in a randomized trial. Typically, an observational study begins by selecting circumstances in which treatment assignment, though not actually random, seems relatively haphazard. With that first step completed, an attempt is made to compare treated and control subjects who look comparable in terms of measured pretreatment covariates, for instance by matching for these covariates. In the absence of random assignment, even very good efforts at these first two tasks may fail to produce treated and control groups that were comparable prior to treatment in terms of covariates that were not measured, so differing outcomes in these groups may not be effects of the treatment.

It is often said (e.g., Hill, 1965), and it is equally often denied (e.g., Rothman, 1986, p. 18, #5; Weiss, 1981), that observing a dose–response relationship – that is, larger ostensible effects at larger doses – provides evidence relevant to judging whether an association is causal or whether it is due to bias from covariates not measured. Those who deny the relevance of a dose–response relationship typically do so on the grounds that an unobserved covariate

Causality: Statistical Perspectives and Applications, First Edition. Edited by Carlo Berzuini, Philip Dawid and Luisa Bernardinelli.

correlated with both dose and response can produce a spurious dose–response in the absence of a treatment effect.

19.1.2 Is more chemotherapy for ovarian cancer more effective or more toxic?

Silber *et al.* (2007) asked if more intense chemotherapy for ovarian cancer prolonged survival and if it increased toxicity. Most of the variation in the treatment of patients is a response to variation in the health of patients, but now and then one can find a source of variation in treatment that is not a reaction to the patient. Ovarian cancer is unusual: there is a source of meaningful variation in the intensity of chemotherapy for ovarian cancer that is not primarily a response to the patient herself. Unlike most forms of cancer, chemotherapy for gynecological cancers, including ovarian cancer, may be provided by either a medical oncologist (MO) who treats cancers of all kinds or by a gynecological oncologist (GO), who is an gynecologist (and hence a surgeon) with additional specialized training in the treatment of gynecological cancers. Silber *et al.* (2007) hypothesized correctly that MOs would use chemotherapy more intensively than GOs, and sought to use that variation in the intensity of chemotherapy to study the effects of increased intensity. This is the first step, mentioned in Section 19.1.1: an attempt has been made to find circumstances in which the assignment of treatments, here the intensity or dose of chemotherapy, is determined by something that is somewhat haphazard – here, treatment by an MO rather than a GO.

Using a matching method described in Rosenbaum *et al.* (2007), they matched all 344 ovarian cancer patients in the SEER-Medicare data base from 1991 to 1999 who were treated by a GO to 344 similar patients who were treated by an MO. (Medicare is the US government program that provides health care to people over the age of 65, and SEER is the Surveillance, Epidemiology and End Results program of the US National Cancer Institute, which collects clinical data about cancer patients at several locations in the US.) The matching controlled for 36 covariates, $\mathbf{x}$, specifically surgeon type, clinical stage, tumor grade, age, race, eight SEER sites, year of diagnosis, and numerous comorbid conditions including chronic obstructive pulmonary disease, hypertension, diabetes, and congestive heart failure; see Table 19.1 for balance on a few of the 36 covariates. Notice in Table 19.1 that the two most important clinical variables, stage and grade, were not much out of balance before or after matching, consistent with the notion that treatment by a GO rather than an MO is a somewhat haphazard event not indicative of an unusual type of patient. Conversely, GOs often gave chemotherapy to patients after performing surgery on those same patients, so surgeon type in Table 19.1 was enormously out of balance before matching.

The matching algorithm in Rosenbaum *et al.* (2007) used a propensity score, a covariate distance, and a fine balance constraint. The propensity score tends to balance many observed covariates stochastically, leaving behind chance imbalances, while use of the Mahalanobis distance creates pairs who were individually similar in terms of the most important covariates, such as clinical stage. The fine balance constraint ensured that GO and MO groups had the same number of patients from each SEER site in each time period without trying to match patients individually for these covariates that were clearly less important than clinical stage.

It is, of course, easy to think of covariates, u, that were not measured and not controlled by matching, which might differ in GO and MO groups (e.g., Cannistra, 2007). Chemotherapy was measured by weeks with a chemotherapy administration as recorded by Medicare, and toxicity was measured by weeks with an ICD-9 diagnosis code for chemotherapy associated

Table 19.1 Baseline comparison of a few key covariates for all 344 patients of a gynecologic oncologist (GO), 344 matched patients of a medical oncologist (MO), and all 2011 patients of medical oncologists. Not shown in this table, a total of 36 covariates were similarly balanced.

		GO $n = 344$	Matched-MO $n = 344$	All-MO $n = 2011$
Age	Mean	72.2	72.2	72.8
Stage %	I	9	9	9
	II	11	9	9
	III	51	53	47
	IV	26	26	31
	Missing	3	2	3
Tumor grade %	1	5	4	4
	2	16	13	17
	3	52	55	47
	4	9	8	11
	Missing	18	20	21
Surgeon type %	GO	76	75	33
	Gyn	15	16	39
	General	8	8	28

adverse events, such as anemia, neutropenia, thrombocytopenia (see Table 19.1 in Silber *et al.*, 2007, for detailed definitions and specific codes).

As expected, MOs delivered on average 4.4 more weeks of chemotherapy than did GOs. Survival was virtually identical in the MO and GO matched groups, as was duration of follow-up (see Table 19.2). Figure 19.1 plots the MO-minus-GO matched pair difference in weeks with toxicity against the difference in weeks with chemotherapy, with the medians, quartiles, eights, sixteenths, and extremes indicated in the top and right margins. The median

Table 19.2 Survival in 344 matched pairs.

	GO patients	MO patients
1 Year survival %	86.6	87.5
95 % CI	[83.0, 90.2]	[84.0, 90.1]
5 Year survival %	35.1	34.2
95 % CI	[30.0, 40.2]	[29.2, 39.3]
Median survival (years)	3.04	2.98
95 % CI	[2.50, 3.40]	[2.69, 3.67]
Number at risk year 0	344	344
Number at risk year 2	223	230
Number at risk year 4	133	128

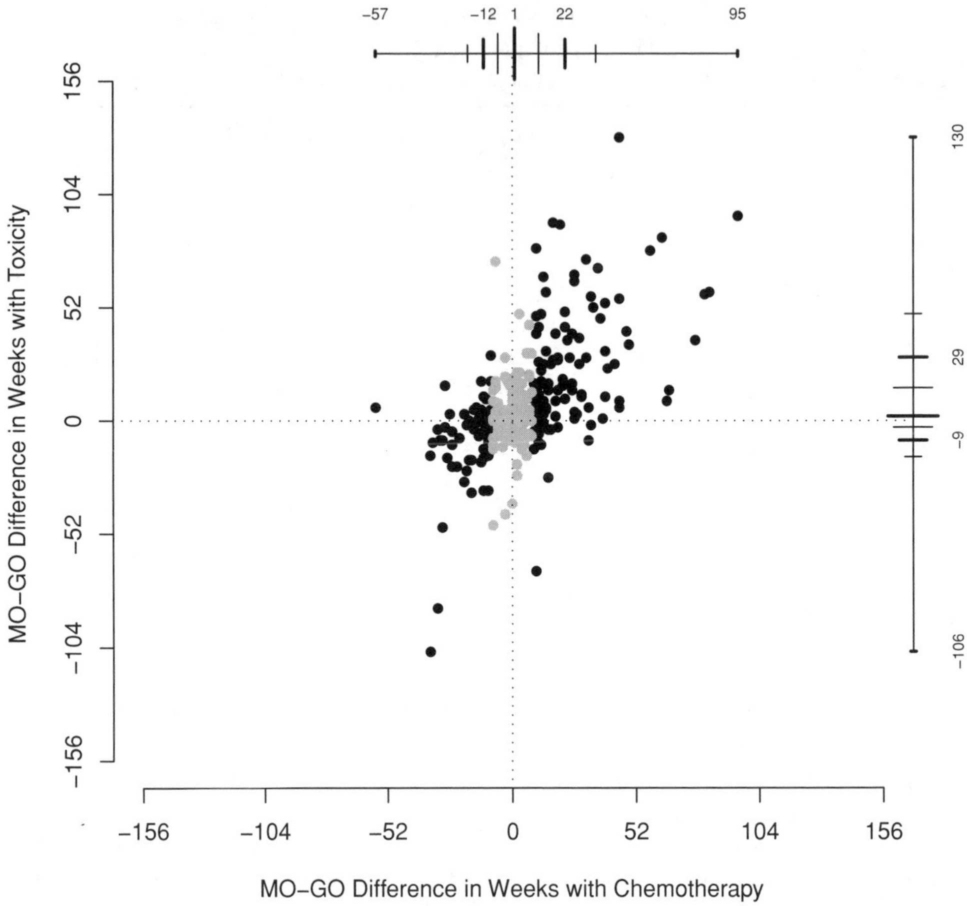

Figure 19.1 *For 344 matched pairs of patients with ovarian cancer, one treated by a medical oncologist (MO) and the other treated by a gynecological oncologist (GO), the MO–GO difference in weeks with toxicity is plotted against the difference in weeks with chemotherapy. The medians, quartiles, eighths, sixteenths, and extremes are indicated in the margins. Points with a difference in chemotherapy weeks between −8 and 8 are shown in grey.*

differences are small, 1 additional week for chemotherapy, 2 additional weeks for toxicity, but at the upper and lower eights the differences are substantial, +22 weeks versus −9 weeks for chemotherapy, +29 versus −9 weeks for toxicity. A difference of 26 weeks is half a year of additional weekly, clinically reportable toxicity, with no apparent compensating difference in survival. There is also a strong dose–response relationship in Figure 19.1: when there is a large difference in chemotherapy weeks, there tends also to be a large difference in the same direction in toxicity weeks.

For reasons to be discussed later, in Figure 19.1, some points appear in grey, others in black. Points with an absolute difference in chemotherapy weeks of 8 weeks or less appear in grey, while those with 9 weeks or more appear in black; i.e., because weeks are recorded as integers, there is an open interval (−9, 9) on the horizontal axis within which points appear in grey. Here, 9 weeks is the median of the absolute differences in chemotherapy weeks and 168 of the 344 pairs have an absolute difference in chemotherapy weeks less than 9 weeks.

The study by Silber *et al.* (2007) viewed the higher dose of chemotherapy given by MOs as an aspect of receiving chemotherapy by an MO rather than a GO; i.e., MOs treat patients more aggressively than do GOs. In the published discussion (e.g., Cannistra, 2007), the possibility was raised that, no, MOs do not treat more aggressively; rather, despite matching for observed covariates such as clinical stage, the patients of MOs are sicker in unmeasured ways, and that is why MOs used more chemotherapy. This alternative view says that, had the patients of MOs gone to GOs instead, the GOs would have recognized them as sicker than their typical patients and would in turn have used more chemotherapy. This alternative view says that the dose reflects the needs of patients typically treated by MOs and GOs, not differing styles of practice of MOs and GOs. Later, this alternative view of the dose will be formalized by saying that the doses are purely reactive. To say that the dose is purely reactive is to say that health care providers all treat the same patient in the same way, and variation in the doses of chemotherapy are simply reactions to variations in the needs of patients. Stated informally, a purely reactive dose is an aspect of the patient, not of the treatment applied to the patient. A counterclaim that a dose is purely reactive creates certain possibilities for analysis that are relevant to the sensitivity of conclusions to biases from unmeasured covariates (see Sections 19.2.5 and 19.3).

19.2 Various concepts of dose

19.2.1 Some notation: Covariates, outcomes, and treatment assignment in matched pairs

There are I matched pairs, $i = 1, \ldots, I$, of two subjects, $j = 1, 2$, one treated, denoted $Z_{ij} = 1$, the other the control, denoted $Z_{ij} = 0$, so $Z_{i1} + Z_{i2} = 1$ for each i. In Section 19.1.1, $Z_{ij} = 1$ for a patient treated by an MO, $Z_{ij} = 0$ for a patient treated by an GO, and $I = 344$. The pairs have been matched for observed covariates, $\mathbf{x}_{ij}$, so $\mathbf{x}_{i1} = \mathbf{x}_{i2} = \mathbf{x}_i$, say for each i, but there is concern that the matching failed to control an unmeasured covariate u_{ij}, so possibly $u_{i1} \neq u_{i2}$ for some or all i. In Section 19.1.1, $\mathbf{x}$ included 36 covariates, including clinical stage, surgeon type, etc., but Cannistra (2007) raised concern about the unrecorded covariate 'residual disease status', that is, the magnitude u of any residual tumors that existed at the time of surgery but that was not removed by surgery. In patients matched for clinical stage, it is not clear whether residual disease status is related to the type of physician, MO or GO, who provided chemotherapy after surgery, but Cannistra suggested this was possible.

Each subject has two potential responses to treatment, namely r_{Tij} if ij receives treatment with $Z_{ij} = 1$ or r_{Cij} if ij receives control with $Z_{ij} = 0$, so the response observed from ij is $R_{ij} = Z_{ij} r_{Tij} + \left(1 - Z_{ij}\right) r_{Cij}$ and the effect caused by the treatment, namely $r_{Tij} - r_{Cij}$, is not observed for any subject (Neyman, 1923; Rubin, 1974). In Figure 19.1, the response is weeks with chemotherapy related toxicity. Fisher's (1935) sharp null hypothesis of no treatment effect asserts that $H_0 : r_{Tij} = r_{Cij}, i = 1, \ldots, I, j = 1, 2$. In pair i, the treated-minus-control difference in observed responses is

$$Y_i = (Z_{i1} - Z_{i2})(R_{i1} - R_{i2}) = Z_{i1}(r_{Ti1} - r_{Ci2}) + Z_{i2}(r_{Ti2} - r_{Ci1}) \tag{19.1}$$

which equals $(Z_{i1} - Z_{i2})(r_{Ci1} - r_{Ci2}) = \pm(r_{Ci1} - r_{Ci2})$ if H_0 is true.

Write $\mathbf{Z} = (Z_{11}, Z_{12}, \ldots, Z_{I2})^T$ for the $2I$-dimensional vector of treatment assignments and write $\mathcal{Z}$ for the set containing the 2^I possible values of $\mathbf{Z}$, so $\mathbf{z} \in \mathcal{Z}$ if $\mathbf{z} = (z_{11}, z_{12}, \ldots, z_{I2})^T$ with each $z_{ij} = 1$ or $z_{ij} = 0$ and $z_{i1} + z_{i2} = 1$. With a slight abuse of notation, in a conditional probability, the event $\mathbf{Z} \in \mathcal{Z}$ is abbreviated to $\mathcal{Z}$; it signifies that reference is made to I pairs, matched for $\mathbf{x}$, with one treated and one control subject.

19.2.2 Reactive and nonreactive doses of treatment

A patient may receive relatively more chemotherapy because her oncologist is aggressive, or because things are going well, or because things are going poorly. If the patient finds chemotherapy tolerable – that is, if things are going well – she may be given more chemotherapy than if she finds chemotherapy intolerable. If her cancer recurs as a consequence of metastases – that is, if things are going poorly – she may receive additional weeks of chemotherapy. If her condition is hopeless, then chemotherapy may be discontinued. Moreover, different individual physicians and patients may judge these issues differently. Receiving care from an MO rather than a GO may have some haphazard aspects related to affiliation or access, but the dose of chemotherapy may be related in a complex way to the patient's health and preferences.

In Figure 19.1, the dose of chemotherapy refers to weeks with a chemotherapy administration. Suppose that patient ij would have received w_{Tij} weeks of chemotherapy if treated by the MO in pair i and would have received w_{Cij} weeks of chemotherapy if treated by the GO in pair i, so this patient actually received $W_{ij} = Z_{ij} w_{Tij} + \left(1 - Z_{ij}\right) w_{Cij}$ weeks of chemotherapy. The horizontal axis in Figure 19.1 is the MO-minus-GO difference in weeks with chemotherapy in pair i, or

$$D_i = (Z_{i1} - Z_{i2})(W_{i1} - W_{i2}) = Z_{i1}(w_{Ti1} - w_{Ci2}) + Z_{i2}(w_{Ti2} - w_{Ci1}) \qquad (19.2)$$

In thinking about D_i in (19.2), three possibilities need to be distinguished.

One possibility is that the variation is weeks of chemotherapy reflects both baseline differences in the condition of patients – stage, grade, and age, for instance – and differences in the styles of practice of different MOs and GOs. If this were true, then perhaps the MO in pair i would deliver the same number of weeks of chemotherapy to any patient of the same age, stage, grade, etc., as the two patients in pair i, as recorded in $\mathbf{x}_i$, and in this case $w_{Ti1} = w_{Ti2} = w_{Ti}$, say. If the GO in pair i also treats patients based solely on clinical characteristics in $\mathbf{x}_i$, then $w_{Ci1} = w_{Ci2} = w_{Ci}$, say. If this is true for both the MO and the GO in all pairs i, then the dose is an attribute of the providers of care when faced with a patient having clinical characteristics $\mathbf{x}_i$. The term 'nonreactive dose' might be used for situations like this. With a nonreactive dose, $D_i = w_{Ti} - w_{Ci}$ in (19.2) and so the treated-minus-control difference in doses D_i does not change with Z_{ij}; that is, it does not react to which patient is treated and which is the control. Speaking informally, a nonreactive dose is purely an aspect of the treatment that is applied to the patient.

A different possibility is that the MO and GO in pair i would give the same number of weeks of chemotherapy to patient $i1$ and they would give the same number of weeks of chemotherapy to patient $i2$; however, patients $i1$ and $i2$ would receive different numbers of weeks of chemotherapy, even though they appeared similar at baseline in terms of stage, grade, age, etc., $\mathbf{x}_{i1} = \mathbf{x}_{i2} = \mathbf{x}_i$. For instance, it might be that patient $i1$ finds chemotherapy difficult to tolerate while $i2$ does not, so $i1$ would receive fewer weeks of chemotherapy no matter who provided the chemotherapy. One might describe this as a 'purely reactive dose', because variation in doses reflects only variations in the patients, not variation in the providers of the treatment. If this were true, then $w_{Ti1} = w_{Ci1} = w_{i1}$, say, $w_{Ti2} = w_{Ci2} = w_{i2}$, say, and $D_i = (Z_{i1} - Z_{i2})(w_{i1} - w_{i2}) = \pm(w_{i1} - w_{i2})$. With a purely reactive dose, the $|D_i| = |w_{i1} - w_{i2}|$ would not change if the two patients in pair i swapped oncologists, but the MO-minus-GO difference in dose would change sign.

The third possibility is that the dose D_i is neither purely reactive nor nonreactive, so neither simplification in D_i occurs. This might be the case if GOs were more likely than MOs to reduce the dose of chemotherapy when patients found chemotherapy intolerable. Alternatively, MOs and GOs might provide similar initial doses of chemotherapy for ovarian cancer, but if the cancer recurs, perhaps MOs will treat the recurrence more aggressively than would GOs. In both cases, MOs and GOs are reacting differently to the reactions of different patients.

A nonreactive dose is an aspect of the treatment. A purely reactive dose is an aspect of the experimental subject. A dose that is neither nonreactive nor purely reactive is an outcome of applying a particular treatment to a particular subject.

To anticipate, if someone were to say that MOs and GOs would treat the same patient in the same way, and the pattern of doses and toxicities in Figure 19.1 is produced by a bias in the way patients select into treatment by MOs and GOs, then what is implicitly being said is that the doses are purely reactive. This conversation is unlikely to occur in an experiment in which patients are randomly assigned to MOs or GOs, because randomization would prevent biased selection into treated and control groups. For this reason, the model of purely reactive doses may have a useful role to play in observational studies that it does not usefully play in randomized experiments.

Write $\mathcal{F} = \{(r_{Tij}, r_{Cij}, w_{Tij}, w_{Cij}, \mathbf{x}_{ij}, u_{ij}),\ i = 1, \ldots, I, j = 1, 2\}$. In all three cases, one simple alternative hypothesis to Fisher's null hypothesis of no effect asserts that the effect $r_{Tij} - r_{Cij}$ of the treatment, MO versus GO, on subject ij's toxicity is proportional to the effect on her dose of chemotherapy, $w_{Tij} - w_{Cij}$, with proportionality constant β_0, or $H_{\beta_0} : r_{Tij} - r_{Cij} = \beta_0 (w_{Tij} - w_{Cij})$. For purely reactive doses, $w_{Tij} = w_{Cij}$ and H_{β_0} is the same as Fisher's hypothesis H_0 of no effect. If H_{β_0} were true, then $r_{Tij} - \beta_0 w_{Tij} = r_{Cij} - \beta_0 w_{Cij}$, so the adjusted responses, $R_{ij} - \beta_0 W_{ij} = r_{Cij} - \beta_0 w_{Cij}$, are determined by β_0 and $\mathcal{F}$; therefore $Y_i - \beta_0 D_i = \epsilon_i$ where $\epsilon_i = (2Z_{i1} - 1)\{(r_{Ci1} - \beta_0 w_{Ci1}) - (r_{Ci2} - \beta_0 w_{Ci2})\}$. These observations can be used to test H_{β_0}, build confidence intervals for β, and obtain point estimates of β whether doses are nonreactive or purely reactive or neither (see, for instance, Rosenbaum, 1999, Section 5; 2010a, Section 5.3). Nonetheless, the issues in Section 19.2.3 arise in the choice of a test statistic.

19.2.3 Three test statistics that use doses in different ways

Consideration will be restricted to three very minor variations on Wilcoxon's signed rank statistic that use the doses in different ways. This restriction reflects my concern with sharply contrasting the three notions of a dose. In some other context, if one had settled on a particular concept of dose, then one might seek a best test in that context, but that is not the current concern. See Rosenbaum (2010a, Section 17.3) for comparisons of statistics involving nonreactive doses and see Rosenbaum (2010b, 2011a) for the relationship between the choice of a test statistic and sensitivity to unmeasured biases.

Let λ_κ be the κ quantile of the $|D_i|$ and let $\mathcal{S}_\kappa \subseteq \{1, 2, \ldots, I\}$ be the set of indices of pairs with $|D_i| \geq \lambda_\kappa$. In Figure 19.1, the black points are pairs with $i \in \mathcal{S}_{1/2}$. Let $q_{i,\kappa}$ be the rank of $|Y_i|$ among the $i \in \mathcal{S}_\kappa$ with average ranks for ties and let q_i be the rank of

$|Y_i|$ for $i \in \{1, 2, \ldots, I\}$, so $q_i = q_{i,0}$. Write $\text{sign}(y) = 1$ if $y > 0$, $\text{sign}(y) = 0$ if $y = 0$, $\text{sign}(y) = -1$ if $y < 0$. The three statistics are

$$T_w = \sum_{i=1}^{I} \text{sign}(Y_i)\, q_i, \quad T_e = \sum_{i \in \mathcal{S}_\kappa} \text{sign}(Y_i)\, q_{i,\kappa}, \quad \text{and} \quad T_n = \sum_{i \in \mathcal{S}_\kappa} \text{sign}(Y_i\, D_i)\, q_{i,\kappa} \qquad (19.3)$$

Here, T_w is simply the centered form of Wilcoxon's statistic (Noether, 1967, p. 46), and it asks whether the patients of MOs experienced more weeks with toxicity than the patients of GOs. The statistic T_e is almost the same, except attention is restricted to the subset $\mathcal{S}_\kappa$ of pairs in which the two patients received a difference in doses that is not extremely small, in the sense that $|D_i| \geq \lambda_\kappa$. The statistic T_n is almost the same as T_e, except that it is concerned with whether the difference in toxicity in pair i has the same sign as the difference in dose of chemotherapy.

Certain expressions are made less repetitive by noting that all three statistics can be written as $T = \sum_{i=1}^{I} \text{sign}(Y_i)\, h_i$. For Wilcoxon's T_w, take $h_i = q_i$, for T_e take $h_i = q_{i,\kappa}$ for $i \in \mathcal{S}_\kappa$ and $h_i = 0$ for $i \notin \mathcal{S}_\kappa$, and for T_n take $h_i = \text{sign}(D_i)\, q_{i,\kappa}$ for $i \in \mathcal{S}_\kappa$ and $h_i = 0$ for $i \notin \mathcal{S}_\kappa$.

As discussed in Section 19.2.4, even in a randomized experiment, the three statistics, T_w, T_e, and T_n, are valid under different assumptions about doses. The Wilcoxon statistic, T_w, was used in Silber *et al.* (2007) to examine toxicity, and in a randomized experiment it requires neither purely reactive nor nonreactive doses. In contrast, in a randomized experiment, T_n is valid with nonreactive doses but not purely reactive doses, while T_e is valid with either nonreactive or purely reactive doses. For this reason, neither T_n nor T_e is generally valid; however, T_e is useful in thinking about a claim that doses are nonreactive vis-à-vis a counterclaim that doses are purely reactive, because T_e is valid in either case. As seen in Sections 19.2.5 and 19.3, the choice of T_w or T_e also affects the sensitivity to unmeasured biases in an observational study.

19.2.4 Randomization inference in randomized experiments

In a paired randomized experiment, the Z_{i1} would be determined by I independent flips of a fair coin. The fairness of this coin requires much more than $\Pr(Z_{i1} = 1) = \frac{1}{2}$; it requires that the fall of this coin is unrelated to the attributes of the patients in pair i, specifically unrelated to their measured and unmeasured covariates and the potential responses $(r_{Tij}, r_{Cij}, w_{Tij}, w_{Cij}, \mathbf{x}_{ij}, u_{ij})$ they might exhibit if assigned to treatment or control. That is, the fair coin has $\Pr(Z_{i1} = 1 | \mathcal{F}, \mathcal{Z}) = \frac{1}{2}$ for each i, and moreover $\Pr(\mathbf{Z} = \mathbf{z} | \mathcal{F}, \mathcal{Z}) = 1/2^I$ for each $\mathbf{z} \in \mathcal{Z}$. See, for instance, Cox (2009) for a survey of the role of randomization and randomization inference in experiments.

Even in a randomized experiment, the three test statistics behave differently depending upon whether the doses are nonreactive, purely reactive, or neither. Simplest is the case of nonreactive doses. Consider testing Fisher's null hypothesis of no effect, $H_0 : r_{Tij} = r_{Cij}$, $i = 1, \ldots, I$, $j = 1, 2$, in a paired randomized experiment when the dose is nonreactive, so that $w_{Ti1} = w_{Ti2} = w_{Ti}$ and $w_{Ci1} = w_{Ci2} = w_{Ci}$ for each i. In this case, $D_i = w_{Ti} - w_{Ci}$ is a function of $\mathcal{F}$ and so is fixed by conditioning on $\mathcal{F}$; hence, the set $\mathcal{S}_\kappa$ is fixed by $\mathcal{F}$. Under H_0, the treated-minus-control difference in outcomes is $Y_i = (Z_{i1} - Z_{i2})(r_{Ci1} - r_{Ci2})$, so when $|Y_i| > 0$, Y_i is $\pm(r_{Ci1} - r_{Ci2})$, each with probability 1/2, and when $|Y_i| = 0$, $Y_i = 0$ with probability 1. Therefore, for each of the three test statistics, the null distribution of $T = \sum_{i=1}^{I} \text{sign}(Y_i)\, h_i$ is simply the distribution of the sum of I independent random variables where $\text{sign}(Y_i)\ h_i = \pm h_i$, each with probability $\frac{1}{2}$ if $|Y_i| > 0$, and $\text{sign}(Y_i)\ h_i = 0$ if $|Y_i| = 0$. Indeed, in the case of nonreactive doses in a randomized experiment, the null distributions of T_w, T_e, and T_n are each the usual null distribution of the centered version of Wilcoxon's statistic, albeit with different sample sizes. In the case of nonreactive doses, the logic of randomization inference permits any of T_w, T_e, and T_n to be used in a standard way to test H_0.

Suppose instead that the doses were purely reactive, so $w_{Ti1} = w_{Ci1} = w_{i1}$, $w_{Ti2} = w_{Ci2} = w_{i2}$, and $D_i = (Z_{i1} - Z_{i2})(w_{i1} - w_{i2})$. In this case, $|D_i|$ is fixed by conditioning on $\mathcal{F}$, so $\mathcal{S}_\kappa$ is also fixed, but D_i varies with Z_{i1}. Under the null hypothesis H_0, the argument in the previous paragraph generates the null distribution of T_w and T_e, but not of T_n. That is, even if the treatment, MO-versus-GO, affects neither the dose nor response, with purely reactive doses the dose and response may be correlated, invalidating the use of T_n. Indeed, under H_0 with purely reactive doses, there is nothing to stop $(w_{i1} - w_{i2})(r_{Ci1} - r_{Ci2}) > 0$ for all i, so that $Y_i D_i = (Z_{i1} - Z_{i2})^2\ (r_{Ci1} - r_{Ci2})\ (w_{i1} - w_{i2}) = (r_{Ci1} - r_{Ci2})\ (w_{i1} - w_{i2}) > 0$ for each i, and $T_\mathrm{n} = \sum_{i \in \mathcal{S}_\kappa} \text{sign}(Y_i\ D_i)\ q_i = \sum_{i \in \mathcal{S}_\kappa} q_i$ takes its maximum possible value in every one of the 2^I possible treatment assignments $\mathbf{z} \in \mathcal{Z}$. One would see $(r_{Ci1} - r_{Ci2})\ (w_{i1} - w_{i2}) > 0$ if in pair i the patient who accepted more weeks of chemotherapy experienced more weeks of toxicity, and that could certainly happen even if MOs and GOs were to treat each patient ij in the same way. If the doses are neither purely reactive nor nonreactive, then H_0 implies only that T_w has its usual randomization distribution.

When applied to Figure 19.1, comparing statistics to their randomization distributions, T_w, T_e, and T_n all yield small P-values, less than 0.000 01. This cannot be the basis for firm conclusions because Figure 19.1 is not from a randomized experiment.

19.2.5 Sensitivity analysis

In an observational study, a sensitivity analysis asks about the magnitude of the departure from random assignment that would need to be present to explain as noncausal an observed association between treatment and response. A simple model for sensitivity analysis in paired observational studies introduces a sensitivity parameter $\Gamma \geq 1$ and asserts that treatment assignments Z_{i1} in distinct pairs are conditionally independent given $(\mathcal{F}, \mathcal{Z})$ with

$$\frac{1}{1+\Gamma} \leq \pi_i = \Pr(Z_{i1} = 1 | \mathcal{F}, \mathcal{Z}) \leq \frac{\Gamma}{1+\Gamma}, \qquad i = 1, \ldots, I \tag{19.4}$$

Then $\Gamma = 1$ yields a random assignment, $\Pr(Z_{i1} = 1 | \mathcal{F}, \mathcal{Z}) = \frac{1}{2}$, whereas each fixed $\Gamma > 1$ implies an unknown but limited departure from a random assignment (see Rosenbaum, 1987; 2002, Section 4; 2010a, Section 3). In (19.4), if $\Gamma = 2$, then two patients, $i1$ and $i2$, matched for observed covariates might differ in their odds $\pi_i / (1 - \pi_i)$ of treatment by an MO rather than a GO by as much as a factor of $\Gamma = 2$; e.g., $\pi_i = \frac{2}{3}$ and $1 - \pi_i = \frac{1}{3}$. For $\Gamma = 1$ in (19.4), $\Pr(Z_{i1} = 1 | \mathcal{F}, \mathcal{Z}) = \frac{1}{2}$, so T_w has its usual randomization distributions under Fisher's null hypothesis H_0, but for $\Gamma > 1$, the null distribution of T_w is unknown to a bounded degree. For several values of $\Gamma \geq 1$, a sensitivity analysis computes the range of possible values of an inference quantity subject to (19.4), for instance the range of possible P-values or point estimates or endpoints for a confidence interval. As $\Gamma \to \infty$ the range of possible P-values goes to [0, 1] – association does not logically imply causation, a sufficiently large bias Γ can explain any observed association – but at some finite value of Γ the range of possible

P-values is long enough to be uninformative, and the inference is sensitive to a bias of that magnitude Γ.

The sensitivity analysis model (19.4) may be extended and interpreted in various ways. The model easily extends to matching with multiple controls, full matching, and unmatched comparisons, and is applicable with continuous, binary, or censored responses; moreover, it may be reexpressed in terms of an explicit failure to match for an unmeasured covariate u_{ij} (Rosenbaum, 2002, Section 4). The one parameter Γ in (19.4) may be replaced by two parameters (Λ, Δ), where Λ controls the relationship between u_{ij} and treatment assignment Z_{ij}, Δ controls the relationship between u_{ij} and response under control r_{Cij}, and values of (Λ, Δ) map into values of Γ, so that a one-dimensional analysis in terms of Γ may be interpreted in terms of the two parameters (Λ, Δ) (see Rosenbaum and Silber, 2009). For instance, suppose treatment assignment Z_{ij} and response under control r_{Cij} would be unrelated if adjustments had been made for both $\mathbf{x}_{ij}$ and u_{ij} (that is, suppose $Z_{ij} \perp\!\!\!\perp r_{Cij} \mid (\mathbf{x}_{ij}, u_{ij})$ in Dawid's (1979) notation), but that failure to match for u_{ij} creates an association between Z_{ij} and r_{Cij} adjusting for $\mathbf{x}_{ij}$ alone. In this case, an unobserved covariate u_{ij} that could double the odds of treatment $(Z_{i1} - Z_{i2} = 1)$ and quadruple the odds of a positive difference $(r_{Ci1} - r_{Ci2} > 0)$ gives exactly the same sensitivity calculations as $\Gamma = 1.5$, but so does a u_{ij} that could quadruple the odds of treatment and double the odds of a positive response difference, and, indeed, so do infinitely many other pairs of values of (Λ, Δ); see Rosenbaum and Silber (2009) who make use of a semiparametric family of distributions introduced by Wolfe (1974).

The above reasoning for Wilcoxon's statistic applies without change to T_{e} and T_{n} providing the doses are such that $\mathcal{F}$ and the null hypothesis of no effect, H_0, fix h_i. That is, if doses are nonreactive, the reasoning applies to both T_{e} and T_{n}. If the doses are purely reactive, the reasoning applies to T_{e} but not generally to T_{n}. In the next paragraph, sensitivity calculations are described for the general statistic T, but these calculations implicitly assume that the doses are such that $\mathcal{F}$ and H_0 fix h_i.

The calculations required by a sensitivity analysis are straightforward. In the current paragraph, for the relevant type of dose, it is assumed that a T in (19.3) has been selected so that h_i is fixed by $\mathcal{F}$ and H_0; this is always true for T_{w}. Let $\overline{\overline{T}}$ be the sum of I independent random variables, $i = 1, \ldots, I$, taking the value 0 if $|Y_i h_i| = 0$, and otherwise the value $|h_i|$ with probability $\Gamma/(1+\Gamma)$ or $-|h_i|$ with probability $1/(1+\Gamma)$. Define $\overline{T}$ similarly with the roles of $\Gamma/(1+\Gamma)$ and $1/(1+\Gamma)$ interchanged. Then, under (19.4) and Fisher's null hypothesis H_0, it is straightforward to show

$$\Pr\left(\overline{T} \geq k \,\middle|\, \mathcal{F}, \mathcal{Z}\right) \leq \Pr(T \geq k \mid \mathcal{F}, \mathcal{Z}) \leq \Pr\left(\overline{\overline{T}} \geq k \,\middle|\, \mathcal{F}, \mathcal{Z}\right) \quad \text{for all } k \tag{19.5}$$

where both bounds in (19.5) equal the randomization distribution when $\Gamma = 1$, and when $\Gamma > 1$ the bounds are sharp in the sense that are attained for a particular π_i satisfying (19.4) (see Rosenbaum, 1987; 2002, Section 4). Then (19.5) yields bounds on P-values and, by inversion, bounds on confidence sets and point estimates. As $I \to \infty$, the central limit theorem yields the approximation to the upper bound in (19.5) as

$$\Pr\left(\overline{\overline{T}} \geq k \,\middle|\, \mathcal{F}, \mathcal{Z}\right) \approx 1 - \Phi\left(\frac{k - \{(\Gamma-1)/(1+\Gamma)\}\sum_{i=1}^{I} h_i}{\sqrt{\{4\Gamma/(1+\Gamma)^2\}\sum_{i=1}^{I} h_i^2}}\right) \tag{19.6}$$

Using (19.6), we find that the approximation to the upper bound (19.5) on the upper tail P-value for T is less than or equal to α if

$$\frac{T - \{(\Gamma - 1)/(1 + \Gamma)\}\sum_{i=1}^{I} h_i}{\sqrt{\{4\Gamma/(1+\Gamma)^2\}\sum_{i=1}^{I} h_i^2}} \geq \Phi^{-1}(1 - \alpha) \qquad (19.7)$$

Alternative methods of sensitivity analysis are discussed by Cornfield *et al.* (1959), Rosenbaum and Rubin (1983), Yanagawa (1984), Gastwirth (1992), Gastwirth *et al.* (1999), Marcus (1997), Robins *et al.* (1999), Copas and Eguchi (2001), and Imbens (2003). For a few applications, see Aakvik (2001), Normand *et al.* (2001), Diprete and Gangl (2004), Silber *et al.* (2005), and Slade *et al.* (2008).

19.2.6 Sensitivity analysis in the example

Table 19.3 is a sensitivity analysis for the data in Figure 19.1. For several values of $\Gamma \geq 1$, Table 19.3 gives the approximate upper bound (19.6) on the one-sided P-value testing the null hypothesis H_0 of no effect based on the three statistics in (19.3) for $\kappa = \frac{1}{2}$ and $\kappa = \frac{2}{3}$. Implicitly, the Wilcoxon statistic has $\kappa = 0$ and uses the difference in toxicity Y_i for all $I = 344$ matched pairs in Figure 19.1. With $\kappa = \frac{1}{2}$, only the black points in Figure 19.1 are used, that is, the pairs with absolute dose differences of $|D_i| \geq 9$ weeks, where $\lambda_{1/2} = 9$ is the median of the $|D_i|$. With $\kappa = \frac{2}{3}$, the analysis uses pairs with absolute dose differences of $|D_i| \geq 13$, or a quarter of a year, where $\lambda_{2/3} = 13$ is the $\frac{2}{3}$ quantile of the $|D_i|$. In Table 19.3, Wilcoxon's statistic T_{w} ignores doses, T_{e} ignores the signs of dose differences but focuses on pairs in which the MO and GO gave quite different doses, and T_{n} takes account of whether the sign of the dose difference agrees with the sign of the toxicity difference.

The three statistics, T_{w}, T_{e}, and T_{n}, give different impressions about the sensitivity of the toxicity difference to bias from nonrandom selection of patients in the MO and GO groups; however, they embody different presumptions about doses. The most basic analysis simply

Table 19.3 Sensitivity analysis without doses and with doses used in various ways. The table gives the upper bound on the one-sided significance level for several values of Γ. In each column, the largest P-value less than or equal to $\alpha = 0.05$ is in **bold**.

	No doses T_{w}	Absolute doses T_{e}		Signed doses T_{n}	
Γ	$\kappa = 0$	$\kappa = 1/2$	$\kappa = 2/3$	$\kappa = 1/2$	$\kappa = 2/3$
1.0	0.00000	0.00000	0.00000	0.00000	0.00000
1.5	**0.0065**	0.00492	0.00080	0.00000	0.00000
1.75	0.08926	**0.03795**	0.00596	0.00000	0.00000
2.1	0.49650	0.20201	**0.03779**	0.00000	0.00000
3.5	0.99992	0.96317	0.59136	0.01231	0.00028
4.0	1.00000	0.99355	0.77431	**0.04381**	0.00108
4.5	1.00000	0.99907	0.88807	0.10769	0.00301
6.5	1.00000	1.00000	0.9965	0.58144	**0.03598**

notes that the 344 patients of MOs were similar at baseline in important respects to the 344 patients of GOs, that the former group received more weeks of chemotherapy, experienced more chemotherapy associated toxicity, but experienced no lengthening of survival. The three components of this story are noted, but not linked. The Wilcoxon statistic T_{w} applied to the difference in toxicity Y_i is a part of this most basic analysis, and it requires no assumptions about the way doses are determined. This analysis is insensitive to small and moderate biases, say $\Gamma = 1.5$, but not to $\Gamma = 2$.

If someone were to claim that MOs and GOs would treat the same patient in the same way and the differences in doses D_i and toxicity Y_i in Figure 19.1 are the result of nonrandom allocation of patients to MOs and GOs, then implicitly that claim asserts that toxicity is unaffected and doses are purely reactive. The statistic T_{e}, which is valid for purely reactive or nonreactive doses, addresses this claim. As seen in Table 19.3 only a larger bias, perhaps $\Gamma > 2.1$ for $\kappa = \frac{2}{3}$, could produce the observed value of T_{e} if doses are purely reactive. This makes sense in light of Figure 19.1 and consideration of an analogous experiment: T_{e} excludes pairs in which $|D_i|$ is small, so it focuses attention on pairs that received meaningfully different doses. No one would design a paired experiment in which nearly the same dose is given to both members of many pairs.

The analysis using T_{n} is least sensitive but depends on the notion that the dose is nonreactive and that the dose-difference D_i in pair i would have been the same had the assignment of patients $i1$ and $i2$ to MO or GO been reversed. If the dose is nonreactive, only a very large bias, perhaps $\Gamma > 6.5$, could produce the observed behavior of T_{n} with $\kappa = \frac{2}{3}$. For comparison, Hammond's (1965) study of the effects of heavy smoking on lung cancer becomes sensitive at about $\Gamma = 6$ (see Rosenbaum, 2002, Section 4.3.2). Although doses might be nonreactive, this is not entirely plausible. It is not uncommon to adjust the dose of chemotherapy to reflect the needs or wishes of individual patients, and both MOs and GOs are likely to do this, perhaps in the same way – a purely reactive dose – or perhaps in different ways – a dose that is neither purely reactive nor nonreactive. Although the analysis using T_{n} is least sensitive, it is also the analysis that rests on the least plausible assumption. An alternative approach separates the assignment of patients to MOs or GOs from the assignment of dose differences within pairs; see Rosenbaum (2010c) for a discussion from this perspective.

Is Table 19.3 an oddity of a particular example? Section 19.3 demonstrates that it is not.

19.3 Design sensitivity

19.3.1 What is design sensitivity?

If an observational study were free of bias from unmeasured covariates u_{ij}, and if the treatment indeed caused its ostensible effects, then we would not know that this was true from the observable data. Call this the 'favorable situation', that is, the situation in which treatment assignment is effectively randomized within pairs, $\Pr(Z_{i1} = 1 \mid \mathcal{F}, \mathcal{Z}) = \frac{1}{2}$ for each i, and Fisher's null hypothesis of no effect is false. In the favorable situation, we might see, for instance, that treated responses are higher than control responses or that they are correlated with doses of treatment, but these observable patterns could also have been produced by a bias from an unmeasured covariate u_{ij}. When we are in the favorable situation, we cannot recognize that this is so from the observable data. The best we could hope to say is that rejection of Fisher's null hypothesis is highly insensitive to bias from unmeasured covariates, so only a large bias Γ could have produced the observed associations between treatment and outcome. The power of a sensitivity analysis is the probability that we will be able to say this.

As with the power of a test, the power of a sensitivity analysis is computed assuming a certain state of affairs, but also assuming that the investigator is ignorant of this state of affairs, and therefore the investigator performs a particular analysis she would not perform if she knew the true state of affairs. In this case, the true state of affairs is a treatment effect without bias from unmeasured covariates. Not knowing this is the true state of affairs; the investigator performs a sensitivity analysis, hoping that the analysis reports that the conclusions are insensitive to a bias of magnitude Γ, for some specific Γ.

The power is computed under a model for $\mathcal{F} = \{(r_{Tij}, r_{Cij}, w_{Tij}, w_{Cij}, \mathbf{x}_{ij}, u_{ij}),\ i = 1, \ldots, I,\ j = 1, 2\}$ in which there is a treatment effect, together with the assumption that $\Pr(Z_{i1} = 1 \mid \mathcal{F}, \mathcal{Z}) = \frac{1}{2}$ for each i; this determines a distribution for (Y_i, D_i). More precisely, the power is defined for a fixed value of the sensitivity parameter $\Gamma \geq 1$ and a fixed level α of the test; it is the probability that the sensitivity analysis will reject the null hypothesis of no effect at level α when the sensitivity analysis is performed with a specific value of Γ. Here, rejection means that the upper bound on the P-value is at most α. For fixed Γ and α, the approximate power is then the probability of the event (19.7). If one knew $\mathcal{F}$, then in principle one could compute the conditional power given $\mathcal{F}$; however, in reality, one does not know $\mathcal{F}$, so instead one computes the unconditional probability of the event (19.7) under a model that generates $\mathcal{F}$. If $\mathcal{F}$ were generated by a certain model, what is the probability of the event (19.7)? For $\Gamma = 1$, this reproduces the power of a randomization test in a randomized experiment. In the calculations here, it is assumed that the $\left(r_{Tij}, r_{Cij}, w_{Tij}, w_{Cij}, \mathbf{x}_{ij}, u_{ij}\right)$ are $2I$ independent observations drawn from the same continuous distribution free of ties of all kinds. For power calculations of this kind, see Rosenbaum (2004, 2005, 2010a, 2010b, 2011a, 2011b), Small and Rosenbaum (2008) and Heller *et al.* (2009).

The design sensitivity, $\widetilde{\Gamma}$, is a number that summarizes the behavior of the power when I is large. It turns out that the power typically tends to 1 as $I \to \infty$ if the sensitivity analysis is performed with the sensitivity parameter set to $\Gamma < \widetilde{\Gamma}$ and the power tends to zero if $\Gamma > \widetilde{\Gamma}$ (see Rosenbaum, 2004, 2010a, Part III, especially Figures 14.2 and 14.3, and 2011b). That is, in large samples, the given sampling model for $\mathcal{F}$ in the favorable situation can be distinguished from the no treatment effect together with a bias of magnitude $\Gamma < \widetilde{\Gamma}$, but it cannot be distinguished from a bias of magnitude $\Gamma > \widetilde{\Gamma}$. The number $\widetilde{\Gamma}$ is a property of a test statistic, say T, when applied in the favorable situation to the given sampling model for $\mathcal{F}$. The design sensitivity may compare different research designs – that is, different sampling models for $\mathcal{F}$ – or it may compare different statistics for a given sampling model.

In the case of Wilcoxon's signed rank statistic, T_{w}, the design sensitivity has a simple formula, namely $\widetilde{\Gamma} = \theta / (1 - \theta)$, where $\theta = \Pr\left(Y_i + Y_j > 0\right), i \neq j$, is the parameter estimated by Hoeffding's (1948) U-statistic that most closely resembles Wilcoxon's signed rank statistic; see Small and Rosenbaum (2008), Heller *et al.* (2009) and Rosenbaum (2010a, Chapter 14, especially Figures 14.2 and 14.3) for specifics. Because T_{e} is the same statistic computed for pairs with $|D_i| \geq \lambda_\kappa$, if the doses are nonreactive or purely reactive, the same formula applies, albeit with a different distribution for Y_i. A few calculations comparing T_{w} and T_{e} with nonreactive doses are given in Section 19.3.2.

19.3.2 Comparison of design sensitivity with purely reactive doses

Table 19.4 calculates the design sensitivity $\widetilde{\Gamma}$ in a few simple situations. To repeat, in these situations, there is a treatment effect and no bias from unmeasured covariates, but the investigator does not know this and cannot deduce this from the data. The best the investigator can hope to say is that the study is insensitive to small and moderate biases, as measured by a sensitivity analysis. The power of the sensitivity analysis is the probability that the upper bound on the significance level will be less than α, conventionally $\alpha = 0.05$. As the sample size I increases, the power tends to 1 for $\Gamma < \widetilde{\Gamma}$ and to 0 for $\Gamma > \widetilde{\Gamma}$.

In Table 19.4, three statistics are compared, namely Wilcoxon's statistic T_w and the statistic T_e for $\kappa = \frac{1}{2}$ and $\kappa = \frac{2}{3}$. Here, T_w does not use doses and requires no assumptions about doses. In contrast, T_e is valid if either the doses are purely reactive or if they are nonreactive.

Table 19.4 calculates the design sensitivity $\widetilde{\Gamma}$ under the model that $D_i \sim_{iid} N(\mu, \sigma)$ and $Y_i = \beta D_i + \epsilon_i$ where the ϵ_i are independent of the D_i and $\epsilon_i \sim_{iid} f_\epsilon$ where f_ϵ is either the standard normal or Cauchy distribution. Then Y_i has a distribution symmetric about $\beta\mu$, where $\beta\mu = \frac{1}{2}$ in the calculations for the normal distribution and $\beta\mu = 1$ for the calculations for the Cauchy distribution. When σ is smaller, the effects βD_i are more nearly constant and less dependent on the dose.

To illustrate the meaning of the design sensitivity, $\widetilde{\Gamma}$, a single sample of $I = 500\,000$ pairs was generated from the first sampling situation in Table 19.4, namely $D_i \sim_{iid} N(1, 1)$ and $Y_i = D_i/2 + \epsilon_i$, $\epsilon_i \sim_{iid} N(0, 1)$. This one sample was then analyzed by two methods in Table 19.4. Because I is large, $\widetilde{\Gamma}$ should forecast quite accurately how these two analyses will turn out. For Wilcoxon's statistic, $\widetilde{\Gamma} = 2.8$ in Table 19.4 and for T_e with $\kappa = \frac{1}{2}$, it is $\widetilde{\Gamma} = 6.5$. Performing the sensitivity analysis on the one large simulated example, the upper bound on the P-value (19.6) for Wilcoxon's statistic is 1.1×10^{-16} for $\Gamma = 2.7$ and it is 1.000 00 for $\Gamma = 2.9$. In parallel, for T_e with $\kappa = \frac{1}{2}$, the upper bound (19.6) is 0.000 27 for $\Gamma = 6.3$ and 0.999 67 for $\Gamma = 6.6$.

Table 19.4 Design sensitivity for purely reactive or nonreactive doses using the Wilcoxon statistic T_w or the statistic T_e.

Sampling distribution			Design sensitivity		
Normal errors			Test statistic		
μ	σ	β	T_w	$T_e, \kappa = \frac{1}{2}$	$T_e, \kappa = \frac{2}{3}$
1	1	$\frac{1}{2}$	2.8	6.5	10.0
1	$\frac{1}{2}$	$\frac{1}{2}$	3.1	5.1	6.2
Cauchy errors			Test statistic		
μ	σ	β	T_w	$T_e, \kappa = \frac{1}{2}$	$T_e, \kappa = \frac{2}{3}$
1	1	1	2.5	4.3	5.3
1	$\frac{1}{2}$	1	2.9	4.0	4.4

In Table 19.4, larger design sensitivities $\widetilde{\Gamma}$ are obtained by restricting attention to pairs in which $|D_i|$ is larger. In Table 19.4, in large samples, T_e with $\kappa = \frac{2}{3}$ is less sensitive to unmeasured bias than is T_w. The pattern in Table 19.4 is similar to that in Table 19.3.

19.4 Summary

The typical analysis using doses D_i views doses as nonreactive, that is, as a part of the treatment that the experimenter inflicts upon the experimental subject. If the doses are nonreactive, the treated-minus-control difference in dose, D_i, is fixed, not changing if treatment assignments are changed in pair i. If the reactions of subjects influence the doses they receive, then the doses are reactive and typical analyses are not appropriate. There is always the option of ignoring the doses. Between these two extremes is the statistic T_e which is valid for testing no effect on the outcome if either the doses are nonreactive or if they are purely reactive. If a dose is either nonreactive or purely reactive, $|D_i|$ is fixed, not changing if the treatment assignments are changed in pair i. The counterclaim that a dose–response relationship, say Figure 19.1, is entirely the product of biased assignment to treatment or control – the counterclaim that experimental subjects would be exactly the same if the treatment assignments were reversed – implicitly includes the notion that doses are purely reactive. In the analysis of a counterclaim, it is fair game to make use of the implicit assumptions of that counterclaim, and in the case of purely reactive doses, this may lead to reduced sensitivity to bias from failure to control for an unmeasured covariate.

References

Aakvik, A. (2001) Bounding a matching estimator: the case of a Norwegian training program. *Oxford Bulletin of Economics and Statistics*, **63**, 115–143.

Cannistra, S.A. (2007) Gynecologic oncology or medical oncology: What's in a name? *Journal of Clinical Oncology*, **25**, 1157–1159.

Copas, J. and Eguchi, S. (2001) Local sensitivity approximations for selectivity bias. *Journal of the Royal Statistical Society B*, **63**, 871–896.

Cornfield, J., Haenszel, W., Hammond, E., Lilienfeld, A., Shimkin, M. and Wynder, E. (1959) Smoking and lung cancer. *Journal of the National Cancer Institute*, **22**, 173–203.

Cox, D.R. (2009) Randomization in the design of experiments. *International Statistical Review*, **77**, 415–429.

Dawid, A.P. (1979) Conditional independence in statistical theory. *Journal of the Royal Statistical Society B*, **41**, 1–31.

Diprete, T.A. and Gangl, M. (2004) Assessing bias in the estimation of causal effects. *Sociological Methodology*, **34**, 271–310.

Fisher, R.A. (1935) *The Design of Experiments*. Edinburgh: Oliver & Boyd.

Gastwirth, J.L. (1992) Methods for assessing the sensitivity of statistical comparisons used in Title VII cases to omitted variables. *Jurimetrics*, **33**, 19–34.

Gastwirth, J.L., Krieger, A.M. and Rosenbaum, P.R. (1998) Dual and simultaneous sensitivity analysis for matched pairs. *Biometrika*, **85**, 907–920.

Hammond, E.C. (1964) Smoking in relation to mortality and morbidity. *Journal of the National Cancer Institute*, **32**, 1161–1188.

Heller, R., Rosenbaum, P.R. and Small, D.S. (2009) Split samples and design sensitivity in observational studies. *Journal of the American Statistical Association*, **104**, 1090–1101.

Hill, A.B. (1965) The environment and disease: association or causation? *Proceedings of the Royal Society of Medicine*, **58**, 295–300.

Hoeffding, W. (1948) A class of statistics with asymptotically normal distribution. *Annals of Mathematical Statistics*, **19**, 293–325.

Imbens, G.W. (2003) Sensitivity to exogeneity assumptions in program evaluation. *American Economic Review*, **93**, 126–132.

Marcus, S.M. (1997) Using omitted variable bias to assess uncertainty in the estimation of an AIDS education treatment effect. *Journal of Educational and Behavioral Statistics*, **22**, 193–201.

Neyman, J. (1923) On the application of probability theory to agricultural experiments: essay on principles, Section 9, reprinted in *Statistical Science*, **5**, 463–480.

Noether, G.E. (1967) *Elements of Nonparametric Statistics*. New York: John Wiley & Sons, Inc.

Normand, S.L.T., Landrum, N.B., Guadagnoli, E., Ayanian, J.Z., Ryan, T.J., Cleary, P.D. and McNeil, B.J. (2001) Validating recommendations for coronary angiography following acute myocardial infarction in the elderly. *Journal of Clinical Epidemiology*, **54**, 387–398.

Pratt, J.W. and Gibbons, J.D. (1981) *Concepts of Nonparametric Theory*. New York: Springer.

Robins, J.M., Rotnitzky, A. and Scharfstein, D. (1999) Sensitivity analysis for selection bias and unmeasured confounding in missing data and causal inference, in *Statistical Models in Epidemiology* (eds E. Halloran and D. Berry). New York: Springer, pp. 1–94.

Rosenbaum, P.R. (1987) Sensitivity analysis for certain permutation inferences in matched observational studies. *Biometrika*, **74**, 13–26.

Rosenbaum, P.R. (1999) Using combined quantile averages in matched observational studies. *Applied Statistics*, **48**, 63–78.

Rosenbaum, P.R. (2002) *Observational Studies*, 2nd ed. New York: Springer.

Rosenbaum, P.R. (2004) Design sensitivity in observational studies. *Biometrika*, **91**, 153–164.

Rosenbaum, P.R. (2005) Heterogeneity and causality: unit heterogeneity and design sensitivity in observational studies. *American Statistician*, **59**, 147–152.

Rosenbaum, P.R. (2010a) *Design of Observational Studies*. New York: Springer.

Rosenbaum, P.R. (2010b) Design sensitivity and efficiency in observational studies. *Journal of the American Statistical Association*, **105**, 692–702.

Rosenbaum, P.R. (2010c) Evidence factors in observational studies. *Biometrika*, **97**, 333–345.

Rosenbaum, P.R. (2011a) A new u-statistic with superior design sensitivity in observational studies. *Biometrics*, **67**, 1017–1027.

Rosenbaum, P.R. (2011b) What aspects of the design of an observational study affect its sensitivity to bias from covariates that were not observed? in *Looking Back: Proceedings of a Conference in Honor of Paul W. Holland* (eds. N.J. Dorans and S. Sinharay). New York: Springer, pp. 87–114, Chapter 6.

Rosenbaum, P. and Rubin, D. (1983) Assessing sensitivity to an unobserved binary covariate in an observational study with binary outcome. *Journal of the Royal Statistical Society B*, **45**, 212–218.

Rosenbaum, P.R. and Silber, J.H. (2009) Amplification of sensitivity analysis in observational studies. *Journal of the American Statistical Association*, **104**, 1398–1405.

Rosenbaum, P.R., Ross, R.N. and Silber, J.H. (2007) Minimum distance matched sampling with fine balance in an observational study of treatment for ovarian cancer. *Journal of the American Statistical Association*, **102**, 75–83.

Rothman, K.J. (1986) *Modern Epidemiology*. Boston, MA: Little, Brown.

Rubin, D.B. (1974) Estimating causal effects of treatments in randomized and nonrandomized studies. *Journal of Educational Psychology*, **66**, 688–701.

Silber, J.H., Rosenbaum, P.R., Trudeau, M.E., Chen, W., Zhang, X., Lorch, S., Rapaport-Kelz, R., Mosher, R.E. and Even-Shoshan, O. (2005) Preoperative antibiotics and mortality in the elderly. *Annals of Surgery*, **242**, 107–114.

Silber, J.H., Rosenbaum, P.R., Polsky, D., Ross, R.N., Even-Shoshan, O., Schwartz, S., Armstrong, K.A. and Randall, T.C. (2007) Does ovarian cancer treatment and survival differ by the specialty providing chemotherapy? *Journal of Clinical Oncology*, **25**, 1169–1175.

Slade, E.P., Stuart, E.A., Alkever, D.S.S., Karakus, M., Green, K.M. and Ialongo, N. (2008) Impacts of age of onset of substance use disorders on risk of adult incarceration among disadvantaged urban youth. *Drug and Alcohol Dependence*, **95**, 1–13.

Small, D. and Rosenbaum, P.R. (2008) War and wages: the strength of instrumental variables and their sensitivity to unobserved biases. *Journal of the American Statistical Association*, **103**, 924–933.

Weiss, N. (1981) Inferring causal relationships: elaboration of the criterion of 'dose–response'. *American Journal of Epidemiology*, **113**, 487–490.

Welch, B.L. (1937) On the z-test in randomized blocks. *Biometrika*, **29**, 21–52.

Wolfe, D.A. (1974) A characterization of population weighted symmetry and related results. *Journal of the American Statistical Association*, **69**, 819–822.

Yanagawa, T. (1984) Case-control studies: assessing the effect of a confounding factor. *Biometrika*, **71**, 191–194.

20

Evaluation of potential mediators in randomised trials of complex interventions (psychotherapies)

Richard Emsley and Graham Dunn
Health Sciences Research Group, School of Community Based Medicine, The University of Manchester, Manchester, UK

20.1 Introduction

In randomised trials in mental health we are often interested in the estimation of mediation effects in order to address issues of noncompliance, identify key processes involved in the intervention, or for testing specific hypothesis relating to scientific theory. In this chapter we focus exclusively on randomised trials of psychological treatments that typically contain many features of a complex intervention: they will almost always collect multivariate outcomes over repeated time points, using scales of measurement that are subject to measurement error; be delivered by a group of heterogeneous therapists to a group of heterogeneous patients; and are likely to contain many interacting components within the intervention.

Readers will be familiar with the intention to treat (ITT) principle in randomised trials, where patients are analysed as belonging to the group to which they were randomly allocated, regardless of the level of intervention they subsequently received (see Chapter 21 in this volume for further details). The pragmatic ITT question asks 'Is there an effect of randomly allocating patients to treatment?', but we can gain deeper insight by asking more subtle questions, such as 'How does the treatment work?', 'Which aspects of the treatment are responsible for the efficacy?', and 'Can the treatment be improved to work more effectively in certain patients?'.

Causality: Statistical Perspectives and Applications, First Edition. Edited by Carlo Berzuini, Philip Dawid and Luisa Bernardinelli.

To answer these questions, we must consider the role of other variables besides random allocation and the outcomes. These variables include mediators, which are defined as intermediate outcomes on the causal pathway between random allocation and the final outcome and are by definition measured post-baseline. By contrast, moderators are pre-baseline characteristics that influence the effect of random allocation on either the intermediate or final outcomes.

We begin by discussing some of the specific characteristics inherent in psychological treatment trials, the causal questions that arise within these and the technical challenges they pose. Our focus is on developing statistical methods to answer these questions and in doing so, we utilise the results and concepts introduced in this volume in Chapter 11 by Vansteelandt and Chapter 12 by Pearl. We outline two statistical approaches, instrumental variables estimation and principal stratification, and after discussing their statistical properties and identification, we illustrate their exposition in a psychotherapy trial in people with an early episode of schizophrenia.

20.2 Potential mediators in psychological treatment trials

In this section, we outline some of the interesting questions that arise in psychotherapy trials and postulate how they may explain treatment effect heterogeneity through mediation. We also discuss some of the technical challenges that need to be considered.

One natural question asks about a patient's adherence to their randomly allocated treatment, which could be formulated as both a binary and continuous measure: for example, does the patient turn up for therapy? Or how many sessions of therapy does she attend? We see in Chapter 21 in this volume how to estimate the complier average causal effect (CACE) (i.e. the ITT effect within the subgroup of compliers) and how a dose–response relationship can be estimated. Here, we formulate this question in terms of mediation. Figure 20.1 shows random allocation as the treatment variable of interest (A), number of sessions of therapy attended as

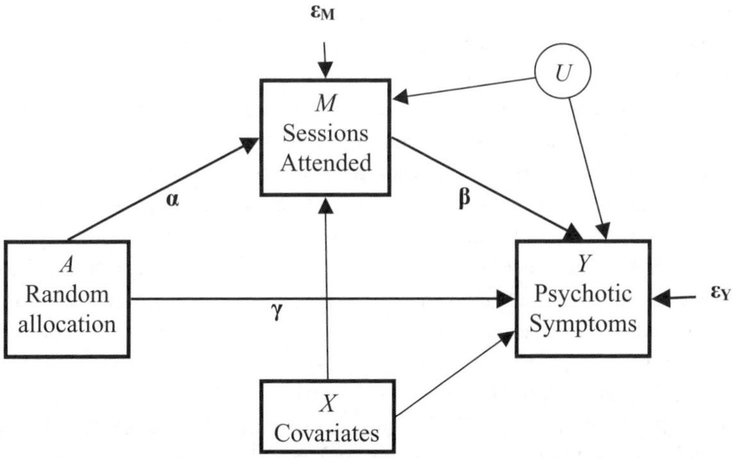

*Figure 20.1 Mediation diagram with random allocation (*A*), the number of sessions attended as the putative mediator (*M*), and psychotic symptoms as the outcome (*Y*). Covariates* X *could influence* M *and/or* Y*, and there are unmeasured confounders* U *between* M *and* Y.

the mediator (M), and a measure of psychotic symptoms (Y) as the outcome. Our main question is what is the effect of attending additional sessions of therapy on psychotic symptoms? And is the effect of sessions the same in all patients? That is, what is the estimate of β and is it constant for all patients?

In a CACE analysis, an assumption is made that there is no direct effect of random allocation on outcome, except through the treatment received (i.e. $\gamma = 0$). This is known as an exclusion restriction. Here, because we are demonstrating a mediation analysis, we do not necessarily impose this assumption, and we demonstrate later in this chapter how identification can still be achieved. The indirect effect is estimated as $\alpha^*\beta$, and observing (rather than setting) γ to be equal to 0 would be an example of complete mediation, such that the entire effect is of A on Y is mediated by M.

Allied to the issue of *quantity* of sessions are issues surrounding the *quality* of the therapeutic relationship, sometimes called the therapeutic alliance. This is a general term for a variety of therapist–client interactional and relational factors that operate in the delivery of treatment, and contains two distinct factors: a personal alliance based on the interpersonal relationship and a task related alliance based on factors of the treatment (Green, 2006). It is measured during the course of therapy, but it is often not observed in the control group if these are receiving some form of treatment as usual (where the concept of an alliance may be undefined). One often observes in the literature a rather naïve examination of the role of therapeutic alliance. Typically, the investigators take data from the treated group (ignoring the controls if the data comes from an experimental trial) and simply look at the association between outcome and alliance. This simplistic approach does not distinguish treatment-free prognosis from the effects of treatment; these associations would reflect an inseparable mix of selection and treatment effects (i.e. the inferred treatment effects would be confounded).

Perhaps more interesting questions pertain to the possible interplay between sessions of therapy and therapeutic alliance, such as whether the quantity (sessions) and quality (alliance) of therapy jointly influence the treatment effect or whether the effect of sessions varies across levels of therapeutic alliance (indicating effect modification by a post-baseline variable). Is there evidence that an increase in the number of sessions attended has a more beneficial effect when the patient has a strong therapeutic alliance with the therapist than a poor alliance?

While we will focus in this chapter on these two 'dose' variables, we illustrate some other examples of potential mediators in psychotherapy trials. The aim is to emphasise that this is a broad topic and that the methods expounded in the later sections are widely applicable to answer a variety of mechanistic questions in complex interventions. Consider the fidelity of therapy, which asks how close is the therapy actually delivered to that described in the treatment manual. For example, is it a true cognitive–behavioural intervention or merely emotional support? Are there subgroups of patients defined by levels of fidelity, such as those receiving full therapy, those receiving partial therapy, and those receiving no therapy? Do these subgroups account for treatment effect heterogeneity between patients? What are the patient's beliefs and does cognitive behaviour therapy (CBT) change beliefs in depressed patients, which, in turn, leads to improvements in symptoms and quality of life? How much of the treatment effect is explained by changes in attributions?

One frequently observes that CBT improves a patient's adherence to concomitant medication, such as antidepressants, and so one could ask whether it is this improved compliance with medication which, in turn, leads to better outcomes? What is the direct effect of CBT after allowing for this or is any observed treatment effect of CBT acting wholly through improved compliance? There are likely to be a range of possible mediation mechanisms pertaining to the psychological model underpinning the intervention that could be further elucidated in this way.

In the context of answering these questions there are some specific technical challenges that arise and have implications for the statistical analysis. Firstly, the potential mediator may be a process measure that is only able to be measured in those patients who are randomly allocated to treatment. That is, it is undefined for those in the control group by definition of the control intervention (the therapeutic alliance or fidelity to a treatment protocol are obvious examples).

Secondly, the mediators and outcomes are likely to be measured with a considerable amount of measurement error (the number of sessions attended, for example, is only a proxy for the 'dose' of therapy; rating scales for strength of the therapeutic alliance will have only modest reliability). This may be in contrast to other subject areas that have 'hard' measures available, such as the presence/absence of a gene, death, or other directly observable measures.

Thirdly, there are also very likely to be hidden selection effects (hidden confounding). A patient may, for example, have a good prognosis under the control condition. If that same person were to receive treatment, however, the factors that predict good outcome in the absence of treatment would also be likely to predict good compliance with the therapy (number of sessions attended or strength of the therapeutic alliance, for example). The severity of symptoms, or level of insight into illness, measured at the time of random allocation, for example, are likely to be predictors of both treatment compliance and treatment outcome. Similarly, the confounders could be post-baseline variables such as life events (e.g. bereavement of a loved one). If these have been measured we can allow for these, but as we shall demonstrate, problems arise when there are unmeasured confounders.

A fourth challenge is that the psychological treatment received is unlikely to be identical for all patients. This variation could be the process measure we wish to investigate, for example, with treatment fidelity, but when this is not the case it leads to a concern that there are, in fact, multiple versions of treatment. In the following section, we make an assumption that allows for this concern. However, related to this is the issue of therapist effects. Sharing therapists is likely to induce shared sources of treatment effect heterogeneity in patients because of therapist characteristics, for example training and experience, working methods, and therapist personality. A standard ITT analysis in this context would involve a model containing random therapist effects (Walwyn and Roberts, 2010). We will ignore this complication in this chapter.

A fifth and final challenge arises from missing outcome data. Such data are unlikely to be missing purely by chance. Prognosis, compliance with allocated treatment, and the treatment outcome, itself, are all potentially related to loss to follow-up. which, in turn, leads to potentially biased estimates of treatment effects and estimates of the influences of treatment–effect modifiers. Similar issues can arise with missing mediator data. For simplicity, we do not consider the extensive challenges of missing data in the remainder of this chapter.

20.3 Methods for mediation in psychological treatment trials

Figure 20.1 shows the standard mediation diagram we consider in this chapter. In general, random allocation of treatment A has an effect (α) on an intermediate outcome M. The intermediate outcome, in turn, has an effect (β) on the final outcome Y. There is also a possible direct effect of A on Y (denoted γ). That part of the influence of A on Y that is explained through the effect of A on M, and of M on Y, is an indirect or mediated effect.

The intermediate variable, M, is a treatment effect mediator and ε_M, ε_Y are the error terms associated with M and Y respectively. Covariates X could predict the mediator and/or the outcome. The U represents unmeasured confounders between M and Y, a critical point we expand on below; it is shown as a latent variable in this diagram. The total effect of A on Y (denoted τ) can be decomposed into the sum of the direct and indirect effects ($\tau = \gamma + \alpha^*\beta$); see Chapters 11 and 12 in this volume for further details.

What are the unmeasured confounders U? Firstly, we can eliminate the possibility of any confounding between random allocation and the mediator, and between random allocation and the final outcome by virtue of randomisation. However, we are still left with possible confounding between the mediator and the final outcome because, unlike random allocation, in most circumstances the mediator is not under the control of the investigator but acts an intermediate outcome on the causal pathway. This is especially true in situations where we are considering process measures of the psychotherapy as potential mediators, for example beliefs about illness or the therapeutic alliance. We can measure and adjust for covariates X, which could be putative confounders, but there will always be the possibility that something important has not been measured or excluded from the model.

We have already alluded to the concern of the unmeasured confounders assumption in Section 20.2, and here we demonstrate its impact on estimation of the direct and indirect effects in Figure 20.1. Consider a situation with complete mediation (i.e. $\gamma = 0$). If there is no unmeasured confounding, then conditional on X and M, A and Y are independent, denoted as $A \perp Y \mid M, X$. If, however, there is unmeasured confounding U, then complete mediation implies $A \perp Y \mid M, X, U$ and it no longer holds that $A \perp Y \mid M, X$. Thus the price for ignoring U will be that the estimates of both β and γ could be biased.

We introduce some further notation, consistent with that used in Chapter 11. For the ith patient we have already defined the following observed variables:

A_i = random treatment allocation ($A_i = 1$ for treatment, 0 for controls)
$X'_i = X1_i, X2_i, \ldots, Xp_i$ = baseline covariates
Y_i = observed outcome
M_i = mediator of the effect of random treatment allocation on outcome (this could be either a quantitative measure or binary)

We now also define the following potential outcomes:

$M_i(a)$ = mediator with treatment level $A = a$
$Y_i(a, m)$ = outcome with treatment level $A = a$ and mediator level $M = m$
$Y_i(0) = Y_i(0, M_i(0))$ = outcome if randomised to the control condition with mediator $M_i(0)$
$Y_i(1) = Y_i(1, M_i(1))$ = outcome if randomised to the treatment with mediator $M_i(1)$

In the control arm where $A = 0$, $Y_i = Y_i(0)$ and $M_i = M_i(0)$, so that $M_i(0)$ and $Y_i(0)$ are the observed values and $M_i(1)$ and $Y_i(1)$ are unobserved. Similarly, in the treatment arm where $A = 1$, $M_i(0)$ and $Y_i(0)$ are unobserved and $M_i = M_i(1)$ and $Y_i = Y_i(1)$ are observed. Throughout this chapter we also assume the stable unit treatment value assumption (Rubin, 1980) or, more accurately if taking into account the variation in A, the treatment-variation irrelevance assumption, which states that even if different versions of $A = 1$ exist they result in the same $Y_i(1)$ (VanderWeele, 2009).

We can partition the total effect of random allocation for patient i as follows:

$$Y_i(1) - Y_i(0) = \{Y_i(1, M_i(0)) - Y_i(0, M_i(0))\} + \{Y_i(1, M_i(1)) - Y_i(1, M_i(0))\}$$

Here, the first element of this decomposition is the direct effect of random allocation given $M(0)$ and the second element is the effect of the change in mediator if randomised to receive treatment (i.e. $A = 1$). The first of these is referred to as the *natural direct effect* and the second is the *natural indirect effect* (Pearl, 2001; Robins and Greenland, 1992). We also define the *controlled direct effect* of random allocation on outcome at the mediator level m as $Y_i(1, m) - Y_i(0, m)$. These effects have been discussed extensively in Chapters 11 and 12 and references therein. We do not intend to add to that discussion in detail here; however, we note that under a set of assumptions, these authors demonstrate how standard regression models can be used to estimate the controlled direct effect and the natural direct and indirect effects, even in the presence of an interaction between A and M.

The methods outlined in these chapters rely on the key assumption that all confounders between M and Y have been measured to ensure that exchangeability (or conditional exchangeability) holds. If we were prepared to assume that this is the case, then the assumptions outlined in Chapter 11 would simplify further in our context because of the role of random allocation. Specifically, the assumption that the exposure–outcome relationship is unconfounded would be true and this would enable direct estimation of the controlled direct effect using a simple regression (VanderWeele and Vansteelandt, 2009). Secondly, the assumption that the exposure–mediator relationship is unconfounded would also hold, and with an additional assumption that no effect of random allocation confounds the M–Y relationship ($Y_i(a, m) \perp M_i(a^*) | X_i$) would enable one to identify the natural direct and indirect effects.

We argue that the no unmeasured confounding assumption is a strong assumption to make in the psychological treatment trial applications being discussed in this chapter. Partly this is a function of investigators not considering this issue at the design stage and so measurement of potential confounders of the mediator–outcome relationship is not given due emphasis, but equally this may be because in the context of complex interventions in mental health it is difficult to conceive and measure all potential confounds where the probable influences on treatment receipt, potential therapy processes, and mental health outcomes are hugely heterogeneous across the patient sample.

Another issue that arises when defining natural and controlled direct effects is that of intervening to set the mediator to a particular value, which may not be possible in some mediators, an obvious example in the psychotherapy context being a patient's beliefs. The best we may be able to manage is to intervene on a variable that manipulates beliefs – this is the essence of the psychological treatment – but there may be other variables that can influence beliefs (or other mediators) and do not have an effect on outcome.

Instead, we propose two alternative approaches to evaluate potential mediators in psychological treatment trials. The first approach does not rely on the unmeasured confounding assumption between the mediator and outcome, but instead makes progress by imposing alternative parametric assumptions on the relationships between variables. This approach uses instrumental variables, which are defined as variables that covary with the mediator, have no direct effect on outcome except that through the mediator (the exclusion restriction), and are unrelated to other causes of the outcome. The concept of using instrumental variables in randomised trials to adjust for unmeasured confounding is not uncommon because it underpins the concept of CACE estimation; the difference in this scenario is that we cannot use random allocation as an instrument because we are explicitly testing the direct effect of this. This requires us to find additional variables to use as instruments, a problem we address in the next section.

Using the instrumental variable approach does, however, rule out the possibility of random allocation by mediator interactions on the outcome; in some circumstances we consider this is restrictive, but in others it is not likely that these interactions are of interest. Consider the situation where we are investigating the role of process variables as possible mediators. Here, the interactions between random allocation and the mediator is unlikely to be of substantive interest and may be undefined if the mediator is unmeasured in the control group.

Alternatively, consider our example of concomitant medication as the putative mediator. In general, patients in the trial will be permitted to continue their usual medication regime (e.g. the intervention is CBT in addition to usual care) since it would be unethical to withhold the medication regime. The argument holds that random allocation to psychotherapy may improve a patient's adherence to their prescribed medication; if this is not the case, then concomitant medication does not meet the criteria for a mediator since the random allocation does not have an effect on medication use. Therefore, assuming this argument holds, the question then is how much of the effect of CBT on outcome is a direct effect and how much may be mediated by medication use. In this context, an interaction between random allocation to CBT and medication use may also be possible and of substantive interest; that is patients who both receive CBT and take concomitant medication may have better outcomes than those who only receive CBT *or* only take concomitant medication. In this case, we are effectively interested in ruling out the role of concomitant medication as a mediator and want to establish that there is a direct effect of CBT, while allowing for possible interactions. In practice we wish to estimate a controlled direct effect where the level of the mediator is controlled at a set level for the population (say, no medication). As outlined above, this can be easily achieved using standard regression models under a no unmeasured confounding assumption and a sensitivity analysis can be performed to assess the validity of this assumption (VanderWeele, 2010).

The second approach to mediation we consider is principal stratification (Frangakis and Rubin, 2002). This is described in detail in Chapter 10 and there is further discussion of its utility and limitations for estimating direct and indirect effects in Chapter 11 and 12. Here, we give a brief overview of the concept and discuss situations we believe to be useful for mediation analysis in psychotherapy trials; for further details see Emsley *et al.* (2010), Gallop *et al.* (2009), Jo (2008), and Elliott *et al.* (2010).

We have highlighted previously how examining variables that are undefined in the control group can be problematic, particularly if one merely examines associations between the mediator and outcome in the treatment group, as this association comprises both treatment and selection effects. Such variables can include compliance with active treatment, since the causal effect of random allocation is likely to be different in compliers and noncompliers (and is often assumed to be zero in the noncompliers) and the strength of the therapeutic alliance between patient and therapist, since the causal effect of random allocation is likely to be more beneficial in the patients who develop a good therapeutic alliance.

Taking the therapeutic alliance example, we would like to compare the effect of random allocation on outcome in those patients who develop a given level of alliance with the effect of random allocation on outcome in the comparative subgroup of control patients who would have developed the same level of alliance if, contrary to fact, they had been allocated to receive the therapy. That is, we wish to make randomised comparisons within subgroups of patients for whom $M_i(1) = M_i(0)$. These subgroups define principal strata; they are latent classes rather than directly observed groups since only one of $M_i(1)$ and $M_i(0)$ can be observed for each patient.

If we take our therapeutic alliance variable to be binary (high versus low), then we can postulate the existence of two strata:

- High alliance stratum – those patients observed to have a high alliance in the therapy group together with those patients in the control group who would have had a high alliance had they been allocated to receive therapy.
- Low alliance stratum – those patients observed to have a low alliance in the therapy group together with those in the control group who would have had a low alliance had they been allocated to receive therapy.

As principal strata are independent of treatment allocation, we can stratify by stratum membership in an analogous way to pre-baseline variables and evaluate the effects of random allocation within them. This within-class effect is known as the principal stratum direct effect $PSDE = E[Y(1) - Y(0)|M(1) = M(0)]$ (see Chapters 11 and 12).

We argue that principal stratification is a useful tool in the context where the mediator is undefined in the control group, because we effectively know the value of $M_i(0)$ is undefined for each patient. In the scenario with a binary mediator there then become only two principal strata, and the identification issue becomes one of identifying which patients in the control group would be in each stratum had they been allocated to the treatment, i.e. predicting $M_i(1)$ when $A_i = 0$. Identification relies on having good predictors of M when $A = 1$. This is conceptually identical to the noncompliance setting when there is no access to treatment in the control arm (no contamination); there, the two principal strata are compliers and noncompliers, and stratum membership is fully observed in the treatment arm. In situations where the mediator is undefined in one group, as with the therapeutic alliance, then we argue that the use of principal stratification in this context is similar to its use in the truncation by death context. There, the issue is that patients may die before the outcome can be measured and so the outcome is undefined, and the interest is in assessing the direct effect of treatment in the principal stratum of those patients who would have survived under either treatment.

The other appeal of principal stratification in psychotherapy trials occurs when it is impossible to intervene directly on the mediator. Intervening on M has been implicitly assumed in the situation where we have counterfactuals of the form $Y(a,m)$. Examples of mediators where experimental manipulation may not be feasible include beliefs and attributions about illness or the fidelity of therapy. Here, the principal stratification method becomes the classification of the patients into homogenous classes and the stratum specific effect could be the explicit focus of one's research. We discuss the concept, identification and limitations of principal stratification in Section 20.5.

20.4 Causal mediation analysis using instrumental variables estimation

In this section we discuss two models that have been proposed for causal mediation analysis, the structural mean model (SMM) and the rank-preserving model (RPM), and demonstrate how these can be estimated using instrumental variable estimation methods. This section draws on previous work in Emsley *et al.* (2010) and Dunn and Bentall (2007), and in Emsley *et al.* (2011).

We consider first the setting where Y and M are both continuous. A structural mean model (SMM) is a model relating the potential outcomes $Y_i(a,m)$ to one another or to $Y_i(0,0)$. If we assume a linear model for the potential outcome $Y_i(0,0)$ in terms of the set of measured baseline covariates, X_i, then we could write

$$Y_i(a,m) = \sum_k \lambda_k X_{ik} + \beta m + \gamma a + \varepsilon_i$$

for all values of a and m, with ε being independent of A but not X and M, i.e. $E[\varepsilon_i|A = a] = 0$. Our aim is to estimate the causal parameters β and γ, which in the presence of hidden confounding between M and Y are not identified using standard linear regression (Robins, 1994). Fischer-Lapp and Goetghebeur (1999, 1997) described the following procedure for estimating the parameters of an SMM using g-estimation:

1. Estimate $E[Y_i \mid \mathrm{X}_i, A_i = 1]$ using linear regression and predict for the whole sample.
2. Estimate $E[Y_i \mid \mathrm{X}_i, A_i = 0]$ using linear regression and predict for the whole sample.
3. Estimate $E[M_i \mid \mathrm{X}_i, A_i = 1]$ using linear regression and predict for the whole sample.
4. Estimate $E[M_i \mid \mathrm{X}_i, A_i = 0]$ using linear regression and predict for the whole sample.
5. Calculate the compliance score, which is defined as the difference between the estimates of $E[M_i \mid \mathrm{X}_i, A_i = 1]$ and $E[M_i \mid \mathrm{X}_i\ A_i = 0]$ (Joffe *et al*., 2003).
6. Regress the difference between the estimates of $E[Y_i \mid \mathrm{X}_i, A_i = 1]$ and $E[Y_i \mid \mathrm{X}_i, A_i = 0]$ on the compliance score.

The whole procedure is bootstrapped to ensure that valid standard errors are obtained. Goetghebeur and Vansteelandt (2005) show how g-estimation consistently estimates mediator effects in the presence of hidden confounding and random errors in the mediators (compliance or dose-response effects in the authors' examples).

It was demonstrated in Dunn and Bentall (2007) and Emsley *et al.* (2010) that this SMM procedure is identical to an instrumental variables model in which the interaction between random allocation A and baseline covariates (X^*A) are instruments, which therefore influence the mediator (M) but have no direct influence on outcome.

A common approach for fitting such an instrumental variables model is the two-stage least squares (2SLS) procedure, which would proceed as follows:

1. Estimate $E[M_i \mid X_i, A_i, X_i^* A_i]$ using linear regression.
2. Estimate $E[Y_i \mid X_i, A_i, E\{M_i \mid X_i, A_i, X_i^* A_i\}]$ using linear regression.

Performing the 2SLS procedure separately yields consistent estimates but the standard errors are invalid, and so a correction needs to be made. In practice, this procedure is straightforward to perform in most statistical software: in Stata (StataCorp, 2010), for example, we could use the following `ivregress` command:

```
ivregress 2sls y a x (m=a*x)
```

The causal relationship is the effect of the compliance score on the difference in expected outcomes (conditional on baseline covariates) between the two randomly allocated groups. Both of these procedures involve 2SLS algorithms to estimate various expected values, and it is the relationship between these two sets of estimated expected values that illustrates the equivalence of the procedures. In the instrumental variables procedure the use of these A by X interactions as instruments (and the implied exclusion restrictions by excluding these from the second-stage regression) is made explicit, whereas it is implicit in the SMM g-estimation procedure. We emphasise that the equivalence between the instrumental variable 2SLS and SMM g-estimation procedures in this linear case arises from our explicit introduction of group by covariate interactions in the prediction of the mediator (i.e. the estimation of the compliance score) together with the assumption that these covariates do not modify the causal effect of sessions attended on outcome.

We now consider the setting with a continuous outcome and a binary mediator. In this context Ten Have, Joffe and colleagues (Ten Have and Joffe, 2012; Ten Have *et al*., 2007) have proposed the use of rank-preserving models (RPM) estimated using g-estimation. The authors define the RPM as $Y(a,m) = g(\tilde{x}) + \theta_M m + \theta_A a + \varepsilon$ where $\theta_M = Y(a,1) - Y(a,0)$ represents the average effect of the mediator on outcome at any fixed level of $A = a$, $\theta_A = Y(1,m) - Y(0,m)$ represents the average controlled direct effect of random allocation on outcome holding the mediator fixed at any level m, and $g(x)$ is an unspecified function, such as $g(X) = \beta^T X$.

Ten Have *et al.* (2007) show how consistent estimates of θ_M and θ_A can be obtained by iteratively solving the estimating equation

$$\sum (A-q)W(x)(Y(0,0)(\theta) - \hat{\beta}^T x) = 0$$

using a Newton–Raphson routine where $q \equiv \Pr(A=1)$; $\hat{\beta}$ is obtained from a linear regression of $Y(0,0)(\theta)$ on X given an estimate of θ from the previous iteration, where

$$W(x)^T = [1, \eta(x)] \quad \text{and} \quad \eta(x) = \Pr(M=1|X, A=1) - \Pr(M=1|X, A=0)$$

Note that again $\eta(x)$ defines the compliance score applied to the binary mediator case.

In Emsley *et al.* (2011), we show that this procedure is identical to an instrumental variable 2SLS estimation procedure using $(A-q)\eta(x)$ as an instrumental variable for M. This procedure can by fitted using the same standard statistical software as previously, for example using the following procedure:

1. Estimate $\Pr[M_i = 1 \mid X_i, A_i = 1]$ using a logistic regression and predict for the sample.
2. Estimate $\Pr[M_i = 1 \mid X_i, A_i = 0]$ using a logistic regression and predict for the sample.
3. Calculate $(A_i - q)\{\Pr[M_i = 1 \mid X_i, A_i = 1] - \Pr[M_i = 1 \mid X_i, A_i = 0]\}$.
4. Fit a 2SLS procedure with $(A_i - q)\{\Pr[M_i = 1 \mid X_i, A_i = 1] - \Pr[M_i = 1 \mid X_i, A_i = 0]\}$ as an instrument for M, for example `ivregress 2sls y a x (m = (a-q)*η)`.

We could distinguish between different sets of covariates to be used in calculating the compliance score and the final 2SLS procedure; however, this has the effect of assuming that the set of covariates used in the compliance score regressions meet the assumptions of instruments (Emsley *et al.*, 2011).

We now illustrate how the use of a baseline covariate by random allocation interactions is used for model identification and parameter estimation. Considering a binary pre-baseline covariate, X, we make the key assumption that X influences the total effect (τ) through its effect on the level of mediation (α), but that X does not modify either the direct effect of the intervention (γ) or the effect of the mediator on the outcome (β). For $X = 1$ we have $\tau_1 = \gamma + \alpha_1\beta$ and similarly for $X = 2$ that $\tau_2 = \gamma + \alpha_2\beta$. Given these expressions, simple algebra gives $\tau_1 - \tau_2 = (\alpha_1 - \alpha_2)\beta$ so that $\beta = (\tau_1 - \tau_2)/(\alpha_1 - \alpha_2)$. Each of the values on the right-hand side are observable by regressing Y and M on A separately in each level of X (possibly while adjusting for other covariates). So β is now identified, as is γ, by back substitution.

In general, if the baseline covariate, X, has many levels then $\tau_x = \gamma + \alpha_x\beta$, and so γ and β are, respectively, the intercept and slope of the straight line relating the total effect (τ) at each level of X to the effect of random allocation on the mediator (α) at that level of X. As the baseline covariate is influencing the size of the effect of random allocation on the mediator, we have an X by A interaction in the structural model for $M(a)$. There is no interaction in the model for $Y(a,m)$ and therefore the interaction is an instrumental variable.

Having demonstrated the use of instrumental variables estimation, we now turn to the issue of finding suitable instruments, in particular finding pre-baseline variables that, when interacted with random allocation, meet the necessary identification assumptions. We note that this also coincides with the concept of mediated moderation (MacKinnon, 2008). Following the lead of other authors who have considered this issue (Gennetian *et al.*, 2005, 2008), we highlight some possibilities in the context of psychotherapy trials:

- Use of a multicentre trial. If random allocation has a differential effect on the mediators within each of the treatment centres then consider centre by random allocation interactions as instruments.
- Randomise to multiple treatments. The number of randomised treatments determines the maximum number of instruments that can be created, which, in turn, determines the maximum number of confounded mediator pathways to be investigated. The nature of the treatments can be chosen such that they are aimed at differentially influencing different mediators.
- Joint analysis of several similar trials. This is conceptually very similar to the use of a single multisite trial but combines several trials and uses trial by random allocation interactions as instruments. In practice, it may be difficult to find such trials since they would require common measures to be collected as the same time points.
- Use of random allocation by baseline covariate interactions.
- Measure baseline variables that predict variation in the mediators but not the outcome, which can then be used as instruments themselves without the interaction, for example genotypes that predict the mediator (phenotype) as applied in mendelian randomisation (Lawlor *et al.*, 2008).

We have demonstrated how instrumental variable estimation procedures can be utilised to estimate the parameters of both SMM and RPMs, for continuous and binary mediators respectively. In Section 20.6, we will illustrate instrumental variable estimation of a structural mean model utilising random allocation by centre interactions, with random allocation by baseline covariate interactions as instruments.

20.5 Causal mediation analysis using principal stratification

In Section 20.3 we proposed two scenarios in which we argue that principal stratification is a useful approach for mediation analysis in psychotherapy trials. In this section, we discuss how the principal strata direct effect (PSDE) can be estimated and address some of the limitations of principal stratification.

Under random allocation A, we consider again a binary mediator ($m = 0,1$), meaning there are two possible values for each potential outcome $M_i(a)$ and there are then four distinct classes for the joint combination of $(M_i(1), M_i(0))$: (0,0), (1,1), (1,0) and (0,1). Within each of these four principal stratum C, we can define the proportion of patients in that class as π^c, and the stratum-specific average treatment effects as τ^c, which can differ between classes. Since class membership is independent of random allocation (i.e. $M_i(1), M_i(0) \perp A_i$), the total effect τ is the weighted average of the within strata effects (i.e. $\tau = \sum_c \pi^c \tau^c$). Recall that the principal strata direct effects of treatment are observed in those classes where $M_i(1) = M_i(0)$ (Vansteelandt, Chapter 11 in this volume).

In this example, a patient's class membership is not known such that if $A_i = 1$ and $M_i = 1$ the patient is only known to belong to the (1,1) or (1,0) classes. Class membership is therefore a latent variable and, viewed in this context, principal stratification ultimately becomes an application of finite mixture modelling, where estimation can be achieved by specifying a full probability model using maximum likelihood, with covariates predicting class membership. It is possible to fit a latent class model for stratum membership and simultaneously a further regression model for the effect of random allocation on outcome within each of the principal strata, usually allowing for the same baseline covariates, for example using the finite mixture model option in Mplus (Muthén and Muthén, 2010). This concept has been developed in the context of noncompliance, using models relating principal stratum membership to baseline covariates, and Bayesian methods or ML/EM algorithms (Jo, 2002b; Jo and Muthén, 2001, 2002; Little and Yau, 1998).

While it is not possible to identify the class membership of a patient, it is possible to determine the most likely classification of stratum for each patient based on the posterior probability of belonging to each class (Muthén, 2002). In the setting we introduced in Section 20.3, where the mediator is only measured in the treatment arm, we know that $M_i(0)$ is undefined for all patients and the identification issue becomes one of estimating $M_i(1)$ when $A_i = 0$, i.e. which of the classes the control arm would belong to, had they been randomly allocated to the treatment arm. In the case of the therapeutic alliance example, we could model the high versus low alliance classes as an outcome using a logistic regression with baseline covariates for the $A_i = 1$ group, and then use this model to calculate posterior probabilities for the $A_i = 0$ group. The identification relies on being able to predict principal strata from baseline covariates, which can also in turn allow the exclusion restriction to be relaxed. This approach is analogous to using the baseline covariate by random allocation interactions as instrumental variables in 2SLS estimation (Emsley *et al.*, 2010).

Chapters 11 by Vansteelandt and 12 by Pearl pointed out some limitations to the principal strata approach. A valid criticism is that this approach can only be conceptualised for binary or categorical mediators, because with continuous mediators there may be unpopulated principal strata where $M_i(0) = M_i(1)$. A further criticism of the principal strata approach is that it cannot correspond to a definition of an indirect effect; however, that is arguably not its purpose if it is being applied to separate the sample into homogeneous classes in order to estimate a PSDE. Furthermore, it is possible to test whether the PSDE is the same between strata by imposing constraints on the model. The essence of the finite mixture model is the evaluation of between-group parameters in models fitted to each group simultaneously, and these model parameters can be left free to vary or be constrained to be equal. If the models are estimated by maximum likelihood and imposing the constraints viewed as model nesting, we can test these equality constraints using, for example, likelihood ratio tests.

When viewing principal stratification as belonging to the mixture model framework, it is also possible to move beyond estimating the principal strata direct effect and investigate other models within principal strata. For example, we have focused on a univariate response, but alternatively we could investigate the advantages of simultaneously estimating direct effects for two or more different outcomes (i.e. multivariate responses). It is also possible to examine multivariate binary outcomes, one of which might be a missing outcome indicator, a notion suggested by Frangakis and Rubin (1999) introducing the concept of latent ignorability. We will illustrate the alternative model idea by looking at the instrumental variables models within principal strata in the next section.

20.6 Our motivating example: The SoCRATES trial

The SoCRATES (Study of Cognitive Re-Alignment Therapy in Early Schizophrenia) trial was designed to evaluate the effects of cognitive behaviour therapy and supportive counselling on the outcomes of an early episode of schizophrenia. It was a three-centre prospective, randomised controlled trial with patients allocated to one of three conditions: cognitive behaviour therapy (CBT) in addition to treatment as usual (TAU), supportive counselling (SC) and TAU, or TAU alone. Further details and the trial outcomes have been reported in detail (Lewis *et al.*, 2002; Tarrier *et al.*, 2004) including aspects of this analysis (Emsley *et al.*, 2010). Briefly, in an ITT analysis of the 18 month follow-up data both the CBT and SC groups had a superior outcome in terms of psychotic symptoms compared to the TAU group, but there were no differences in the effects of CBT and SC. Therefore, for our illustrative purposes we ignore the distinction between CBT and SC, treating both as a combined 'treatment' group.

In summary, 207 patients were allocated to our treatment group and 102 to TAU alone, of which 104 patients in the treatment arm and 71 in the TAU alone arm were interviewed at 18 month follow-up. The primary outcome measure at 18 months was the PANSS (Positive and Negative Syndromes Schedule), an objective interview-based scale for rating psychotic and nonpsychotic symptoms, with scores ranging from 30 to 210, where low scores indicated an improved outcome. The potential mediators are the total number of sessions of therapy actually attended and the therapeutic alliance measured at the fourth session of therapy using a patient rated measure of alliance, with total scores ranging from 0 (low alliance) to 7 (high alliance). The analyses reported in the present article are based on all of the TAU group patients but only includes those from the treatment group who provided both an alliance score at the fourth session of therapy and the total number of sessions attended ($n = 114$).

There were large differences between the three centres in the mean PANSS scores at baseline and 18 months. There were also large differences in the ITT effects within the three centres, with one centre having a significant ITT effect, unlike the other two. The treated group in this centre also attended the highest number of therapy sessions and had higher levels of therapeutic alliance. Therefore, the specific questions we illustrate using SoCRATES are:

1. What are the joint effects of sessions attended and therapeutic alliance on the PANSS score at 18 months?
2. What is the direct effect of random allocation on the PANSS score at 18 months and how is this influenced by the therapeutic alliance?
3. Is the direct effect of the number of sessions attended on the PANSS score at 18 months influenced by therapeutic alliance?

20.6.1 What are the joint effects of sessions attended and therapeutic alliance on the PANSS score at 18 months?

Recall that we can define the individual treatment effect as a comparison of the outcome actually observed with the corresponding treatment free outcome, i.e. $\tau_i = Y_i(s,m) - Y_i(0)$. The first question asks how is τ_i influenced jointly by the number of sessions (s) and the strength of the therapeutic alliance (m)? We postulate the following structural mean model, with a linear dose–response relationship for s and m acting solely as an effect modifier:

$$E[\tau_i | S_i = \text{s and } M_i = m] = \beta_s s + \beta_{sm} sm$$

The absence of a constant in the model implies that receiving no sessions of therapy produces no treatment effect; this is the exclusion restriction, expressed as $E[\tau_i \mid S_i = 0] = 0$. The effect of alliance is multiplicative and there is no main effect of alliance on outcome since alliance is undefined unless $s > 0$. This follows as a corollary to the exclusion restriction that there is no modification by M: $E[\tau_i \mid M_i = m, S_i = 0] = E[\tau_i \mid S_i = 0] = 0$. Note also that $s = 0$ when $A = 0$ because patients in the TAU arm had no access to therapy. This model also takes into account the undefined m when $s = 0$, if we assume that the interaction term $sm = 0$ when $s = 0$, for all m.

We rescale the therapeutic alliance to have a maximum value of 0, which means that β_s is the effect of one additional session of therapy at the maximum level of alliance. We include the following baseline covariates in the model: PANSS total score at baseline, dummy variables for treatment centre, logarithm of the duration of untreated psychosis and number of years of education. The analysis is performed with 1000 bootstrap samples.

We estimate the parameters β_s and β_{sm} of the structural mean model using the instrumental variable approach outlined in Section 20.4. As we are not assessing the direct effect of random allocation in this model, we assume that this acts as an instrumental variable for the effect of sessions on outcome. Our model is still not identified without additional instruments since sm is also confounded and so, as outlined in Section 20.4, we use random allocation by covariate interactions as instruments, using interactions with the complete set of covariates in this example. Here, we are especially utilising the variation in the number of sessions attended and therapeutic alliance between the treatment centres. In effect, we assume that the entire centre effect observed in the ITT analysis is due to this variation in sessions and alliance between centres, but that the effects of sessions and its interaction with alliance on outcome is homogeneous across centres. This analysis is performed in Stata with the following command, replacing the variable names as required:

```
ivregress 2sls Y A X1 X2 X3 (S SM = AX1 AX2 AX3)
```

We compare these estimates with those obtained from a standard regression analysis estimated using OLS, performing a complete case analysis. A more detailed analysis including adjustment for missing outcome data is presented in Dunn and Bentall (2007). The instrumental variable estimates for β_s and β_{sm} are –2.40 (standard error 0.70) and –1.28 (s.e. 0.48). Since both coefficients are negative and significant (recalling that a low PANSS outcome is beneficial), this implies that increasing the number of sessions improves the effect of therapy on outcome, and this effect is greater when it is associated with a stronger therapeutic alliance. Conversely, at the minimum level of therapeutic alliance, the effect of attending additional sessions of therapy appears detrimental (the expected treatment effect is $-2.40s + (-7 \times -1.28)s = +6.55s$). The corresponding OLS estimates β_s and β_{sm} are –0.95 (s.e. 0.22) and –0.39 (s.e. 0.11).

20.6.2 What is the direct effect of random allocation on the PANSS score at 18 months and how is this influenced by the therapeutic alliance?

To answer this question we illustrate the principal stratification approach using the dichotomised alliance variable ($M = 1$ when alliance >5, 0 otherwise). We form two principal strata, the low alliance and the high alliance classes, and are interested in estimating the direct effect of random allocation within each of these two classes; these are the principal stratum direct effects. These classes are predicted by the same set of baseline covariates as previously, and we use these same covariates plus the effect of random allocation to model outcome within principal strata. We first allow the ITT effect in the low alliance class to be freely estimated and then repeat the analysis imposing an exclusion restriction by forcing this effect to be 0, in order to evaluate the impact of this constraint on the ITT effect in the high alliance class. We show results for a missing at-random mechanism assumption for the missing PANSS outcomes. We perform bootstrapping with 250 replications to obtain standard errors. Further details, including models allowing for latently ignorable outcome data, and the Mplus code used to generate these results, are available in Emsley *et al.* (2010).

We find that the ITT effect in the low alliance class is +7.50 (s.e. 8.18), which although this is nonsignificant suggests that psychological treatment could be detrimental in this class. The corresponding ITT effect estimate in the high alliance class is –15.46 (s.e. 4.61), which shows a significant beneficial effect of psychological treatment. Imposing the zero ITT constraint (exclusion restriction) in the low alliance class decreases the estimated effect in the high alliance class to –12.73 (s.e. 4.75). We would therefore conclude that the direct effect of random allocation is influenced by the therapeutic alliance, and while random allocation to psychological therapy is beneficial in the high alliance class, it may be detrimental in the low alliance class.

Table 20.1 Estimated effect of sessions (estimate (SE)) attended on PANSS at 18 months using instrumental variables SEM under MAR assumption in SoCRATES; α is the effect of random allocation on number of sessions and β is the effect of sessions on outcome.

	Low alliance class	High alliance class
α: effect of A on S	14.90 (0.97)	16.95 (0.46)
β: effect of S on Y	+0.37 (0.47)	−0.80 (0.29)
α: effect of A on S	14.84 (0.98)	16.94 (0.46)
β: effect of S on Y	0 (constraint)	−0.71 (0.26)

20.6.3 Is the direct effect of the number of sessions attended on the PANSS score at 18 months influenced by therapeutic alliance?

To answer this final question, we now illustrate the use of instrumental variable models within principal strata to estimate the effect of number of sessions attended on the PANSS score at 18 months within each of the two principal strata as estimated previously. To allow for the likely selection effects we use random allocation as an instrument for the number of sessions. Within each stratum we are interested in a regression of PANSS at 18 months on the number of sessions and the set of covariates (this estimates β), with a zero intercept implying that attending no sessions of therapy implies there is no treatment effect. We can also introduce an exclusion restriction (zero constraint) for the effect of sessions on outcome in the low alliance class. For a more expansive analysis involving standard regression estimates and latent ignorability assumptions, see Table 5 in Emsley *et al.* (2010).

Table 20.1 presents the results from this analysis. Since there are no sessions of therapy in the control group, the estimate of α is simply the mean number of sessions attended in the treatment group within each of the classes; we observe that, on average, more sessions are attended in those patients in the high alliance class than the low alliance class (16.95 to 14.90). More pertinent are the estimates of β, that is the effect of attending one additional session of therapy on outcome within each class. In the low alliance class this is +0.37 (s.e. 0.47), whereas in the high alliance class the estimate is –0.80 (0.29). Introducing the exclusion restriction in the low alliance class again reduces the corresponding estimate in the high alliance class. As previously with the $S + SM$ model, we show that in the low alliance patients each additional session of the psychological intervention may be detrimental, whereas in the high alliance patients there is a beneficial effect.

20.7 Conclusions

Causal extensions to the traditional regression-based approaches for mediation have been demonstrated elsewhere in this volume (Chapter 11) but rely on assumptions that are open to challenge. As an alternative approach, by making different assumptions, we have illustrated the use of instrumental variable estimation and principal stratification to evaluate potential mediators in psychological treatments, and demonstrated their ease of implementation in standard statistical software.

Perhaps inevitably, in order to make progress we have also utilised assumptions that are difficult to verify. Specifically these are:

1. The effect of random allocation on the mediator is moderated by one or more baseline covariates.
2. The direct effect of random allocation on the final outcome is not moderated by these baseline covariates (moderation by other covariates is still possible).
3. The effect of the putative mediator on the final outcome is neither moderated by random allocation nor these baseline covariates (moderation by other covariates is still possible).

Assumptions 1 and 2 above encode the instrumental variable assumptions, with assumption 2 defining the exclusion restriction. We cannot test this empirically in the data. In psychotherapy trials, even the use of random allocation may not meet the exclusion restriction. One concern is that there is no blinding and so patients are fully aware of their random allocation, and this in turn could lead to resentful demoralisation in patients in the control arm where the demoralisation from not being offered the intervention could negatively influence the patient's outcome (Shadish *et al.*, 2002). Jo (2002a) has described alternative model assumptions for identifiability when using instrumental variables for noncompliance and gave two settings where the exclusion restriction can be relaxed with additional covariates.

Alternatively, in the case of overidentification (there are more instruments than confounded mediators) we can relax each of the constraints in turn using a specification test (Hausman, 1978). There is also an extensive literature on Bayesian analysis for instrumental variables, including sensitivity for violations of the exclusion restriction (Imbens and Rubin, 1997; Kraay, 2010).

Finally, we have limited our discussion in this chapter to the mediation diagram in Figure 20.1 that contains only one mediator and one outcome, each measured on a single occasion. Of course, the more realistic scenario is likely to have been repeated measures of outcomes with multiple mediators, which themselves could be measured repeatedly, and there will be missing data. Readers will have observed that the challenges of the 'simple' setting are considerable and these challenges only increase when we consider the more realistic scenario. In order to make progress in estimating direct and mediated effects, regardless of the statistical method used, we ultimately have to impose unverifiable assumptions and argue for the validity of these in each individual application.

Acknowledgements

Both R.E. and G.D. are members of the UK Mental Health Research Network (MHRN) Methodology Research Group, which is led by G.D. Research funding is provided by the UK Medical Research Council (Grant Numbers G0600555 and G0900678) and R.E. holds a UK Medical Research Council Career Development Award in Biostatistics (G0802418). We thank Shôn Lewis, Nick Tarrier and the rest of the SoCRATES team for the use of their trial data.

References

Dunn, G. and Bentall, R. (2007) Modelling treatment-effect heterogeneity in randomized controlled trials of complex interventions (psychological treatments). *Statistics in Medicine*, **26** (26), 4719–4745.

Elliott, M.R., Raghunathan, T.E. and Li, Y. (2010) Bayesian inference for causal mediation effects using principal stratification with dichotomous mediators and outcomes. *Biostatistics*, **11** (2), 353–372.

Emsley, R., Dunn, G. and White, I.R. (2010) Mediation and moderation of treatment effects in randomised controlled trials of complex interventions. *Statistical Methods in Medical Research*, **19** (3), 237–270.

Emsley, R., Windmeijer, F., Liu, H., Clarke, P., White, I.R. and Dunn, G. (2011) *Equivalence of Rank Preserving Structural Models and Instrumental Variable Models for Causal Mediation Analysis*, Biostatistics Group Working Paper, University of Manchester.

Fischer-Lapp, K. and Goetghebeur, E. (1999) Practical properties of some structural mean analyses of the effect of compliance in randomized trials. *Controlled Clinical Trials*, **20** (6), 531–546.

Frangakis, C.E. and Rubin, D.B. (1999) Addressing complications of intention-to-treat analysis in the combined presence of all-or-none treatment-noncompliance and subsequent missing outcomes. *Biometrika*, **86** (2), 365–379.

Frangakis, C.E. and Rubin, D.B. (2002) Principal stratification in causal inference. *Biometrics*, **58** (1), 21–29.

Gallop, R., Small, D.S., Lin, J.Y., Elliott, M.R., Joffe, M. and Ten Have, T.R. (2009) Mediation analysis with principal stratification. *Statistics in Medicine*, **28** (7), 1108–1130.

Gennetian, L.A., Morris, P.A., Bos, J.M. and Bloom, H.S. (2005) Constructing instrumental variables from experimental data to explore how treatments produce effects, in *Learning More From Social Experiments: Evolving Analytic Approaches* (ed. H.S. Bloom), 1st edn. New York: Russell Sage Foundation, pp. 75–114.

Gennetian, L.A., Magnuson, K. and Morris, P.A. (2008) From statistical associations to causation: what developmentalists can learn from instrumental variables techniques coupled with experimental data. *Developmental Psychology*, **44** (2), 381–394.

Goetghebeur, E. and Lapp, K. (1997) The effect of treatment compliance in a placebo-controlled trial: regression with unpaired data. *Applied Statistics – Journal of the Royal Statistical Society Series C*, **46** (3), 351–364.

Goetghebeur, E. and Vansteelandt, S. (2005) Structural mean models for compliance analysis in randomized clinical trials and the impact of errors on measures of exposure. *Statistical Methods in Medical Research*, **14** (4), 397–415.

Green, J. (2006) Annotation: the therapeutic alliance – a significant but neglected variable in child mental health treatment studies. *Journal of Child Psychology and Psychiatry*, **47** (5), 425–435.

Hausman, J.A. (1978) Specification tests in econometrics. *Econometrica*, **46** (6), 1251–1271.

Imbens, G.W. and Rubin, D.B. (1997) Bayesian inference for causal effects in randomized experiments with noncompliance. *Annals of Statistics*, **25** (1), 305–327.

Jo, B. (2002a) Estimation of intervention effects with noncompliance: alternative model specifications. *Journal of Educational and Behavioral Statistics*, **27** (4), 385–409.

Jo, B. (2002b) Model misspecification sensitivity analysis in estimating causal effects of interventions with non-compliance. *Statistics in Medicine*, **21** (21), 3161–3181.

Jo, B. (2008) Causal inference in randomized experiments with mediational processes. *Psychological Methods*, **13** (4), 314–336.

Jo, B. and Muthén, B.O. (2001) Modeling of intervention effects with noncompliance: a latent variable approach for randomized trials, in *New Developments and Techniques in Structural Equation Modeling* (eds G.A. Marcoulides and R.E. Schumacker). Mahwah, NJ: Lawrence Erlbaum Associates, pp. 57–87.

Jo, B. and Muthén, B.O. (2002) Longitudinal studies with intervention and noncompliance: estimation of causal effects in growth mixture modeling, in *Multilevel Modeling: Methodological Advances, Issues, and Applications* (eds N. Duan and S. Reise). Mahwah, NJ: Lawrence Erlbaum Associates, pp. 112–139.

Joffe, M.M., Ten Have, T.R. and Brensinger, C. (2003) The compliance score as a regressor in randomized trials. *Biostatistics*, **4** (3), 327–340.

Kraay, A. (2010) Instrumental variables regressions with uncertain exclusion restrictions: a Bayesian approach. *Journal of Applied Econometrics*.

Lawlor, D.A., Harbord, R.M., Sterne, J.A.C., Timpson, N. and Smith, G.D. (2008) Mendelian randomization: using genes as instruments for making causal inferences in epidemiology. *Statistics in Medicine*, **27** (8), 1133–1163.

Lewis, S., Tarrier, N., Haddock, G., Bentall, R., Kinderman, P., Kingdon, D., Siddle, R., Drake, R., Everitt, J., Leadley, K., Benn, A., Grazebrook, K., Haley, C., Akhtar, S., Davies, L., Palmer, S., Faragher, B. and Dunn, G. (2002) Randomised controlled trial of cognitive-behavioural therapy in early schizophrenia: acute-phase outcomes. *British Journal of Psychiatry*, **181**, S91–S97.

Little, R. J. and Yau, L. (1998) Statistical techniques for analysing data from prevention trials: treatment of no-shows using Rubin's causal model. *Psychological Methods*, **3** (2), 147–159.

MacKinnon, D.P. (2008) *Introduction to Statistical Mediation Analysis*, 1st edn. New York: Taylor & Francis Group.

Muthén, B.O. (2002) Beyond SEM: general latent variable modeling. *Behaviormetrika*, **29** (1), 81–117.

Muthén and Muthén (2010) *MPlus Version 6.1*, Computer Program. Muthén and Muthén.

Pearl, J. (2001) Direct and indirect effects, in *Proceedings of the Seventeenth Conference on Uncertainty in Artificial Intelligence*, pp. 411–420.

Robins, J.M. (1994) Correcting for non-compliance in randomised trials using structural nested mean models. *Communications in Statistics – Theory and Methods*, **23** (8), 2379–2412.

Robins, J.M. and Greenland, S. (1992) Identifiability and exchangeability for direct and indirect effects. *Epidemiology*, **3** (2), 143–155.

Rubin, D.B. (1980) Randomization analysis of experimental-data – the Fisher randomization test – comment. *Journal of the American Statistical Association*, **75** (371), 591–593.

Shadish, W.R., Cook, T.D. and Campbell, D.T. (2002) *Experimental and Quasi-Experimental Designs*. Boston, MA: Houghton Mifflin Company.

StataCorp. (2010) *Intercooled Stata Statistical Software: Release 11.1*, Computer Program. College Station, TX: Stata Corporation.

Tarrier, N., Lewis, S., Haddock, G., Bentall, R., Drake, R., Kinderman, P., Kingdon, D., Siddle, R., Everitt, J., Leadley, K., Benn, A., Grazebrook, K., Haley, C., Akhtar, S., Davies, L., Palmer, S. and Dunn, G. (2004) Cognitive-behavioural therapy in first-episode and early schizophrenia – 18-month follow-up of a randomised controlled trial. *British Journal of Psychiatry*, **184**, 231–239.

Ten Have, T.R. and Joffe, M. (2012) A review of causal estimation of effects in mediation analyses. *Statistical Methods in Medical Research*, **21** (1), 77–107.

Ten Have, T.R., Joffe, M.M., Lynch, K.G., Brown, G.K., Maisto, S.A. and Beck, A.T. (2007) Causal mediation analyses with rank preserving models. *Biometrics*, **63** (3), 926–934.

VanderWeele, T.J. (2009) Concerning the consistency assumption in causal inference. *Epidemiology*, **20** (6), 880–883.

VanderWeele, T.J. (2010) Bias formulas for sensitivity analysis for direct and indirect effects. *Epidemiology*, **21** (4), 540–551.

VanderWeele, T.J. and Vansteelandt, S. (2009) Conceptual issues concerning mediation, interventions and composition. *Statistics and Its Interface*, **2** (4), 457–468.

Walwyn, R. and Roberts, C. (2010) Therapist variation within randomised trials of psychotherapy: implications for precision, internal and external validity. *Statistical Methods in Medical Research*, **19** (3), 291–315.

21

Causal inference in clinical trials

Krista Fischer[1,2] and Ian R. White[1]
[1]MRC Biostatistics Unit, Cambridge, UK
[2]Estonian Genome Center, University of Tartu, Estonia

21.1 Introduction

In observational studies in medicine, the main complication in assessing the causal effect of exposures is the presence of possible unobserved confounding. This means that people at different exposure levels can also be different with respect to unobserved factors that influence the outcome (e.g. health status) of interest. Therefore the observed outcome differences between exposed and unexposed individuals cannot necessarily be attributed to the exposure of interest.

Such problems can be avoided in randomized controlled trials (RCTs), where random allocation of subjects to different exposure arms guarantees that any systematic differences in the outcomes across arms can only be explained by the different exposure levels. One may well say that RCTs are designed to answer causal questions about the effect of certain medical treatments or other interventions. This chapter focuses on clinical RCTs, hereafter simply called 'randomized trials' or just 'trials'.

In an ideal randomized trial there is one-to-one correspondence between the randomized treatment assignment and actually received treatment exposure for all study subjects. Outcomes of such a study would allow one to estimate the average effect size of the specific exposure in the exact way it is administered (dose, timing, etc.) in the randomized subjects and possibly generalize to the population they represent.

In reality, randomized trials conducted on humans rarely turn out to be ideal. The main discrepancy from the ideal randomized trial is often the presence of protocol *nonadherence*. In trials comparing an experimental treatment with a control regime (a standard treatment,

Causality: Statistical Perspectives and Applications, First Edition. Edited by Carlo Berzuini, Philip Dawid and Luisa Bernardinelli.

placebo or no treatment), there are two main kinds of nonadherence. The first is *noncompliance*, where some subjects randomized to the experimental treatment do not receive it or receive it in a different way than assigned (e.g. a smaller dose). The second is *contamination*, where some control arm individuals actually receive the experimental treatment.

The primary analysis approach in such trials is *intention to treat (ITT)* analysis, which compares average outcomes on the randomized arms regardless of the actual exposure levels. Such comparisons provide valid estimates of the effect of treatment *assignment*. One may argue that ITT analysis estimates the average health impact of making the treatment generally used, as patients outside randomized trials also do not always comply with their assigned treatments. On the other hand, it is unlikely that treatment adherence in a trial would closely mimic adherence in practice, as knowledge of the experimental status of the treatment (and, sometimes, awareness of the possibility of actually receiving a placebo) could have a strong impact on compliance. One may still see the ITT effect in the presence of nonadherence as a diluted estimate of the actual treatment efficacy and wish to adjust the analysis for adherence, to learn more about the possible 'true' effect of the treatment.

Naive approaches would either restrict the analysis to adherent patients in a so-called per-protocol analysis or compare the patients with respect to their actual (rather than assigned) exposure in an as-treated analysis. This would mean regrouping the patients to nonrandom and noncomparable subgroups, leading to possible biases caused by unobserved confounding. In other words, by conducting such analyses one would lose the advantage of randomization and be back in the situation of an observational study.

Recent decades have seen considerable progress in the field of adherence-adjusted analysis of randomized trials. For simple 'one-shot' treatments where exposure could be characterized as a binary variable, *principal stratification* approaches have been recommended that aim to estimate the effect of treatment in the latent subgroup of *compliers* – individuals who would potentially adhere under either assignment. One of the first publications on this approach was by Sommer and Zeger (1991), illustrating a simple principal stratification approach on a vitamin A trial in rural Indonesia. In their example, villages were randomized either to receive or not to receive vitamin A for all infants to reduce infant mortality. Some of the villages randomized to treatment did not receive the vitamin supplies, but no contamination was observed within the control villages. Angrist *et al.* (1996), Cuzick *et al.* (1997) and others have extended the methodology to allow for contamination.

For more general settings with possibly continuous observed dosage levels (or other forms of compliance), instrumental variable (IV) approaches have been proposed. Robins (1994) proposed a general framework for structural nested mean models, allowing for repeated outcome measurements. A simplification of these ideas for univariate outcomes, structural mean models, has been explored more in practical settings (Goetghebeur and Lapp, 1997; Fischer-Lapp and Goetghebeur, 1999), with generalized structural mean models extending these ideas to nonnormal outcomes (Vansteelandt and Goetghebeur, 2003).

In this chapter, we describe various causal models that may be used in randomized trials, focusing on their estimands, their assumptions and estimation procedures. Our discussion will be based on the example of the Estonian postmenopausal hormone therapy (EPHT) trial (Veerus *et al.*, 2006a), which started in 1999 and was stopped prematurely in the beginning of 2004 (due to negative results of large international trials). The EPHT trial had two subtrials: a double-blind placebo-controlled trial and an open-label trial. Here we analyse the effect of hormone therapy (HT) on average healthcare costs per year for randomized women in the open-label subtrial (as in this trial the participants' health behaviour would have a better resemblance to a nontrial situation than in the blind trial). As there were some women (3.5 %) on the HT arm who never initiated HT and a considerable proportion (26.4 %) on the control arm who did initiate HT, it is of interest to adjust the analysis for noncompliance and contamination.

21.2 Causal effect of treatment in randomized trials

21.2.1 Observed data and notation

Consider a trial where n subjects ($i = 1, \ldots, n$) have been randomized to either an experimental treatment or control. Suppose the following data have been observed:

R_i: assignment indicator, with $R_i = 1$ for experimental and $R_i = 0$ for control arm subjects.

Z_i: a summary of actually received experimental treatment (depending on R, Z measures either compliance or contamination). We will first consider only univariate summaries, but later discuss possible generalizations for multivariate cases.

Y_i: an outcome variable. Here we will confine our discussion to binary or continuous outcome types.

$\mathbf{X}_i$: a vector of baseline covariates.

21.2.2 Defining the effects of interest via potential outcomes

The *individual causal effect* of a treatment assignment can be seen as a difference between the outcome the individual would have under the experimental treatment assignment and his/her potential outcome under the control assignment. Although the two outcomes are never jointly observed, we will use these potential or *counterfactual* outcomes to define the effects of interest.

Therefore we define:

$Y_i(0)$: the potential outcome of individual i under the control assignment (if R_i were set to 0).

$Y_i(1)$: the potential outcome of individual i under the experimental treatment assignment (if R_i were set to 1).

Depending on the actual value of R_i, either $Y_i(0)$ or $Y_i(1)$ is actually observed, so $Y_i = (1 - R_i)Y_i(0) + R_iY_i(1)$. Now we define:

$Y_i(1) - Y_i(0)$: the *individual causal effect of the experimental treatment assignment* compared to placebo.

Throughout this chapter we always use the term *treatment effect* to denote the effect of treatment assignment. The aim of the methodology presented here is to investigate the heterogeneity in individual treatment (assignment) effects, assuming the effect to be a function of an individual's treatment adherence. This is not equivalent to a dose–response analysis, which requires either much stronger assumptions (Efron and Feldman, 1991; Jin and

Rubin, 2008) or, preferably, a dose–response experiment where treatment doses are fixed by investigators.

Next, we study how the following summary statistics of the individual causal effect of treatment assignment can be estimated in the context of randomized trials: the overall average, the average conditional on baseline covariates and the average conditional on adherence and baseline covariates. The first two are estimated in the context of conventional intention to treat (ITT) analysis, but the third requires instrumental variables techniques.

21.2.2.1 The crude (unadjusted) ITT analysis via potential outcomes

The unadjusted intention to treat (ITT) analysis will test the null hypothesis of the (overall) average effect being zero:

$$H_0: \ \mathrm{E}[Y_i(1) - Y_i(0)] = 0$$

The ITT null hypothesis can be tested by comparing the average outcomes on the two randomized arms, given *the randomization assumption for potential outcomes:*

$$Y_i(0), Y_i(1) \perp\!\!\!\perp R_i \tag{21.1}$$

Now

$$\mathrm{E}[Y_i(1) - Y_i(0)] = \mathrm{E}[Y_i(1)|R_i = 1] - \mathrm{E}[Y_i(0)|R_i = 0] = \mathrm{E}[Y_i|R_i = 1] - \mathrm{E}[Y_i|R_i = 0]$$

and an appropriate test for the ITT null hypothesis is a test of

$$\mathrm{E}[Y_i|R_i = 1] = \mathrm{E}[Y_i|R_i = 0] = \mathrm{E}[Y_i]$$

(the standard ITT analysis).

21.2.2.2 Covariate-adjusted ITT analysis

One may wish to adjust the analysis for a set of measured baseline covariates $\mathbf{X}_i$ that are likely to be associated with the outcome, for the following reasons:

1. To correct for possible imbalances on the randomized arms with respect to some important covariates.
2. To reduce error variability in the two-sample comparison.
3. To assess whether some of the covariates are *treatment effect modifiers*.

For the adjusted analysis to be valid, one needs the *covariate-adjusted randomization assumption:*

$$Y_i(0), Y_i(1) \perp\!\!\!\perp R_i|\mathbf{X}_i \tag{21.2}$$

stating that the two randomized arms are comparable within levels of baseline covariates. In the adjusted analysis one assumes (21.2) while relaxing the unconditional assumption (21.1). The intention to treat (ITT) null hypothesis conditional on $\mathbf{X}$ is

$$H_0: \ \mathrm{E}[Y_i(1) - Y_i(0)|\mathbf{X}_i] = 0$$

Similarly to the unconditional analysis, under the randomization assumption (21.2), an equivalent hypothesis in terms of observed data is

$$H_0: \ \mathrm{E}[Y_i|R_i = 0, \mathbf{X}_i] = \mathrm{E}[Y_i|R_i = 1, \mathbf{X}_i] = \mathrm{E}[Y_i|\mathbf{X}_i]$$

and it can usually be tested by classical or generalized regression methods.

When the purpose of the adjusted analysis is to test for effect modification (reason 3), one needs to test whether $\mathrm{E}[Y_i(1) - Y_i(0)|\mathbf{X}_i]$ is a function of $\mathbf{X}_i$. A proper analysis approach here would be a regression model for $\mathrm{E}[Y_i|R_i, \mathbf{X}_i]$ that allows for interactions of the components of $\mathbf{X}_i$ with R_i and estimates the corresponding effect sizes. In other words, adjusted analysis would allow for *treatment–effect heterogeneity* between levels of $\mathbf{X}_i$.

21.2.3 Adherence-adjusted ITT analysis

21.2.3.1 Need for adherence adjustment and complications

To allow for treatment–effect heterogeneity in individuals at different adherence levels, one needs to adjust the analysis for treatment nonadherence.

Adjustment for treatment adherence cannot be done as easily as adjustment for baseline covariates, as adherence is a post-randomization characteristic, being itself influenced by randomization and possibly also by components of outcome. To understand the latter, consider a situation where both outcome and adherence are formed as time-dependent processes (the patient is told to take the antihypertensive medication every day, blood pressure is changing constantly), but only summaries of the final state are used for analysis (percentage of the prescribed drug taken is used as a compliance summary, final blood pressure after 2 months of therapy is used as an outcome variable). As the two processes are likely to influence each other, both of them can at the same time be caused by the other one. Besides such direct mutual influence of adherence and outcome processes, one can also think of third processes that are influencing both adherence and outcome.

21.2.3.2 Counterfactual exposures

As the observed adherence–outcome association is a mixture of different causal effects, it is easier to disentangle the effects of interest via *potential (counterfactual) exposures*, defined as:

$Z_i(1)$: the potential experimental treatment exposure of individual i under the experimental treatment assignment (if R_i were set to 1).

$Z_i(0)$: the potential experimental treatment exposure of individual i under the control assignment (if R_i were set to 0).

Here $Z_i(1)$ summarizes an individual's treatment compliance and would be observed together with outcome $Y_i(1)$ if the individual is assigned to the experimental arm: $R_i = 1$. Similarly,

$Z_i(0)$ and $Y_i(0)$ are observed for the control arm subjects and $Z_i = (1 - R_i)Z_i(0) + R_i Z_i(1)$. If randomization creates two comparable groups with respect to all observed and unobserved quantities, the following assumption is satisfied.

The *randomization assumption* for potential outcomes and exposure levels is

$$[Y_i(0), Y_i(1), Z_i(0), Z_i(1)] \perp\!\!\!\perp R_i|\mathbf{X}_i \tag{21.3}$$

Although the four-way distribution of the vector $[Y_i(0), Y_i(1), Z_i(0), Z_i(1)]$ is not observed, the marginal distribution of each of the four variables is observed in either the experimental treatment or the control arms.

21.2.3.3 Adherence-adjusted ITT effect and a range of feasible models

As $Y_i(1)$ is likely to be associated with $Z_i(1)$ and $Y_i(0)$ with $Z_i(0)$, the individual effect of assignment $Y_i(1) - Y_i(0)$ may depend on both $Z_i(1)$ and $Z_i(0)$. Therefore the *adherence-adjusted treatment (assignment) effect* or, equivalently, the adherence-adjusted ITT effect can be defined as

$$\mathrm{E}[Y_i(1) - Y_i(0)|Z_i(1), Z_i(0)]$$

or, if one additionally adjusts for $\mathbf{X}_i$,

$$\mathrm{E}[Y_i(1) - Y_i(0)|Z_i(1), Z_i(0), \mathbf{X}_i]$$

A *structural mean model (SMM)* for the adherence-adjusted treatment effect assumes a functional form for this dependence:

$$\mathrm{E}[Y_i(1) - Y_i(0)|Z_i(1), Z_i(0), \mathbf{X}_i] = \gamma(Z_i(1), Z_i(0), \mathbf{X}_i; \boldsymbol{\psi}) \tag{21.4}$$

where γ is a known function of potential treatment exposures under the two assignments, baseline characteristics and an unknown parameter vector $\boldsymbol{\psi}$. The aim of the analysis is to estimate the parameters in $\boldsymbol{\psi}$.

A *linear SMM* would assume a linear form for γ. For instance:

$$\mathrm{E}[Y_i(1) - Y_i(0)|Z_i(1), Z_i(0), \mathbf{X}_i] = \psi^{(1)}Z_i(1) - \psi^{(0)}Z_i(0) \tag{21.5}$$

A further simplification assumes the effects of $Z_i(1)$ and $Z_i(0)$ to be the same:

$$\mathrm{E}[Y_i(1) - Y_i(0)|Z_i(1), Z_i(0), \mathbf{X}_i] = \psi[Z_i(1) - Z_i(0)] \tag{21.6}$$

Model (21.5) implies that the average assignment effect would be 0 for individuals who would not receive any experimental treatment, regardless of assignment ($Z_i(1) = Z_i(0) = 0$). Model (21.6) implies that the average assignment effect is 0 for all individuals who would receive the same dose of treatment under either assignment ($Z_i(1) = Z_i(0)$).

These two assumptions refer to different forms of an assumption called *exclusion restriction*, basically stating that the effect of assignment is completely mediated by the treatment exposure (there is no direct effect of assignment).

We list three possible versions of the exclusion restriction, starting from the most restrictive:

Strong exclusion restriction:

$$P[Y_i(0) = Y_i(1)|Z_i(0) = Z_i(1)] = 1$$

Mean exclusion restriction:

$$E[Y_i(0) - Y_i(1)|Z_i(0) = Z_i(1)] = 0 \quad \text{or} \quad E[Y_i(0) - Y_i(1)|Z_i(0) = Z_i(1), \mathbf{X}_i] = 0$$

Mean zero-exposure exclusion restriction:

$$E[Y_i(0) - Y_i(1)|Z_i(0) = Z_i(1) = 0] = 0 \quad \text{or} \quad E[Y_i(0) - Y_i(1)|Z_i(0) = Z_i(1) = 0, \mathbf{X}_i] = 0$$

All three versions would be violated if assignment to the experimental treatment systematically differs from the control assignment in other aspects, possibly influential for the outcome, than the quantity measured by Z_i. In such cases the actual treatment and also the outcome of a subject with $Z_i(0) = Z_i(1) = 0$ would still depend on the value of R_i.

The analysis presented here does not require the strong exclusion restriction to be valid. The mean exclusion restriction is implied by model (21.6) and the mean zero-exposure exclusion restriction by model (21.5).

The mean exclusion restriction and mean zero-exposure exclusion restriction are equivalent in trials where the experimental treatment is not available on the control arm, as there $Z_i(0) \equiv 0$. In such trials the SMM can be defined as a model for $\mathrm{E}[Y_i(1) - Y_i(0)|Z_i(1), \mathbf{X}_i]$. See Goetghebeur and Lapp (1997) and Fischer-Lapp and Goetghebeur (1999) for SMM estimation algorithms and practical properties of SMM in such trials.

To assess the validity of the mean exclusion restriction in trials with contamination, one should find out whether the experimental treatment received by control arm subjects is identical (the same treatment, same dose-timing, etc.) to the experimental treatment received by the treatment arm subjects. If the treatments are not identical, the assumption can only be valid if the treatment differs in aspects that do not affect the outcome. The latter cannot be tested, but should be assessed by using the expert knowledge.

Next we will show how an estimator for the parameters in models (21.5) and (21.6) can be derived, assuming either mean or mean zero-exposure exclusion restriction, depending on the chosen model.

21.3 Estimation for a linear structural mean model

21.3.1 A general estimation procedure

For any SMM of form (21.4), parameter estimation is possible without any additional assumptions (other than the randomization assumption and those implied by the model), provided that the function $\gamma(Z_i(1), Z_i(0), \mathbf{X}_i; \boldsymbol{\psi})$ can be written as a difference of two components:

$$\gamma(Z_i(1), Z_i(0), \mathbf{X}_i; \boldsymbol{\psi}) = \gamma^{(1)}(Z_i(1), \mathbf{X}_i; \boldsymbol{\psi}^{(1)}) - \gamma^{(0)}(Z_i(0), \mathbf{X}_i; \boldsymbol{\psi}^{(0)})$$

where $\gamma^{(1)}$ is a function of variables that are observable on arm $R_i = 1$ and $\gamma^{(0)}$ is a function of variables that are observable on arm $R_i = 0$, with $\boldsymbol{\psi} = (\boldsymbol{\psi}^{(1)}, \boldsymbol{\psi}^{(0)})'$. Now the model (21.4) can be written as

$$\mathrm{E}[Y_i(1) - \gamma^{(1)}|Z_i(1), Z_i(0), \mathbf{X}_i] = \mathrm{E}[Y_i(0) - \gamma^{(0)}|Z_i(1), Z_i(0), \mathbf{X}_i]$$

(we omit the arguments of $\gamma^{(1)}$ and $\gamma^{(0)}$ for ease of presentation), implying

$$\mathrm{E}[Y_i(1) - \gamma^{(1)}|\mathbf{X}_i] = \mathrm{E}[Y_i(0) - \gamma^{(0)}|\mathbf{X}_i]$$

and, given the randomization assumption,

$$\mathrm{E}[Y_i(1) - \gamma^{(1)}|\mathbf{X}_i, R_i = 1] = \mathrm{E}[Y_i(0) - \gamma^{(0)}|\mathbf{X}_i, R_i = 0] \tag{21.7}$$

Now, for given functional forms of $\gamma^{(1)}$ and $\gamma^{(0)}$ and given $\boldsymbol{\psi}$, the expectation on the left-hand side of (21.7) is estimable from the experimental treatment arm and the expectation on the right-hand side from the control arm data.

The idea of *G-estimation* is to find values of the components of $\boldsymbol{\psi}$ that make (21.7) satisfied in the sample.

If the equality (21.7) does not involve baseline covariates (no conditioning on $\mathbf{X}_i$), it implies just one estimating equation, where the expectations can be replaced by sample averages on both arms. This single equation can uniquely identify only one parameter, such as the parameter ψ in (21.6). If a realistic model involves more parameters, sensitivity analysis approaches may be useful. With baseline covariates involved, one may be able to estimate more parameters in $\boldsymbol{\psi}$, provided that (21.7) implies two or more *independent* estimating equations. We will illustrate these issues next for linear models.

21.3.2 Identifiability and closed-form estimation of the parameters in a linear SMM

21.3.2.1 A simple one-parameter SMM

First we consider the model (21.6). The estimating equation (21.7) becomes in this case:

$$\mathrm{E}[Y_i - \psi Z_i|\mathbf{X}_i, R_i = 1] = \mathrm{E}[Y_i - \psi Z_i|\mathbf{X}_i, R_i = 0]$$

If there are no baseline covariates involved, all variables in the equality

$$\mathrm{E}[Y_i - \psi Z_i|R_i = 1] = \mathrm{E}[Y_i - \psi Z_i|R_i = 0] \tag{21.8}$$

can be replaced by their sample averages on the corresponding arm and a simple closed-form estimator for ψ would be

$$\hat{\psi} = \frac{\bar{y}_1 - \bar{y}_0}{\bar{z}_1 - \bar{z}_0} \tag{21.9}$$

where $\bar{y}_1$ and $\bar{z}_1$ are sample averages of the outcome and treatment exposure variables on the experimental treatment arm and $\bar{y}_0$ and $\bar{z}_0$ are the corresponding averages on the control arm.

The estimator cannot be used in the unlikely situation when treatment allocation fails to create different average exposure levels on the two arms.

If the equations condition on $\mathbf{X}_i$, a semiparametrically efficient estimator can be derived, following the general ideas of Robins (1994), using the derivation in Goetghebeur and Lapp (1997), which easily extends to allow for contamination. Mostly, as shown in Fischer-Lapp and Goetghebeur (1999), the estimator will simplify to a two-stage least squares estimator. At the first stage, one regresses outcome and exposure variables on the baseline covariates in each arm separately, obtaining predicted values $\hat{Y}_i(0)$, $\hat{Y}_i(1)$, $\hat{Z}_i(0)$ and $\hat{Z}_i(1)$ for all individuals in the study. At the second stage, one regresses $\hat{Y}_i(1) - \hat{Y}_i(0)$ on $\hat{Z}_i(1) - \hat{Z}_i(0)$ (without an intercept) to estimate ψ. A sandwich variance estimator can be derived (Fischer-Lapp and Goetghebeur, 1999).

Note that the estimator is valid under the mean exclusion restriction, validity of which can only partially be assessed.

21.3.2.2 A two-parameter SMM

The one-parameter SMM (21.6) relies on the mean exclusion restricton, assuming no assignment effect for subjects who would receive similar treatment exposure on the control arm than they would do on the experimental arm, so that for them $Z_i(1) = Z_i(0)$. This can be violated in trials where the variable Z_i captures just one aspect of the treatment exposure – so for subjects with $Z_i(1) = Z_i(0)$, the potential treatment exposures on control and treatment arms can still differ in some other aspects, possibly influencing the outcome.

One way to relax this assumption would be to assume model (21.5) instead, where estimating equations are derived from

$$\mathrm{E}[Y_i - \psi^{(1)} Z_i | \mathbf{X}_i, R_i = 1] = \mathrm{E}[Y_i - \psi^{(0)} Z_i | \mathbf{X}_i, R_i = 0] \tag{21.10}$$

Note that an equivalent model can be used for trials that compare two active treatments, allowing $Z_i(1)$ and $Z_i(0)$ to measure exposures to two different treatments, respectively. The model and estimation in the context of active-treatment comparisons have been discussed in detail by Fischer *et al.* (2011). If there are no baseline covariates available, the two distinct parameters cannot be identified. However, a sensitivity analysis is possible. We reparameterize the model so that $\psi^{(1)} = \psi$ and $\psi^{(0)} = \psi + \delta$. Now with $\mu_y^{(1)} = \mathrm{E}[Y_i | R_i = 1]$, $\mu_y^{(0)} = \mathrm{E}[Y_i | R_i = 0]$, $\mu_z^{(1)} = \mathrm{E}[Z_i | R_i = 1]$ and $\mu_z^{(0)} = \mathrm{E}[Z_i | R_i = 0]$, and omitting $\mathbf{X}_i$ from (21.10), we get

$$\frac{\mu_y^{(1)} - \mu_y^{(0)}}{\mu_z^{(1)} - \mu_z^{(0)}} = \psi + \delta \frac{\mu_z^{(0)}}{\mu_z^{(1)} - \mu_z^{(0)}}$$

As the left-hand side of this expression is estimated from (21.9), the quantity on the right-hand side is the one that we actually estimate in cases where the model (21.5) is true with $\psi^{(1)} \neq \psi^{(0)}$, but (21.6) is assumed and used instead for the estimation. As a kind of sensitivity analysis one could use this equation to find the value of ψ for a range of realistic values of δ.

If baseline covariates are used, the first stage of the estimation procedure is the same as in Section 21.3.2.1, but in the second stage the difference $\hat{Y}_i(1) - \hat{Y}_i(0)$ is regressed on

Table 21.1 Summary of outcomes and adherence in the EPHT trial (open-label subtrial only).

	HT arm	Control arm
Total healthcare costs per year (EUR): median (range)	201.4 (0 . . . 3356)	122.3 (0 . . . 7687)
Healthcare costs, excluding the cost of HT (EUR): median (range)	132.5 (0 . . . 3356)	110.3 (0 . . . 7687)
Number (%) of women initiating HT	477 (96.5 %)	134 (26.4 %)
Number (%) of women taking HT for more than a year	291 (58.9 %)	43 (8.5 %)

the two predicted variables, $\hat{Z}_i(1)$ and $\hat{Z}_i(0)$ (without intercept), to obtain the two parameter estimates. Clearly the two distinct parameters are not estimable if the expected compliance and contamination levels, $E[Z_i(1)|\mathbf{X}_i]$ and $E[Z_i(0)|\mathbf{X}_i]$, are proportional. The precision of estimation therefore depends on the existence of covariates $\mathbf{X}_i$, which make these two quantites far from proportional. The identifiability conditions are more formally presented in Fischer *et al.* (2011).

21.3.3 Analysis of the EPHT trial

We analyse the data of 1001 women randomized to either open-label HT ($n = 494$) or untreated control arm ($n = 507$). We consider two possible outcomes for the HT trial: the average total healthcare cost per year, including the cost of the hormone therapy (outcome 1) and the total healthcare cost excluding the cost of the HT (outcome 2). Both outcomes are log-transformed, due to highly skewed distribution. The ITT analysis of the outcomes (for both subtrials) on an untransformed scale can be found in Veerus *et al.* (2006b). For the adjusted analysis, baseline age, body weight, smoking and education levels are used.

Table 21.1 summarizes the outcomes and adherence levels in this trial. The ITT analysis finds a difference of 0.66 ($SE = 0.084$) between log total healthcare costs on HT and control arms, after adjustment for baseline covariates. We can therefore conclude that HT arm women had significantly more healthcare costs than control arm women. When the HT costs are subtracted from the total costs, the adjusted ITT difference becomes 0.22 ($SE = 0.11$), slightly favouring the existence of a real nonzero difference.

To adjust additionally for adherence, one cannot simply use a crude as-treated or per-protocol analysis, as the decision to actually initiate HT (after the initial assignment) is certainly not a random decision. It is possible that women with certain health complaints relate their problems to menopause and seek HT, but at the same time HT can be contraindicated when some other treatment is taken. Therefore HT initiation itself is likely to be influenced by a woman's health status, which in turn is directly related to healthcare costs. If a crude adherence–outcome association parameter is estimated (e.g. a difference in outcomes between women who initiated HT and those who did not), this is likely to be a confounded (biased) estimate of the effect of HT on healthcare costs.

Table 21.2 Structural mean model estimates in the EPHT trial.

Outcome	Model	Parameter	Estimate (SE)
Total healthcare costs	One-parameter SMM	ψ	0.94 (0.12)
	Two-parameter SMM	$\psi^{(1)}$	0.09 (0.61)
		$\psi^{(0)}$	−2.15 (2.18)
		$\psi^{(1)} - \psi^{(0)}$	2.23 (1.58)
Healthcare costs excluding HT	One-parameter SMM	ψ	0.32 (0.16)
	Two-parameter SMM	$\psi^{(1)}$	−0.53 (0.74)
		$\psi^{(0)}$	−2.76 (2.65)
		$\psi^{(1)} - \psi^{(0)}$	2.23 (1.93)

21.3.3.1 Adherence-adjusted analysis: one-parameter SMM

The participating women were encouraged to report to their physician if they decided to stop the trial treatment. For those who did not report, the number of received HT pill packages is known (they had to come to clinic to receive a next package every 7 months). For women in the control arm, the prescriptions database from the Estonian Health Insurance Fund was used to retrieve information on their HT usage. Based on these data, various adherence variables can be derived. Here we use one of the simplest (and least measurement-error-prone) measures – the binary indicator of HT initiation.

We fit model (21.5) for the two outcomes and two adherence measures. The results are shown in Table 21.2. The parameter ψ in one-parameter SMM can be interpreted as the effect in the subset who would initiate HT on the HT arm, but would not initiate on the control arm (complier average causal effect or CACE). We see it is clearly nonzero for total healthcare costs that include HT. If we subtract the costs of HT from the total costs and fit the model again, the parameter estimate decreases, as expected, but still remains positive. Similarly to the ITT analysis we see a borderline significant effect of HT increasing the healthcare costs.

The one-parameter model and its interpretation are valid only when the underlying assumptions are satisfied. Here the main concern lies in the validity of the mean exclusion restriction. First, similarly to the mean-zero exclusion restriction, this states that the average healthcare costs for women who would not initiate HT regardless of assignment would not depend on their treatment assignment. As this is an open-label trial, this assumption might be violated if the healthcare seeking behaviour of the women (or the attitude of their physicians) is influenced by knowledge of their treatment assignment. The mean exclusion restriction also states that women who would initiate HT whether assigned to control or HT arm do not experience any assignment effect on their healthcare costs. As HT arm women are likely to take HT for a longer period than those taking HT on the control arm, and the effect of HT can depend on the length of time spent on HT, this assumption may well be violated here.

There are two possible ways to relax the mean exclusion restriction. One possibility would be to use a continuous adherence measure, such as the proportion of pills taken during the follow-up period. This would bring in an extra modelling assumption on linearity (or other specified functional form).

A second solution is to fit a two-parameter SMM, using (21.10). The results of such analysis are also shown in Table 21.2. Although the two distinct parameters are identified here, they are estimated with high uncertainty. The negative point estimates in the two-parameter model offer a different interpretation of the ITT result: overall cost was greater in the HT arm because HT use in the control arm (but not in the HT arm) dramatically reduced costs. This does not seem to be a plausible finding and is likely to be due to the wide uncertainty arising from estimating a two-parameter model. The estimated difference between the two parameters has exactly the same point estimate for the two outcomes and is positive – although nonsignificant, it supports the findings from the one-parameter model: the costs are higher on the HT arm and this difference cannot be fully explained by the higher cost of HT only.

A third approach would be to conduct a sensitivity analysis for the possible value for $\psi^{(1)}$ while varying the values of $\psi^{(0)}$. As a special case we fit here the model where $\psi^{(0)} = 0$, assuming no effect on healthcare costs on the control arm. We get $\hat{\psi}^{(1)} = 1.37$ ($SE = 0.17$) for total healthcare costs and $\hat{\psi}^{(1)} = 0.49$ ($SE = 0.23$) for healthcare costs excluding HT. These parameters can be interpreted as the average effects of HT assignment on the healthcare costs for the subset who initiate HT on the HT arm, allowing for the possibility that some of them would also initiate HT on the control arm.

Although not explicitly required by the models here, it eases the interpretation if we assume that women who initiate HT on the control arm would also initiate HT on the HT arm. Such a *no-defiers* assumption is formally stated in the next section, where we introduce the principal stratification approach to estimation.

21.4 Alternative approaches for causal inference in randomized trials comparing experimental treatment with a control

21.4.1 Principal stratification

With binary compliance and contamination, the bivariate distribution of the potential exposures $(Z_i(0), Z_i(1))$ divides the target population into four latent classes or *principal strata* (Frangakis and Rubin, 2002). The classes are commonly named as follows:

Compliers:

$$(Z_i(0), Z_i(1)) = (0, 1)$$

Always-takers:

$$(Z_i(0), Z_i(1)) = (1, 1)$$

Never-takers:

$$(Z_i(0), Z_i(1)) = (0, 0)$$

Defiers:

$$(Z_i(0), Z_i(1)) = (1, 0)$$

In a randomized trial with perfect adherence, all subjects would be compliers. In a trial with no contamination, there would be two classes of subjects, compliers and never-takers, and the classes would be identifiable on the experimental treatment arm. However, in a trial with both noncompliance and contamination, all four classes may be present, and class membership is not observable for any of the individual subjects.

Within the class of compliers, the randomized trial is perfect – treatment assignment corresponds to the actual treatment exposure. The aim is therefore to estimate the causal effect of treatment exposure in compliers – the so-called *complier average causal effect (CACE)* (Imbens and Rubin, 1997). Estimation of this parameter in a trial without contamination was proposed by Sommer and Zeger (1991) and allowing for contamination by Cuzick *et al.* (1997). Identifiability of the CACE parameter is achieved by assuming that *there are no defiers* and that *there is no effect of assignment in never-takers and always-takers.*

One can easily see that the second assumption is identical to the mean exclusion restriction and so CACE can be estimated as the parameter ψ in model (21.6). However, the SMM (21.6) does not require the absence of defiers. Instead it requires that if defiers are present, the effect of treatment exposure in defiers and compliers is the same: $\mathrm{E}[Y_i(1) - Y_i(0)|Z_i(1) = 0, Z_i(0) = 1] = -\psi$.

Therefore the principal stratification approach is not a fundamentally different approach from the SMM for an adherence-adjusted ITT effect, but rather a way to interpret the parameters and assumptions involved.

21.4.2 SMM for the average treatment effect on the treated (ATT)

An alternative approach for causal inference in randomized trials would be to state a model for the effect of observed exposure in the population of interest. Therefore one defines a *potential exposure-free outcome* $Y_i(z = 0)$ as the outcome that one would have observed had the individual i received no experimental treatment. A linear model could be proposed as

$$\mathrm{E}[Y_i - Y_i(z = 0)|Z_i, R_i, \mathbf{X}_i] = \beta Z_i \tag{21.11}$$

The parameter β is the *average treatment effect on the treated (ATT)*. In the absence of contamination this model is identical to the SMM (21.6), as $Z_i(0)$ would be constantly 0 and $Z_i(1)$ is the observed treatment exposure Z_i for treatment arm individuals.

Estimation proceeds by assuming that $Y_i(z = 0)$ is independent of R_i (the randomization assumption) and thus

$$\mathrm{E}[Y_i - \beta Z_i|Z_i, R_i] = \mathrm{E}[Y_i(z = 0)|Z_i, R_i] = \mathrm{E}[Y_i(z = 0)|Z_i]$$

implying that $\mathrm{E}[Y_i - \beta Z_i|Z_i, R_i]$ should not depend on R_i and so

$$\mathrm{E}[Y_i - \beta Z_i|R_i = 1] = \mathrm{E}[Y_i - \beta Z_i|R_i = 0] \tag{21.12}$$

This is the same estimating equation as (21.8), so any estimate for ψ is also a valid estimate for β. However, the interpretation of the parameters is different as the parameter β in model (21.11) measures *the average effect of treatment exposure on all exposed subjects*, regardless of their randomization status, whereas parameter ψ in (21.6) measures *the effect of assignment in subjects whose exposure level depends on the assignment.*

21.4.2.1 Conceptual differences between modelling ATT and modelling the adherence-adjusted ITT effect

Modelling the ATT in a randomized trial is motivated by the ideas of instrumental variables (IV) estimation of causal effects in observational studies (Hogan and Lancaster, 2004). As in observational settings one usually does not have a particular intervention in mind (at least not at the time when the effect of exposure is assessed), the main interest lies in assessing the effect of hypothetical 'perfect' interventions that would remove the exposure entirely. One wishes to estimate the average effect of exposure in all exposed individuals, where the effect of exposure is defined as the difference between the observed outcome and the unobserved exposure-free outcome. An *instrument* would here be a variable that is associated with the exposure, but is believed to have no direct effect on the outcome. The observed statistical association between the instrument and the outcome can now be split into two components – the instrument–exposure association and the exposure–outcome (causal) association. As the first component is estimable, the second one, involving the causal parameter(s) of interest, can also be identified.

As randomization provides a perfect instrument – a variable that is strongly correlated with the exposure, but is guaranteed (in properly conducted blind trials) to have no other association with the outcome – it is tempting to use the IV methods here in a similar fashion. However, one should keep in mind the exact nature of randomization. Random assignment, by definition, is not just a variable associated with the exposure, but for the majority of the patients it is the factor that completely determines the exposure. In most trials, nonadherent individuals can be considered as an outlying minority that may differ from the rest of the sample in many ways. There is no reason to believe that the effect of exposure on nonadherent individuals would be the same as on adherent subjects.

As an example, consider a hypothetical trial on the effect of aspirin (or another mild painkiller) on stress headaches, where the control group is untreated. In such a trial it is likely that noncompliers on the treatment arm are those with mildest headaches – they do not feel the need for medication and as their pain level is low, any possible medication-related reduction is likely to be small. At the same time the contaminators on the control arm might be those with worst headaches, as they, although told not to take the medication, still decide to take it. Thus their possible effect from the treatment may be large. Although this example is hypothetical, similar phenomena have been supported by the data of real trials. As the model for ATT assumes *no exposure–effect heterogeneity* across all individuals, regardless of assignment, conditional on the observed exposure, it does not allow for such differences. Although it does not require generalization to unexposed individuals, it assumes that the exposure effect among contaminators on the control arm is the same as among compliers on the treatment arm.

Therefore, if there is a subset of individuals with identical nonzero experimental treatment exposure level z regardless of assignment, model (21.11) assumes that their effect of exposure is βz. However, as one could express the expectations in (21.12) as weighted sums of expectations within different subcategories of the sample, such subcategories will contribute an identical component to the sum on both sides of equality, which will cancel out: the solution for β will not depend on their actual exposure effect. Thus the information about β is provided only by the individuals whose exposure level depends on the assignment. This means that in the presence of such heterogeneity the model would provide a biased estimate of the ATT, as a subset of the sample is not accounted for while estimating the effect.

21.5 Discussion

A number of different causal questions can be addressed in randomized trials, and it is important to identify the question which is of greatest clinical interest. First, it may only be of interest to adjust for some forms of noncompliance. It might be of interest to adjust for noncompliance that arose from the circumstances of the trial, perhaps due to an ethical requirement to provide experimental treatment to control individuals whose disease progresses. On the other hand, it might not be of interest to adjust for inevitable forms of noncompliance such as noncompliance due to treatment toxicity (White *et al.*, 1999).

Different conceptualizations are also possible. Often it is desirable to avoid making assumptions about possible treatment effects within subgroups whose treatment exposure was not affected by randomization. We therefore favour the adherence-adjusted ITT effect as a natural extension of the ITT analysis.

The ideas presented have implications for the design of randomized trials. First, it is always desirable to minimize the degree of nonadherence, except in 'pragmatic' trials, which aim to estimate the effects of interventions as applied in practice and in which adjustment for nonadherence may not be necessary. Second, baseline covariates have a crucial role, not just in improving precision (as they do in any randomized trial), but also in facilitating the identification of more complex models with multiple causal parameters. Few covariates have been found with the desirable property of predicting adherence differently in the two arms.

The models considered here may be extended by allowing the treatment effect to vary with baseline covariates, or even to vary with the placebo response $Y_i(0)$ (Fischer and Goetghebeur, 2004; Vansteelandt and Goetghebeur, 2004).

We have focused on singly-measured quantitative outcomes. Other models available include the structural nested mean models for repeatedly measured quantitative outcomes (Robins, 1994), multiplicative SMMs and other generalized SMMs for binary outcomes (Vansteelandt and Goetghebeur, 2003), and rank-preserving structural nested failure time models for censored survival outcomes (Robins and Tsiatis, 1991; Korhonen and Palmgren, 2002). However, despite the methodology being available, few practical applications of these methods have been published. More work is needed to bring the methodology closer to practical data analysts in clinical trials – including clarification of all underlying assumptions, approaches for sensitivity analysis and user-friendly software packages.

An area of medicine where causal modelling is particularly needed is in equivalence and noninferiority trials. There the aim is to show that an experimental treatment is equivalent to or no worse than a standard treatment, usually because it is superior in other ways, such as cost or acceptability to patients. In such trials, cross-over from experimental treatment to standard treatment tends to dilute the ITT difference and hence leads to anti-conservative ITT analyses (Sheng and Kim, 2006). It is conventional to support the ITT analysis with per-protocol analyses (Jones *et al.*, 1996), but 'methodological approaches are needed that preserve more effectively the integrity of randomization than is achieved by the popular "per-protocol" analyses' (Fleming, 2008). Such trials usually include patients who stop all treatment, so the two-parameter causal models of Section 21.3.2.2 are needed, with their attendant estimation difficulties.

References

Angrist, J.D., Imbens, G.W. and Rubin, D.B. (1996) Identification of causal effects using instrumental variables. *Journal of the American Statistical Association*, **91**, 444–455.

Cuzick, J., Edwards, R. and Segnan, N. (1997) Adjusting for noncompliance and contamination in randomized controlled trials. *Statistics in Medicine*, **16**, 1017–1029.

Efron, B. and Feldman, D. (1991) Compliance as an explanatory variable in clinical trials. *Journal of the American Statistical Association*, **86**, 9–17.

Fischer, K. and Goetghebeur, E. (2004) Structural mean effects of noncompliance: estimating interaction with baseline prognosis and selection effects. *Journal of the American Statistical Association*, **99**, 918–928.

Fischer, K., Goetghebeur, E., Vrijens, B. and White, I.R. (2011) A structural mean model to analyze the effect of compliance when comparing two active treatments. *Biostatistics*, **12**, 247–257.

Fischer-Lapp, K. and Goetghebeur, E. (1999) Practical properties of some structural mean analyses of the effect of compliance in randomized trials. *Controlled Clinical Trials*, **20**, 531–546.

Fleming, T. (2008) Current issues in non-inferiority trials. *Statistics in Medicine*, **27**, 317–332.

Frangakis, C.E. and Rubin, D.B. (2002) Principal stratification in causal inference. *Biometrics*, **58**, 21–29.

Goetghebeur, E. and Lapp, K. (1997) The effect of treatment compliance in a placebo-controlled trial: regression with unpaired data. *Journal of the Royal Statistical Society, Series C*, **46**, 351–364.

Hogan, J.W. and Lancaster, T. (2004) Instrumental variables and inverse probability weighting for causal inference from longitudinal observational studies. *Statistical Methods in Medical Research*, **13**, 17–48.

Imbens, G.W. and Rubin, D.B. (1997) Bayesian inference for causal effects in randomized experiments with noncompliance. *Annals of Statistics*, **25**, 305–327.

Jin, H. and Rubin, D. (2008) Principal stratification for causal inference with extended partial compliance. *Journal of the American Statistical Association*, **103**, 101–111.

Jones, B., Jarvis, P., Lewis, J. and Ebbutt, A. (1996) Trials to assess equivalence: the importance of rigorous methods. *British Medical Journal*, **313**, 36–39.

Korhonen, P. and Palmgren, J. (2002) Effect modification in a randomized trial under non-ignorable non-compliance: an application to the alpha-tocopherol beta-carotene study. *Journal of the Royal Statistical Society, Series C*, **51**, 115–133.

Robins, J.M. (1994) Correcting for non-compliance in randomized trials using structural nested mean models. *Communications in Statistics – Theory and Methods*, **23**, 2379–2412.

Robins, J.M. and Tsiatis, A.A. (1991) Correcting for non-compliance in randomized trials using rank preserving structural failure time models. *Communications in Statistics – Theory and Methods*, **20** (8), 2609–2631.

Sheng, D. and Kim, M. (2006) The effects of non-compliance on intent-to-treat analysis of equivalence trials. *Statistics in Medicine*, **25**, 1183.

Sommer, A. and Zeger, S.L. (1991) On estimating efficacy from clinical trials. *Statistics in Medicine*, **10**, 45–52.

Vansteelandt, S. and Goetghebeur, E. (2003) Causal inference with generalized structural mean models. *Journal of the Royal Statistical Society, Series B*, **65**, 817–835.

Vansteelandt, S. and Goetghebeur, E. (2004) Using potential outcomes as predictors of treatment activity via strong structural mean models. *Statistica Sinica*, **14**, 907–925.

Veerus, P., Fischer, K., Hovi, S., Hakama, M., Rahu, M. and Hemminki, E. (2006a) Postmenopausal hormone therapy increases use of health services: experience from the Estonian postmenopausal hormone therapy trial [isrctn35338757]. *American Journal of Obstetrics and Gynecology*, **195**, 62–71.

Veerus, P., Hovi, S., Fischer, K., Rahu, M., Hakama, M. and Hemminki, E. (2006b) Results from the Estonian postmenopausal hormone therapy trial [ISRCTN35338757]. *Maturitas*, **55** (2), 162–173.

White, I.R., Babiker, A.G., Walker, S. and Darbyshire, J.H. (1999) Randomisation-based methods for correcting for treatment changes: examples from the Concorde trial. *Statistics in Medicine*, **18**, 2617–2634.

22

Causal inference in time series analysis

Michael Eichler
Department of Quantitative Economics, Maastricht University, Maastricht, The Netherlands

22.1 Introduction

The identification of causal relationships is an important part of scientific research and essential for understanding the consequences when moving from empirical findings to actions. At the same time, the notion of causality has shown to be evasive when trying to formalize it. Among the many properties a general definition of causality should or should not have, there are two important aspects that are of practical relevance:

> *Temporal precedence:* causes precede their effects.
>
> *Physical influence:* manipulation of the cause changes the effects.

The second aspect is central to most of the recent literature on causal inference (e.g. Pearl, 2000; Lauritzen, 2001; Dawid, 2002; Spirtes *et al.*, 2001), which is also demonstrated by previous chapters in this volume. Here, causality is defined in terms of the effect of interventions, which break the symmetry of association and thus give a direction to the association between two variables.

In time series analysis, most approaches to causal inference make use of the first aspect of temporal precedence. One the one hand, controlled experiments are often not feasible in many time series applications and researchers may be reluctant to think in these terms. On the other hand, temporal precedence is readily available in time series data.

Causality: Statistical Perspectives and Applications, First Edition. Edited by Carlo Berzuini, Philip Dawid and Luisa Bernardinelli.

Among these approaches, the definition introduced by Granger (1969, 1980, 1988) is probably the most prominent and most widely used concept. This concept of causality does not rely on the specification of a scientific model and thus is particularly suited for empirical investigations of cause–effect relationships. On the other hand, it is commonly known that Granger causality basically is a measure of association between the variables and thus can lead to so-called spurious causalities if important relevant variables are not included in the analysis (Hsiao, 1982). Since in most analyses involving time series data the presence of latent variables that affect the measured components cannot be ruled out, this raises the question whether and how the causal structure can be recovered from time series data.

The objective of this chapter is to embed the concept of Granger causality in the broader framework of modern graph-based causal inference. It is based on a series of papers by the author (Eichler, 2005, 2006, 2007, 2009, 2010; Eichler and Didelez, 2007, 2010). We start in Section 22.2 by comparing four possible definitions of causality that have been used in the context of time series. In Section 22.3, we sketch approaches for representing the dependence structure of a time series graphically. Furthermore, we discuss in more detail Markov properties associated with Granger causality as these are most relevant for the purpose of this chapter. This is continued in Section 22.4, where graphical representations of systems with latent variables are considered. Section 22.5 covers the identification of causal effects from observational data while Sections 22.6 and 22.7 present approaches for learning causal structures based on Granger causality. Section 22.8 concludes.

22.2 Causality for time series

Suppose that $X = (X_t)_{t\in\mathbb{Z}}$ and $Y = (Y_t)_{t\in\mathbb{Z}}$ are two stationary time series that are statistically dependent on each other. When is it justified to say that the one series X causes the other series Y? Questions of this kind are important when planning to devise actions, implementing new policies or subjecting patients to a treatment. Nonetheless, the notion of causality has been evasive and formal approaches to define causality have been much debated and criticized. In this section, we review four approaches to formalize causality in the context of multivariate time series. We start by introducting some notation that we will use throughout this chapter.

Consider a multivariate stationary time series $X = (X_t)_{t\in\mathbb{Z}}$ consisting of random vectors $X_t = (X_{1,t}, \ldots, X_{n_x,t})'$ defined on a joint probability space $(\Omega, \mathscr{L}, \mathbb{P})$. The history of X up to time t will be denoted by $X^t = (X_s, s \leq t)$; furthermore, $\sigma\{X^t\}$ denotes the corresponding σ-algebra generated by X^t. The σ-algebra $\sigma\{X^t\}$ represents the information that is obtained by observing the time series X. Finally, for every $1 \leq a \leq n_x$, we decompose X_t into its ath component $X_{a,t}$ and all other components $X_{-a,t}$.

22.2.1 Intervention causality

The previous chapters of this book have shown how important and fruitful the concept of interventions has been for the understanding of causality. This approach formalizes the understanding that causal relationships are fundamental and should persist if certain aspects of the system are changed. In the context of time series, this idea of defining a causal effect as the effect of an intervention in such a system has first been proposed by Eichler and Didelez (2007).

We start by introducing the concept of intervention indicators already seen in Chapters 3, 4, 14, and 16 of this volume. Such indicators allow us to distinguish formally between the 'natural' behaviour of a system and its behaviour under an intervention (e.g. Pearl, 1993, 2000; Lauritzen, 2001; Dawid, 2002; Spirtes *et al.*, 2001).

Definition 1 Regimes *Consider a set of indicators $\sigma = \{\sigma_t; t \in \tau\}$ denoting interventions in X_t at points $t \in \tau$ in time. Each σ_t takes values in some set $\mathcal{S}$ augmented by one additional state $\emptyset$. Different values of σ indicate different distributions of the time series $V_t = (X_t, Y_t, Z_t)$ in the following way:*

(i) *Idle regime. If $\sigma_t = \emptyset$ no intervention is performed and the process X_t arises naturally. The corresponding probability measure will be denoted as $\mathbb{P}_{\sigma_t=\emptyset}$ (often abbreviated to $\mathbb{P}_\emptyset$ or even just $\mathbb{P}$). This regime is also called the* observational regime.

(ii) *Atomic interventions. Here σ_t takes values in the domain of X_t such that $\sigma_t = x^*$ means we intervene and force X_t to assume the value x^*. Hence $\mathbb{P}_{\sigma_t=x^*}$ (or shorter $\mathbb{P}_{x^*}$) denotes the probability measure under such an atomic intervention with*

$$\mathbb{P}_{\sigma_t=x^*}(X_t = x|V^{t-1}) = \delta_{\{x^*\}}(x)$$

where $\delta_A(x)$ is one if $x \in A$ and zero otherwise.

(iii) *Conditional intervention: here $\mathcal{S}$ consists of functions $g_{x,t}(C^{t-1})$, where C is a subseries of the process V, such that $\sigma_t = g$ means X_t is forced to take on a value that depends on past observations C^{t-1}. With $\mathbb{P}_{\sigma_t=g}$ denoting the distribution of the time series V under such a conditional intervention, we have*

$$\mathbb{P}_{\sigma_t=g}(X_t = x|V^{t-1}) = \mathbb{P}_{\sigma_t=g}(X_t = x|C^{t-1}) = \delta_{\{g_{x,t}(C^{t-1})\}}(x).$$

(iv) *Random intervention. Here $\mathcal{S}$ consists of distributions meaning that X_t is forced to arise from such a distribution; that is the conditional distribution $\mathbb{P}_{\sigma_t=s}(X_t|V^{t-1})$ is* known *and is possibly a function of C^{t-1} for a subseries C of V.*

With this definition, we are considering a family of probability measures $\mathbb{P}_\sigma$ on $(\Omega, \mathscr{F})$ indexed by the possible values that σ can take. While $\mathbb{P} = \mathbb{P}_{\sigma=\emptyset}$ describes the natural behaviour of the time series under the observational regime, any implementation of an intervention strategy $\sigma = s$ will change the probability measure to $\mathbb{P}_{\sigma=s}$. We note that the assumption of stationarity is only made for the natural probability measure $\mathbb{P}$. In the following, we write $\mathbb{E}_\emptyset$ (or shorter just $\mathbb{E}$) and $\mathbb{E}_{\sigma=s}$ to distinguish between expectations with respect to $\mathbb{P}_\emptyset$ and $\mathbb{P}_{\sigma=s}$; the shorthand $\mathbb{P}_s$ and $\mathbb{E}_s$ is used when it is clear from the context what variables are intervened in.

One common problem in time series analysis is that controlled experiments cannot be carried out. Therefore, in order to assess the possible success of a meditated intervention, the effect of this intervention must be predicted from data collected under the observational regime. This is only possible if the intervention affects only the distribution of the target variable X_t at times $t \in \tau$ according to Definition 1 (ii) to (iv), whereas all other conditional distributions remain the same as under the idle regime. To formalize this invariance under different regimes, we say that a random variable Y is independent of σ_t conditionally on some σ-algebra $\mathscr{H} \subseteq \mathscr{F}$ if the conditional distributions of Y given $\mathscr{H}$ under the probability measures $\mathbb{P}_{\sigma_t=s}$ and $\mathbb{P}_\emptyset$ are almost surely equal for all $s \in \mathcal{S}$. With this notation, the required link between the probability measures $\mathbb{P}$ and $\mathbb{P}_\sigma$ under the observational and the interventional regime, respectively, is established by the following assumptions, which are analogous to those of (extended) stability in Dawid and Didelez (2005). See also Chapter 8 in this volume by Berzuini, Dawid and Didelez for a discussion of this issue.

Assumption 1 Stability *Let $V = (X, Y, Z)$ be a multivariate time series that is stationary under the idle regime. The interventions $\sigma = \{\sigma_t, t \in \tau\}$ from Definition 1 are assumed to have the following properties:*

(I1) for all $t \notin \tau$: $V_t \perp\!\!\!\perp \sigma | V^{t-1}$;

(I2) for all $t \in \tau$: $V_t \perp\!\!\!\perp \{\sigma_{t'} | t' \in \tau, t' \neq t\} | V^{t-1}, \sigma_t$;

(I3) for all $t \in \tau$: $Y_t, Z_t \perp\!\!\!\perp \sigma_t | V^{t-1}$;

(I4) for all $t \in \tau$ and all $a = 1, \ldots, n_X$: $X_{a,t} \perp\!\!\!\perp X_{-a,t}, \sigma_{-a,t} | V^{t-1}, \sigma_{a,t} = s$.

When working with such (conditional) independence relations, it is important to remember that the intervention indicators σ_t are not random variables and that the above notion of independence – although being one of statistical independence – is not symmetric. For a more detailed discussion of conditional independence involving nonrandom quantities we refer to Dawid (2002) and Chapter 4 by Dawid in this volume.

With the above assumptions the distribution of the time series $V = (X, Y, Z)$ under the interventional regime is fully specified by its natural distribution described by $\mathbb{P} = \mathbb{P}_\emptyset$ and the conditional distributions given in Definition 1 (ii) to (iv), under the chosen intervention. As it is usually not possible or feasible to collect data under the interventional regimes, all model assumptions such as stationarity are supposed to apply to the idle regime.

In the case of a single intervention, that is in one variable at one point in time, the above assumptions (I1) to (I3) simplify to

$$X^{t-1}, Y^t, Z^t \perp\!\!\!\perp \sigma_t \tag{22.1}$$

and

$$\{X_{t'}, Y_{t'}, Z_{t'} | t' > t\} \perp\!\!\!\perp \sigma_t | X^t, Y^t, Z^t. \tag{22.2}$$

Let us now consider effects of interventions. In general, this can be any function of the post-intervention distribution of $\{V_{t'} | t' > t\}$ given an individual intervention $\sigma_t = s$, for instance. It will often involve the comparison of setting X_t to different values, e.g. setting $X_t = x^*$ as compared to setting it to $X_t = x^0$, which could be a baseline value in some sense. One may also want to use the idle case as baseline for the comparison. Typically, one is interested in the mean difference between interventions or between an intervention and the idle case. This leads to the following definition of the average causal effect.

Definition 2 Average causal effect *The* average causal effect (ACE) *of interventions in $X_{t_1}, \ldots, X_{t_m}$ according to strategy* s *on the response variable $Y_{t'}$ with $t' > t$ is given by*

$$\text{ACE}_s = \mathbb{E}_{\sigma_t = s} Y_{t'} - \mathbb{E} Y_{t'}.$$

In the stationary case, we may assume without loss of generality that $\mathbb{E}Y_{t'} = 0$ and thus $\mathrm{ACE}_s = \mathbb{E}_{\sigma_t=s}Y_{t'}$. Furthermore, different intervention strategies can be compared by considering the difference $\mathrm{ACE}_{s_1} - \mathrm{ACE}_{s_2}$. Finally, we note that the effect of an intervention need not be restricted to the mean. For example, in a financial time series, an intervention might aim at reducing the volatility of the stock market. In general, one can consider any functional of the post-intervention distribution $\mathbb{P}_s(Y_{t'})$.

22.2.2 Structural causality

In a recent article, White and Lu (2010) proposed a new concept of so-called direct structural causality for the discussion of causality in dynamic structural systems. The approach is based on the assumption that the data-generating process (DGP) has a recursive dynamic structure in which predecessors structurally determine successors. For ease of notation, the following definitions are slightly modified and simplified.

Suppose that we are interested in the causal relationship between two processes, the 'cause of interest' X and the 'response of interest' Y. We assume that X and Y are structurally generated as

$$\begin{aligned} X_t &= q_{x,t}(X^{t-1}, Y^{t-1}, Z^{t-1}, U_{x,t}) \\ Y_t &= q_{y,t}(X^{t-1}, Y^{t-1}, Z^{t-1}, U_{y,t}) \end{aligned} \tag{22.3}$$

for all $t \in \mathbb{Z}$. Here, the process Z includes all relevant observed variables while the realizations of $U = (U_x, U_y)$ are assumed to be unobserved. The functions $q_{x,t}$ and $q_{y,t}$ are also assumed to be unknown.

Definition 3 *The process X* does not directly structurally cause *the process Y if the function $q_{y,t}(x^{t-1}, y^{t-1}, z^{t-1}, u_{y,t})$ is constant in x^{t-1} for all admissible values for y^{t-1}, z^{t-1}, and $u_{y,t}$. Otherwise, X is said to* directly structurally cause *Y.*

We note that similar approaches of defining causality by assuming a set of structural equations have been considered before by a number of authors (e.g. Pearl and Verma, 1991; Pearl, 2000; see also Chapter 3 by Shpitser in this volume). However, in contrast to this strand of literature, White and Lu (2010) make no reference to interventions or to graphs.

It is clear from the definition that an intervention σ_t on X_t results in replacing the generating equation by the corresponding equation under the interventional regime. For instance, in the case of a conditional intervention, we have $X_t = g_{x,t}(C^{t-1})$, where the subprocess C denotes the set of conditioning variables. Consequently, as the generating equation for the response variable $Y_{t'}$, $t' > t$, is unaffected by the intervention, we immediately obtain the following result.

Corollary 1 *Suppose that (X, Y) is generated by (22.3). If X does not directly structurally cause Y, then*

$$\mathbb{E}_{\sigma_t=s}\big(h(Y_{t+1})\big) = \mathbb{E}_{\emptyset}\big(h(Y_{t+1})\big)$$

for all measurable functions h.

White and Lu (2010) also propose a definition of total structural causality. With this, the above corollary can be generalized to responses at arbitrary times $t' > t$. We omit the details and refer the reader to White and Lu (2010). Furthermore, we note that the converse of the above result is generally not true. This is due to the fact that the function $q_{y,t}$ might depend on x^{t-1} only on a set having probability zero. In that case, the dependence on x^{t-1} will not show up in the ACE.

22.2.3 Granger causality

In time series analysis, inference about cause–effect relationships is commonly based on the concept of Granger causality (Granger 1969, 1980). Unlike the two previous approaches, this probabilistic concept of causality does not rely on the specification of a scientific model and thus is particularly suited for empirical investigations of cause–effect relationships. For his general definition of causality, Granger (1969, 1980) evokes the following two fundamental principles:

(i) The effect does not precede its cause in time.

(ii) The causal series contains unique information about the series being caused that is not available otherwise.

The first principle of temporal precedence of causes is commonly accepted and has been also the basis for other probabilistic theories of causation (e.g. Good, 1961, 1962; Suppes 1970). In contrast, the second principle is more subtle as it requires the separation of the special information provided by the former series X from any other possible information. To this end, Granger considers two information sets:

(i) $\mathcal{I}^*(t)$ is the set of all information in the universe up to time t.

(ii) $\mathcal{I}^*_{-X}(t)$ contains the same information set except for the values of series X up to time t.

Here it is assumed that all variables in the universe are measured at equidistant points in time, namely $t \in \mathbb{Z}$. Now, if the series X causes series Y, we expect by the above principles that the conditional probability distributions of Y_{t+1} given the two information sets $\mathcal{I}^*(t)$ and $\mathcal{I}^*_{-X}(t)$ differ from each other. The following equivalent formulation is chosen to avoid measure-theoretic subtleties.

Granger's definition of causality (1969, 1980) The series X *does not cause* the series Y if

$$Y_{t+1} \perp\!\!\!\perp \mathcal{I}^*(t) | \mathcal{I}^*_{-X}(t) \tag{22.4}$$

for all $t \in \mathbb{Z}$; otherwise the series X is said to *cause* the series Y.

Besides measure-theoretic subtleties and the obviously abstract nature of the set $\mathcal{I}^*(t)$, a problem of the above definition is whether $\mathcal{I}^*_{-X}(t)$ contains truely less information than $\mathcal{I}^*(t)$. Implicitly, such a separation of the two information sets $\mathcal{I}^*(t)$ and $\mathcal{I}^*_{-X}(t)$ seems to be based on the assumption that the universe considered is discretized not only in time (as we consider time-discrete processes) but also in space.

Leaving aside such theoretical problems, it is interesting to note that Granger's definition is also specified in terms of the DGP (cf. Granger, 2003) and is closely related to the direct structural causality. Like direct structural causality, Granger's definition covers only direct causal relationships. For example, if X affects Y only via a third series Z, then $\mathcal{I}^*_{-X}(t)$ comprises the past values of Z and Y_{t+1} is independent from the past values of X given $\mathcal{I}^*_{-X}(t)$. The following result by White and Lu (2010) shows that in the absence of latent variables Granger causality can be interpreted as direct structural causality.

Proposition 1 *Suppose that X and Y are generated by the DGP in (22.3) and assume additionally that* $U_{y,t+1} \perp\!\!\!\perp X^t | Y^t, Z^t$. *If X does not directly structurally cause Y, then* $Y_{t+1} \perp\!\!\!\perp X^t | Y^t, Z^t$.

Since the process Y_{t+1} is supposed to be generated solely from the variables in Y^t, X^t and Z^t, the conditioning set coincides with $\mathcal{I}^*_{-X}(t)$. Thus, direct structural noncausality implies (22.4). As in the case of intervention causality, the converse implication is generally untrue. As an example, suppose that the process Y is generated by

$$Y_t = q_t(X_{t-1}, U_{1,t}, U_{2,t}) = \frac{X_{t-1}}{\sqrt{1 + X_{t-1}^2}} U_{1,t} + \frac{1}{\sqrt{1 + X_{t-1}^2}} U_{2,t}$$

while $U_{1,t}$, $U_{2,t}$ and X_t are all independent and normally distributed with mean zero and variance σ^2. Furthermore, we assume that $U_{1,t}$ and $U_{2,t}$ are unobservable. Then $Y_t | X^{t-1} \sim \mathcal{N}(0, \sigma^2)$, which implies that X does not cause Y in the meaning of Granger. However, this argument raises the question of what information should belong to $\mathcal{I}^*(t)$. Namely, if we add at least one of the variables U_1 and U_2 to the information set, Y_t becomes dependent on X^t. Since Y_t is thought to be generated by its equation, the variables $U_{1,t}$ and $U_{2,t}$ must exist and therefore should belong to $\mathcal{I}^*(t)$.

If we – like White and Lu (2010) – are not willing to accept $U_{1,t}$ and $U_{2,t}$ as separate entities that in principle should belong to $\mathcal{I}^*(t)$, does the difference between the two definitions of causality matter in practice? We think not, as the difference basically can be viewed as counterfactual. More precisely, let $Y_t^* = q_t(x^*, U_{1,t}, U_{2,t})$ be the value we would have got for Y_t if we had set the value of X_{t-1} to x^* (so we use the same realizations of $U_{1,t}$ and $U_{2,t}$). Then $Y_t - Y_t^* \neq 0$ almost surely. However, this cannot be tested from data as Y^* is counterfactual and cannot be observed together with Y_t; furthermore, the average $\mathbb{E}(Y_t) - \mathbb{E}(Y_t^*)$ is zero.

We end our discussion of the relationship between the two concepts by remarking that White and Lu (2010) showed that Granger's notion of causality is equivalent to a slightly weaker notion of direct almost sure causality. For further details, we refer to White and Lu (2010).

It is clear that the above definition of causality usually cannot be used with actual data. In practice, only the background knowledge available at time t can be incorporated into an analysis. Therefore, the definition must be modified to become operational. Suppose that the process $V = (X, Y, Z)$ has been observed. Substituting the new information sets $\sigma\{X^t, Y^t, Z^t\}$

and $\sigma\{Y^t, Z^t\}$ for $\mathcal{I}_V(t)$ and $\mathcal{I}^*_{-X}(t)$, respectively, we obtain the following modified version of the above definition (Granger 1980, 1988).

Definition 4 Granger causality *The series X is* Granger-noncausal *for the series Y with respect to* $V = (X, Y, Z)$ *if*

$$Y_{t+1} \perp\!\!\!\perp X^t | Y^t, Z^t \tag{22.5}$$

Otherwise we say that X Granger-causes Y with respect to V.

Granger (1980, 1988) used the term '*X* is a prima facie cause of *Y*' to emphasize the fact that a cause in the sense of Granger causality must be considered only as a potential cause. This, however, is not to say that the concept of Granger causality is completely useless for causal inference. Indeed, as we discuss in Section 22.6, it can be an essential tool for recovering (at least partially) the causal structure of a time series *X* if used in the right way.

Furthermore, we note that the above definition in terms of conditional independence is usually referred to as *strong Granger causality*; other existing notions are *Granger causality in mean* (Granger, 1980, 1988) and *linear Granger causality* (Hosoya, 1977; Florens and Mouchart, 1985).

It is clear from the general definition given above that Granger intended the information to be chosen as large as possible including all available and possibly relevant variables. Despite this, most (econometric) textbooks (e.g. Lütkepohl, 1993) introduce Granger causality only in the bivariate case. This has led to some confusion about a multivariate definition of Granger causality (e.g. Kamiński *et al.* 2001).

With the above definition of a causal effect in terms of interventions a first connection between Granger noncausality and the effect of an intervention can be established as follows.

Corollary 2 *Consider a multivariate time series* $V = (X, Y, Z)$ *and an individual intervention* $\sigma_a(t) = s$ *satisfying (22.1) and (22.2). If X is Granger-noncausal for Y with respect to V, that is* $Y_{t+1} \perp\!\!\!\perp X^t | Y^t, Z^t$*, then there is no causal effect of intervening in* X_t *on* Y_{t+1}.

The proof of the corollary, which can be found in Eichler and Didelez (2010), relies on both conditions (22.1) and (22.2), underpinning the fact that Granger noncausality on its own is not enough to make statements about the effect of interventions. However, we do not need the whole *V* to be observable in the above corollary; the system *V* with respect to which *X* is Granger-noncausal for *Y* can therefore include latent time series if this helps to justify the stability assumptions.

22.2.4 Sims causality

The econometric literature features other less well known probabilistic notions of causality that are related to Granger causality. Among these, the concept introduced by Sims (1972) seems of most interest. In contrast to Granger causality, it takes in account not only direct but also indirect causal effects. Thus it can be seen as a concept for total causality. The following definition is a slight variation proposed by Florens and Mouchart (1982).

Definition 5 Sims noncausality *The process X is* Sims-noncausal *for another process Y with respect to the process $V = (X, Y, Z)$ if*

$$\{Y_{t'}|t' > t\} \perp\!\!\!\perp X_t | X^{t-1}, Y^t, Z^t$$

for all $t \in \mathbb{Z}$.

Now suppose that we are interested in the causal effect of an intervention $\sigma = s$ in X_t on $Y_{t'}$ for some $t' > t$. Let $V = (X, Y, Z)$ be a process such that the stability assumptions (I1) to (I4) in Section 22.2.1 are satisfied. If X is Sims-noncausal for Y with respect to V, it follows from (I1) that

$$Y_{t+h} \perp\!\!\!\perp \sigma_t, X_t | X^{t-1}, Y^t, Z^t$$

and furthermore, by (I1) and (I3) that

$$\begin{aligned}\mathbb{E}_s\big(g(Y_{t+h})\big) &= \mathbb{E}_s\big[\mathbb{E}_s\big(g(Y_{t+h})|X^t, Y^t, Z^t\big)\big] = \mathbb{E}_s\big[\mathbb{E}_\emptyset\big(g(Y_{t+h})|X^{t-1}, Y^t, Z^t\big)\big] \\ &= \mathbb{E}_\emptyset\big[\mathbb{E}_\emptyset\big(g(Y_{t+h})|X^{t-1}, Y^t, Z^t\big)\big] = \mathbb{E}_\emptyset\big(g(Y_{t+h})\big)\end{aligned}$$

This suggests the following result which relates the concepts of intervention causality and Sims causality. The details of the proof are omitted.

Proposition 2 *Consider a multivariate time series $V = (X, Y, Z)$ and an individual intervention $\sigma_a(t) = s$ satisfying (22.1) and (22.2). Then X is Sims-noncausal for Y with respect to V if and only if the average causal effect of $\sigma = s$ on $g(Y_{t'})$ is zero for all measurable functions g and all $t' > t$.*

22.3 Graphical representations for time series

In this section, we briefly review the two main approaches for representing dependences among multiple time series by graphs. For simplicity, we consider only the case of stationary Gaussian processes. Therefore, throughout this section, we assume that $X = (X_t)_{t\in\mathbb{Z}}$ with $X_t = (X_{1,t}, \ldots, X_{d,t})'$ is a stationary Gaussian process with mean zero and covariances $\Gamma(u) = \mathbb{E}(X_t X_{t-u}')$; furthermore, we make the following assumption.

Assumption 2 *The spectral density matrix*

$$f(\lambda) = \frac{1}{2\pi} \sum_{u=-\infty}^{\infty} \Gamma(u) \, e^{-i\lambda u}$$

of X exists, and its eigenvalues are bounded and bounded away from zero uniformly for all $\lambda \in [-\pi, \pi]$.

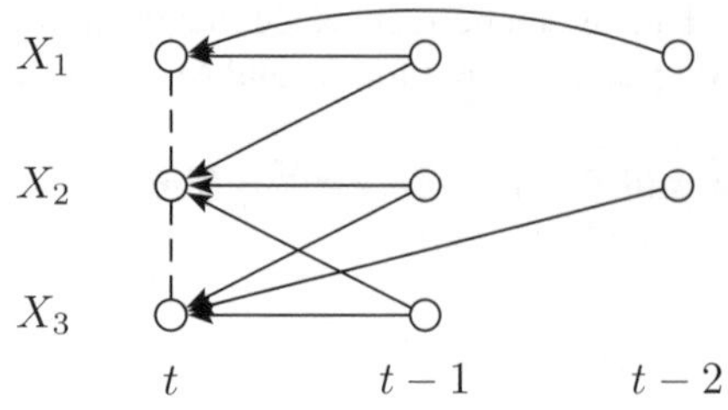

Figure 22.1 Graphical representation of the conditional distribution of X_t given its past X^{t-1} for the process in Example 1.

This technical assumption ensures that the process has a mean-square convergent autoregressive representation

$$X_t = \sum_{u=1}^{\infty} \Phi(u)\, X_{t-u} + \varepsilon_t, \tag{22.6}$$

where $\Phi(u)$ is a square summable sequence of $d \times d$ matrices and $\varepsilon = (\varepsilon_t)_t$ is a Gaussian white noise process with non-singular covariance matrix Σ.

22.3.1 Conditional distributions and chain graphs

The simplest approach to visualize the dependence structure of a vector autoregressive process is to construct a graph from the conditional distribution of X_t given its past values X^{t-1} for fixed t. More precisely, representing the variables $X_{a,t}$ by vertices a_t, two distinct vertices a_t and b_t are connected by an undirected (dashed) edge whenever $\Sigma_{ab} \neq 0$. Additionally, vertices b_{t-k} represent the lagged instances $X_{b,t-k}$ of the variable X_b and are connected to nodes a_t whenever $\Phi_{ab}(k) \neq 0$. Any vertices b_{t-k} for which $\Phi_{ab}(k) = 0$ for all a are omitted from the graph. The resulting *conditional distribution graph* gives a concise picture of the autoregressive structure of the process X.

Example 1 For an illustration, we consider the trivariate process X given by

$$\begin{aligned}
X_{1,t} &= \phi_{11}(1)\, X_{1,t-1} + \phi_{11}(2)\, X_{1,t-2} + \varepsilon_{1,t} \\
X_{2,t} &= \phi_{22}(1)\, X_{2,t-1} + \phi_{21}(1)\, X_{1,t-1} + \phi_{23}(1)\, X_{3,t-1} + \varepsilon_{2,t} \\
X_{3,t} &= \phi_{33}(1)\, X_{3,t-1} + \phi_{32}(1)\, X_{2,t-1} + \phi_{32}(2)\, X_{2,t-2} + \varepsilon_{1,t}
\end{aligned}$$

where ε_1, ε_2 and ε_3 are white noise processes with $\mathrm{corr}(\varepsilon_{1,t}, \varepsilon_{3,t}) = 0$. The corresponding conditional distribution graph is depicted in Figure 22.1.

As can be seen from the example, graphs of this type give a concise picture of the autoregressive structure of the process X. The disadvantage of such graphs is that they encode only a very limited set of conditional independences. In fact, the only conditional dependences that can be derived from such graphs are concerned with the relationships among the variables $X_{a,t}$ conditionally on the complete history X^{t-1} of the process. Therefore, such or similar graphs

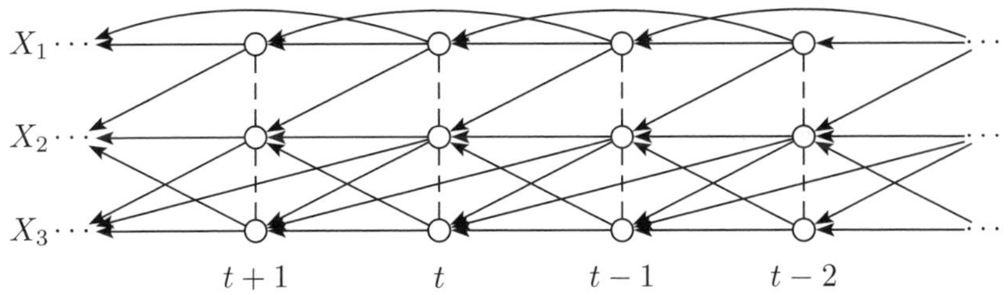

Figure 22.2 *Time series chain graph for the vector autoregressive process X_V in Example 1.*

have been used mainly for investigating the causal ordering of the variables $X_{1,t}, \ldots, X_{d,t}$ (Swanson and Granger, 1997; Reale and Tunnicliffe Wilson, 2001; Demiralp and Hoover, 2003; Moneta and Spirtes, 2006; Moneta, 2007; Hyvärinen *et al.*, 2010).

In general, we also want to be able to investigate, for instance, the conditional independences that hold among the variables of a subprocess $Y \subseteq X$. For this, we must also include the dependences among lagged variables in the graph. This leads to graphs with an infinite number of variables a_t with $a \in \{1, \ldots, d\}$ and $t \in \mathbb{Z}$. Because of stationarity, edges in this graph are translation invariant and can be obtained by using the above conditional distribution graph as a blueprint. The resulting graph is called *time series chain graph* (Lynggaard and Walther, 1993; Dahlhaus and Eichler, 2003).

While time series chain graphs (or time series DAGs in the case of structural vector autoregressions) allow a full discussion of the Markov properties that are entailed by the autoregressive structure of the process, the graphs become complicted and infeasible even for small numbers of variables unless the dependence structure is very sparse. Figure 22.2 shows the time series chain graph for the process in Example 1.

22.3.2 Path diagrams and Granger causality graphs

In order to avoid the complexity of time series chain graphs, Eichler (2007) proposed to consider path diagrams $G = (V, E)$ with $V = \{1, \ldots, d\}$ in which every vertex $v \in V$ corresponds to one complete component series $(X_{v,t})$ while edges – arrows and lines – between vertices indicate nonzero coefficients in the autoregressive representation of the process.

Definition 6 Path diagram *Let $X = X_V$ be a stationary Gaussian process with the autoregressive representation (22.6). Then the* path diagram *associated with X is the graph $G = (V, E)$ with vertex set V and edge set E such that for distinct vertices $a, b \in V$:*

(i) $a \longrightarrow b \notin E \iff \Phi_{ba}(u) = 0 \quad \forall u \in \mathbb{N}.$

(ii) $a \dashrightarrow b \notin E \iff \Sigma_{ab} = \Sigma_{ba} = 0.$

Figure 22.3 depicts the path diagram for the autoregressive process X_V in Example 1. Compared for the graphs in Figures 22.1 and 22.2, this path diagram contains less vertices and less edges. This reduction in complexity compared to time series chain graphs is paid for by a loss of information. While path diagrams only encode whether or not one variable X_a depends on the past of another variable X_b, the time series chain graphs additionally yield the lags at

Figure 22.3 Path diagram for the vector autoregressive process X_V in Example 1.

which past instances of X_b have an effect on $X_{a,t}$. However, when concerned with problems of causal inference, algorithms for identifying causal relationships become infeasible if the level of detail is too high. Therefore, path diagrams provide a reasonable compromise between time series chain graphs and even simpler graphs such as partial correlation graphs (Dahlhaus, 2000).

The use of path diagrams for causal inference is based on the following lemma, which links edges in the graph with Granger-causal relationships between the corresponding components.

Lemma 1 *Let $G = (V, E)$ be the path diagram associated with a stationary Gaussian process* X *satisfying Assumption 2. Then:*

(i) The directed edge $a \longrightarrow b$ is absent in the graph if and only if X_a is Granger-noncausal for X_b with respect to the full process X, that is $X_{b,t+1} \perp\!\!\!\perp X_a^t | X_{-a}^t$.

(ii) The undirected edge a --- b is absent in the graph if and only if X_a and X_b are not contemporaneously correlated with respect to the full process, that is $X_{a,t+1} \perp\!\!\!\perp X_{b,t+1} | X^t$.

Because of this result, we will call the path diagram associated with a process X also the *Granger causality graph* of X. We note that in the more general case of nonlinear non-Gaussian processes X, the definition of such graphs is entirely based on the concept of Granger causality. For details we refer to Eichler (2011).

22.3.3 Markov properties for Granger causality graphs

Under the assumptions imposed on the process X_V, more general Granger-causal relationships than those in Lemma 1 can be derived from the path diagram associated with X_V. This global Markov interpretation is based on a path-oriented concept of separating subsets of vertices in a mixed graph, which has been used previously to represent the Markov properties of linear structural equation systems (e.g. Spirtes *et al.*, 1998; Koster, 1999). Following Richardson (2003), we will call this notion of separation in mixed graphs *m*-separation.

More precisely, let $G = (V, E)$ be a mixed graph and $a, b \in V$. A *path* π in G is a sequence $\pi = \langle e_1, \dots, e_n \rangle$ of edges $e_i \in E$ with an associated sequence of nodes $v_0, \dots, v_n$ such that e_i is an edge between v_{i-1} and v_i. The vertices v_0 and v_n are the *endpoints* while $v_1, \dots, v_{n-1}$ are the *intermediate vertices* of the path. Notice that paths may be self-intersecting since we do not require that the vertices v_i are distinct.

An intermediate vertex c on a path π is said to be an *m-collider* on π if the edges preceding and suceeding c both have an arrowhead or a dashed tail at c (i.e. $\longrightarrow c \longleftarrow$, $\longrightarrow c$ ---, --- $c \longleftarrow$, --- c ---); otherwise c is said to be an *m-noncollider on π. A path π between a* and b is said to be *m-connecting* given a set C if

(i) every m-noncollider on π is not in C and

(ii) every m-collider on π is in C;

otherwise we say that π is *m-blocked* given C. If all paths between a and b are m-blocked given C, then a and b are said to be *m-separated* given C. Similarly, two sets A and B are said to be m-separated given C if for every pair $a \in A$ and $b \in B$, a and b are m-separated given C.

With this notion of separation, it can be shown that path diagrams for multivariate time series have a similar Markov interpretation as path diagrams for linear structural equation systems (cf. Koster, 1999). But since each vertex $v \in V$ corresponds to a complete process X_v, separation in the path diagram encodes a conditional independence relation among complete subprocesses of X_V.

Proposition 3 *Let X_V be a stationary Gaussian process that satisfies Assumption 2, and let G be its path diagram. Then, for all disjoint $A, B, S \subseteq V$,*

$$A \bowtie_m B|S \text{ in } G \Longrightarrow X_A \perp\!\!\!\perp X_B | X_S.$$

Derivation of such conditional independence statements requires that all paths between two sets are m-blocked. For the derivation of Granger-causal relationships, it suffices to consider only a subset of these paths, namely those having an arrowhead at one endpoint. For a formal definition, we say that a path π between a and b is *b-pointing* if it has an arrowhead at the endpoint b; furthermore, a path between sets A and B is said to be *B-pointing* if it is b-pointing for some $b \in B$. Then, to establish Granger noncausality from X_A to X_B, it suffices to consider only all B-pointing paths between A and B. Similarly, a graphical condition for contemporaneous correlation can be obtained based on a *bi-pointing* path, which have an arrowhead at both endpoints.

Definition 7 *A stationary Gaussian process X_V is Markov for a graph $G = (V, E)$ if, for all disjoint subsets $A, B, C \subseteq V$, the following two conditions hold:*

(i) If every B-pointing path between A and B is m-blocked *given $B \cup C$, then X_A is Granger-noncausal for X_B with respect to $X_{A \cup B \cup C}$.*

(ii) If the sets A and B are not connected by an undirected edge (- - -) and every bi-pointing path between A and B is m-blocked given $A \cup B \cup C$, then X_A and X_B are contemporaneously uncorrelated with respect to $X_{A \cup B \cup C}$.

With this definition, it can be shown that path diagrams for vector autoregressions can be interpreted in terms of such global Granger-causal relationships.

Theorem 1 *Let X_V be a stationary Gaussian process that satisfies Assumption 2, and let G be the associated path diagram. Then X_V is Markov for G.*

22.4 Representation of systems with latent variables

The notion of Granger causality is based on the assumption that all relevant information is included in the analysis (Granger, 1969, 1980). It is well known that the omission of important variables can lead to temporal correlations among the observed components that are falsely detected as causal relationships. The detection of such so-called spurious causalities (Hsiao,

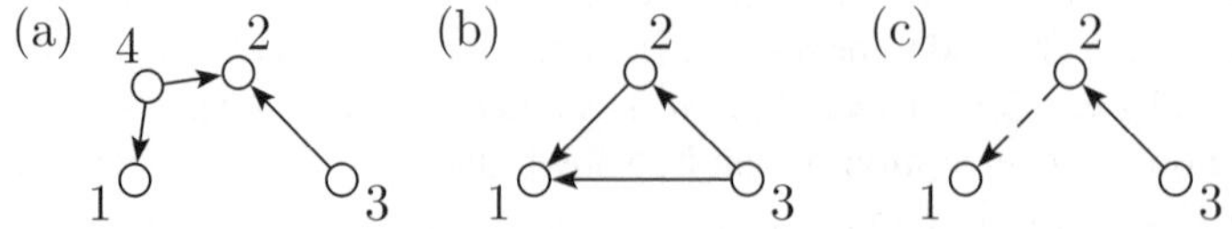

Figure 22.4 Graphical representations of the four-dimensional VAR(2) process in (4.1): (a) path diagram associated with $X_{\{1,2,3,4\}}$; (b) path diagram associated with $X_{\{1,2,3\}}$; (c) general path diagram for $X_{\{1,2,3\}}$.

1982) becomes a major problem when identifying the structure of systems that may be affected by latent variables.

Of particular interest will be spurious causalities of type I, where a Granger-causal relationship with respect to the complete process vanishes when only a subprocess is considered. Since Granger causality graphs are defined in terms of the pairwise Granger-causal relationships with respect to the complete process, they provide no means to distinguish such spurious causalities of type I from true causal relationships. To illustrate this remark, we consider the four-dimensional vector autoregressive process X with components

$$\begin{aligned} X_{1,t} &= \alpha\, X_{4,t-2} + \varepsilon_{1,t}, & X_{3,t} &= \varepsilon_{3,t} \\ X_{2,t} &= \beta\, X_{4,t-1} + \gamma\, X_{3,t-1} + \varepsilon_{2,t}, & X_{4,t} &= \varepsilon_{4,t} \end{aligned} \tag{22.7}$$

where $\varepsilon_i(t), i = 1, \ldots, 4$, are uncorrelated white noise processes with mean zero and variance one. The true dynamic structure of the process is shown in Figure 22.4(a). In this graph, the 1-pointing path $3 \longrightarrow 2 \longleftarrow 4 \longrightarrow 1$ is m-connecting given $S = \{2\}$, but not given the empty set. By Theorem 1, we conclude that X_3 is Granger-noncausal for X_1 in a bivariate analysis, but not necessarily in an analysis based on $X_{\{1,2,3\}}$.

Now suppose that variable X_4 is latent. Simple derivations show (cf. Eichler, 2005) that the autoregressive representation of $X_{\{1,2,3\}}$ is given by

$$\begin{aligned} X_{1,t} &= \frac{\alpha\,\beta}{1+\beta^2}\, X_{2,t-1} + \frac{\alpha\,\beta\,\gamma}{1+\beta^2}\, X_{3,t-2} + \tilde{\varepsilon}_{1,t} \\ X_{2,t} &= \gamma\, X_{3,t-1} + \tilde{\varepsilon}_{2,t} \\ X_{3,t} &= \varepsilon_{3,t} \end{aligned}$$

where $\tilde{\varepsilon}_{2,t} = \varepsilon_{2,t} + \beta\, X_{4,t-1}$ and

$$\tilde{\varepsilon}_{1,t} = \varepsilon_{1,t} - \frac{\alpha\beta}{1+\beta^2}\, \varepsilon_{2,t-1} + \frac{\alpha}{1+\beta^2}\, X_{4,t-2}$$

The path diagram associated with $X_{\{1,2,3\}}$ is depicted in Figure 22.4(b). In contrast to the graph in Figure 22.4(a), this path diagram contains an edge $3 \longrightarrow 1$ and, thus, does not encode that X_3 is Granger-noncausal for X_1 in a bivariate analysis.

As a response to such situations, two approaches have been considered in the literature. One approach suggests to include all latent variables explicitly as additional nodes in the graph (e.g. Pearl, 2000, Eichler, 2007); this leads to models with hidden variables, which can be estimated, for example, by application of the EM algorithm (e.g. Boyen *et al.*, 1999). For a list of possible problems with this approach, we refer to Richardson and Spirtes (2002, Section 1).

The alternative approach focuses on the conditional independence relations among the observed variables; examples of this approach include linear structural equations with correlated errors (e.g. Pearl, 1995; Koster, 1999) and the maximal ancestral graphs by Richardson and Spirtes (2002). In the time series setting, this approach has been discussed by Eichler (2005), who considered path diagrams in which dashed edges represent associations due to latent variables. For the trivariate subprocess $X_{\{1,2,3\}}$ in the above example, such a path diagram is depicted in Figure 22.4(c).

Following this latter approach, we consider mixed graphs that may contain three types of edges, namely undirected edges (- - -), directed edges ($\longrightarrow$), and dashed directed edges ($\dashrightarrow$). For the sake of simplicity, we also use $a \longleftrightarrow b$ as an abbreviation for the triple edge $a \overset{\dashleftarrow}{\underset{\dashrightarrow}{- - -}} b$. Unlike path diagrams for autoregressions, these graphs in general are not defined in terms of pairwise Granger-causal relationships, but only through the global Markov interpretation according to Definition 7. To this end, we simply extend the concept of m-separation introduced in the previous section by adapting the definition of m-noncolliders and m-colliders. Let π be a path in a mixed graph G. Then an intermediate vertex n is called an m-noncollider on π if at least one of the edges preceding and suceeding c on the path is a directed edge ($\longrightarrow$) and has its tail at c. Otherwise, c is called an m-collider on π. With this extension, we leave all other definitions such as m-separation or pointing paths unchanged.

22.4.1 Marginalization

The main difference between the class of mixed graphs with directed ($\longrightarrow$) and undirected (- - -) edges and the more general class of mixed graphs that has been just introduced is that the latter class is closed under marginalization. This property makes it suitable for representing systems with latent variables.

Let $G = (V, E)$ be a mixed graph and $i \in V$. For every subpath $\pi = \langle e_1, e_2\rangle$ of length 2 between vertices $a, b \in V\backslash\{i\}$ such that i as an intermediate vertex and an m-noncollider on π, we define an edge e_π according to Table 22.1. Let $A^{\{i\}}$ the set of all such edges e_π. Furthermore, let $E^{\{i\}}$ be the subset of edges in E that have both endpoints in $V\backslash\{i\}$. Then we define $G^{\{i\}} = \big(V\backslash\{i\}, E^{\{i\}} \cup A^{\{i\}}\big)$ as the graph obtained by marginalizing over $\{i\}$. Furthermore, for $L = \{i_1, \ldots, i_n\}$ we set $G^L = ((G^{\{i_1\}})^{\{i_2\}} \cdots)^{\{i_n\}}$; that is we proceed iteratively by marginalizing over i_j, for $j = 1, \ldots, n$. Similarly, as in Koster (1999), it can be shown that the order of the vertices does not matter and that the graph G^L is indeed well defined.

Table 22.1 Creation of edges by marginalizing over i.

Subpath π in G	Associated edge e_π in $G^{\{i\}}$
$a \longrightarrow i \longrightarrow b$	$a \longrightarrow b$
$a \dashrightarrow i \longrightarrow b$	$a \dashrightarrow b$
a - - - $i \longrightarrow b$	$a \dashrightarrow b$
$a \longleftarrow i \longrightarrow b$	$a \longleftrightarrow b$
$a \dashleftarrow i \longrightarrow b$	$a \longleftrightarrow b$

We note that the graph G^L obtained by marginalizing over the set L in general contains self-loops. Simple considerations, however, show that G^L is Markov-equivalent to a graph $\tilde{G}^L$ with all subpaths of the form a --- b --→ b and a ←-- b --→ b replaced by a ⇢ b and $a \longleftrightarrow b$, respectively, with all self-loops deleted. It therefore suffices to consider mixed graphs without self-loops. We omit the details.

Now suppose that, for some subsets $A, B, C \subseteq V \backslash L$, π is an m-connecting path between A and B given S. Then all intermediate vertices on π that are in L must be m-noncolliders. Removing these vertices according to Table 22.1, we obtain a path π' in G^L that is still m-connecting. Since the converse is also true, we obtain the following result.

Proposition 4 *Let $G = (V, E)$ be a mixed graph and $L \subseteq V$. Then it holds that, for all distinct $a, b \in V \backslash L$ and all $C \subseteq V \backslash L$, every path between a and b in G is m-blocked given C if and only if the same is true for the paths in G^L. Furthermore, the same equivalence holds for all pointing paths and for all bi-pointing paths.*

It follows that, if a process X_V is Markov for a graph G, the subprocess $X_{V \backslash L}$ is Markov for G^L, which encodes all relationships about $X_{V \backslash L}$ that are also encoded in G.

We note that insertion of edges according to Table 22.1 is sufficient but not always necessary for representing the relations in the subprocess $X_{V \backslash L}$. This applies in particular to the last two cases in Table 22.1. For an example, we consider again the process (22.7) with the associated path diagram in Figure 22.4(a). By Table 22.1, the subpath $1 \longleftarrow 4 \longrightarrow 2$ should be replaced by $1 \longleftrightarrow 2$, which suggests that X_1 Granger-causes X_2 (as does the path $1 \longleftarrow 4 \longrightarrow 2$ in the original path diagram), while in fact the structure can be represented by the graph in Figure 22.4(c).

22.4.2 Ancestral graphs

For systems with latent variables, the set of Granger-causal relationships and contemporaneous independencies that hold for the observed process does not uniquely determine a graphical representation within the class of general path diagrams. As an example, Figure 22.5 displays two graphs that are Markov equivalent; that is they encode the same set of Granger-causal and contemporaneous independence relations among the variables. Therefore, the corresponding graphical models – models that obey the conditional independence constraints imposed by the graph – are statistically indistinguishable. This suggests that one unique representative should be chosen for each Markov equivalence class and model selection should be restricted to these.

Following Richardson and Spirtes (2002), one suitable choice are maximal ancestral graphs. For vertices $a, b \in V$, we say that a is an ancestor of b if $a = b$ or there exists a directed path $a \longrightarrow \cdots \longrightarrow b$ in G. The set of ancestors of b is denoted by an(b). Then

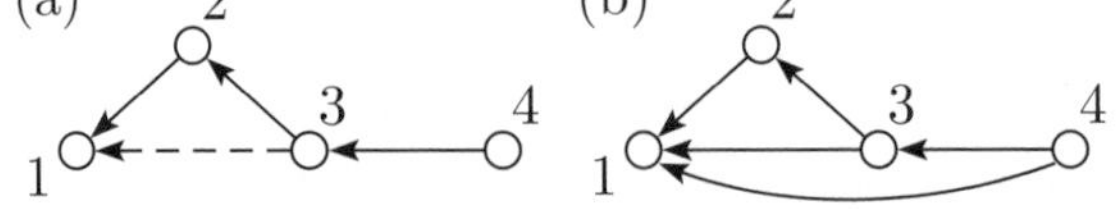

Figure 22.5 Two Markov equivalent graphs: (a) nonancestral graph and (b) corresponding ancestral graph.

$G = (V, E)$ is an ancestral graph if

$$a \in \mathrm{an}(b) \Longrightarrow a \dashrightarrow b \notin E \tag{22.8}$$

for all distinct $a, b \in V$. We note that, in contrast to Richardson and Spirtes (2002), we do not require acyclicity (which is hidden in the time ordering). Furthermore, an ancestral graph G is maximal if addition of further edges changes the independence models; for details, we refer to Richardson and Spirtes (2002).

22.5 Identification of causal effects

Suppose that we are interested in studying the causal effect of an intervention in X on future instances of Y. As we want to be sure about the effect of the intervention before actually intervening, the effect must be predicted from data obtained under the observational regime.

Following the definition of causal effects in Section 22.2.1, let Z be a set of other relevant variables such that the process $V = (X, Y, Z)$ satisfies the stability assumptions (I1) to (I4). For simplicity, we consider the case of a simple intervention, that is an intervention in one variable at a single point in time. Then, using the law of iterated expectation, the ACE can be written as

$$\mathrm{ACE}_s = \mathbb{E}_s\mathbb{E}_s\big[\mathbb{E}_s\big(Y_{t+h}\big|X^t, Y^t, Z^t\big)\big|X^{t-1}, Y^t, Z^t\big].$$

Noting that under the simplifying assumptions conditions (22.1) and (22.2) hold, we can use invariance under a change of regime to get

$$\mathrm{ACE}_s = \mathbb{E}_\emptyset\mathbb{E}_s\big[\mathbb{E}_\emptyset\big(Y_{t+h}\big|X^t, Y^t, Z^t\big)\big|X^{t-1}, Y^t, Z^t\big].$$

If the full process V is observable, the causal effect can be estimated from data. For instance, if V is a Gaussian stationary process, the causal effect of setting X_t to x^* on $Y_{t'}$ is given by the corresponding coefficient in a linear regression of $Y_{t'}$ on V^t.

The problem is that usually not all variables that are deemed as relevant and included in Z are observable. Therefore, suppose that only the subprocess $\tilde{Z}$ of Z is observed. In the case where the desired intervention is conditional as in Definition 1 (iii) and (iv), we assume that the conditioning set C is a subset of the variable in $(X, Y, \tilde{Z})$. The following result states sufficient conditions under which the above derivation of the ACE with Z replaced by $\tilde{Z}$ continues to hold. This so-called back-door criterion has been established first by Eichler and Didelez (2007) and reflects what is known in epidemiology as adjusting for confounding. The name is due to the graphical way of checking this criterion. For a more detailed discussion including proofs of the presented results, we refer to Eichler and Didelez (2010) as well as to the related material in Chapters 3 and 8 in this volume.

Theorem 2 Back-door criterion *Suppose that the process $V = (X, Y, Z)$ satisfies assumptions (22.1) and (22.2) and that for some $S \subseteq Z$*

$$Y_{t+h} \perp\!\!\!\perp \sigma_t | X^t, Y^t, S^t \tag{22.9}$$

for all $h > 0$. *Then* $\tilde{V} = (X, Y, S)$ *identifies the effect of* $\sigma_t = s$ *on* Y_{t+h} *for all* $h > 0$, *and the average causal effect* ACE_s *is given by*

$$\mathbb{E}_s Y_{t+h} = \mathbb{E}_\emptyset \mathbb{E}_s \left[\mathbb{E}_\emptyset \left(Y_{t+h} \middle| X^t, Y^t, S^t \right) \middle| X^{t-1}, Y^t, S^t \right]. \tag{22.10}$$

In (22.10) we can estimate $\mathbb{E}_\emptyset(Y_{t+h}|X^t, Y^t, S^t)$ from observational data, while the second expectation is with respect to the interventional distribution, which is fully known. The outer expectation is again observational. Hence, provided that (X, Y, S) has been observed, we can use the above to estimate the causal effect ignoring any variables that are in Z but not in S. Dawid (2002) calls such a set (X, Y, S) 'sufficient covariates' or 'unconfounder' (see also Chapter 4 by Dawid in this volume). Note that under the stability assumptions, $V = (X, Y, Z)$ always identifies the causal effect due to condition (22.2). In this sense we could say that the whole system V contains all 'relevant' variables or components to identify an individual causal effect of X_t on Y_{t+h}. Note, however, that if an intervention in a different variable $\tilde{X}$ is considered, a different system $\tilde{V}$ might be required to justify the stability assumptions with respect to this intervention.

As in the case of ordinary multivariate distributions, the back-door criterion for time series has an intuitive graphical representation, which can be used to check if a set $S \subset V$ exists that satisfies the back-door or front-door criteria.

Theorem 3 Graphical back-door criterion *Consider a multivariate time series* X_V *that obeys the global Markov properties for a graph* G. *Furthermore, assume that* a *is an ancestor of* b ($a \in \mathrm{an}(b)$).

(i) Assumption (22.9) of Theorem 2 is satisfied if all $\mathrm{an}(b)$*-pointing back-door paths between* a *and* $\mathrm{an}(b)$ *are m-blocked given* S.

(ii) The minimal set S *satisfying (i) is given by* $S = \{a, b\} \cup \mathrm{pa}(a) \cup D$, *where* D *is the set of all nodes* v *such that there is a back-door path from node* a *to* v *for which all intermediate nodes are m-colliders and all intermediate nodes as well as* v *itself are ancestors of* b.

The following example illustrates the application of the graphical back-door criterion. A more complex example with sequential interventions can be found in Eichler and Didelez (2010).

Example 2 Consider the following trivariate Gaussian process X with

$$\begin{aligned} X_{1,t} &= \alpha_1 Z_{t-2} + \beta_{12} X_{2,t-1} + \varepsilon_{1,t} \\ X_{2,t} &= \alpha_2 Z_{t-1} + \beta_{23} X_{3,t-1} + \varepsilon_{2,t} \\ X_{3,t} &= \beta_{32} X_{2,t-1} + \varepsilon_{3,t} \end{aligned}$$

where Z and ε_i, $i = 1, 2, 3$, are independent Gaussian white noise processes with mean 0 and variance σ^2. The corresponding graph is shown in Figure 22.6.

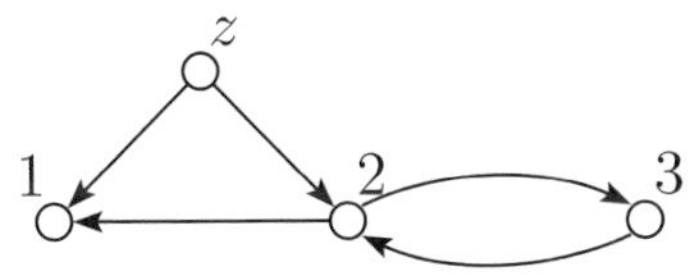

Figure 22.6 Mixed graph associated with the processes X and Z in Example 2.

Now suppose that we are interested in the effect of an intervention s setting $X_{3,t}$ to x_3^* on $X_{1,t+2}$. If both X and Z have been observed, the equations for the full model immediately yield

$$\mathbb{E}_s X_{1,t+2} = \beta_{12}\beta_{23}x_3^*.$$

If only X has been observed while the process Z takes the role of an unobserved variable, the ACE can be still computed. First, straightforward calculations show that X has the autoregressive representation

$$\begin{aligned} X_{1,t} &= \left(\frac{\alpha_1\alpha_2}{1+\alpha_2^2} + \beta_{12}\right) X_{2,t-1} - \frac{\alpha_1\alpha_2\beta_{23}}{1+\alpha_2^2} X_{3,t-2} + \tilde{\varepsilon}_{1,t} \\ X_{2,t} &= \beta_{23}\, X_{3,t-1} + \tilde{\varepsilon}_{2,t} \\ X_{3,t} &= \beta_{32}\, X_{2,t-1} + \tilde{\varepsilon}_{3,t} \end{aligned}$$

where $\tilde{\varepsilon}_i$, $i = 1, 2, 3$, are again independent zero mean Gaussian white noise processes and independent of the other X components. To apply the back-door criterion in Theorem 3, we note that in Figure 22.6 every pointing back-door path between 3 and some other node v is bi-pointing starting with the edge $3 \longleftarrow 2$ and hence is m-blocked given $S = \{1, 2, 3\}$ because 2 is a noncollider. Thus, S identifies the effect of $X_{3,t}$ on $X_{1,t+2}$ and the average causal effect can be obtained from the above autoregressive representation of X as

$$\mathbb{E}_s X_{1,t+2} = \phi_{13}^{(2)}(1) = \phi_{12}(1)\phi_{23}(1) + \phi_{13}(2) = \beta_{12}\beta_{23}x_3^*.$$

Similarly, the example also shows that the effect of $X_{3,t}$ on $X_{1,t+2}$ is not identified if Z instead of X_2 is observed as there exists a back-door path $3 \longleftarrow 2 \longrightarrow 1$ that is not blocked by z.

We close our discussion of the identification of causal effects by an example that points out some limitations of the theory due to serial correlation usually present in time series.

Example 3 Consider a four-dimensional autoregressive process of the form

$$\begin{aligned} X_{1,t} &= \phi_{11}\, X_{1,t-1} + \phi_{12}\, X_{2,t-1} + \varepsilon_{1,t} \\ X_{2,t} &= \phi_{22}\, X_{2,t-1} + \varepsilon_{2,t} \\ X_{3,t} &= \phi_{32}\, X_{2,t-1} + \phi_{33}\, X_{3,t-1} + \varepsilon_{3,t} \\ X_{4,t} &= \phi_{41}\, X_{1,t-1} + \phi_{43}\, X_{3,t-1} + \phi_{44}\, X_{4,t-1} + \varepsilon_{4,t} \end{aligned}$$

where $(\varepsilon_1, \varepsilon_2, \varepsilon_3, \varepsilon_4)$ is a Gaussian white noise process with mean zero and covariance matrix $\sigma^2 I$. The path diagram is shown in Figure 22.7(a).

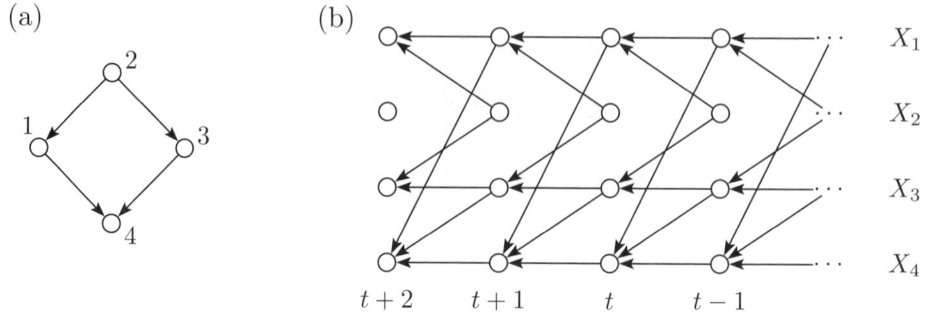

Figure 22.7 *(a) Granger causality graph and (b) time series chain graph (with an additional imposed constraint $\phi_{22} = 0$) for the process* X *in Example 3.*

Suppose that we are interested in the effect of an intervention s setting $X_{1,t}$ to x^* on $X_{4,t+2}$. If all variables are observed, the causal effect of $X_{1,t}$ on $X_{4,t+2}$ is identified and given by

$$\mathbb{E}_s X_{4,t+2} = (\phi_{41}\,\phi_{11} + \phi_{44}\,\phi_{41})\,x^*.$$

Similarly, knowledge of X_2 is sufficient for identification of the causal effect as every back-door path is m-blocked given $S = \{1, 2, 4\}$. In contrast, knowledge of only (X_1, X_3, X_4) is not sufficient as the path $1 \longleftarrow 2 \longrightarrow 1 \longrightarrow 4$ is unblocked.

To understand why this path induces a confounding association between $X_{1,t}$ and $X_{4,t+2}$, we consider explicitly the instances at the relevant points in time along the path. Thus, the above path $1 \longleftarrow 2 \longrightarrow 1 \longrightarrow 4$ corresponds to the path

$$X_{1,t} \longleftarrow X_{2,t-1} \longrightarrow X_{2,t+1} \longrightarrow X_{1,t+1} \longrightarrow X_{4,t+2}.$$

The problem is that every variable that directly influences a variable of interest acts as a confounder for all causal links from this variable of interest to other variables due to serial correlation.

The problem can be resolved if additional information about the process constraints the dynamic dependence structure further. For example, if $\phi_{22} = 0$ for the above process X, the causal effect of $X_{1,t}$ on $X_{4,t+1}$ is identified by (X_1, X_3, X_4). However, a graphical criterion for identifiability must be based on the full time series chain graph depicted in Figure 22.7(b) as the Granger causality graph cannot encode such additional specific constraints.

22.6 Learning causal structures

There are two major approaches for learning causal structures: One approach utilizes constraint-based search algorithms such as the fast causal inference (FCI) algorithm (Spirtes *et al.*, 2001) while the other consists of score-based model selection. Both approaches have been adapted to the time series case (Eichler, 2009, 2010). In this and the next section, we briefly discuss the two approaches.

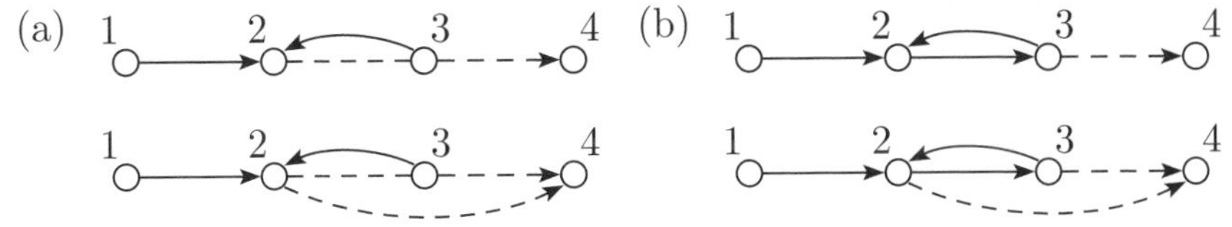

Figure 22.8 Inducing paths: (a) dynamic ancestral graph with an inducing 4-pointing path between 2 and 4 and its corresponding Markov equivalent maximal dynamic ancestral graph; (b) maximal dynamic ancestral graph with an inducing 4-pointing path between 2 and 4. The graph with additional edge $2 \dashrightarrow 4$ *is not Markov equivalent.*

The first approach requires as input the set of all Granger-noncausal relations and contemporaneous independences that hold for the process. Usually, this will be accomplished by fitting vector autoregressions to all subseries X_S of X_V. Then the constraint-based search tries to find a graph that matches the empirically determined conditional independences. It usually consists of two steps:

(1) identification of adjacencies of the graph;

(2) identification of the type and the orientation of edges whenever possible.

In the case of ancestral graphs, the first step makes use of the fact that for every ancestral graph there exists a unique Markov-equivalent maximal ancestral graph (MAG), in which every missing edge corresponds to a conditional independence relation among the variables. Here an ancestral graph G is said to be maximal if addition of further edges would change the Markov equivalence class. MAGs are closely related to the concept of inducing paths; in fact, Richardson and Spirtes (2002) used inducing paths to define MAGs and then showed the maximality property.

Definition 8 *In a (dynamic) ancestral graph, a path π between two vertices a and b is called an inducing path if every intermediate vertex on π is, first, a collider on π and, second, an ancestor of a or b.*

Figures 22.8(a) and (b) give two examples of dynamic ancestral graphs, in which 2 --- 3 $\dashrightarrow$ 4 respectively 2 $\longrightarrow$ 3 $\dashrightarrow$ 4, are inducing 4-pointing paths. The graph in (b) shows that – unlike in the case of ordinary ancestral graphs – inducing paths may start with a tail at one of two vertices. As a consequence, insertion of an edge 2 $\dashrightarrow$ 4 or 2 $\longrightarrow$ 4 changes the encoded Granger-causal relationships. While the upper graph implies that X_1 is Granger-noncausal for X_4 with respect to $X_{\{1,2,3,4\}}$, this is not true for the lower graph. It follows that the method used for identifying adjacencies in ordinary MAGs does not apply to dynamic ancestral graphs.

The problem can be solved by observing that m-connecting pointing paths encode not only Granger-causal relationships but, depending on whether they start with $a \longrightarrow c$ or $a \dashrightarrow c$, also a related type of conditional independences. More precisely, we have the following result.

Proposition 5 *Suppose that* a *and* b *are not connected by an edge $a \longrightarrow b$ or $a \dashrightarrow b$ or by a* b*-pointing inducing path starting with $a \longleftarrow c$, $a \dashleftarrow c$ or a --- c. Then there exist disjoint* *subsets* S_1, S_2 *with* $b \in S_1$ *and* $a \notin S_1 \cup S_2$ *such that*

$$X_{a,t-k} \perp\!\!\!\perp X_{b,t+1} \mid X^{t}_{S_1}, X^{t-k}_{S_2}, X^{t-k-1}_{a}$$

for all $k \in \mathbb{N}$ *and all* $t \in \mathbb{Z}$.

The proof is based on the fact that inducing paths starting with an edge $a \longrightarrow c$ or $a \dashrightarrow c$ only induce an association between $X_{a,t-k}$ and $X_{b,t+1}$ if one conditions on $X_{c,t-k+1}, \ldots, X_{c,t}$. To block any other paths, we set S_2 to be the set of all intermediate vertices on all b-pointing inducing paths connecting a and b, and S_1 to be the set of all ancestors of a and b except a and S_2.

This leads us to the following algorithm for the identification of the Markov equivalence classes of dynamic ancestral graphs. Here, we use dotted directed edges $\cdots\!\!\rightarrow$ to indicate that the tail of the directed edge is (yet) undetermined.

Algorithm

Identification of adjacencies:

1. Insert $a\ {-}\,{-}\,{-}\ b$ whenever X_a and X_b are not contemporaneously independent with respect to X_V.
2. Insert $a \cdots\!\!\rightarrow b$ whenever
   (i) X_a Granger-causes X_b with respect to X_S for all $S \subseteq V$ with $a, b \in S$ and
   (ii) $X_{a,t-k}$ and $X_{b,t+1}$ are not conditionally independent given $X^{t}_{S_1}, X^{t-k}_{S_2}, X^{t-k-1}_{a}$ for some $k \in \mathbb{N}$, all $t \in \mathbb{Z}$, and all disjoint $S_1, S_2 \subseteq V$ with $b \in S_1$ and $a \notin S_1 \cup S_2$.

Identification of tails:

1. *Colliders*. Suppose that G does not contain $a \cdots\!\!\rightarrow b$, $a \longrightarrow b$ or $a \dashrightarrow b$. If $a \cdots\!\!\rightarrow c \cdots\!\!\rightarrow b$ and X_a is Granger-noncausal for X_b with respect to X_S for some set S with $c \notin S$, replace $c \cdots\!\!\rightarrow b$ by $c \dashrightarrow b$.
2. *Noncolliders*. Suppose that G does not contain $a \cdots\!\!\rightarrow b$, $a \longrightarrow b$, or $a \dashrightarrow b$. If $a \cdots\!\!\rightarrow c \cdots\!\!\rightarrow b$ and X_a is Granger-noncausal for X_b with respect to X_S for some set S with $c \in S$, replace $c \cdots\!\!\rightarrow b$ by $c \longrightarrow b$.
3. *Ancestors*. If $a \in \text{an}(b)$ replace $a \cdots\!\!\rightarrow b$ by $a \longrightarrow b$.
4. *Discriminating paths*. A fourth rule is based on the concept of discriminating paths. For details, we refer to Ali *et al.* (2004).

We note that in contrast to the case of ordinary ancestral graphs only the tails of the dotted directed edges need to be identified. The positions of the arrowheads are determined by the time ordering of the Granger-causal relationships. The above algorithm probably can be complemented by further rules; see also Zhang and Spirtes (2005).

To illustrate the identification algorithm, we consider the graphs in Figure 22.9. The original general path diagram is depicted in (a). Since 4 is an ancestor of 5, this graph is not ancestral. The adjacencies determined by the algorithm are shown in (b). Since X_1 does not

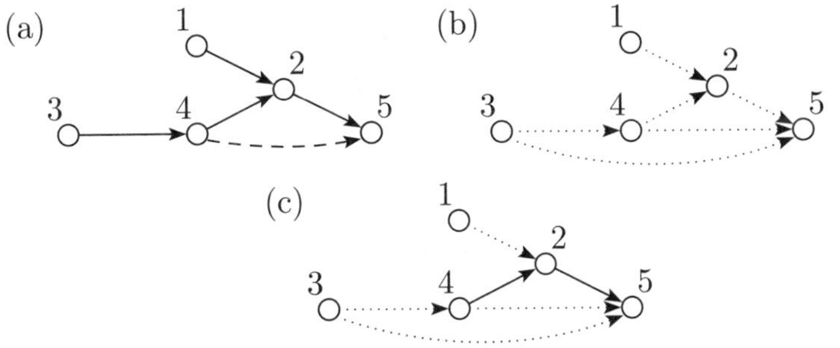

Figure 22.9 Identification of dynamic ancestral graphs: (a) underlying structure; (b) adjacencies; (c) identification of tails.

Granger-cause X_5 with respect to X_V, we find that 2 and 5 are connected by $2 \longrightarrow 5$. Similarly, X_3 is Granger-noncausal for X_2 with respect to $X_{\{2,3,4\}}$, which implies that the graph also contains the edge $4 \longrightarrow 2$. No further tails can be identified; the final graph is given in (c).

22.7 A new parametric model

The second major approach for identifying causal structures requires fitting suitably constrained models that are Markov with respect to a given path diagram to the data. Searching over all possible path diagrams, the fit is evaluated by a model selection criterion such as AIC or BIC. The path diagram with the lowest score is taken as an estimate of the true causal structure.

In the case of simple path diagrams with no dashed directed edges, Definition 6 states the constraints that are required for the model being Markov with respect to the graph. In particular, we note that the undirected edges in the path diagram correspond to dependences in the innovation process ε. Eichler (2010) showed that by allowing correlations between lagged instances of the innovation process ε dashed directed edges can be incorporated into the framework of graphical vector autoregressive models.

More precisely, we consider multivariate stationary Gaussian processes $X = X_V$ that are given by

$$X_t = \sum_{u=1}^{p} \Phi(u) X_{t-u} + \varepsilon_t \tag{22.11}$$

where $\varepsilon = \varepsilon_V$ is a stationary Gaussian process with mean zero and covariances

$$\operatorname{cov}\big(\varepsilon_t, \varepsilon_{t-u}\big) = \begin{cases} \Omega(u) & \text{if } |u| \leq q \\ 0 & \text{otherwise} \end{cases} \tag{22.12}$$

for some $q \in \mathbb{N}$. The distribution of these processes can be parametrized by the vector $\theta = (\phi, \omega)$ with

$$\phi = \operatorname{vec}(\Phi(1), \ldots, \Phi(p))$$

and

$$\omega = \begin{pmatrix} \mathrm{vech}\Omega(0) \\ \mathrm{vec}\big(\Omega(1), \ldots, \Omega(q)\big) \end{pmatrix}$$

where as usual the vec operator stacks the columns of a matrix and the vech operator stacks only the elements contained in the lower triangular submatrix. The parameter ϕ corresponds to the autoregressive structure of the process X while ω parametrizes the dependence structure of the innovation process ε. The following result states that suitable zero constraints on the parameters ϕ and ω ensure that the model is Markov with respect to a given mixed graph G.

Theorem 4 *Let $X = X_V$ be a stationary Gaussian process of the form (22.11) and (22.12), and suppose that Assumption 2 holds. Furthermore, let $G = (V, E)$ be a mixed graph such that the model parameters ϕ and ω satisfy the following constraints:*

(i) If $a \longrightarrow b \notin E$ then $\Phi_{ba}(u) = 0$ for all $u > 0$.

(ii) If $a \dashrightarrow b \notin E$ then $\Omega_{ba}(u) = 0$ for all $u > 0$.

(iii) If a --- $b \notin E$ then $\Omega_{ba}(0) = \Omega_{ab}(0) = 0$.

Then X is Markov for G.

Sufficient conditions for X to satisfy Assumption 2 can be formulated in terms of the parameters ϕ and ω. For this, let $\Phi(z) = I - \Phi(1)\,z - \cdots - \Phi(p)\,z^p$ be the characteristic polynomial of the autoregressive part of X and let $g_\omega(\lambda)$ be the spectral matrix of ε. Then, if $\det \Phi(z) \neq 0$ for all $z \in \mathbb{C}$ such that $|z| \leq 1$ and the eigenvalues of $g_\omega(\lambda)$ are bounded and bounded away from zero uniformly for all $\lambda \in [-\pi, \pi]$, Assumption 2 holds for the process X. For details on this and the proof of the above result we refer to Eichler (2010).

Autoregressive models with an innovation process having a correlation structure of the form (22.12) are uncommon in time series analysis. However, it has been shown by Eichler (2010) that models of this form can be viewed as graphical multivariate ARMA models that satisfy the conditional independence constraints encoded by general path diagrams. We note that under the assumptions the error process ε has a moving average representation. Since the covariance function $\Omega(u)$ vanishes for $|u| > q$, it follows that the moving average representation is of order q and we have

$$\varepsilon_t = \sum_{u=0}^{q} \Theta(u)\,\eta_{t-u}$$

where $\Theta(0) = I$ and η is a Gaussian white noise process with mean zero and covariance matrix Σ. The coefficients of this moving average representation are uniquely determined by the equation system

$$\Omega(v) = \sum_{u=0}^{q-v} \Theta(u)'\,\Sigma\,\Theta(u+v) \qquad (22.13)$$

which can be iteratively solved for $\Theta(u)$ and Σ (Tunnicliffe Wilson, 1972). It follows that the process X can be represented as a multivariate ARMA(p,q) process

$$X_t = \sum_{u=1}^{p} \Phi(u)\, X_{t-u} + \sum_{u=0}^{q} \Theta(u)\, \eta_{t-u}.$$

We note that, because of (22.13), the zero constraints on the matrices $\Omega(u)$ do not translate into equally simple constraints on the parameters $\Theta(1), \ldots, \Theta(q)$ and Σ.

22.8 Concluding remarks

In time series analysis, inference about causal relationships is still predominantly based on the concept of Granger causality as it can be simply formulated and tested in many standard time series models. Nevertheless, Granger causality has always been criticized as being not a true notion of causality due to the fact that it can lead to spurious causalities when confounding variables are not included in the analysis.

In this chapter, we have pointed out the usefulness of Granger causality as an empirical concept for causal inference by reviewing recent results that link the traditional concept of Granger causality to the modern graph-based approach to causality. Here it is helpful to distinguish between the theoretical notion of Granger causality as orginally defined by Granger (1969, 1980, 1988) and its various empirical versions, which are usually referred to as 'Granger causality'. As Granger (2003) stated himself, these empirical versions should be seen as implications of Granger causality and not mistaken as the definition itself.

With this in mind, we have compared four definitions of causality, namely intervention, direct structural, Granger and Sims causality used in the context of time series. We found that all four concepts are closely related and differ only in whether they are concerned with the total or only the direct causal effect. The subtle differences that do exist we think are not of practical relevance.

Although Granger causality does not require a scientific model – that is no pre-knowledge about the system to analysed – it is limited in its scope by the statistical model used for the empirical analysis. The traditional framework to study Granger causality are vector autoregressive models (cf. Eichler, 2005, 2006). As a consequence, only linear relationships are covered by the model and any nonlinear cause–effect relationship might stay undetected. Although there exist general tests for nonlinear Granger causality that are based on nonparametric estimation methods (e.g. Su and White, 2007; Bouezmarni *et al.*, 2009; Huang, 2009; Marinazzo *et al.*, 2011), applying this on large-scale systems usually is not feasible. Inference can be greatly facilitated if additional background knowledge about the system can be incorporated in the model. As an example, we mention the recent article by Valdes-Sosa *et al.* (2011), which lists various complex state-space models used for causal inference from brain-imaging data.

An important topic that we have not raised in our discussion is model uncertainty. The score-based approach yields a single model – specified by a path diagram – that minimizes some model selection criterion, but at the same time there are other models that have only a slightly larger score and therefore should be deemed competitive. Such models usually do not give a significantly worse fit if tested by an appropriate test (e.g. Vuong, 1989). One possible solution to this model uncertainty could be model averaging, but it is not immediately clear how to 'average' path diagrams and how to draw conclusions from such an averaged path diagram. In the constraint-based approach, model uncertainty enters through the set of Granger noncausality statements that serve as an input to the identification algorithm. As the set of statements is determined by a multitude of statistical tests, the empirical set deviates from the true set of statements that hold for the process. The uncertainty introduced in this way in the identification algorithm has so far been neglected.

Finally, we note that there is no reason to restrict the discussion to time-discrete stochastic processes. Indeed, Comte and Renault (1996) and Florens and Fougère (1996) extended the notion of Granger causality to time-continuous processes. Closely related to these definitions is the notion of local independence introduced earlier by Schweder (1970) and Aalen (1987) and used by Didelez (2008) to define graphical models for marked point processes and by Commenges and Gègout-Petit (2009) for causal inference.

References

Aalen, O.O. (1987) Dynamic modeling and causality. *Scandinavian Actuarial Journal*, 177–190.

Ali, R.A., Richardson, T.S. and Spirtes, P. (2004) Markov equivalence for ancestral graphs. Technical Report 466, Department of Statistics, University of Washington.

Bouezmarni, T., Rombouts, J.V.K. and Taamouti, A. (2009) A nonparametric copula based test for conditional independence with applications to Granger causality. Economics Working Papers we093419, Departamento de Economía, Universidad Carlos III.

Boyen, X., Friedman, N. and Koller, D. (1999) Discovering the hidden structure of complex dynamic systems, in *Proceedings of the 15th Conference on Uncertainty in Artificial Intelligence*, Morgan Kaufmann, San Francisco, pp. 91–100.

Commenges, D. and Gègout-Petit, A. (2009) A general dynamical statistical model with causal interpretation. *Journal of the Royal Statistical Society, Series B*, **71**, 1–18.

Comte, F. and Renault, E. (1996) Noncausality in continuous time models. *Econometric Theory*, **12**, 215–256.

Dahlhaus, R. (2000) Graphical interaction models for multivariate time series. *Metrika*, **51**, 157–172.

Dahlhaus, R. and Eichler, M. (2003) Causality and graphical models in time series analysis, in *Highly Structured Stochastic Systems* (eds P. Green, N. Hjort and S. Richardson). Oxford: University Press, pp. 115–137.

Dawid, A.P. (2002) Influence diagrams for causal modelling and inference. *International Statistical Review*, **70**, 161–189.

Dawid, A.P. and Didelez, V. (2005) Identifying the consequences of dynamic treatment strategies. Technical Report 262, Department of Statistical Science, University College London.

Demiralp, S. and Hoover, K.D. (2003) Searching for the causal structure of a vector autoregression. *Oxford Bulletin of Economics and Statistics*, **65**(Supplement), 745–767.

Didelez, V. (2008) Graphical models for marked point processes based on local independence. *Journal of the Royal Statistical Society, Series B*, **70**, 245–264.

Eichler, M. (2005) A graphical approach for evaluating effective connectivity in neural systems. *Philosophical Transactions of The Royal Society, Series B*, **360**, 953–967.

Eichler, M. (2006) Graphical modelling of dynamic relationships in multivariate time series, in *Handbook of Time Series Analysis* (eds M. Winterhalder, B. Schelter and J. Timmer). Wiley-VCH, pp. 335–372.

Eichler, M. (2007) Granger causality and path diagrams for multivariate time series. *Journal of Econometrics*, **137**, 334–353.

Eichler, M. (2009) Causal inference from multivariate time series: What can be learned from Granger causality, in *Proceedings from the 13th International Congress of Logic, Methodology and Philosophy of Science* (eds C. Glymour, W. Wang and D. Westerståhl). London: College Publications, pp. 481–496.

Eichler, M. (2010) Graphical Gaussian modelling of multivariate time series with latent variables, in *Proceedings of the 13th International Conference on Artificial Intelligence and Statistics*, Journal of Machine Learning Research W&CP 9.

Eichler, M. (2011) Graphical modelling of multivariate time series. *Probability Theory and Related Fields*. DOI:10.1007/s00440-011-0345-8.

Eichler, M. and Didelez, V. (2007) Causal reasoning in graphical time series models, in *Proceedings of the 23rd Conference on Uncertainty in Artificial Intelligence*.

Eichler, M. and Didelez, V. (2010) On Granger-causality and the effect of interventions in time series. *Life Time Data Analysis*, **16**, 3–32.

Florens, J. P. and Fougère, D. (1996) Noncausality in continuous time. *Econometrica*, **64**, 1195–1212.

Florens, J.P. and Mouchart, M. (1982) A note on noncausality. *Econometrica*, **50**, 583–591.

Florens, J.P. and Mouchart, M. (1985) A linear theory for noncausality. *Econometrica*, **53**, 157–175.

Good, I. J. (1961) A causal calculus (I). *British Journal for the Philosophy of Science*, **11**, 305–318.

Good, I. J. (1962) A causal calculus (II). *British Journal of the Philosophy of Science*, **12**, 43–51.

Granger, C.W.J. (1969) Investigating causal relations by econometric models and cross-spectral methods. *Econometrica*, **37**, 424–438.

Granger, C. W. J. (1980) Testing for causality, a personal viewpoint. *Journal of Economic Dynamics and Control*, **2**, 329–352.

Granger, C.W.J. (1988) Some recent developments in a concept of causality. *Journal of Econometrics*, **39**, 199–211.

Granger, C.W.J. (2003) Some aspects of causal relationships. *Journal of Econometrics*, **112**, 69–71.

Hosoya, Y. (1977) On the Granger condition for non-causality. *Econometrica*, **45**, 1735–1736.

Hsiao, C. (1982) Autoregressive modeling and causal ordering of econometric variables. *Journal of Economic Dynamics and Control*, **4**, 243–259.

Huang, M. (2009) *Essays on testing conditional independence*. Doctoral Thesis, University of California, San Diego.

Hyvärinen, A., Zhang, K., Shimizu, S. and Hoyer, P. (2010) Estimation of a structural vector autoregression model using non-Gaussianity. *Journal of Machine Learning Research*, **11**, 1709–1731.

Kamiński, M., Ding, M., Truccolo, W.A. and Bressler, S.L. (2001) Evaluating causal relations in neural systems: Granger causality, directed transfer function and statistical assessment of significance. *Biological Cybernetics*, **85**, 145–157.

Koster, J.T.A. (1999) On the validity of the Markov interpretation of path diagrams of Gaussian structural equations systems with correlated errors. *Scandinavian Journal of Statistics*, **26**, 413–431.

Lauritzen, S.L. (2001) Causal inference from graphical models, in *Complex Stochastic Systems* (eds O.E. Barndorff-Nielsen, D.R. Cox and C. Klüppelberg). London: CRC Press, pp. 63–107.

Lütkepohl, H. (1993) *Introduction to Multiple Time Series Analysis*. New York: Springer.

Lynggaard, H. and Walther, K.H. (1993) *Dynamic modelling with mixed graphical association models*. Master's Thesis, Aalborg University.

Marinazzo, D., Liao, W., Chen, H. and Stramaglia, S. (2011) Nonlinear connectivity by Granger causality. *NeuroImage*, **58**, 330–338.

Moneta, A. (2007) Graphical causal models and VARs: an empirical assessment of the real business cycles hypothesis. *Empirical Economics*. DOI 10.1007/s00181-007-0159-9.

Moneta, A. and Spirtes, P. (2006) Graphical models for the identification of causal structures in multivariate time series models, in *Joint Conference on Information Sciences Proceedings*. Atlantis Press.

Pearl, J. (1993) Graphical models, causality and interventions. *Statistical Science*, **8**, 266–269.

Pearl, J. (1995) Causal diagrams for empirical research (with discussion). *Biometrika*, **82**, 669–710.

Pearl, J. (2000) *Causality*. Cambridge, UK: Cambridge University Press.

Pearl, J. and Verma, T. (1991) A theory of inferred causation, in *Principles of Knowledge Representation and Reasoning: Proceedings of the 2nd International Conference* (eds J.A. Allen, F. Fikes and E. Sandewall), San Mateo, CA: Morgan Kaufmann, pp. 441–452.

Reale, M. and Tunnicliffe Wilson, G. (2001) Identification of vector AR models with recursive structural errors using conditional independence graphs. *Statistical Methods and Applications*, **10**, 49–65.

Richardson, T. (2003) Markov properties for acyclic directed mixed graphs. *Scandinavian Journal of Statistics*, **30**, 145–157.

Richardson, T. and Spirtes, P. (2002) Ancestral graph Markov models. *Annals of Statistics*, **30**, 962–1030.

Schweder, T. (1970) Composable Markov processes. *Journal of Applied Probability*, **7**, 400–410.

Sims, C.A. (1972) Money, income and causality. *American Economic Review*, **62**, 540–552.

Spirtes, P., Richardson, T.S., Meek, C., Scheines, R. and Glymour, C. (1998) Using path diagrams as a structural equation modelling tool. *Society of Methods Research*, **27**, 182–225.

Spirtes, P., Glymour, C. and Scheines, R. (2001) *Causation, Prediction, and Search*. 2nd edn. Cambridge, MA: MIT Press, with additional material by David Heckerman, Christopher Meek, Gregory F. Cooper and Thomas Richardson.

Su, L. and White, H. (2007) A consistent characteristic function based test for conditional independence. *Journal of Econometrics*, **141**, 807–834.

Suppes, P. (1970) *A Probabilistic Theory of Causality*. Amsterdam: North-Holland.

Swanson, N.R. and Granger, C.W.J. (1997) Impulse response functions based on a causal approach to residual orthogonalization in vector autoregressions. *Journal of the American Statistical Association*, **92**, 357–367.

Tunnicliffe Wilson, G. (1972) The factorization of matricial spectral densities. *SIAM Journal of Applied Mathematics*, **23**, 420–426.

Valdes-Sosa, P.A., Roebroeck, A., Daunizeau, J. and Friston, K. (2011) Effective connectivity: influence, causality and biophysical modeling. *NeuroImage*, **58**, 339–361.

Vuong, Q.H. (1989) Likelihood ratio tests for model selection and non-nested hypotheses. *Econometrica*, **57**, 307–333.

White, H. and Lu, X. (2010) Granger causality and dynamic structural systems. *Journal of Financial Econometrics*, **8**, 193–243.

Zhang, J. and Spirtes, P. (2005) A characterization of Markov equivalence classes for ancestral graphical models. Technical Report 168, Department of Philosophy, Carnegie Mellon University.

23

Dynamic molecular networks and mechanisms in the biosciences: A statistical framework

Clive G. Bowsher
School of Mathematics, University of Bristol, Bristol, UK

23.1 Introduction

Problems of causal inference often arise in connection with biomedical data, and many of the developments in the field have been motivated by biomedical and epidemiological applications. Scientific knowledge of underlying molecular mechanisms in genetics, biochemistry and cell biology has burgeoned in recent decades, and there is often a need to incorporate such mechanistic knowledge when making causal inferences about disease outcomes. For example, valid application of Mendelian randomisation in epidemiology requires the conditional independence of the instrumental variable (genotype) and the disease outcome, given the intermediate phenotype and unobserved (confounding) effects (Didelez and Sheehan, 2007). Detailed knowledge of the associated biological science and mechanisms is extremely helpful in assessing the likely validity of such conditional independences, which cannot be tested statistically.

Dynamic stochastic models formulated at the level of molecular mechanisms are currently of considerable importance in biological science, especially in cellular systems biology. The construction and analysis of such models is an area to which causal inference (and graphical modelling) can contribute and from which it can benefit. Systems biology is an increasingly influential, interdisciplinary approach that aims to build dynamic, causal models of biomolecular networks in order to understand and ultimately control the behaviour of the cell. The networks involved are often very large indeed, and their stochastic dynamic properties are difficult to

Causality: Statistical Perspectives and Applications, First Edition. Edited by Carlo Berzuini, Philip Dawid and Luisa Bernardinelli.

understand. Putative causal models of these networks must be constructed by integrating the results of many experiments and are subject to substantial uncertainty. The models must be tested by designing and performing appropriate experimental interventions, making reliable causal inferences on the basis of these and then refining the network model accordingly.

Well-specified dynamic models of molecular mechanisms can readily accommodate interventions such as pharmacological treatments by incorporating the reactions in which the drug participates – other aspects of (is reactions in) the model remain 'stable' features of the model because the underlying biochemistry does not change. The issues of stability under intervention and invariance of conditional distributions (Dawid, 2010) are no longer paramount due to an increased mechanistic understanding, although one must still beware the presence of unknown biomolecules and/or unknown biochemical reactions.

The field of causal inference would most likely benefit from a renewed engagement with experimental design, with the endeavour of drawing inferences from experimental data and with the experimental biosciences. Hypothesised causal models derived from observational data (graphical or otherwise) should always be tested if possible using data collected under the posited interventional regimes. In systems biology, testing the validity of causal inferences drawn from a hypothesised causal model is often feasible using direct experimental intervention under controlled conditions in the laboratory.

The above discussion motivates the focus of this chapter on *stochastic kinetic models* (SKMs), in particular their graphical representation, local and global dynamic conditional independence properties and coarse-grained architecture. The structure of the present chapter is as follows. Section 23.2 introduces SKMs and biochemical reaction networks. Section 23.3 is concerned with kinetic independence graphs of SKMs – their local independence properties and notions of causal influence. Section 23.4 explores the notion of modules (quasi-autonomous groups of vertices) which is important in systems biology, and describes the MIDIA algorithm for identification of modularisations based simultaneously on local and global dynamic independence of the constituent modules. Section 23.5 applies the concepts discussed to the important MAPK cell signalling mechanism, which is activated by many cancer genes, and Section 23.6 concludes.

23.2 SKMs and biochemical reaction networks

A stochastic kinetic model (Bowsher, 2010) is a continuous-time jump process modelling the state of a biochemical system, $X(t) = [X_1(t), \ldots, X_n(t)]'$, where $X_i(t)$ is interpreted as the nonnegative, integer number of molecules of type i present at time t. The set of different types of molecule or the *species set* is given by $\mathcal{V} := \{1, \ldots, n\}$. There are a finite number of possible values of jump in $X(t)$ that may take place, corresponding to the different types of possible *reaction*, $m \in \mathcal{M} := \{1, \ldots, M\}$. It is particularly useful for our purposes to view an SKM as a marked point process in which the points or 'events' correspond to the jump times of the process $X(t)$ and each mark indicates the value of the jump associated with the corresponding jump time. Mathematically, a particular reaction can then be identified with an element of the finite mark space.

An SKM is denoted here by $\{T_s, Z_s\}_{s \geq 1}$, where T_s is the sth jump time. The mark $Z_s \in \{S_m | m \in \mathcal{M}\}$ is the value of the jump and is interpreted as the *changes* in the number of molecules of each species. The matrix $S := [S_1, S_2, \ldots, S_M]$ is usually known as the stoichiometric matrix. Any two columns of S are taken to be nonequal; hence there is a bijection between the mark space and the reaction space, $\mathcal{M}$. The following linear equation determines the dynamic evolution of $X(t)$:

$$X(t) = X(0) + SN(t), \quad t \geq 0 \tag{23.1}$$

where $N(t) = [N_1(t), \ldots, N_M(t)]'$ is the M-variate counting process associated with the marked point process $\{T_s, Z_s\}_{s\geq 1}$. Thus, $N_m(t)$ is interpreted as counting the number of reactions of type m during $(0, t]$. Denote by $\mathcal{F}_t^N := \sigma(N(s); 0 \leq s \leq t)$ the internal history of the entire M-variate process and by $\mathcal{F}_t^m := \sigma(N_m(s); 0 \leq s \leq t)$ the internal history of the mth counting process.

A basic familiarity with the chemical representation and interpretation of reactions is helpful for what follows (see also Wilkinson, 2009, for an accessible introduction). Each reaction $m \in \mathcal{M}$ has the chemical representation

$$\sum_{i \in R[m]} \alpha_i X_i \to^m \sum_{j \in P[m]} \beta_j X_j \tag{23.2}$$

which is read as follows: when the event called 'reaction m' takes place, α_i molecules of type i are consumed for each i in the subset $R[m] \subset \mathcal{V}$ and β_j molecules of type j are produced for each j in the subset $P[m] \subset \mathcal{V}$. The species $R[m]$ are called the *reactants* (or inputs) of the reaction m and the species $P[m]$ are called the *products* (or outputs) of m. The integer coefficients $[\{\alpha_i\}, \{\beta_j\}]$ are known as the stoichiometries of the reaction. If a species k is a reactant but not a product, then its corresponding entry in the stoichiometric matrix S of Equation (23.1) – i.e. the change in the level of k caused by reaction m – is given by $S_{km} = -\alpha_k$. Alternatively, if species k is a product but not a reactant, then $S_{km} = \beta_k$. There is no assumption that $R[m] \cap P[m] = \emptyset$, and if k is both a product and a reactant then $S_{km} = \beta_k - \alpha_k$. A common situation in this case is $\beta_k = \alpha_k$, that is k acts as a 'catalyst', increasing the rate of the reaction but not itself being 'changed' by the reaction – i.e. not itself being net consumed or produced when reaction m takes place. Formally, the sets $R[m]$ and $P[m]$ are defined by allowing zero stoichiometries and writing the mth reaction as $\sum_{i\in\mathcal{V}} \alpha_i X_i \to^m \sum_{j\in\mathcal{V}} \beta_j X_j$. Then $R[m] := \{i \in \mathcal{V} | \alpha_i > 0\}$ and $P[m] := \{j \in \mathcal{V} | \beta_j > 0\}$.

In systems biology, a living cell is often viewed as a network of interacting biomolecules of different types, with n and M both large (and often $M > n$). The interaction is selective – only species that are reactants for some reaction m can together react to give products. Each reaction involves only a few species, so the cardinality of $R[m] \cup P[m]$ is small. Certain reactions are 'coupled' in that a product of one reaction is also a reactant of another reaction. From a stochastic process perspective, the specification of the list of component reactions as in Equation (23.2) for all $m \in \mathcal{M}$ implies dependences between the levels (or concentrations) of the different biomolecules.

The following example of an SKM provides a simple but nonetheless biochemically meaningful illustration.

Example 1 Consider the SKM with the five different species $\mathcal{V} = \{P, R, g, P_2, gP_2\}$ and the six reactions

$$g \to^{trc} g + R, \qquad R \to^{trl} R + P, \qquad 2P \to^{d} P_2$$
$$P_2 \to^{rd} 2P, \qquad g + P_2 \to^{b} gP_2, \qquad gP_2 \to^{ub} g + P_2$$

The gene (g) is responsible for the production of molecules of protein (P) via the intermediate (mRNA) species (R). In this simplified representation, g and R act as simple catalysts in the reactions *trc* ('transcription') and *trl* ('translation') respectively. The third reaction d consists of the binding of two molecules of P (the sole reactant) to form the new molecule P_2 (the sole product). The fourth reaction *rd* is the reverse of the third. The fifth reaction sets up a 'negative feedback cycle' whereby the production of P is negatively self-regulated by the binding of P_2 to g to form the distinct species gP_2. Genes bound in this way to P_2 are not then available to participate in the *trc* reaction, thus preventing overproduction of the protein. Notice that the quantity $[X^{gP_2}(t) + X^g(t)]$ is 'conserved', that is constant over time. We shall return later to the same example (see Figure 23.1).

Returning to Equation (23.1) for a general SKM, the probability measure P generating the process $N(t)$, and hence generating $X(t)$, is uniquely determined by what are known as the $\mathcal{F}_t^N$-conditional intensities, $[\lambda_m(t); m \in \mathcal{M}]$. When $N(t)$ has finite expectation for all $t > 0$, this means that $[N_m(t) - \int_0^t \lambda_m(s)ds]$ is an $\mathcal{F}_t^N$-martingale for all m. That the intensities satisfy

$$\lim_{h\downarrow 0} \frac{1}{h}\mathsf{P}(N_m(t+h) - N_m(t) = 1|\mathcal{F}_t^N) = \lambda_m(t+),\ \ m \in \mathcal{M} \tag{23.3}$$

$$\lim_{h\downarrow 0} \frac{1}{h}\mathsf{P}(\bar{N}(t+h) - \bar{N}(t) > 1|\mathcal{F}_t^N) = 0$$

where $\bar{N}(t) := \sum_{m\in\mathcal{M}} N_m(t)$, is in fact a principle conclusion of the arguments of stochastic kinetic theory (Gillespie, 1992). The assumptions of the theory are that the system is spatially homogeneous, confined to a fixed volume and held at constant temperature. The theory further implies that the $\mathcal{F}_t^N$-intensities, $\lambda_m(t)$, have the form

$$\lambda_m(t) = c_m g_m\{X^{R[m]}(t-)\} \tag{23.4}$$

where $c_m > 0$ is a deterministic ('rate') constant and $g_m\{\cdot\} \geq 0$ is a continuous function depending *only on the levels of the reactants* of the reaction, $R[m]$.

The interpretation of the conditional intensities, $\lambda_m(t)$, is that each one determines the local (or instantaneous) dependence of $N_m(t)$ on the internal history of the entire, M-variate process, $\mathcal{F}_t^N$. Confining attention to a finite interval of time $\mathcal{T}$, provided that $N(t)$ has finite expectation $\forall t \in \mathcal{T}$ (and that $[\lambda_m(t); t \in \mathcal{T}]$ is bounded by an integrable random variable), each intensity is a local rate of reaction in exactly the biochemical sense – that is $\lambda_m(t+) = \lim_{h\downarrow 0} \mathsf{E}[h^{-1}\{N_m(t+h) - N_m(t)\}|\mathcal{F}_t^N]$, the conditionally expected number of reactions of type m per unit time in the limit as h goes to zero. A technical subtlety is that $\lambda_m(t)$ is defined to have sample paths that are left-continuous (with limits from the right), compared to the right-continuous sample paths of $X(t)$.

23.3 Local independence properties of SKMs

23.3.1 Local independence and kinetic independence graphs

A counting process treatment of SKMs (Bowsher, 2010) greatly facilitates the construction of a directed graph encoding its local independence structure. For any subset of molecular species $A \subseteq \mathcal{V}$, let the vector process $\{X^A(t)\} := \{X_i(t); i \in A\}$ denote the corresponding subprocess of X. Now X^A may be identified with the multivariate counting process, $N^A(t)$, each element of which counts the number of occurrences during $(0, t]$ of reactions that result in a given (nonzero) change in A. Putting these elements together for all possible types of change in A to form a sample path of $N^A(t)$ captures exactly the 'information' given by the corresponding sample path of $X^A(t)$. Indeed, there is a bijection between the sample paths of $N^A(t)$ and those of $X^A(t)$, and the internal history of the counting process $N^A(t)$ is identical to that of $X^A(t)$.

Viewing subprocesses as counting processes allows one to derive straightforwardly their $\mathcal{F}_t^N$-intensities, $\lambda^A(t)$, and to interpret each intensity in the usual manner as determining the local or instantaneous dependence of A on $\mathcal{F}_t^N$. Each element of the vector $\lambda^A(t)$ is the sum of the conditional intensities of all those reactions causing a given (nonzero) change in the levels of A. Kinetic independence graphs (Bowsher, 2010) or KIGs are directed, cyclic graphs with vertex set equal to the set of biochemical species, $\mathcal{V}$. For a given reaction network the KIG is constructed so that, for each species k, all species other than k and its parents are known to be irrelevant for the conditional intensity of k, $\lambda^k(t)$. Formally, the KIG is defined as follows.

Definition 1 *The directed graph G with vertex set $\mathcal{V}$ is the kinetic independence graph (KIG) of the SKM* $[N, S, \mathsf{P}]$ *if and only if*

$$pa(k) = R[\Delta(k)]\backslash\{k\}, \quad \forall k \in \mathcal{V} \tag{23.5}$$

where $pa(k) = \{i \in \mathcal{V} | i \to k\}$ is the set of parents of vertex k, $\Delta(k) = \{m \in \mathcal{M} | S_{km} \neq 0\}$, and $R[\Delta(k)] := \bigcup_{m \in \Delta(k)} R[m]$ is the set of reactants of all reactions that change the level of species k.

The motivation for Definition 1 is that the local evolution of species k depends only on the intensity (or stochastic rate) of reactions that change the number of molecules (or level) of k, which in turn depend only on the levels of their reactants (see Equation(23.4)). To make this exact, the concept of local independence (Didelez, 2008) is needed. Let $A, B \subset \mathcal{V}$. We will say that N^A is *locally independent* of N^B (given $N^{\mathcal{V}\backslash B}$) if and only if the $\mathcal{F}_t^N$-intensity, $\lambda^A(t)$, is measurable $\mathcal{F}_t^{\mathcal{V}\backslash B}$ for all t – that is the internal history of X_t^B is irrelevant for the $\mathcal{F}_t^N$ -intensity of the species in A. Only intensities of subprocesses conditional on the history of the whole system, $\mathcal{F}_t^N$, are considered here (as opposed to $\mathcal{G}_t$-intensities where $\mathcal{G}_t \subset \mathcal{F}_t^N$).

As a consequence of Definition 1 one can read off from the KIG, for any collection of vertices A, those subprocesses with respect to which N^A is locally independent, that is which are irrelevant for the instantaneous evolution of A. Denote the closure of A by $cl(A) := pa(A) \cup A$.

Proposition 1 *Let G be the KIG of the SKM* $[N, S, \mathsf{P}]$ *and let $A, B \subset \mathcal{V}$. Then the $\mathcal{F}_t^N$-intensity $\lambda^A(t)$ is measurable $\mathcal{F}_t^{cl(A)}$ for all t, that is N^A is* locally independent *of $N^{\mathcal{V}\backslash cl(A)}$ (given $N^{cl(A)}$). Suppose that $B \cap cl(A) = \emptyset$. Then $\lambda^A(t)$ is measurable $\mathcal{F}_t^{\mathcal{V}\backslash B}$.*

Proposition 1 can be understood intuitively as follows (see Bowsher, 2010, for its proof). Given the internal history of $X^{cl(A)}$ at time t, the levels of the species $R[\Delta(A)]$ just prior to t

are 'known'. These are exactly the species levels that determine the local dynamics of A since, as reactants, they determine the rate of all reactions that change the levels of A. Therefore, any further information about species histories, including the internal history of $N^{\mathcal{V}\setminus cl(A)}$, is irrelevant for the local dynamics of A.

Separations in the undirected version of the KIG, $G^{\sim}$, can be related to both local and global dynamic independence properties of its SKM. The notation $A \perp_{G^{\sim}} B|D$ stands for the *graphical separation* of A and B by D, that is the property that every sequence of edges (or path) in $G^{\sim}$ that begins with some $a \in A$ and (without any repetition of vertices) ends with some $b \in B$ includes a vertex in D. With $[A, B, D]$ a partition of $\mathcal{V}$, such a separation in $G^{\sim}$ is equivalent to the nonexistence of $(a \in A, b \in B)$ such that there is an edge $a \rightarrow b$ or an edge $b \rightarrow a$ in G, and hence equivalent to the property $A \cap cl(B) = B \cap cl(A) = \emptyset$. This graphical separation then implies the following mutual local independence property (by Proposition 1).

Proposition 2 *Let G be the KIG of an SKM $[N, S, \mathsf{P}]$ and let $[A, B, D]$ be a partition of $\mathcal{V}$. If $A \perp_{G^{\sim}} B|D$, then N^B is* locally independent *of N^A (given $N^{B\cup D}$) and N^A is locally independent of N^B (given $N^{A\cup D}$) or, equivalently, $\lambda^B(t)$ is measurable $\mathcal{F}_t^{B\cup D}$ and $\lambda^A(t)$ is measurable $\mathcal{F}_t^{A\cup D}$.*

Note that it follows from the definition of the KIG that the graphical separation in Proposition 2 is equivalent to the biochemical property $A \cap R[\Delta(B)] = B \cap R[\Delta(A)] = \emptyset$; that is A does not participate as a reactant in any reaction that changes B, and vice versa. Therefore, for example, $R[\Delta(B)] \subseteq B \cup D$ and hence, given the levels of B and D (which fully determine the rate of reactions that change B), the levels of A are irrelevant for the instantaneous evolution of B, and vice versa. Section 23.4.2 (Theorem 1) will establish conditions under which the separation $A \perp_{G^{\sim}} B|D$ in $G^{\sim}$ implies not only mutual local independence but also global conditional independence of the internal histories of A and B given that of D.

Only partial information about the SKM is required for construction of its KIG – for each $m \in \mathcal{M}$, it is required to know the reactants $R[m]$ and the species (reactants and products) changed by the reaction, that is $\{i \in \mathcal{V}|S_{im} \neq 0\}$. This information is currently available for many biochemical reaction networks. Full knowledge of the stoichiometric matrix S is neither necessary nor sufficient for construction of the KIG. Note that the possible presence of a catalyst among the reactants $R[m]$ implies that $R[m]$ cannot be reliably reconstructed from S. No knowledge of the rate parameters c_m is required for construction of the KIG, which is important since their measurement is difficult experimentally. Nor is it necessary to assume a particular functional form for the functions $g_m\{\cdot\}$ in Equation (23.4).

As an illustration of the concepts discussed so far, consider again the SKM of Example 1. The corresponding KIG is shown in Figure 23.1. Note the presence of cycles in the KIG, including $g \rightarrow R \rightarrow P \rightarrow P_2 \rightarrow g$, which might be termed the 'negative feedback cycle'. Clearly $\{P, R\} \perp_{G^{\sim}} \{gP_2\}|\{g, P_2\}$. Let $D := \{g, P_2\}$. Note also that subprocesses of the SKM are given by $N^{gP_2} = [N_b, N_{ub}]'$ and $N^D = [N_d, N_{rd}, N_b, N_{ub}]'$. Hence $\mathcal{F}_t^{gP_2} \subset \mathcal{F}_t^D$ (owing to the deterministic relationship between $X^{gP_2}(t)$ and $X^g(t)$ in this particular example) and therefore the global independence $\mathcal{F}_t^{P,R} \perp\!\!\!\perp \mathcal{F}_t^{gP_2}|\mathcal{F}_t^D$ holds immediately here. It is apparent from Figure 23.1 that D does not separate $\{P, R\}$ and $\{gP_2\}$ in the moralised KIG (where moralisation is understood in the usual way familiar from the DAG literature). The reason for this will become clear later from our discussion of Theorem 1.

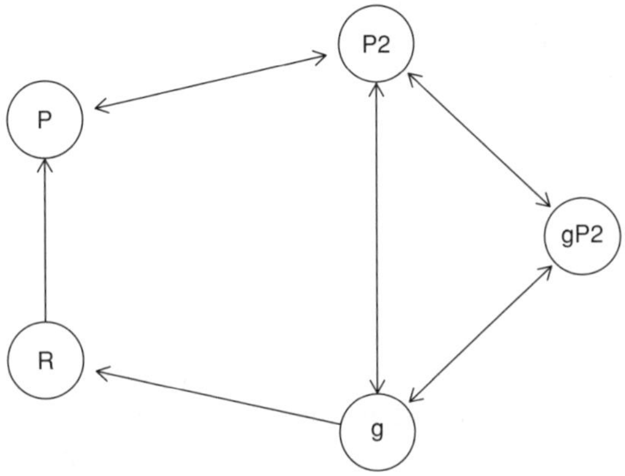

Figure 23.1 Kinetic independence graph of the SKM in Example 1.

23.3.2 Local independence and causal influence

SKMs constitute an appealing model class for possible causal models of cellular systems.[1] We propose below a definition of direct and indirect causal influence in dynamic biomolecular networks based on the KIG of a sufficiently large SKM. This is in line with a mechanistic, physical process account of causality, and with a conserved quantity theory in particular (Dowe, 2000). The biochemical reactions of the SKM conserve both mass and charge. A possible objection to such a mechanistic account is its restricted scope and limited applicability, for example in the social sciences (Williamson, 2005). However, in the context that concerns us here, a mechanistic approach seems both useful and appropriate. Decades of hard-won biochemical and chemical scientific understanding should not go unheeded in the very domain for which it is intended.

Also in a mechanistic vein, Commenges Gégout-Petit (2009) introduced a general dynamical model as a framework for causal interpretation, adopting an approach to causality based on 'physical laws in sufficiently large systems'. In this framework, component *j* has a (direct) causal influence on *k* if *j* (directly) influences *k* in a 'perfect system'. Such a system is a sufficiently high-dimensional stochastic process whose generating probability measure (the 'physical law') coincides with the 'true' probability law when restricted to the σ-field of events of interest. The direct influence $j \to k$ in Commenges and Gégout-Petit (2009) corresponds to the failure of the statement that *k* is locally independent of *j* (given all components other than *j*).

One might ask if the framework of Commenges Gégout-Petit (2009) can be applied to an observable cellular system with a sufficiently large SKM acting as the 'perfect system'. Unfortunately, as the jump processes followed by biochemical species and SKMs do not belong to the class ($\mathcal{D}$) of special semimartingales to which Commenges and Gégout-Petit confine attention – the components $X^k(t)$ are neither counting processes, nor are they continuous with a

[1] Note, however, that use of an SKM assumes that the system is modelled as spatially homogeneous. When this is not appropriate, e.g. in calcium signalling systems where spatial effects are important, alternative model classes must be sought.

deterministic bracket process (hence Assumption 2 of Commenges and Gégout-Petit, 2009, fails) – some modification of the framework would be required.

These things being said, we (tentatively) propose the following definition of direct and indirect causal influence in dynamic biomolecular networks. Assume that there exists a sufficiently large SKM, $[N^*, S^*, \mathsf{P}^*]$, such that $\mathsf{P}^*_{\mathcal{E}}$ coincides with the 'true' probability law (where $\mathcal{E}$ is the σ-field of observable cellular events). Then it seems natural to say that species j has a direct causal influence on species k if $j \in pa(k)$ in the KIG, G^*, of the SKM $[N^*, S^*, \mathsf{P}^*]$ and to say that j has a causal influence on species k if there is a directed path from j to k in G^*. Recall (Definition 1) that $pa(k) = R[\Delta(k)]\backslash\{k\}$. Thus all reactants (excluding k itself) in reactions that net produce or consume k, i.e. the species $R[\Delta(k)]\backslash\{k\}$, then have a direct causal influence on k. These include species that react with k as a reactant in some reaction $m \in \Delta(k)$. This definition seems appealing from a biochemical perspective since any species $i \notin pa(k)$ $(i \neq k)$ has no effect on the instantaneous kinetics of any reaction that changes the level of species k. In terms of local independence, N^k is locally independent of N^i (given $N^{\mathcal{V}\backslash i}$) since $\lambda^k(t)$ is measurable $\mathcal{F}_t^{\mathcal{V}\backslash i}$.

23.4 Modularisation of SKMs

Biomolecular networks in cells are increasingly thought of as having a 'modular' structure (Jeong *et al.*, 2001; Ravasz *et al.*, 2002; Guimerà and Amaral, 2005; Kashtan and Alon, 2005) a separable architecture that can be decomposed into units (groups of species) that perform in some sense independently of one another. It has been recognised both that understanding such modularity in biomolecular networks requires a dynamic approach (Alexander *et al.*, 2009) and that rigorous definition and identification of such dynamic modularisations is difficult (Szallasi *et al.*, 2006, Chapter 3). In recent work (Bowsher, 2011a), we addressed this problem by defining modularisations in terms of the local and global dynamic independence of the constituent modules. The usual sense of the term modularity in the biological literature is retained here. It should be noted that this is distinct from the various uses of the term in the causality literature.

23.4.1 Modularisations and dynamic independence

Denote by $\mathcal{F}_t^A := \sigma(X^A(s); s \leq t)$ the internal history of any subset of species $A \subset \mathcal{V}$. We write $\mathcal{F}_t^A \perp\!\!\!\perp \mathcal{F}_t^B | \mathcal{F}_t^D$ to mean that the histories (at time t) of X^A and X^B are conditionally independent given that of X^D. This dynamic independence is said to be *global* (as opposed to local) when it holds for all time intervals $(0, t]$ of nonzero length. The conditional independence $\mathcal{F}_t^A \perp\!\!\!\perp \mathcal{F}_t^B | \mathcal{F}_t^D$ implies that the expectation of (suitably measurable) functions of the path of X^A (respectively X^B) over the time interval $(0, t]$, conditional on the internal history of X^D over the same interval, are unchanged when the conditioning also includes the internal history of X^B (respectively X^A).

A *modularisation* of an SKM here is a hypergraph of its species set (i.e. a collection of subsets of its species) satisfying the following local and global dynamic independence properties.

Definition 2 *Let $\mathcal{V}$ be the species set of an SKM. The finite collection of subsets of $\mathcal{V}$, $\{M_d | M_d \subset \mathcal{V}\}$, is a* modularisation of the *SKM if and only if all species in $\mathcal{V}$ are in at* *least one module, and the following two conditions both hold: (1) each module satisfies the graphical separation* $M_d \perp_{G^\sim} \{\cup_{e \neq d} M_e\} | S_d$ *in the undirected KIG of the SKM, where the module intersection* $S_d := M_d \cap \{\cup_{e \neq d} M_e\}$ $\forall d$*, and (2) each module history satisfies the conditional independence*

$$\mathcal{F}_t^{M_d} \perp\!\!\!\perp \mathcal{F}_t^{\cup_{e \neq d} M_e} | \mathcal{F}_t^{S_d} \quad \forall d, \forall t > 0 \tag{23.6}$$

By Proposition 2, condition (1) above implies that each *module residual*, $M_d \backslash S_d$, is locally independent of $\mathcal{V} \backslash M_d$ (given M_d), i.e. each conditional intensity $\lambda^{M_d \backslash S_d}(t)$ is measurable $\mathcal{F}_t^{M_d}$. Notice also that Equation (23.6) of condition (2) above is equivalent to the statement $\mathcal{F}_t^{M_d \backslash S_d} \perp\!\!\!\perp \mathcal{F}_t^{\mathcal{V} \backslash M_d} | \mathcal{F}_t^{S_d}$.

Suppose that $\{M_d^*\}$ is a modularisation of a 'perfect system' SKM $[N^*, S^*, \mathsf{P}^*]$ (see Section 23.3.2). Condition (1) of Definition 2 then implies the absence, for all d, of any direct causal influence from species not in the dth module, $\mathcal{V} \backslash M_d^*$, to the module residual $M_d^* \backslash S_d^*$ and vice versa (since the graphical separation is equivalent to $[\{\mathcal{V} \backslash M_d\} \nrightarrow \{M_d \backslash S_d\}$ and $\{M_d \backslash S_d\} \nrightarrow \{\mathcal{V} \backslash M_d\}]$ in $G^\sim$). Thus any direct causal influence from other modules, $\bigcup_{e \neq d} M_e^*$, to the residual $M_d^* \backslash S_d^*$ must be a direct causal influence of the intersection S_d^* on the residual. (Of course, there may be direct causal influences on S_d^* from modules other than d.) Condition (2) of Definition 2 requires further that, roughly speaking, the information contained in the internal history at t of the intersection S_d renders any information in the internal history at t of $\cup_{e \neq d} M_e$ irrelevant for that of the module M_d (and vice versa). This condition has a biochemical motivation in terms of *information transfer* or *'processing'* within the network (Bowsher, 2011a) – for each d, the species in the intersection S_d plays the role of informational intermediaries or 'information carriers' between the module residual and the rest of the network. As explained in (Bowsher, 2011a), imposing this condition in addition to the first results in modularisations whose properties are more interesting from a biological perspective, especially when the SKM constitutes a cell signalling network (see Section 23.5).

23.4.2 MIDIA Algorithm

Since the biomolecular networks studied are often very large, automated and semi-automated methods for identifying SKM modularisations are a necessary, valuable tool for analysing and understanding their structure. We review below our Modularisation Identification by Dynamic Independence Algorithm (MIDIA) (Bowsher, 2010, 2011a, 2011b), which uses graphical decomposition methods to identify SKM modularisations according to Definition 2. MIDIA relies on the following result linking graphical separation in the undirected KIG to global dynamic independence of species histories (see Bowsher, 2010, for its proof). The changes in the species in D when reaction m occurs are written as S_m^D.

Theorem 1 *Let* G *be the KIG of a standard SKM,* X*, and let* $[A, B, D]$ *be a partition of* $\mathcal{V}$ *such that the separation* $A \perp_{G^\sim} B | D$ *holds in the undirected KIG. Suppose also that Condition 1 (see the Appendix) holds for the set of reactions that change both* A *and* B *(i.e. for the set* $\Delta(A) \cap \Delta(B)$*); and that for any two reactions that change* D *the same (that is* $m, \tilde{m} \in \Delta(D)$ *with* $S_m^D = S_{\tilde{m}}^D$*), the reaction* $\tilde{m}$ *has the same membership of the two sets* $[\Delta(A), \Delta(B)]$ *as does the reaction* m*. Then the global dynamic independence* $\mathcal{F}_t^A \perp\!\!\!\perp \mathcal{F}_t^B | \mathcal{F}_t^D$ *holds* $\forall t > 0$.

The regularity conditions imposed by Theorem 1 are weak ones – for an explanation of Condition 1 and what is meant by a standard SKM see the Appendix. Unsurprisingly, it is possible to show (Bowsher, 2010, Theorem 4.7) that under these regularity conditions, the separation $A \perp_{G^{\sim}} B | D$ is not sufficient for the global dynamic independence $\mathcal{F}_t^A \perp\!\!\!\perp \mathcal{F}_t^B | \mathcal{F}_t^D$. Nor is the separation $A \perp_{G^{\mathrm{m}}} B | D$ in the moralised KIG, G^{m}, sufficient for the same global dynamic independence. The questions involved are intricate and the interested reader should consult the exposition in Bowsher (2010, Sections 4.2 and 4.3). The crucial aspect is the condition of Theorem 1 that, for $m, \tilde{m} \in \Delta(D)$ with $S_m^D = S_{\tilde{m}}^D$, the reaction $\tilde{m}$ has the same membership of the two sets $[\Delta(A), \Delta(B)]$ as does the reaction m. Put briefly, if this condition (alone) were dropped from Theorem 1, then a *weaker* conclusion $\mathcal{F}_t^A \perp\!\!\!\perp \mathcal{F}_t^B | \mathcal{F}_t^{D^*}$ always holds, where $\mathcal{F}_t^D \subset \mathcal{F}_t^{D^*}$ $\forall t$ (and $\mathcal{F}_t^{D^*}$ is defined in Bowsher, 2010), but the global dynamic independence $\mathcal{F}_t^A \perp\!\!\!\perp \mathcal{F}_t^B | \mathcal{F}_t^D$ may fail. This comes as no surprise – for separation in the undirected KIG (together with Condition 1) to guarantee global dynamic independence, certain 'additional' conditioning information must be included along with the internal history of D.

The MIDIA algorithm consists of two main steps (see Bowsher, 2011a, 2011b, for full details and for the associated R software package). The first step (*clique decomposition and aggregation*) ensures that the separation $M_d \perp_{G^{\sim}} \cup_{e \neq d} M_e | S_d$ holds in the undirected KIG for all d. The step involves obtaining a junction tree of the clique decomposition of a minimal triangulation of $G^{\sim}$, followed by the iterated, pairwise aggregation of selected, neighbouring vertices in the tree. For large networks this selection may be automated, allowing the degree of coarse-graining of the network to be controlled by setting a minimum size for the module residuals, $M_d \backslash S_d$. The resultant junction tree, $\mathcal{T}_{M,L}$, constitutes a modularisation in which each module residual is locally independent of all the network species outside the module. (A junction tree is a connected, acyclic, undirected graph in which the vertices of the graph are the modules. It has the property that the intersection of any two modules of the tree, $M_d \cap M_e$ $(d \neq e)$, is contained in every module on the unique path in the tree between M_d and M_e.)

The second step (*species copying*) involves enlarging the intersection of neighbouring modules, S_{de}, by 'copying' certain species from a module containing them, M_g say, both to M_d (which lies on the unique path between g and e) and to M_e. The species are also copied to all other modules on that path. The species copied are chosen to ensure that the key condition of Theorem 1 holds: namely that, for any two reactions $m, \tilde{m} \in \Delta(S_{de})$ with $S_m^{S_{de}} = S_{\tilde{m}}^{S_{de}}$, the reaction $\tilde{m}$ has the same membership of the two sets $[\Delta(V_{de} \backslash S_{de}), \Delta(V_{ed} \backslash S_{de})]$ as does the reaction m. (Together with the graphical separation, this then implies that the dynamic independence $\mathcal{F}_t^{V_{de}} \perp\!\!\!\perp \mathcal{F}_t^{V_{ed}} | \mathcal{F}_t^{S_{de}}$ stated in Theorem 2 below holds.) The algorithm works 'backwards' through the junction tree $\mathcal{T}_{M,L}$, imposing this condition on each edge S_{de} so as to preserve the junction tree property and to leave edges treated in previous steps unchanged. The resultant junction tree is denoted by $\mathcal{T}_{M,G}$. It has three important properties, stated in Theorems 2 and 3 below (see Bowsher, 2011a, for proofs).

Theorem 2 *Let $\mathcal{T}_{M,G}$ be the junction tree returned by the algorithm MIDIA. Denote the modules of $\mathcal{T}_{M,G}$ by $\{M_d\}$ and associate each edge in the tree between adjacent modules with the separator of these modules, $S_{de} := M_d \cap M_e$. Define V_{de} (V_{ed}) as the union of the species in the modules in $\mathcal{T}_{M,G}^{de}$ $(\mathcal{T}_{M,G}^{ed})$, where the $\mathcal{T}_{M,G}^{\bullet}$ are the two subtrees obtained by cutting the edge $M_d \sim M_e$ in $\mathcal{T}_{M,G}$, and $M_d \subset V_{de}$ $(M_e \subset V_{ed})$.*

Then $S_{de} = V_{de} \cap V_{ed}$ *and the separation* $V_{de} \perp_{G^{\sim}} V_{ed}|S_{de}$ *holds in the undirected KIG. Furthermore, provided that Condition 1 holds for the set of reactions* $\Delta(V_{de} \backslash S_{de}) \cap \Delta(V_{ed} \backslash S_{de})$, *the global dynamic independence* $\mathcal{F}_t^{V_{de}} \perp\!\!\!\perp \mathcal{F}_t^{V_{ed}} | \mathcal{F}_t^{S_{de}}$ *holds for all pairs of adjacent modules in* $\mathcal{T}_{M,G}$.

Theorem 2 says that if we take the junction tree of the modularisation of the SKM returned by the MIDIA algorithm and cut one of its edges, say the one between the modules M_d and M_e, then two important properties hold: (1) the sets of species present in each of the subtrees obtained by the cutting procedure are separated in $G^{\sim}$ by the intersection of M_d and M_e and (2) under the stated Condition 1, there is global dynamic independence of the histories of the two species sets given the history of their intersection. The conditional independences $\mathcal{F}_t^{V_{de}} \perp\!\!\!\perp \mathcal{F}_t^{V_{ed}} | \mathcal{F}_t^{S_{de}}$ then imply those required for the global independence property of the modularisation in Equation (23.6).

Theorem 3 *Suppose that Condition 1 holds for the set of reactions* $\Delta(V_{de} \backslash S_{de}) \cap \Delta(V_{ed} \backslash S_{de})$, *for all pairs of adjacent modules in the junction tree* $\mathcal{T}_{M,G}$ *returned by the MIDIA algorithm. Then the modules* $\{M_d\}$ *of* $\mathcal{T}_{M,G}$ *are a modularisation of the SKM according to Definition 2 and* $S_d = M_d \cap \{\cup_{e \in ne(d)} M_e\}$, *where* $ne(d)$ *are the indices of the modules adjacent to* $\mathrm{M_d}$ *in the junction tree.*

Note that the intersection of each module with all the other modules present is given by its intersection just with those modules immediately adjacent to it in the junction tree $\mathcal{T}_{M,G}$. MIDIA thus returns a powerful visualisation, $\mathcal{T}_{M,G}$, of the modularisation obtained, which reveals the organisation of the molecular network and allows the direct reading of local and global dynamic independences from the junction tree.

23.5 Illustrative example – MAPK cell signalling

This section aims to illustrate the foregoing material in a biologically interesting setting. A little background biochemistry will therefore be helpful for appreciation of its contents. *Cell signalling* refers to the processes by which a cell converts a (chemical) signal or message arriving at the cell surface into a response. The focus here is on a widespread signalling mechanism found in (nucleated or 'eukaryotic') cells from yeast to mammals and involved in cellular proliferation, differentiation, surivival and death. The mechanism is known as the MAPK (Mitogen activated protein kinase) signalling cascade (Seger and Krebs, 1995; Levchenko *et al.*, 2000). Many genes involved in cancer (oncogenes) result in the transmission of growth-promoting signals via this mechanism. A full description of the species and reactions comprising the MAPK signalling network analysed here may be found in Novère (2009).[2]

The MAPK cascade involves, among others, three species of biomolecule, which we write as $\{K_1, K_2, K_3\}$, all of which are 'protein kinases'. Activation of K_3 by the species that acts as the signal, S, results in increased levels of the active form of K_2, which then results in increased levels of the active form of K_1.[3] For biochemical reasons, the active forms are

[2] The network analysed corresponds to the MAPK cascade described by Levchenko *et al.* (2000) in the absence of scaffold protein. This cytoplasmic cascade is considered here for simplicity.

[3] The biochemical names for $\{K_1, K_2, K_3, S\}$ are {MAPK, MEK, RAF, RAFK} respectively.

written as K_3P, K_2P_2 and K_1P_2. The last of these constitutes the biochemically active MAPK, which may be regarded as the output of the signalling process here and subsequently results in changes in cell behaviour (in response to the original signal S). As we shall see, the cascade after which the mechanism is named is present in the KIG of the network as the following path of direct causal influences:

$$S \to S \cdot K_3 \to K_3P \to K_2 \cdot K_3P \to K_2P \to K_3P \cdot K_2P \to$$
$$K_2P_2 \to K_1 \cdot K_2P_2 \to K_1P \to K_2P_2 \cdot K_1P \to K_1P_2 \qquad (23.7)$$

where $\cdot$ always indicates the (physical) binding of two molecules to form a single molecule (which then constitues a distinct species). The presence during $(0, t]$ of each type of molecule in the path shown is necessary for the presence of all the subsequent species in the path.

Kinases may be regarded as having the effect of 'adding' P (a phosphate group) to a biomolecule, while so-called phosphatases have the effect of removing that P (thus preventing permanent effects of a transient input signal). The phosphatase molecule that acts on (the phosphoforms of) K_i is written as K_iase ($i = 1, 2, 3$). The SKM corresponding to the MAPK network is taken to include a (possibly inhomogeneous) Poisson counting process, N^{S^*}, which inputs signal molecules S into the system. Its intensity may sometimes take the value zero, thus allowing for transient 'stimulation'. (Similarly, a second Poisson counting process for degradation of S could be included without altering the results reported here in any way.)

Figure 23.2 shows the KIG of the SKM for the MAPK network. The reader may trace in the KIG the presence of the 'cascade' or path of direct causal influences in Equation (23.2). Graphical separations immediately apparent from inspection of the KIG suggest a modularisation $\{M_1, M_2, M_3\}$ based on local independence with $M_1 \cap M_2 = K_2P_2$ and $M_2 \cap M_3 = K_3P$. Roughly speaking, the three modules correspond to sets of vertices in the 'north-west', 'central' and 'southern' regions of the KIG. Understanding the function of each of the modules is straightforward. For $d \in \{1, 2, 3\}$, the module M_d consists (exactly) of those species involved in the activation (phosphorylation) and deactivation (dephosphorylation) of K_d. The kinase K_3P (respectively K_2P_2) is the result of one activation and the agent of another, and is therefore present in both modules M_3 and M_2 (respectively M_2 and M_1). Clearly, $M_3 \perp_G \sim \{M_2 \cup M_1\} | K_3P$ and $M_1 \perp_{G\sim} \{M_2 \cup M_3\} | K_2P_2$. The only direct causal influences of M_3 on M_2 (and vice versa) are those from K_3P (there are no direct causal influences between M_3 and M_1). Similarly, the only direct causal influences of M_2 on M_1 (and vice versa) are those from K_2P_2. Biochemists thus speak of the separators K_3P and K_2P_2 as 'signalling intermediaries'. It is worth noting that the direct causal influences arising from these vertices run in both directions – there are 'retroactivities' going in the opposite direction to that of signal 'propagation'.

The modularisation $\{M_1, M_2, M_3\}$ is not a modularisation according to Definition 2, as a result of the need to ensure global dynamic independence. A modularisation $\mathcal{T}_{M,G}$ identified by the MIDIA algorithm is shown in Figure 23.3 (where modules are labelled with the corresponding module residual). Notice that the species $\{K_2ase, K_2P \cdot K_3P\}$ have been added ('copied') to the first (uppermost) module and therefore appear in its separator together with K_2P_2. Similarly, the species $\{K_3ase, K_3 \cdot S\}$ have been added to the second (central) module and therefore appear in its separator together with K_3P. Denote this modularisation by $\{M_{1,G}, M_{2,G}, M_{3,G}\}$.

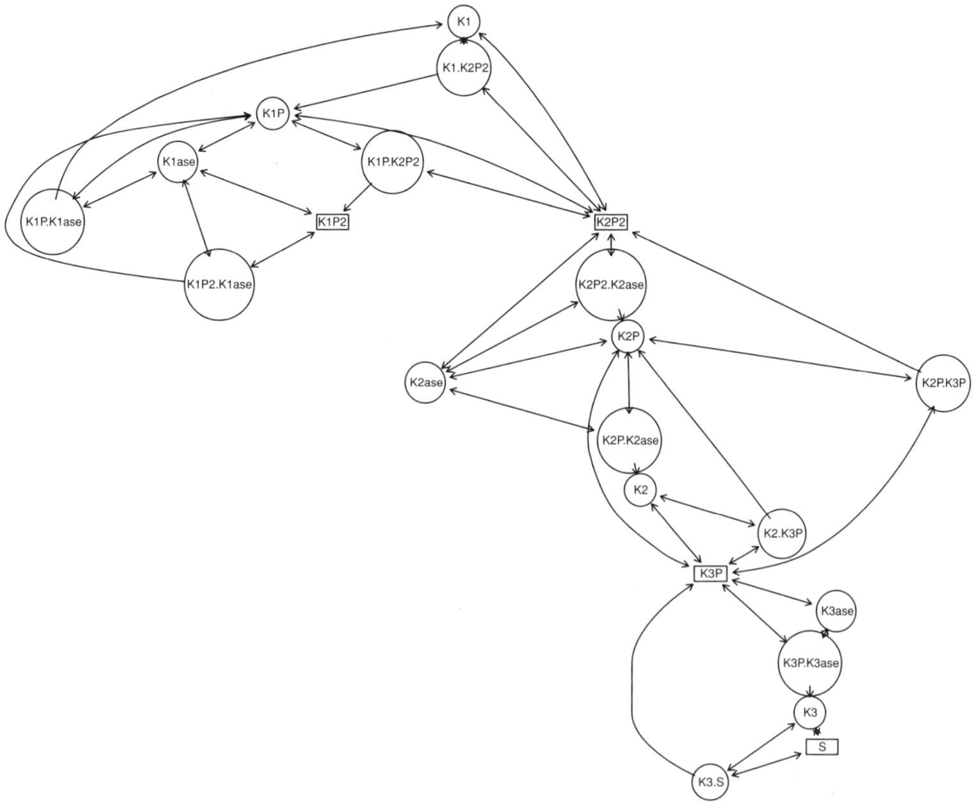

Figure 23.2 Kinetic independence graph of the MAPK signalling network of Levchenko et al. (2000). Species names are defined in the text. Boxes denote particularly important vertices: the input S, the output K_1P_2 and the separators (active, intermediary kinases) K_3P, K_2P_2.

As a result of the structure of the SKM here, $\mathcal{F}_t^{S^*} \subset \mathcal{F}_t^{\{S, K_3 \cdot S\}}$, which is the natural filtration of the external input signal, is contained in that of the species $\{S, K_3 \cdot S\}$. Also recall that the activated MAPK, K_1P_2, constitutes the output of the signalling network. From the perspective of information transfer or 'processing' within the cell, the following global dynamic independences between the input signal and output that may be read from $\mathcal{T}_{M,G}$ (using Theorem 2 and properties of conditional independence) are of particular interest:

$$\mathcal{F}_t^{K_1P_2} \perp\!\!\!\perp \mathcal{F}_t^{S^*} | \mathcal{F}_t^{\{K_3P, K_3ase, K_3 \cdot S\}}, \qquad \text{and} \qquad \mathcal{F}_t^{K_1P_2} \perp\!\!\!\perp \mathcal{F}_t^{S^*} | \mathcal{F}_t^{\{K_2ase, K_2P_2, K_2P \cdot K_3P\}}$$

Thus, over any time interval $(0, t]$, the 'encoding' of information by the dynamic evolution of either one of the (relatively small) separators $M_{2,G} \cap M_{3,G}$ or $M_{1,G} \cap M_{2,G}$ renders the information in the external input signal irrelevant for the dynamic evolution of the output. Furthermore, the junction tree tells us about the process of sequential information transfer by the signalling network. The junction tree implies that $\mathcal{F}_t^{K_1P_2} \perp\!\!\!\perp \mathcal{F}_t^{S^*} \vee \mathcal{F}_t^{\{K_3P, K_3ase, K_3 \cdot S\}} | \mathcal{F}_t^{\{K_2ase, K_2P_2, K_2P \cdot K_3P\}}$; hence we see that the 'second stage encoders' $\{K_2ase, K_2P \cdot K_3P,$

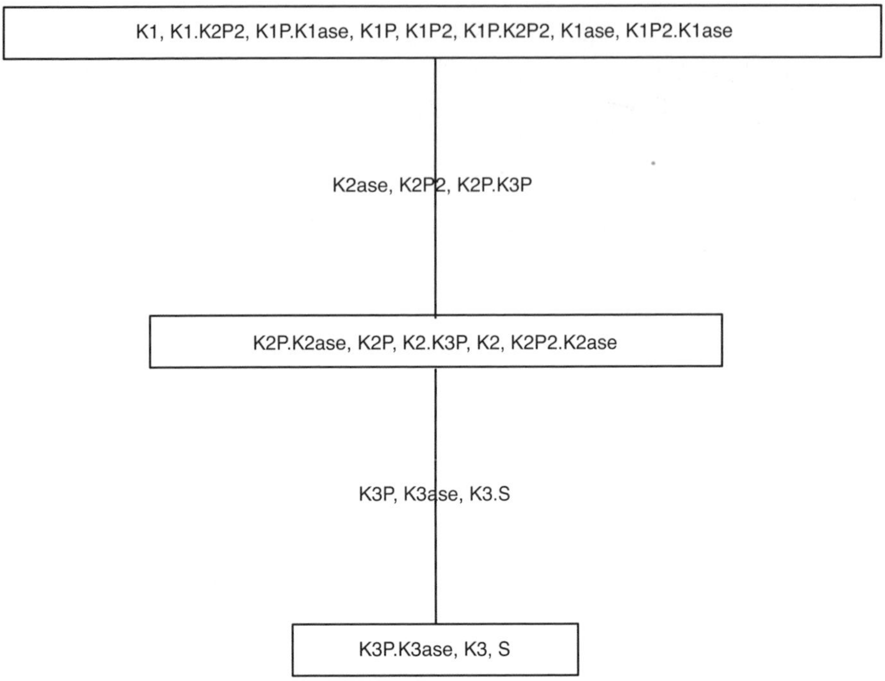

Figure 23.3 Junction tree of the modularisation for the MAPK signalling network. Modules are labelled with the contents of the corresponding module residual and edges are labelled with the corresponding separator, $S_{de} = M_d \cap M_e$.

$K_2P_2\}$ make the information in both the input signal and the 'first stage encoders' $\{K_3P, K_3ase, K_3 \cdot S\}$ irrelevant for the outputs.

This example illustrates the insights that may be gained into the organisation and functioning of biomolecular networks by identifying modularisations explicitly based on the dynamic properties of the SKM. The informational properties of the MAPK network we have derived are not obvious and provide biological insight – information is encoded not by the dynamic evolution of the active intermediary kinase(s) alone but by the dynamics of the distribution of each kinase across certain of its possible forms (e.g. for K_3, its distribution across its active form, its complex with S and its complex with K_3ase). The MIDIA algorithm continues to be feasible for very large networks (Bowsher, 2011a), a setting in which a modular view is particularly useful for gaining understanding and automated approaches become essential.

SKM modularisations should prove useful in identifying potential targets for pharmacological intervention (drug treatment), especially where there are complicated interactions between different 'pathways' in the network. In the MAPK example, K_2P_2 (MEK) is an obvious target and indeed MEK-inhibitors have undergone clinical trials in cancer treatment (Emerya *et al.*, 2009). The analysis of SKMs subject to interventions is an important area for future research. Suppose that a (selective) MEK-inhibitor D is known to bind to K_2P_2 to form the (inactive) complex $K_2P_2 \cdot D$, but D does not react with any of the other species in the KIG in Figure 23.2. Then the KIG of the (augmented) SKM for the network including the drug treatment is identical to that in Figure 23.2 except for the addition of the two vertices $\{D, K_2P_2 \cdot D\}$ and the addition of the six edges in $\{K_2P_2, D, K_2P_2 \cdot D\} \times \{K_2P_2, D, K_2P_2 \cdot D\}$ but excluding loops (self-edges).

The effects of pharmacological interventions suggested by analysis of KIGs and their associated modularisations can be investigated initially *in silico* using Monte Carlo simulation. The causal effects of interventions that appear promising may then be estimated using controlled laboratory experiments.

23.6 Conclusion

Consideration of molecular mechanisms is of widespread importance for understanding causality in the biosciences. A stochastic kinetic model (SKM) is a mechanistic description of the dynamics of the molecular networks inside living cells that is firmly grounded in biochemical kinetics. The kinetic independence graph of an SKM encodes its local independence structure. Putative causal influences in a cellular system may be defined in terms of the directed edges of the KIG of a sufficiently large SKM. Given a partition $[A, B, D]$ of its vertices, the graphical separation $A \perp B | D$ in the undirected KIG has an intuitive biochemical interpretation and implies that A is locally independent of B given $A \cup D$. A condition is given under which this separation also results in global independence of the internal histories of A and B conditional on that of D. The MIDIA algorithm allows automated and semi-automated identification of SKM modularisations based simultaneously on local and global dynamic independence of the constituent modules. The modularisation may be visualised as a junction tree encoding the independences. A modular approach explicitly based on the dynamics of the system has the potential to significantly aid causal understanding of biomolecular networks.

23.7 Appendix: SKM regularity conditions

The MIDIA algorithm applies to *standard* SKMs. For the straightforward case where the levels of all reactants of any reaction in the SKM are changed by that reaction, this means simply that zeroth-order reactions (those with no reactants) are specified to each have only one product. The MAPK signalling network is such a case and is a standard SKM. (A slightly more general definition of a standard SKM is given in Bowsher, 2010.) The following condition is used in Theorem 1.

Condition 1 *A subset of reactions Γ of an SKM is said to be identified by consumption of reactants if and only if: (i) for all $m \in \Gamma$, $S_{im} \leq 0$ for all $i \in R[m]$ and $S_{km} < 0$ for some $k \in R[m]$ (provided that $R[m]$ is not empty), and $S_{im} \geq 0$ for all $i \in P[m]$, and (ii) there does not exist a pair of distinct reactions $m, \tilde{m} \in \Gamma$ such that $S_m^- = S_{\tilde{m}}^-$, where S_m^- denotes the vector formed by setting all positive elements of S_m to zero.*

Condition 1 implies that no two reactions in Γ change reactants identically and will be satisfied for all reactions in the network by most SKMs, possibly after explicit inclusion of enzymes, etc., in reaction mechanisms. See Bowsher (2010, Remark 4.1) for a discussion. For the MAPK network junction tree, $\mathcal{T}_{M,G}$, Condition 1 holds for the subset of reactions $\Delta(V_{de} \backslash S_{de}) \cap \Delta(V_{ed} \backslash S_{de})$, for $(d, e) \in \{(1, 2), (2, 3)\}$.

Acknowledgements

This research was jointly funded by the EPSRC and MRC (United Kingdom). The author is grateful to the Editors and referee, and also to the Statistical Laboratory and the Cambridge Statistics Initiative (University of Cambridge, UK) for the research environment provided.

References

Alexander, R.P., Kim, P.M., Emonet, T. and Gerstein, M.B. (2009) Understanding modularity in molecular networks requires dynamics. *Science Signaling*, **2** (81), 44.

Bowsher, C.G. (2010) Stochastic kinetic models: dynamic independence, modularity and graphs. *Annals of Statistics*, **38** (4), 2242–2281.

Bowsher, C.G. (2011a) Information processing by biochemical networks: a dynamic approach. *Journal of the Royal Society Interface*, **8** (55), 186–200.

Bowsher, C.G. (2011b) Automated analysis of information processing, kinetic independence and modular architecture in biochemical networks using MIDIA. *Bioinformatics*, **27**.

Commenges, D. and Gégout-Petit, A. (2009) A general dynamical statistical model with causal interpretation. *Journal of the Royal Statistical Society, Series B*, **71** (3), 719–736.

Dawid, A.P. (2010) Beware of the DAG! *Journal of Machine Learning Research Workshop and Conference Proceedings*, **6**, 59–86.

Didelez, V. (2008) Graphical models for marked point processes based on local independence. *Journal of the Royal Statistical Society, Series B*, **70**, 245–264.

Didelez, V. and Sheehan, N.A. (2007) Mendelian randomisation as an instrumental variable approach to causal inference. *Statistical Methods in Medical Research*, **16**, 309–330.

Dowe, P. (2000) *Physical Causation*. Cambridge University Press, Cambridge.

Emerya, C.M. *et al.* (2009) Mek1 mutations confer resistance to Mek and B-Raf inhibition. *Proceedings of the National Academy of Science*, **106** (48), 20411–20416.

Gillespie, D.T. (1992) A rigorous derivation of the chemical master equation. *Physica A*, **188**, 404–425.

Guimerà, R. and Amaral, L.A.N. (2005) Functional cartography of complex metabolic networks. *Nature*, **433**, 895–900.

Jeong, H., Mason, S., Barabási, A.L. and Oltvai, Z. (2001) Lethality and centrality in protein networks. *Nature*, **411**, 41–42.

Kashtan, N. and Alon, U. (2005) Spontaneous evolution of modularity and network motifs. *Proceedings of the National Academy of Science*, **102** (39), 13773–13778.

Levchenko, A. Bruck, J. and Sternberg, P.W. (2000) Scaffold proteins may biphasically affect the levels of mitogen-activated protein kinase signaling and reduce its threshold properties. *Proceedings of the National Academy of Science*, **97** (11), 5818–5823.

Novère, N.L. (2009) Biomodels database: model number BIOMD0000000011. http://www.ebi.ac.uk/biomodels-main/BIOMD0000000011. Version as at 2009-01-05.

Ravasz, E. Somera, A.L. Mongru, D.A. Oltvai, Z.N. and Barabási, A.L. (2002) Hierarchical organization of modularity in metabolic networks. *Science*, **297**, 1551–1555.

Seger, R. and Krebs, E.C. (1995) The MAPK signaling cascade. *FASEB Journal*, **9**, 726–735.

Szallasi, Z., Periwal, V. and Stelling, J. (2006) System Modeling in Cellular Biology. Cambridge: MIT Press. Chapter 3.

Williamson, J. (2005) *Bayesian Nets and Causality*. Oxford: Oxford University Press.

Wilkinson, D.J. (2009) Stochastic modelling for quantitative description of heterogeneous biological systems. *Nature Reviews Genetics*, **10**, 122–133.

Index

Note: Page numbers printed in bold face indicate the location in the book where the term is defined, or where the primary discussion of it is located.

Causality: Statistical Perspectives and Applications, First Edition. Edited by Carlo Berzuini, Philip Dawid and Luisa Bernardinelli.

WILEY SERIES IN PROBABILITY AND STATISTICS
ESTABLISHED BY WALTER A. SHEWHART AND SAMUEL S. WILKS

The ***Wiley Series in Probability and Statistics*** is well established and authoritative. It covers many topics of current research interest in both pure and applied statistics and probability theory. Written by leading statisticians and institutions, the titles span both state-of-the-art developments in the field and classical methods.

Reflecting the wide range of current research in statistics, the series encompasses applied, methodological and theoretical statistics, ranging from applications and new techniques made possible by advances in computerized practice to rigorous treatment of theoretical approaches.

This series provides essential and invaluable reading for all statisticians, whether in academia, industry, government, or research.

† ABRAHAM and LEDOLTER • Statistical Methods for Forecasting
AGRESTI • Analysis of Ordinal Categorical Data, *Second Edition*
AGRESTI • An Introduction to Categorical Data Analysis, *Second Edition*
AGRESTI • Categorical Data Analysis, *Second Edition*
ALTMAN, GILL, and McDONALD • Numerical Issues in Statistical Computing for the Social Scientist
AMARATUNGA and CABRERA • Exploration and Analysis of DNA Microarray and Protein Array Data
ANDĚL • Mathematics of Chance
ANDERSON • An Introduction to Multivariate Statistical Analysis, *Third Edition*
* ANDERSON • The Statistical Analysis of Time Series
ANDERSON, AUQUIER, HAUCK, OAKES, VANDAELE, and WEISBERG • Statistical Methods for Comparative Studies
ANDERSON and LOYNES • The Teaching of Practical Statistics
ARMITAGE and DAVID (editors) • Advances in Biometry
ARNOLD, BALAKRISHNAN, and NAGARAJA • Records
* ARTHANARI and DODGE • Mathematical Programming in Statistics
* BAILEY • The Elements of Stochastic Processes with Applications to the Natural Sciences
BAJORSKI • Statistics for Imaging, Optics, and Photonics
BALAKRISHNAN and KOUTRAS • Runs and Scans with Applications
BALAKRISHNAN and NG • Precedence-Type Tests and Applications
BARNETT • Comparative Statistical Inference, *Third Edition*
BARNETT • Environmental Statistics
BARNETT and LEWIS • Outliers in Statistical Data, *Third Edition*
BARTHOLOMEW, KNOTT, and MOUSTAKI • Latent Variable Models and Factor Analysis: A Unified Approach, *Third Edition*
BARTOSZYNSKI and NIEWIADOMSKA-BUGAJ • Probability and Statistical Inference, *Second Edition*
BASILEVSKY • Statistical Factor Analysis and Related Methods: Theory and Applications
BATES and WATTS • Nonlinear Regression Analysis and Its Applications

*Now available in a lower priced paperback edition in the Wiley Classics Library.
†Now available in a lower priced paperback edition in the Wiley–Interscience Paperback Series.

BECHHOFER, SANTNER, and GOLDSMAN • Design and Analysis of Experiments for Statistical Selection, Screening, and Multiple Comparisons
BEIRLANT, GOEGEBEUR, SEGERS, TEUGELS, and DE WAAL • Statistics of Extremes: Theory and Applications
BELSLEY • Conditioning Diagnostics: Collinearity and Weak Data in Regression
† BELSLEY, KUH, and WELSCH • Regression Diagnostics: Identifying Influential Data and Sources of Collinearity
BENDAT and PIERSOL • Random Data: Analysis and Measurement Procedures, *Fourth Edition*
BERNARDO and SMITH • Bayesian Theory
BERZUINI • Causality: Statistical Perspectives and Applications
BHAT and MILLER • Elements of Applied Stochastic Processes, *Third Edition*
BHATTACHARYA and WAYMIRE • Stochastic Processes with Applications
BIEMER, GROVES, LYBERG, MATHIOWETZ, and SUDMAN • Measurement Errors in Surveys
BILLINGSLEY • Convergence of Probability Measures, *Second Edition*
BILLINGSLEY • Probability and Measure, *Anniversary Edition*
BIRKES and DODGE • Alternative Methods of Regression
BISGAARD and KULAHCI • Time Series Analysis and Forecasting by Example
BISWAS, DATTA, FINE, and SEGAL • Statistical Advances in the Biomedical Sciences: Clinical Trials, Epidemiology, Survival Analysis, and Bioinformatics
BLISCHKE AND MURTHY (editors) • Case Studies in Reliability and Maintenance
BLISCHKE AND MURTHY • Reliability: Modeling, Prediction, and Optimization
BLOOMFIELD • Fourier Analysis of Time Series: An Introduction, *Second Edition*
BOLLEN • Structural Equations with Latent Variables
BOLLEN and CURRAN • Latent Curve Models: A Structural Equation Perspective
BOROVKOV • Ergodicity and Stability of Stochastic Processes
BOSQ and BLANKE • Inference and Prediction in Large Dimensions
BOULEAU • Numerical Methods for Stochastic Processes
* BOX • Bayesian Inference in Statistical Analysis
BOX • Improving Almost Anything, *Revised Edition*
* BOX and DRAPER • Evolutionary Operation: A Statistical Method for Process Improvement
BOX and DRAPER • Response Surfaces, Mixtures, and Ridge Analyses, *Second Edition*
BOX, HUNTER, and HUNTER • Statistics for Experimenters: Design, Innovation, and Discovery, *Second Editon*
BOX, JENKINS, and REINSEL • Time Series Analysis: Forcasting and Control, *Fourth Edition*
BOX, LUCEÑO, and PANIAGUA-QUIÑONES • Statistical Control by Monitoring and Adjustment, *Second Edition*
* BROWN and HOLLANDER • Statistics: A Biomedical Introduction
CAIROLI and DALANG • Sequential Stochastic Optimization
CASTILLO, HADI, BALAKRISHNAN, and SARABIA • Extreme Value and Related Models with Applications in Engineering and Science
CHAN • Time Series: Applications to Finance with R and S-Plus®, *Second Edition*
CHARALAMBIDES • Combinatorial Methods in Discrete Distributions
CHATTERJEE and HADI • Regression Analysis by Example, *Fourth Edition*
CHATTERJEE and HADI • Sensitivity Analysis in Linear Regression
CHERNICK • Bootstrap Methods: A Guide for Practitioners and Researchers, *Second Edition*
CHERNICK and FRIIS • Introductory Biostatistics for the Health Sciences
CHILÈS and DELFINER • Geostatistics: Modeling Spatial Uncertainty, *Second Edition*
CHOW and LIU • Design and Analysis of Clinical Trials: Concepts and Methodologies, *Second Edition*
CLARKE • Linear Models: The Theory and Application of Analysis of Variance
CLARKE and DISNEY • Probability and Random Processes: A First Course with Applications, *Second Edition*
* COCHRAN and COX • Experimental Designs, *Second Edition*
COLLINS and LANZA • Latent Class and Latent Transition Analysis: With Applications in the Social, Behavioral, and Health Sciences

*Now available in a lower priced paperback edition in the Wiley Classics Library.
†Now available in a lower priced paperback edition in the Wiley–Interscience Paperback Series.

CONGDON • Applied Bayesian Modelling
CONGDON • Bayesian Models for Categorical Data
CONGDON • Bayesian Statistical Modelling, *Second Edition*
CONOVER • Practical Nonparametric Statistics, *Third Edition*
COOK • Regression Graphics
COOK and WEISBERG • An Introduction to Regression Graphics
COOK and WEISBERG • Applied Regression Including Computing and Graphics
CORNELL • A Primer on Experiments with Mixtures
CORNELL • Experiments with Mixtures, Designs, Models, and the Analysis of Mixture Data, *Third Edition*
COX • A Handbook of Introductory Statistical Methods
CRESSIE • Statistics for Spatial Data, *Revised Edition*
CRESSIE and WIKLE • Statistics for Spatio-Temporal Data
CSÖRGŐ and HORVÁTH • Limit Theorems in Change Point Analysis
DAGPUNAR • Simulation and Monte Carlo: With Applications in Finance and MCMC
DANIEL • Applications of Statistics to Industrial Experimentation
DANIEL • Biostatistics: A Foundation for Analysis in the Health Sciences, *Eighth Edition*
* DANIEL • Fitting Equations to Data: Computer Analysis of Multifactor Data, *Second Edition*
DASU and JOHNSON • Exploratory Data Mining and Data Cleaning
DAVID and NAGARAJA • Order Statistics, *Third Edition*
DE ROCQUIGNY • Modelling Under Risk and Uncertainty: An Introduction to Statistical, Phenomenological and Computational Models
* DEGROOT, FIENBERG, and KADANE • Statistics and the Law
DEL CASTILLO • Statistical Process Adjustment for Quality Control
DEMARIS • Regression with Social Data: Modeling Continuous and Limited Response Variables
DEMIDENKO • Mixed Models: Theory and Applications
DENISON, HOLMES, MALLICK and SMITH • Bayesian Methods for Nonlinear Classification and Regression
DETTE and STUDDEN • The Theory of Canonical Moments with Applications in Statistics, Probability, and Analysis
DEY and MUKERJEE • Fractional Factorial Plans
DILLON and GOLDSTEIN • Multivariate Analysis: Methods and Applications
* DODGE and ROMIG • Sampling Inspection Tables, *Second Edition*
* DOOB • Stochastic Processes
DOWDY, WEARDEN, and CHILKO • Statistics for Research, *Third Edition*
DRAPER and SMITH • Applied Regression Analysis, *Third Edition*
DRYDEN and MARDIA • Statistical Shape Analysis
DUDEWICZ and MISHRA • Modern Mathematical Statistics
DUNN and CLARK • Basic Statistics: A Primer for the Biomedical Sciences, *Fourth Edition*
DUPUIS and ELLIS • A Weak Convergence Approach to the Theory of Large Deviations
EDLER and KITSOS • Recent Advances in Quantitative Methods in Cancer and Human Health Risk Assessment
* ELANDT-JOHNSON and JOHNSON • Survival Models and Data Analysis
ENDERS • Applied Econometric Time Series, *Third Edition*
† ETHIER and KURTZ • Markov Processes: Characterization and Convergence
EVANS, HASTINGS, and PEACOCK • Statistical Distributions, *Third Edition*
EVERITT, LANDAU, LEESE, and STAHL • Cluster Analysis, *Fifth Edition*
FEDERER and KING • Variations on Split Plot and Split Block Experiment Designs
FELLER • An Introduction to Probability Theory and Its Applications, Volume I, *Third Edition*, Revised; Volume II, *Second Edition*
FITZMAURICE, LAIRD, and WARE • Applied Longitudinal Analysis, *Second Edition*
* FLEISS • The Design and Analysis of Clinical Experiments
FLEISS • Statistical Methods for Rates and Proportions, *Third Edition*
† FLEMING and HARRINGTON • Counting Processes and Survival Analysis

*Now available in a lower priced paperback edition in the Wiley Classics Library.
†Now available in a lower priced paperback edition in the Wiley–Interscience Paperback Series.

FUJIKOSHI, ULYANOV, and SHIMIZU • Multivariate Statistics: High-Dimensional and Large-Sample Approximations
FULLER • Introduction to Statistical Time Series, *Second Edition*
† FULLER • Measurement Error Models
GALLANT • Nonlinear Statistical Models
GEISSER • Modes of Parametric Statistical Inference
GELMAN and MENG • Applied Bayesian Modeling and Causal Inference from Incomplete-Data Perspectives
GEWEKE • Contemporary Bayesian Econometrics and Statistics
GHOSH, MUKHOPADHYAY, and SEN • Sequential Estimation
GIESBRECHT and GUMPERTZ • Planning, Construction, and Statistical Analysis of Comparative Experiments
GIFI • Nonlinear Multivariate Analysis
GIVENS and HOETING • Computational Statistics
GLASSERMAN and YAO • Monotone Structure in Discrete-Event Systems
GNANADESIKAN • Methods for Statistical Data Analysis of Multivariate Observations, *Second Edition*
GOLDSTEIN • Multilevel Statistical Models, *Fourth Edition*
GOLDSTEIN and LEWIS • Assessment: Problems, Development, and Statistical Issues
GOLDSTEIN and WOOFF • Bayes Linear Statistics
GREENWOOD and NIKULIN • A Guide to Chi-Squared Testing
GROSS, SHORTLE, THOMPSON, and HARRIS • Fundamentals of Queueing Theory, *Fourth Edition*
GROSS, SHORTLE, THOMPSON, and HARRIS • Solutions Manual to Accompany Fundamentals of Queueing Theory, *Fourth Edition*
* HAHN and SHAPIRO • Statistical Models in Engineering
HAHN and MEEKER • Statistical Intervals: A Guide for Practitioners
HALD • A History of Probability and Statistics and their Applications Before 1750
† HAMPEL • Robust Statistics: The Approach Based on Influence Functions
HARTUNG, KNAPP, and SINHA • Statistical Meta-Analysis with Applications
HEIBERGER • Computation for the Analysis of Designed Experiments
HEDAYAT and SINHA • Design and Inference in Finite Population Sampling
HEDEKER and GIBBONS • Longitudinal Data Analysis
HELLER • MACSYMA for Statisticians
HERITIER, CANTONI, COPT, and VICTORIA-FESER • Robust Methods in Biostatistics
HINKELMANN and KEMPTHORNE • Design and Analysis of Experiments, Volume 1: Introduction to Experimental Design, *Second Edition*
HINKELMANN and KEMPTHORNE • Design and Analysis of Experiments, Volume 2: Advanced Experimental Design
HINKELMANN (editor) • Design and Analysis of Experiments, Volume 3: Special Designs and Applications
HOAGLIN, MOSTELLER, and TUKEY • Fundamentals of Exploratory Analysis of Variance
* HOAGLIN, MOSTELLER, and TUKEY • Exploring Data Tables, Trends and Shapes
* HOAGLIN, MOSTELLER, and TUKEY • Understanding Robust and Exploratory Data Analysis
HOCHBERG and TAMHANE • Multiple Comparison Procedures
HOCKING • Methods and Applications of Linear Models: Regression and the Analysis of Variance, *Second Edition*
HOEL • Introduction to Mathematical Statistics, *Fifth Edition*
HOGG and KLUGMAN • Loss Distributions
HOLLANDER and WOLFE • Nonparametric Statistical Methods, *Second Edition*
HOSMER and LEMESHOW • Applied Logistic Regression, *Second Edition*
HOSMER, LEMESHOW, and MAY • Applied Survival Analysis: Regression Modeling of Time-to-Event Data, *Second Edition*
HUBER • Data Analysis: What Can Be Learned From the Past 50 Years
HUBER • Robust Statistics
† HUBER and RONCHETTI • Robust Statistics, *Second Edition*
HUBERTY • Applied Discriminant Analysis, *Second Edition*

*Now available in a lower priced paperback edition in the Wiley Classics Library.
†Now available in a lower priced paperback edition in the Wiley–Interscience Paperback Series.

HUBERTY and OLEJNIK • Applied MANOVA and Discriminant Analysis, *Second Edition*
HUITEMA • The Analysis of Covariance and Alternatives: Statistical Methods for Experiments, Quasi-Experiments, and Single-Case Studies, *Second Edition*
HUNT and KENNEDY • Financial Derivatives in Theory and Practice, *Revised Edition*
HURD and MIAMEE • Periodically Correlated Random Sequences: Spectral Theory and Practice
HUSKOVA, BERAN, and DUPAC • Collected Works of Jaroslav Hajek—with Commentary
HUZURBAZAR • Flowgraph Models for Multistate Time-to-Event Data
INSUA, RUGGERI and WIPER • Bayesian Analysis of Stochastic Process Models
JACKMAN • Bayesian Analysis for the Social Sciences
† JACKSON • A User's Guide to Principle Components
JOHN • Statistical Methods in Engineering and Quality Assurance
JOHNSON • Multivariate Statistical Simulation
JOHNSON and BALAKRISHNAN • Advances in the Theory and Practice of Statistics: A Volume in Honor of Samuel Kotz
JOHNSON, KEMP, and KOTZ • Univariate Discrete Distributions, *Third Edition*
JOHNSON and KOTZ (editors) • Leading Personalities in Statistical Sciences: From the Seventeenth Century to the Present
JOHNSON, KOTZ, and BALAKRISHNAN • Continuous Univariate Distributions, Volume 1, *Second Edition*
JOHNSON, KOTZ, and BALAKRISHNAN • Continuous Univariate Distributions, Volume 2, *Second Edition*
JOHNSON, KOTZ, and BALAKRISHNAN • Discrete Multivariate Distributions
JUDGE, GRIFFITHS, HILL, LÜTKEPOHL, and LEE • The Theory and Practice of Econometrics, *Second Edition*
JUREK and MASON • Operator-Limit Distributions in Probability Theory
KADANE • Bayesian Methods and Ethics in a Clinical Trial Design
KADANE AND SCHUM • A Probabilistic Analysis of the Sacco and Vanzetti Evidence
KALBFLEISCH and PRENTICE • The Statistical Analysis of Failure Time Data, *Second Edition*
KARIYA and KURATA • Generalized Least Squares
KASS and VOS • Geometrical Foundations of Asymptotic Inference
† KAUFMAN and ROUSSEEUW • Finding Groups in Data: An Introduction to Cluster Analysis
KEDEM and FOKIANOS • Regression Models for Time Series Analysis
KENDALL, BARDEN, CARNE, and LE • Shape and Shape Theory
KHURI • Advanced Calculus with Applications in Statistics, *Second Edition*
KHURI, MATHEW, and SINHA • Statistical Tests for Mixed Linear Models
* KISH • Statistical Design for Research
KLEIBER and KOTZ • Statistical Size Distributions in Economics and Actuarial Sciences
KLEMELÄ • Smoothing of Multivariate Data: Density Estimation and Visualization
KLUGMAN, PANJER, and WILLMOT • Loss Models: From Data to Decisions, *Third Edition*
KLUGMAN, PANJER, and WILLMOT • Solutions Manual to Accompany Loss Models: From Data to Decisions, *Third Edition*
KOSKI and NOBLE • Bayesian Networks: An Introduction
KOTZ, BALAKRISHNAN, and JOHNSON • Continuous Multivariate Distributions, Volume 1, *Second Edition*
KOTZ and JOHNSON (editors) • Encyclopedia of Statistical Sciences: Volumes 1 to 9 with Index
KOTZ and JOHNSON (editors) • Encyclopedia of Statistical Sciences: Supplement Volume
KOTZ, READ, and BANKS (editors) • Encyclopedia of Statistical Sciences: Update Volume 1
KOTZ, READ, and BANKS (editors) • Encyclopedia of Statistical Sciences: Update Volume 2
KOWALSKI and TU • Modern Applied U-Statistics
KRISHNAMOORTHY and MATHEW • Statistical Tolerance Regions: Theory, Applications, and Computation
KROESE, TAIMRE, and BOTEV • Handbook of Monte Carlo Methods
KROONENBERG • Applied Multiway Data Analysis
KULINSKAYA, MORGENTHALER, and STAUDTE • Meta Analysis: A Guide to Calibrating and Combining Statistical Evidence
KULKARNI and HARMAN • An Elementary Introduction to Statistical Learning Theory
KUROWICKA and COOKE • Uncertainty Analysis with High Dimensional Dependence Modelling

*Now available in a lower priced paperback edition in the Wiley Classics Library.
†Now available in a lower priced paperback edition in the Wiley–Interscience Paperback Series.

KVAM and VIDAKOVIC • Nonparametric Statistics with Applications to Science and Engineering
LACHIN • Biostatistical Methods: The Assessment of Relative Risks, *Second Edition*
LAD • Operational Subjective Statistical Methods: A Mathematical, Philosophical, and Historical Introduction
LAMPERTI • Probability: A Survey of the Mathematical Theory, *Second Edition*
LAWLESS • Statistical Models and Methods for Lifetime Data, *Second Edition*
LAWSON • Statistical Methods in Spatial Epidemiology, *Second Edition*
LE • Applied Categorical Data Analysis, *Second Edition*
LE • Applied Survival Analysis
LEE • Structural Equation Modeling: A Bayesian Approach
LEE and WANG • Statistical Methods for Survival Data Analysis, *Third Edition*
LEPAGE and BILLARD • Exploring the Limits of Bootstrap
LESSLER and KALSBEEK • Nonsampling Errors in Surveys
LEYLAND and GOLDSTEIN (editors) • Multilevel Modelling of Health Statistics
LIAO • Statistical Group Comparison
LIN • Introductory Stochastic Analysis for Finance and Insurance
LITTLE and RUBIN • Statistical Analysis with Missing Data, *Second Edition*
LLOYD • The Statistical Analysis of Categorical Data
LOWEN and TEICH • Fractal-Based Point Processes
MAGNUS and NEUDECKER • Matrix Differential Calculus with Applications in Statistics and Econometrics, *Revised Edition*
MALLER and ZHOU • Survival Analysis with Long Term Survivors
MARCHETTE • Random Graphs for Statistical Pattern Recognition
MARDIA and JUPP • Directional Statistics
MARKOVICH • Nonparametric Analysis of Univariate Heavy-Tailed Data: Research and Practice
MARONNA, MARTIN and YOHAI • Robust Statistics: Theory and Methods
MASON, GUNST, and HESS • Statistical Design and Analysis of Experiments with Applications to Engineering and Science, *Second Edition*
McCULLOCH, SEARLE, and NEUHAUS • Generalized, Linear, and Mixed Models, *Second Edition*
McFADDEN • Management of Data in Clinical Trials, *Second Edition*
* McLACHLAN • Discriminant Analysis and Statistical Pattern Recognition
McLACHLAN, DO, and AMBROISE • Analyzing Microarray Gene Expression Data
McLACHLAN and KRISHNAN • The EM Algorithm and Extensions, *Second Edition*
McLACHLAN and PEEL • Finite Mixture Models
McNEIL • Epidemiological Research Methods
MEEKER and ESCOBAR • Statistical Methods for Reliability Data
MEERSCHAERT and SCHEFFLER • Limit Distributions for Sums of Independent Random Vectors: Heavy Tails in Theory and Practice
MENGERSEN, ROBERT, and TITTERINGTON • Mixtures: Estimation and Applications
MICKEY, DUNN, and CLARK • Applied Statistics: Analysis of Variance and Regression, *Third Edition*
* MILLER • Survival Analysis, *Second Edition*
MONTGOMERY, JENNINGS, and KULAHCI • Introduction to Time Series Analysis and Forecasting
MONTGOMERY, PECK, and VINING • Introduction to Linear Regression Analysis, *Fifth Edition*
MORGENTHALER and TUKEY • Configural Polysampling: A Route to Practical Robustness
MUIRHEAD • Aspects of Multivariate Statistical Theory
MULLER and STOYAN • Comparison Methods for Stochastic Models and Risks
MURTHY, XIE, and JIANG • Weibull Models
MYERS, MONTGOMERY, and ANDERSON-COOK • Response Surface Methodology: Process and Product Optimization Using Designed Experiments, *Third Edition*
MYERS, MONTGOMERY, VINING, and ROBINSON • Generalized Linear Models. With Applications in Engineering and the Sciences, *Second Edition*
NATVIG • Multistate Systems Reliability Theory With Applications
† NELSON • Accelerated Testing, Statistical Models, Test Plans, and Data Analyses
† NELSON • Applied Life Data Analysis

*Now available in a lower priced paperback edition in the Wiley Classics Library.
†Now available in a lower priced paperback edition in the Wiley–Interscience Paperback Series.

NEWMAN • Biostatistical Methods in Epidemiology
NG, TAIN, and TANG • Dirichlet Theory: Theory, Methods and Applications
OKABE, BOOTS, SUGIHARA, and CHIU • Spatial Tesselations: Concepts and Applications of Voronoi Diagrams, *Second Edition*
OLIVER and SMITH • Influence Diagrams, Belief Nets and Decision Analysis
PALTA • Quantitative Methods in Population Health: Extensions of Ordinary Regressions
PANJER • Operational Risk: Modeling and Analytics
PANKRATZ • Forecasting with Dynamic Regression Models
PANKRATZ • Forecasting with Univariate Box-Jenkins Models: Concepts and Cases
PARDOUX • Markov Processes and Applications: Algorithms, Networks, Genome and Finance
PARMIGIANI and INOUE • Decision Theory: Principles and Approaches
* PARZEN • Modern Probability Theory and Its Applications
PEÑA, TIAO, and TSAY • A Course in Time Series Analysis
PESARIN and SALMASO • Permutation Tests for Complex Data: Applications and Software
PIANTADOSI • Clinical Trials: A Methodologic Perspective, *Second Edition*
POURAHMADI • Foundations of Time Series Analysis and Prediction Theory
POWELL • Approximate Dynamic Programming: Solving the Curses of Dimensionality, *Second Edition*
POWELL and RYZHOV • Optimal Learning
PRESS • Subjective and Objective Bayesian Statistics, *Second Edition*
PRESS and TANUR • The Subjectivity of Scientists and the Bayesian Approach
PURI, VILAPLANA, and WERTZ • New Perspectives in Theoretical and Applied Statistics
† PUTERMAN • Markov Decision Processes: Discrete Stochastic Dynamic Programming
QIU • Image Processing and Jump Regression Analysis
* RAO • Linear Statistical Inference and Its Applications, *Second Edition*
RAO • Statistical Inference for Fractional Diffusion Processes
RAUSAND and HØYLAND • System Reliability Theory: Models, Statistical Methods, and Applications, *Second Edition*
RAYNER, THAS, and BEST • Smooth Tests of Goodnes of Fit: Using R, *Second Edition*
RENCHER • Linear Models in Statistics, *Second Edition*
RENCHER • Methods of Multivariate Analysis, *Second Edition*
RENCHER • Multivariate Statistical Inference with Applications
RIGDON and BASU • Statistical Methods for the Reliability of Repairable Systems
* RIPLEY • Spatial Statistics
* RIPLEY • Stochastic Simulation
ROHATGI and SALEH • An Introduction to Probability and Statistics, *Second Edition*
ROLSKI, SCHMIDLI, SCHMIDT, and TEUGELS • Stochastic Processes for Insurance and Finance
ROSENBERGER and LACHIN • Randomization in Clinical Trials: Theory and Practice
ROSSI, ALLENBY, and McCULLOCH • Bayesian Statistics and Marketing
† ROUSSEEUW and LEROY • Robust Regression and Outlier Detection
ROYSTON and SAUERBREI • Multivariate Model Building: A Pragmatic Approach to Regression Analysis Based on Fractional Polynomials for Modeling Continuous Variables
* RUBIN • Multiple Imputation for Nonresponse in Surveys
RUBINSTEIN and KROESE • Simulation and the Monte Carlo Method, *Second Edition*
RUBINSTEIN and MELAMED • Modern Simulation and Modeling
RYAN • Modern Engineering Statistics
RYAN • Modern Experimental Design
RYAN • Modern Regression Methods, *Second Edition*
RYAN • Statistical Methods for Quality Improvement, *Third Edition*
SALEH • Theory of Preliminary Test and Stein-Type Estimation with Applications
SALTELLI, CHAN, and SCOTT (editors) • Sensitivity Analysis
SCHERER • Batch Effects and Noise in Microarray Experiments: Sources and Solutions
* SCHEFFE • The Analysis of Variance
SCHIMEK • Smoothing and Regression: Approaches, Computation, and Application

*Now available in a lower priced paperback edition in the Wiley Classics Library.
†Now available in a lower priced paperback edition in the Wiley–Interscience Paperback Series.

SCHOTT • Matrix Analysis for Statistics, *Second Edition*
SCHOUTENS • Levy Processes in Finance: Pricing Financial Derivatives
SCOTT • Multivariate Density Estimation: Theory, Practice, and Visualization
* SEARLE • Linear Models
† SEARLE • Linear Models for Unbalanced Data
† SEARLE • Matrix Algebra Useful for Statistics
† SEARLE, CASELLA, and McCULLOCH • Variance Components
SEARLE and WILLETT • Matrix Algebra for Applied Economics
SEBER • A Matrix Handbook For Statisticians
† SEBER • Multivariate Observations
SEBER and LEE • Linear Regression Analysis, *Second Edition*
† SEBER and WILD • Nonlinear Regression
SENNOTT • Stochastic Dynamic Programming and the Control of Queueing Systems
* SERFLING • Approximation Theorems of Mathematical Statistics
SHAFER and VOVK • Probability and Finance: It's Only a Game!
SHERMAN • Spatial Statistics and Spatio-Temporal Data: Covariance Functions and Directional Properties
SILVAPULLE and SEN • Constrained Statistical Inference: Inequality, Order, and Shape Restrictions
SINGPURWALLA • Reliability and Risk: A Bayesian Perspective
SMALL and McLEISH • Hilbert Space Methods in Probability and Statistical Inference
SRIVASTAVA • Methods of Multivariate Statistics
STAPLETON • Linear Statistical Models, *Second Edition*
STAPLETON • Models for Probability and Statistical Inference: Theory and Applications
STAUDTE and SHEATHER • Robust Estimation and Testing
STOYAN • Counterexamples in Probability, *Second Edition*
STOYAN, KENDALL, and MECKE • Stochastic Geometry and Its Applications, *Second Edition*
STOYAN and STOYAN • Fractals, Random Shapes and Point Fields: Methods of Geometrical Statistics
STREET and BURGESS • The Construction of Optimal Stated Choice Experiments: Theory and Methods
STYAN • The Collected Papers of T. W. Anderson: 1943–1985
SUTTON, ABRAMS, JONES, SHELDON, and SONG • Methods for Meta-Analysis in Medical Research
TAKEZAWA • Introduction to Nonparametric Regression
TAMHANE • Statistical Analysis of Designed Experiments: Theory and Applications
TANAKA • Time Series Analysis: Nonstationary and Noninvertible Distribution Theory
THOMPSON • Empirical Model Building: Data, Models, and Reality, *Second Edition*
THOMPSON • Sampling, *Third Edition*
THOMPSON • Simulation: A Modeler's Approach
THOMPSON and SEBER • Adaptive Sampling
THOMPSON, WILLIAMS, and FINDLAY • Models for Investors in Real World Markets
TIERNEY • LISP-STAT: An Object-Oriented Environment for Statistical Computing and Dynamic Graphics
TSAY • Analysis of Financial Time Series, *Third Edition*
UPTON and FINGLETON • Spatial Data Analysis by Example, Volume II: Categorical and Directional Data
† VAN BELLE • Statistical Rules of Thumb, *Second Edition*
VAN BELLE, FISHER, HEAGERTY, and LUMLEY • Biostatistics: A Methodology for the Health Sciences, *Second Edition*
VESTRUP • The Theory of Measures and Integration
VIDAKOVIC • Statistical Modeling by Wavelets
VIERTL • Statistical Methods for Fuzzy Data
VINOD and REAGLE • Preparing for the Worst: Incorporating Downside Risk in Stock Market Investments
WALLER and GOTWAY • Applied Spatial Statistics for Public Health Data
WEISBERG • Applied Linear Regression, *Third Edition*
WEISBERG • Bias and Causation: Models and Judgment for Valid Comparisons
WELSH • Aspects of Statistical Inference
WESTFALL and YOUNG • Resampling-Based Multiple Testing: Examples and Methods for p-Value Adjustment

*Now available in a lower priced paperback edition in the Wiley Classics Library.
†Now available in a lower priced paperback edition in the Wiley–Interscience Paperback Series.

* WHITTAKER • Graphical Models in Applied Multivariate Statistics
WINKER • Optimization Heuristics in Economics: Applications of Threshold Accepting
WOODWORTH • Biostatistics: A Bayesian Introduction
WOOLSON and CLARKE • Statistical Methods for the Analysis of Biomedical Data, *Second Edition*
WU and HAMADA • Experiments: Planning, Analysis, and Parameter Design Optimization, *Second Edition*
WU and ZHANG • Nonparametric Regression Methods for Longitudinal Data Analysis
YIN • Clinical Trial Design: Bayesian and Frequentist Adaptive Methods
YOUNG, VALERO-MORA, and FRIENDLY • Visual Statistics: Seeing Data with Dynamic Interactive Graphics
ZACKS • Stage-Wise Adaptive Designs
* ZELLNER • An Introduction to Bayesian Inference in Econometrics
ZELTERMAN • Discrete Distributions—Applications in the Health Sciences
ZHOU, OBUCHOWSKI, and McCLISH • Statistical Methods in Diagnostic Medicine, *Second Edition*

*Now available in a lower priced paperback edition in the Wiley Classics Library.
†Now available in a lower priced paperback edition in the Wiley–Interscience Paperback Series.